Understanding and Preventing Violence

Volume 2
Biobehavioral Influences

Albert J. Reiss, Jr., Klaus A. Miczek, and
Jeffrey A. Roth, eds.

Panel on the Understanding and Control
of Violent Behavior
Committee on Law and Justice
Commission on Behavioral and Social Sciences
and Education
National Research Council

NATIONAL ACADEMY PRESS
Washington, D.C. 1994

NATIONAL ACADEMY PRESS   2101 Constitution Avenue, N.W.   Washington, D.C.   20418

NOTICE: The project that is the subject of this report was approved by the Governing Board of the National Research Council, whose members are drawn from the councils of the National Academy of Sciences, the National Academy of Engineering, and the Institute of Medicine. The members of the committee responsible for the report were chosen for their special competences and with regard for appropriate balance.

This report has been reviewed by a group other than the authors according to procedures approved by a Report Review Committee consisting of members of the National Academy of Sciences, the National Academy of Engineering, and the Institute of Medicine.

The panel study on understanding and preventing violence was supported by grants from the National Science Foundation, the Centers for Disease Control and Prevention of the U.S. Department of Health and Human Services, and the National Institute of Justice of the U.S. Department of Justice. Additional funding to support publication of the commissioned papers was provided by the John D. and Catherine T. MacArthur Foundation, the National Institute of Mental Health of the U.S. Department of Health and Human Services, and the National Institute of Justice.

Library of Congress Cataloging-in-Publication Data
(Revised for vol. 2)

Understanding and preventing violence.

"Panel on the Understanding and Control of Violent Behavior, Committee on Law and Justice, Commission on Behavioral and Social Sciences and Education, National Research Council."
   Vol. 2 edited by Albert J. Reiss, Jr.,   Klaus A. Miczek, and Jeffrey A. Roth.
   Includes bibliographical references and index.
   Contents: v. [1]. [without special title] — v. 2. Biobehavioral influences.
   1. Violence—United States. 2. Violence—United States—Prevention. 3. Violent crimes—United States. I. Reiss, Albert J. II. Roth, Jeffrey A., 1945-III. Miczek, Klaus A. IV. National Research Council (U.S.). Panel on the Understanding and Control of Violent Behavior.
HN90.V5U53   1993   303.6   92-32137
ISBN 0-309-04594-0 (v. 1)
ISBN 0-309-04649-1 (v. 2)

First Printing, May 1994
Second Printing, November 1994

PANEL ON THE UNDERSTANDING AND CONTROL OF VIOLENT BEHAVIOR

ALBERT J. REISS, JR., *Chair*, Department of Sociology, Yale University

DAVID P. FARRINGTON, *Vice Chair*, Institute of Criminology, Cambridge University

ELIJAH ANDERSON, Department of Sociology, University of Pennsylvania

GREGORY CAREY, Institute of Behavior Genetics, University of Colorado

JACQUELINE COHEN, School of Urban and Public Affairs, Carnegie Mellon University

PHILIP J. COOK, Institute of Policy Sciences, Duke University

FELTON EARLS, Department of Behavioral Sciences, Harvard University

LEONARD ERON, Department of Psychology, University of Illinois

LUCY FRIEDMAN, Victim Services Agency, New York

TED ROBERT GURR, Department of Government and Politics, University of Maryland

JEROME KAGAN, Department of Psychology, Harvard University

ARTHUR KELLERMANN, Emergency Department, Regional Medical Center, Memphis, and Department of Internal Medicine and Preventive Medicine, University of Tennessee

RON LANGEVIN, Juniper Psychological Services, Toronto, and Department of Psychiatry, University of Toronto

COLIN LOFTIN, Institute of Criminal Justice and Criminology, University of Maryland

KLAUS A. MICZEK, Department of Psychology, Tufts University

MARK H. MOORE, Kennedy School of Government, Harvard University

JAMES F. SHORT, JR., Social and Economic Sciences Research Center, Washington State University

LLOYD STREET, College of Human Ecology, Cornell University

FRANKLIN E. ZIMRING, Law School, University of California, Berkeley

JEFFREY A. ROTH, *Principal Staff Officer*

Contents

The National Academy of Sciences is a private, nonprofit, self-perpetuating society of distinguished scholars engaged in scientific and engineering research, dedicated to the furtherance of science and technology and to their use for the general welfare. Upon the authority of the charter granted to it by the Congress in 1863, the Academy has a mandate that requires it to advise the federal government on scientific and technical matters. Dr. Bruce M. Alberts is president of the National Academy of Sciences.

The National Academy of Engineering was established in 1964, under the charter of the National Academy of Sciences, as a parallel organization of outstanding engineers. It is autonomous in its administration and in the selection of its members, sharing with the National Academy of Sciences the responsibility for advising the federal government. The National Academy of Engineering also sponsors engineering programs aimed at meeting national needs, encourages education and research, and recognizes the superior achievements of engineers. Dr. Robert M. White is president of the National Academy of Engineering.

The Institute of Medicine was established in 1970 by the National Academy of Sciences to secure the services of eminent members of appropriate professions in the examination of policy matters pertaining to the health of the public. The Institute acts under the responsibility given to the National Academy of Sciences by its congressional charter to be an adviser to the federal government and, upon its own initiative, to identify issues of medical care, research, and education. Dr. Kenneth I. Shine is president of the Institute of Medicine.

The National Research Council was organized by the National Academy of Sciences in 1916 to associate the broad community of science and technology with the Academy's purposes of furthering knowledge and advising the federal government. Functioning in accordance with general policies determined by the Academy, the Council has become the principal operating agency of both the National Academy of Sciences and the National Academy of Engineering in providing services to the government, the public, and the scientific and engineering communities. The Council is administered jointly by both Academies and the Institute of Medicine. Dr. Bruce M. Alberts and Dr. Robert M. White are chairman and vice chairman, respectively, of the National Research Council.

Foreword

In cities, suburban areas, and even small towns, Americans are fearful and concerned that violence has permeated the fabric of their communities and degraded the quality of their lives. This anxiety is not unfounded. In recent years, murders have killed about 23,000 people annually, while upward of 3,000,000 nonfatal but serious violent victimizations have occurred each year. These incidents are sources of chronic fear and public concern over the seeming inability of public authorities to prevent them.

Because of this concern, three federal agencies requested the National Research Council to carry out a comprehensive review of research applicable to the understanding and control of violence. Within the general topic of violence, the three sponsors expressed somewhat different sets of priorities. The National Science Foundation's Law and Social Science Program sought a review of current knowledge of the causes of violent behavior and recommendations about priorities in funding future basic research. The other two sponsors were more concerned with the application of that knowledge to the prevention and control of violence. The National Institute of Justice sought advice on how to prevent and control violent crimes, using the combined resources of criminal justice and other agencies. The National Center for Injury Prevention and Control of the Centers for Disease Control and Prevention sought assistance in setting priorities in efforts to prevent injuries and deaths from violent events.

In response, the Commission on Behavioral and Social Sciences and Education, through its Committee on Law and Justice, established the Panel on the Understanding and Control of Violent Behavior and took primary responsibility for shaping the specific mandate and composition of the panel. Two features of its mandate carried particular weight. First, to draw implications from past research and to chart its future course, the perspectives and models of biological, psychological, and social science research on violence should be integrated. Second, as a matter of science policy, the panel's work should orient the future allocation of research and evaluation resources toward the development and refinement of promising strategies for reducing violence and its consequences.

Early on, the panel recognized that the extraordinary breadth of its mandate demanded the mobilization of expertise beyond that of its own members and staff. Therefore, in addition to preparing a number of internal review memoranda, it commissioned a number of reviews and analyses by experts in certain specialized topics. Although the commissioned papers reflect the views of their authors and not necessarily those of the panel, all were valuable resources for the panel. From the entire set, the panel selected 15 for publication in supplementary volumes because it found them particularly useful. The panel is grateful to all the authors and to the discussants who prepared comments for the panel's Symposium on Understanding and Preventing Violence.

This volume contains the commissioned reviews of research on biological influences on violent or aggressive behavior. Panel member Gregory Carey reviews the available statistical evidence on genetic contributions to the probability of violent and related behaviors. Allan Mirsky and Allan Siegel review available research on features of brain structure and functioning that have been implicated in aggressive behavior. Paul Brain's review concerns the roles of hormonal and neurological interactions in violent behavior. Panel member Klaus Miczek and his colleagues reviewed research on animal and human subjects on the neurochemistry of violence and aggression and its implications for the management of those behaviors. Robin Kanarek reviewed existing research on dietary influences on violent behavior. To increase the accessibility of these papers, they are introduced by an overview prepared by Klaus Miczek and his colleagues Allan Mirsky, Gregory Carey, Joseph DeBold, and Adrian Raine. The panel members believe that, like themselves, others will find these reviews to be extremely helpful resources.

Understanding and Preventing

Violence

Volume 2

Biobehavioral Influences

An Overview of Biological Influences on Violent Behavior

Klaus A. Miczek, Allan F. Mirsky, Gregory Carey, Joseph DeBold, and Adrian Raine

Even the most complex social environmental influences on an individual's propensity to engage in violent behavior may eventually be traced to their biologic bases. In order to sketch such an interactive model, it is useful to begin with a consideration of the genetic influences on violent behavior as studied in animals as well as humans. Steroid and peptide hormones as well as peptides and biogenic amines are critically important in the neural and physiologic mechanisms initiating, executing, and coping with violent behavior. It is here that important endocrine and pharmacologic interventions are targeted. These neurochemical systems mediating violent behavior are specific to discrete neuroanatomic networks. Indirect measures of neural mechanisms of violent behavior may be obtained via neuroimaging and functional neuropsychologic assessments.

GENETIC MECHANISMS

Behavioral genetic research has shown that genes influence individual differences in a wide range of human behaviors—cogni-

---

Klaus Miczek is at the Department of Psychology, Tufts University; Allan Mirsky is at the National Institutes of Health; Gregory Carey is at the Institute of Behavior Genetics, University of Colorado; Joseph DeBold is at the Department of Psychology, Tufts University; and Adrian Raine is at the University of Southern California.

1

tion, academic achievement, personality and temperament (including such traits as aggression and hostility), psychopathology, and even vocational interests and social attitudes (Plomin et al., 1989). Hence, a research finding that criminal or violent behavior had some heritable component would come as no surprise—especially since violent and criminal behaviors are themselves correlated with some of the other behaviors for which genetic relationships have been established. Beyond confirming the existence of heritability in violent behavior, the more interesting intellectual challenges are

(1) isolating the precise nature of the mechanisms through which an individual's propensity to engage in or refrain from violent behavior may be inherited;

(2) using quantitative methodology to control for heritable influences so that conclusions about environmental influences on violent behavior can be clarified; and

(3) quantifying the genetic effect in terms of its importance or triviality in explaining human behavior and the magnitude of its correlation with risk factors for violence.

On the first challenge, quantitative genetic studies have not isolated any simple genetic syndrome, either Mendelian or chromosomal, that is invariably associated with violence or, more broadly, with antisocial behavior. Like inherited propensities for other behaviors, a genetic liability toward violence is likely to involve many genes and substantial environmental variation. The existence of such mechanisms may well be confirmed by future quantitative genetic research, but knowledge of their precise nature must await progress in detecting genes—and markers linked to them—that account for small variations in behavior, a problem in molecular biology that lies beyond the scope of this paper.

The second challenge suggests a more promising line of research than the reiteration of long-standing, sterile "nature versus nurture" debates—that genetic research designs may clarify environmental effects. This can best be illustrated by a hypothetical example. Suppose that a propensity toward violent behavior is transmitted from parent to offspring by two mechanisms: one operating through the genes and the other through social learning. How can these two mechanisms be detected and quantified in a study of intact nuclear families? If the parent-offspring correlation is interpreted solely in terms of social learning, then the environmental transmission will be overestimated. On the other hand, if the correlation is interpreted solely in terms of genetic transmission, then the social learning of aggression will be overlooked. Twin studies, studies of adoptive parent-offspring pairs, and studies of the biologic parents of adoptees are required to untangle the joint effects of genetic and family environmental transmission. Although such designs are becoming routine in the study of cognitive development, they are rare in the study of violent behavior.

The third challenge aims at determining whether a genetic propensity to violence is substantial or trivial and the extent to which it is correlated with other behaviors. Does the genetic influence on intelligence and on alcohol abuse explain genetic liability toward aggression?

STUDIES OF HUMANS

Studies of the inheritance of violent behavior in humans rely on adoption or twin designs to tease apart the effects of shared family environment from those of shared genes. The adoption design capitalizes on the fact that an adoptee does not receive environmental transmission from a biological parent or genetic transmission from an adoptive parent. Similarly, adoptive siblings share environments but not genes, whereas biological siblings raised apart share genes but not environments. Twin studies rely on the fact that identical twins have all genes in common, whereas fraternal pairs share on average only half their genes (plus a small effect from assortative mating). In both kinds of studies, heritability coefficients—the proportion of observed variation due to genetic variation—may be calculated. As with all human behaviors, the interpretation of these coefficients may be confounded by several factors—selective placement in adoptions, and imitation and co-offending in twins. In addition, the study of violence presents problems of its own. Not only is the base rate for violence low, but it is also more poorly measured than most behaviors studied by behavior geneticists.

Despite methodologic weaknesses in the early twin studies, later twin and adoption research suggests important heritability for *adult* antisocial behavior with perhaps a smaller genetic influence on *juvenile* criminality (Bohman et al., 1982; Christiansen, 1977; Cloninger and Gottesman, 1987; Mednick et al., 1984). Heritability estimates range from a low of about .20 to a high of almost .70 in Danish samples. Needless to say, such estimates cannot be easily extrapolated to other cultures.

In contrast to these data, the evidence for a genetic basis to *violent* offending is much weaker. Only three samples permit one to assess the role of genetics in violent offending, and two of the three produced nonsignificant results.

These findings suggest at most a weak role for heredity in violent behavior. But studies that use samples at high risk for violent behavior or that measure violent behavior through self-reports rather than arrest records may yet discover genetic relationships that have so far remained hidden—or underestimated—because arrests for violent offenses are rare in samples of the general population. One positive lead is the correlation between violence in biological parents and alcohol abuse in adopted-away sons. This suggests a genetic relationship between the two, an important link given the well-established correlation between alcohol and violent behavior as discussed below (see Miczek et al., Volume 3). Because many violent offenders also commit nonviolent offenses, heritability for criminality per se provides another possible link between genes and violence. Finally and perhaps most importantly, the gene-environment interaction reported for antisocial behavior (Cloninger et al., 1982; Cadoret et al., 1983) may also extend to violence.

The principles of quantitative genetics raise strong cautions about the extrapolation of empirical research findings on violence. First, evidence for the heritability of individual differences *within* a population cannot be used to explain average differences *between* populations or even within the same population *over time.* It is unlikely that genetic differences could account for anything but a small fraction of the change in violence over the twentieth century, differences in violent crimes among nations, or variance in rates among certain subgroups within a nation. Second, heritability cannot predict or explain an individual's culpability in a particular violent event. Third, many estimates of heritability are based on data from the Scandinavian countries, where the necessary data are routinely collected in national registries. Because the environmental variance relevant to violence may not be the same in the Scandinavian countries and in the United States, for example, the heritability estimates cannot be readily extrapolated.

STUDIES OF ANIMALS

A large number of strain comparisons and the successful establishment of selected lines demonstrate significant heritability for rodent aggression (Ebert and Sawyer, 1980; Lagerspetz and Lagerspetz, 1975; Scott, 1942, 1966; van Oortmerssen and Bakker,

1981). Although there is controversy over the extent to which the genetic mechanisms are the same for male and female aggression in *Mus*, the testing situation can change the rank order of selected lines (Hood and Cairns, 1988)—females from high-aggression lines exhibit their agonistic behavior mostly in sex-appropriate settings (e.g., postpartum tests). Similarly, studies of selected lines show that aggression may be modified by experience. Thus, although agonistic behavior shows *some* developmental continuity and cross-situational generality in inbred or selected strains, it is clearly not a single genetic phenomenon that can be studied in isolation from specific contextual cues, social environment, and development (e.g., Cairns et al., 1990; Jones and Brain, 1987).

The recent trend in behavior genetic research is less toward demonstrating the fact of the heritability of aggression and more on elucidating its genetic correlates and identifying genetic loci that underlie agonistic behavior. Here, the Y chromosome may contribute to individual differences in male aggression in the mouse, at least in some strains (Carlier et al., 1990; Maxson et al., 1989; Selmanoff et al., 1975). There also appears to be genetic sensitivity to the effects of early neonatal androgens on aggression in mice (e.g., Vale et al., 1972; Michard-Vanhee, 1988).

NEUROCHEMICAL MECHANISMS

ENDOCRINE MECHANISMS AND VIOLENT BEHAVIOR

Steroids

Research suggests that testosterone and its androgenic and estrogenic metabolites influence the probability of aggressive responding to environmental events and stimuli through *organizational* as well as *activational* mechanisms. Organizational effects are traditionally those exerted, generally permanently, by a hormone during some sensitive period of development. This type of mechanism appears to explain sex differences in anatomy and some aspects of sex differences in behavior. For example, testosterone present during a particular period of fetal development in mammals induces the development of the male reproductive tract and genitalia. If androgen levels are low, as is normally the case in females, this development does not occur and female genitalia develop instead. A similar control for male aggression has been demonstrated in a number of laboratory animal species. For example, female mice given a single injection of testosterone at birth become much more sensitive to the aggression-enhancing effects of androgens as adults. Prenatal treatment of female rhesus monkeys with testosterone results in females that are male-like in their higher level of "rough and tumble" play as juveniles and more aggressive as adults. In humans, there is evidence for a similar, but reduced in magnitude, modulation of aggression by androgens. This research uses children that were accidentally exposed to inappropriate steroids during fetal development and assesses their behavior though observations, interviews, and psychologic assessments. In girls prenatally exposed to heightened levels of androgens, there is a trend for increased levels of aggression. In boys exposed to estrogens or antiandrogenic steroids during pregnancy, there is a trend for decreased aggressiveness (see Brain, Table 4 in this volume). However, generally these steroids also have had some effect on genital development and the behavioral differences may be due to altered body image or to the affected children being treated differently by parents or peers. Interestingly, prenatal testosterone also alters the development of parts of the preoptic area of the brain. Preoptic area structure and neurochemistry are sexually dimorphic in animals and in humans, and this brain area is also thought to have a role in aggressive behavior. However, a direct link between the sexual dimorphism of the preoptic area and human violent behavior remains elusive.

In animals, testosterone (or its metabolites) has effects on the probability of aggressive response to conspecifics or other environmental events. This is frequently referred to as an activational effect although, mechanistically, androgens are not stimulating aggressive behavior *in vacuo*; more accurately, they appear to be altering the response to aggression-provoking stimuli. In laboratory animals, particularly rodents, there is research that demonstrates the brain sites involved in this action and the importance of the biochemical mechanisms by which testosterone can alter neural activity. The strength of the modulation that testosterone exerts on aggressive behavior seems to decrease in more complex social animals. In nonhuman primates, the correlation between testosterone and aggressiveness or dominance frequently, but not in all studies, exists, but the activational effect of testosterone is more variable and harder to demonstrate. This trend is perhaps more exaggerated in humans. Positive correlations have been reported between androgen levels and aggressive or violent behavior in adolescent boys and in men, but these correlations are not high, they are sometimes difficult to replicate, and importantly, they do not demonstrate causation. In fact there is better evidence for the reverse relationship (behavior altering hormonal levels). Stress (e.g., from being subject to aggression or being defeated) decreases androgen levels, and winning—even in innocuous laboratory competitions—can increase testosterone.

The results of manipulating androgens with antiandrogen therapy in violent offenders are also mixed and difficult to interpret because of confounding influences on the data collection. Some critical reviews have concluded that antiandrogens show promise as an adjunct therapy for violent sex offenders. These may be relying more on the clearer relationship between testosterone and sexual motivation than between testosterone and violence. Another interesting approach will be to study the effects of anabolic steroids, but these studies have just begun and face very difficult methodologic problems. In general, most investigators conclude that there can be an influence of androgens on violence but that it is only one component accounting for a small amount of the variance.

Gonadal steroids have also been postulated to be involved in the increased irritability and hostility seen in some women with premenstrual syndrome (PMS). However, the endocrine evidence in support of this view is weak, and most recent papers find that individual differences in estrogens, progestins, and other hormones across the menstrual cycle do not explain the variability in intensity of PMS symptoms.

Adrenal steroids (glucocorticoids, such as cortisol and corticosterone) and the pituitary hormone ACTH (adrenocorticotropic hormone) have also been found to be related to aggressive behavior in animals. However, the strongest relationship is a negative one. Chronically increased levels of corticosteroids decrease aggressiveness, and ACTH increases submissiveness and avoidance of attack. These two effects are difficult to separate endocrinologically, but they appear to be mediated by different mechanisms. Correlations between dominance and corticosteroid levels in primates may more directly reflect variations in the ability to adapt to stress.

In summary, there is no simple relationship between steroids and aggression, much less violence. The strongest conclusion is that in humans, androgens can influence and be influenced by aggressive behavior. However, they are only one of many influences and *not* the determining factor. The opposite relationship (i.e., the environment and behavior influencing hormone secretion) is the stronger of the two linkages.

Other Hormones

Steroids are not the only hormones that have been related to aggression and violence, but other hormones appear to have less direct or less specific effects. For example, adrenal norepinephrine secretion has been related to the commission of violent crime, but norepinephrine and epinephrine are released in response to a wide variety of arousing or emotional conditions and are important in coping with stress. Animal and clinical studies have found evidence for a role of central nervous system (CNS) norepinephrine in aggressiveness—but when it acts as a neurotransmitter, not as a hormone. Because hormones can alter many aspects of cellular activity and because aggressive behavior involves so many areas of the brain, the potential for indirect or secondary effects of hormones is high.

In summary, there is no simple relationship between hormones and aggression, much less violence. The strongest conclusion is that in humans, steroids can influence and be influenced by aggressive behavior. However, they are only one influence of many and *not* the determining factor. The opposite relationship (i.e., the environment and behavior influence on hormone secretion) is the stronger of the two linkages.

NEUROTRANSMITTERS AND RECEPTORS

Dopamine

Evidence from animal studies points to large changes in brain dopamine systems during aggressive or defensive behavior. At present, evidence for similarly altered dopamine activity in brain regions of violent humans is not available. It is possible that brain dopamine systems are particularly significant in the rewarding aspects of violent and aggressive behavior. However, at present, a "marker" for some aspect of brain dopamine activity that is selective to a specific kind of aggression or violent behavior has not been identified in any of the accessible bodily fluids or via imaging methods in the brain.

The most frequently used treatment of violent outbursts in emergency situations and also in long-term medication of violence-prone individuals employs drugs that act principally at dopamine receptors. Particularly, drugs that antagonize the D2 subtype of dopamine receptors represent widely used antipsychotics with frequent application to violent patients. Evidence from animal and human studies emphasizes the many debilitating side effects of these drugs that render them problematic as treatment options, representing little more than a form of "chemical restraint." Antipsychotic drugs that are antagonists at D2 dopamine receptors show a wide range of behavioral activities and, when used chronically, lead to various neurologic problems.

Cocaine and amphetamine activate behavior and engender euphoria in all likelihood via action on brain dopamine receptors. The broad spectrum of behavioral and mood-elevating effects of these drugs may also include the aggression-enhancing effects that are seen in animals under certain conditions and that may be relevant to the occasional incidence of human violence after psychomotor stimulants. More important, however, are the psychopathic conditions that *precede* chronic amphetamine or cocaine use in predicting violent outbursts. Whether the paranoid psychosis due to amphetamine or cocaine use represents the causative condition for occasional violent behavior or the psychopathology preceding drug use is unclear at present. The relatively infrequent occurrences of violent activities in stimulant abusers appear to result from brain dopamine changes that are counteracted by treatment with antipsychotic drugs.

Norepinephrine

Behavioral events involving intense affect are accompanied by adrenergic activity, in both the peripheral and the central nervous system. For several decades, the adrenergic contribution to the "flight-fight" syndrome in the form of increased sympathetic innervation as well as adrenal output has been well established. More recently, large changes in noradrenergic neurotransmitter activity in limbic, diencephalic, and mesencephalic regions, while preparing for, executing, and recovering from highly arousing activities—among them aggressive and violent behavior—have been documented. So far, in neither animal nor human studies have noradrenergic "markers" emerged that selectively identify the propensity to engage in an aggressive or violent act. Rather, noradrenergic activity, either measured in the form of metabolite levels in a bodily fluid or indirectly assessed in a sympathetically innervated end organ, is correlated with the level of general arousal, degree of behavioral exertion and activation, and either positive or negative affect, but not with a specific behavior or mood change such as a violent act.

The most significant development during the past dozen years in applying noradrenergic drugs in the management and treatment of retarded, schizophrenic, or autistic patients with a high rate of violent behavior is the use of beta-blockers primarily for their effects on the central nervous system rather than for their autonomic effects. Drugs that block adrenergic beta-receptors also act on certain subtypes of serotonin receptors, and their aggression-reducing effects may be derived from their action on these latter sites. Beta-blockers have not been compared in effectiveness and side effects, particularly during long-term treatment, with other therapeutic agents that reduce aggressive and violent activities.

Clonidine, an adrenergic drug that targets a specific alpha-receptor subtype, has been used with success in managing withdrawal from alcohol, nicotine, and opiate addiction. Evidence from animal and human studies demonstrates that withdrawal states are often associated with irritability and a higher incidence of aggressive and defensive acts.

The application of therapeutic agents with increasing selectivity for adrenergic receptor subtypes to managing and treating patients with violent outbursts represents an important therapeutic alternative to the classic antipsychotics.

Serotonin (5-Hydroxytryptamine)

For the past 30 years, the most intensively studied amine in violent individuals has been serotonin. Evidence from studies ranging from invertebrates to primates highlights marked changes in aspects of serotonin activity in bodily fluids or neural tissue in individuals that have engaged in violent and aggressive behavior on repeated occasions. There is considerable evolutionary variation in the role of serotonin in mediating aggressive or violent behavior across animal species, functionally divergent roles even being represented at the nonhuman primate level. In psychiatric studies, deficits in serotonin synthesis, release, and metabolism have been explored as potential "markers" for certain types of alcoholic and personality disorders with poor impulse control. It is very difficult to extract, from single measures of whole brain serotonin or blood levels, activity information that is specific to past violent behavior, or represents a risk for future propensity, without also considering seasonal and circadian rhythmicity, level of arousal, nutritional status, or past drug history, particularly alcohol abuse. No single type or class of violent activity has emerged as being specifically linked to a "trait" serotonin metabolite level. However, challenges with pharmacologic probes and physiologic or environmental stresses begin to reveal an important profile of serotonin-mediated response patterns.

During the past decade, remarkable advances in serotonin receptor pharmacology have promised to yield important new therapeutic options. Evidence from animal studies suggest that drugs with specific actions at certain serotonin receptors selectively decrease several types of aggressive behavior. A new class of antianxiety drugs that target certain serotonin receptors is currently finding acceptance in clinical practice. However, specific antiaggressive effects have not been demonstrated for the serotonin anxiolytics. In humans, brain imaging of serotonin receptors begins to point to distinct alterations in serotonin receptor populations in subgroups of affectively disordered patients. These ongoing developments promise to be significant for diagnostic and therapeutic applications to violent individuals.

Sensational incidents of violence have been linked to the use of hallucinogens that act at distinct serotonin receptor subtypes. However, little is known as to whether or not the action at serotonin receptors is the actual mechanism by which these substances engender violent outbursts in rare, possibly psychopathic individuals.

Gamma-Aminobutyric Acid—Benzodiazepine Receptors

Thirty percent of all synapses in the brain use gamma-aminobutyric acid (GABA), and many of the GABA-containing neurons are inhibitory in nature. This cellular inhibitory role has been postulated to apply also to the physiologic and behavioral levels, including aggression. However, the present neurochemical evidence from animal studies finds inhibitory as well as excitatory influences of GABA manipulations on different aggressive patterns in discrete brain regions.

Interest in GABA is currently intense because one subtype of GABA receptor (GABA-A) forms a supramolecular complex that also localizes benzodiazepine receptors and that also is the site of action for certain alcohol effects. These receptors are the site of action for the most important antianxiety substances that have also been used for their antiaggressive properties. Evidence from animal and human studies documents the effectiveness of benzodiazepine anxiolytic for their calming and quieting effects. However, under certain pharmacologic and physiologic conditions, at low doses benzodiazepine anxiolytics may increase aggressive behavior in animals and humans, leading sometimes to violent outbursts that are termed "paradoxical rage."

The study of the benzodiazepine-GABA-A receptor complex in

individuals with a high rate of violent behavior promises to enhance the currently available diagnostic and therapeutic tools for the management of violence.

BRAIN MECHANISMS AND VIOLENT BEHAVIOR

NEUROANATOMIC APPROACH

Physiologic research on aggression in animals has discovered that different neural circuits appear to underlie "predatory attack" behavior as opposed to "affective defense" in animals (Siegel and Edinger, 1981, 1983; Siegel and Brutus, 1990). Sites in which electrical stimulation elicits predatory attack behavior in the cat include midbrain periaqueductal gray matter, the locus ceruleus, substantia innominata, and central nucleus of the amygdala. Brain sites that mediate affective defense reactions in the cat include the medial hypothalamus and the dorsal aspect of periaqueductal gray matter. In general, limbic structures such as the amygdala, hypothalamus, midbrain periaqueductal gray, and septal area, as well as cortical areas such as the prefrontal cortex and the anterior cingulate gyrus, contain networks of excitatory and inhibitory processes of different kinds of aggressive and defensive behavior.

Aggression in animals largely reflects an adaptive response when viewed within an evolutionary framework. Whether or not violent offending in humans constitutes an instrumental act that can be viewed as adaptive is open to question. Nevertheless, the distinction between quiet, predatory, planned attack, on the one hand, and affective, explosive aggression occurring in the context of high autonomic arousal may be of heuristic value for understanding human violence. Possibly, cat (or rat) models could serve as effective screening methods to identify new drugs—some for control of "cold calculated" aggression, others for control of "explosive" aggression. Neuroanatomic and neuropsychologic studies are needed, however, to determine whether disruption of different brain mechanisms is indeed implicated in these two forms of aggression in humans.

Data on the neuroanatomy of violence in humans stem largely from clinical studies of the effects of epileptic activity and other forms of brain damage on behavior, as well as from reports of the effects of brain resections on control of violent behavior (i.e., psychosurgery). Psychosurgical studies in Japan, India, and the United States have aimed at destroying portions of the limbic system, especially the amygdala and medial hypothalamus, in cases of patients with uncontrollable violence. Other symptoms have been targeted for psychosurgical treatment, as well. Favorable outcomes are reported, but clinical improvement has been variable; moreover, the basis for assessing success is controversial (see Mirsky and Siegel, in this volume; O'Callaghan and Carroll, 1987). Several studies suggest a link between violence and temporal lobe epilepsy, although violence occurring during a seizure is extremely rare (Mirsky and Siegel, in this volume). The question remains unanswered as to whether some patients with seizure disorders are more violence prone (because of their putative heightened emotionality) than other persons. Another important question that remains unanswered to date concerns whether these limbic system structures (portions of the temporal lobe, hippocampus, amygdala, hypothalamus) are also implicated in ostensibly "normal" criminally violent offenders who are not preselected under the suspicion of neural abnormalities. Brain imaging techniques constitute one relatively new methodology for addressing such questions.

Clinical case studies of patients with damage to the prefrontal lobes provide some support for a link between this area and features of psychopathic behavior; however, the overlap between these two syndromes is only partial. Given the animal data implicating the prefrontal cortex in the inhibition of aggression, together with neuropsychologic data on frontal dysfunction in violence, it would seem important to pursue research linking this brain site with violence. Studies are needed that combine brain imaging and social, cognitive, emotional or affective measures in order to assess both direct and indirect relationships between the prefrontal cortex and violence in humans.

NEUROPSYCHOLOGIC APPROACH

A large number of studies have found that violent offenders have brain dysfunction as reflected in deficits on neuropsychologic tests (e.g., Bryant et al., 1984; see review in Mirsky and Siegel, in this volume). Although the etiological implications of these deficits are not fully understood, there is converging evidence that cognitive deficits may underlie early school failure, dropouts, alcohol and drug use, and ultimately, encounters with the legal system as violent offenders.

An important issue requiring resolution concerns whether neuropsychologic disturbances are a cause or an effect of violence. Left-hemisphere dysfunction that disrupts linguistic processing may be causal with respect to violence in that poorer verbal comprehension and communication may contribute to a misinterpretation of events and motives in an interpersonal encounter; this in turn could precipitate a violent encounter. Similarly, poor verbal abilities and communication skills could contribute to peer rejection in childhood, which in combination with other later social and situational factors could predispose to alienation and, ultimately, to violence. Alternatively, left-hemisphere dysfunction could result in verbal deficits that lead to school failure, which in turn could lead to violence. Left-hemisphere dysfunction may, however, be a result (rather than the cause) of violent behavior, since blows to the head and falls may result in concussion and damage to the cortex.

Another major source of damage to the brain that may have profound and irreversible consequences for adaptive behavior is in environmental toxins. Maternal use of ethanol (as in beverage alcohol) has effects on the fetus that may persist for many years, and are manifest in poorer attention at ages 4, 7, and older (Streissguth et al., 1984, 1986, 1989). The effects of lead on cognitive and social adaptation have been the focus of investigation by Needleman and collaborators (1990); even relatively "small" elevations of lead in the body are associated with poor attention, academic failure, and other impairment in life success. Maternal use of cocaine, opiates, and tobacco has also been shown to have a deleterious effect on the neurobehavioral capacities of the infant and developing child. These early effects may be associated, as well, with long-term academic and social failures (summarized in Mirsky and Siegel, in this volume).

Large-scale epidemiologic and prospective studies are required in order to help elucidate the etiologic significance of neuropsychologic impairment for violence. One limitation of neuropsychologic studies, however, is that they are only indirect measures of brain dysfunction; additional statements regarding brain dysfunction in violence can be made on the basis of future studies that combine neuropsychologic testing with electroencephalogram (EEG) and positron emission tomography (PET) measures of brain activity.

PSYCHOPHYSIOLOGIC APPROACH

Neurochemical, neuroanatomic, and neurophysiologic research on violent behavior faces formidable difficulties. Measurement of the central neurologic processes is costly, often invasive, and difficult to implement so as to observe the processes during reactions to transitory situations in the social environment. To cope with these difficulties, an alternative approach to the study of some types of violent offenders has been provided by the measurement of psychophysiologic variables (i.e., assaying autonomic and CNS functions by means of recordings from the periphery of the body). Included among these variables are heart rate and skin conductance (autonomic nervous system variables), as well as EEG and event-related brain potentials (central nervous system variables). Differences among criminals, delinquents and conduct-disordered children on the one hand, and control subjects, on the other, have been shown to exist in resting heart rate (lower in offenders and in persons characterized as fearless) (Raine et al., 1990a; Raine and Jones, 1987; Venables and Raine, 1987; Kagan, 1989). Some offenders have also been shown to have lower skin conductance responses to orienting stimuli than controls, although the reverse may be true for criminal offenders designated as psychopathic (Siddle et al., 1973; Raine et al., 1990b).

With respect to EEG studies, many have reported an excess of slow wave activity in the records of incarcerated criminal offenders. It is unclear whether this is best interpreted as the effects of underarousal in the prison setting, developmental anomalies, or the sequela of brain damage (Williams, 1969; Hare, 1980).

Event-related brain potentials (ERPs), in particular the P300 component, have been studied in a number of disordered populations. The P300 wave, which is an index of the allocation of attention to a stimulus (Duncan, 1990), is an example of a "cognitive" component of the ERP. These components vary as a function of some information processing requirement or task administered to the subjects. The P300 has been found to be larger in some groups of psychopathic criminals (Raine and Venables, 1988). The interpretation of this finding is unclear, although it suggests that these persons process information differently from normal subjects.

Neuroimaging Approach

Perhaps the most recent technical development in research into the antecedents of violence involves the application of new brain imaging techniques. Positron emission tomography and regional cerebral blood flow (RCBF) techniques allow direct and indirect assessments of glucose metabolism (or blood flow in the case of RCBF) throughout the brain either during a resting state or during performance of a certain task. As such, PET and RCBF techniques assess brain *function*. Conversely, computerized tomography and magnetic resonance imaging (MRI) techniques, while providing detailed images, assess brain *structure* only.

Although studies suggest differences between violent and nonviolent offenders, sample sizes have been relatively small, and findings should be viewed as preliminary. Nevertheless, brain imaging is clearly a new field that has enormous potential for addressing questions concerning altered brain structure and function in violent offenders. For example, PET studies would be capable of directly assessing differential effects of alcohol administration on brain glucose metabolism in violent and nonviolent offender groups, and could help address the issue of whether some violent offenders constitute a subgroup that is particularly susceptible to the disinhibitory effects of alcohol on specific brain areas. Studies that combine both MRI and PET techniques are clearly desirable in that assessments of both structure and function would allow more complete statements to be made with regard to brain dysfunction in violence. Studies that combine brain imaging assessment with neuropsychologic, cognitive-psychophysiologic, and hormonal assessments in violent and nonviolent subjects would allow us to address the potentially important interactions between different biologic systems in predisposing to violence.

HYPOGLYCEMIA, DIET, AND VIOLENT BEHAVIOR

Some studies have observed that violent offenders, particularly those with a history of alcohol abuse, are characterized by reactive hypoglycemia (Virkkunen, 1986). Although there have been no demonstrations to date that violent individuals are hypoglycemic at the time of the commission of violence, it is possible that low blood glucose levels (hypoglycemia) could be conducive to aggressive behavior. Increased irritability is one symptom of hypoglycemia (Marks, 1981), and this could be the first step in the development of a full-blown aggressive outburst. Anthropologic studies, studies of aggressive personality in "normal" subjects, and experimental studies in animals all support a link between hypoglycemia and aggression (Venables and Raine, 1987). Acute symptoms of hypoglycemia are reported as maximal at 11.00-11.30 a.m. (Marks, 1981), and this time corresponds to peaks in assaults on both staff and other inmates in prison, both of which reach their maximum at 11.00-11.30 a.m. (Davies, 1982).

A number of studies have claimed that dietary changes aimed at reducing sugar consumption reduce institutional antisocial behavior in juvenile offenders, but these studies have methodological weaknesses that preclude drawing any firm conclusions at the present time (see Kanarek, in this volume). There is also some limited evidence that food additives may contribute to hyperactivity, although the data on sugar intake and hyperactivity are inconclusive (Kanarek, in this volume). There have been reports however that home environment mediates dietary effects on behavior.

Although data are limited at the present time, further double-blind studies into the effects of dietary manipulation on aggression and violence in institutions seem warranted. Furthermore, investigation of the interconnections between hypoglycemia or diet and other factors at both biological and social levels seems warranted. Since alcohol increases the susceptibility to hypoglycemia through its capacity to increase insulin secretion (Marks, 1981), it may well be that predispositions to both hypoglycemia and alcohol abuse would make an individual particularly predisposed to violence. Hypoglycemia has also been theoretically linked to both low heart rate and EEG slowing, factors that have been found to characterize violent offenders (Venables, 1988). The fact that children from a supportive home environment show more dietary improvement than those from an unsupportive home (Rumsey and Rapoport, 1983) also suggests an interaction between diet and family environment in antisocial behavior. Clearly, diet and hypoglycemia should not be studied independently of interactions with factors at other levels.

REFERENCES

Bohman, M., C.R. Cloninger, S. Sigvardsson, and A.L. von Knorring
  1982 Predisposition to petty criminality in Swedish adoptees: I. Genetic and environmental heterogeneity. *Archives of General Psychiatry* 39:1233-1241.
Bryant, E.T., M.L. Scott, C.D. Tori, and C.J. Golden
  1984 Neuropsychological deficits, learning disability, and violent behavior. *Journal of Consulting and Clinical Psychology* 52:323-324.
Cadoret, R.J., C. Cain, and R.R. Crowe
  1983 Evidence for a gene-environmental interaction in the development of adolescent antisocial behavior. *Behavior Genetics* 13:301-310.
Cairns, R.B., J.-L. Gariepy, and K.E. Hood
  1990 Development, microevolution and social behavior. *Psychological Review* 97:49-65.
Carlier, M., P.L. Roubertoux, M.L. Kottler, and H. Degrelle
  1990 Y chromosome and aggression in strains of laboratory mice. *Behavior Genetics* 20:137-156.

Christiansen, K.O.
1977   A review of studies of criminality among twins. In S.A. Mednick and K.O. Christiansen, eds., *Biosocial Bases of Criminal Behavior*. New York: Gardner.
Cloninger, C.R., and I.I. Gottesman
1987   Genetic and environmental factors in antisocial behavioral disorders. Pp. 92-109 in S.A. Mednick, T.E. Moffitt, and S.A. Stack, eds., *The Causes of Crime: New Biological Approaches*. New York: Cambridge University Press.
Cloninger, C.R., S. Sigvardsson, M. Bohman, and A.L. von Knorring
1982   Predisposition to petty criminality in Swedish adoptees: II. Cross-fostering analysis of gene-environmental interaction. *Archives of General Psychiatry* 39:1242-1247.
Davies, W.
1982   Violence in prisons. In P. Feldman, ed., *Developments in the Study of Criminal Behavior. Vol. 2: Violence*. London: Wiley.
Duncan, C.C.
1990   Current issues in the application of P300 to research on schizophrenia. In E.R. Straube and K. Hahlweg, eds., *Schizophrenia: Concepts, Vulnerability, and Intervention*. New York: Springer-Verlag.
Ebert, P.D., and R.G. Sawyer
1980   Selection for agonistic behavior in wild female *Mus musculus*. *Behavior Genetics* 10:349-360.
Gorenstein, E.
1982   Frontal lobe functions in psychopaths. *Journal of Abnormal Psychology* 91:368-379.
Hare, R.D.
1980   *Psychopathy: Theory and Practice*. New York: Wiley.
Hood, K.E., and R.B. Cairns
1988   A developmental-genetic analysis of aggressive behavior in mice: II. Cross-sex inheritance. *Behavior Genetics* 18:605-619.
Jones, S.E., and P.F. Brain
1987   Performances of inbred and outbed laboratory mice in putative tests of aggression. *Behavior Genetics* 17:87-96.
Kagan, J.
1989   Temperamental contributions to social behavior. *American Psychologist* 44:668-674.
Lagerspetz, K.M.J., and K.Y.H. Lagerspetz
1975   The expression of the genes of aggressiveness in mice: The effect of androgen on aggression and sexual behavior in females. *Aggressive Behavior* 1:291-296.
Marks, V.
1981   The regulation of blood glucose. In V. Marks and F.C. Rose, eds., *Hypoglycemia*. Oxford: Blackwell.
Maxson, S.C., A. Didier-Erickson, and S. Ogawa
1989   The Y chromosome, social signals, and offense in mice. *Behavioral and Neural Biology* 52:251-259.

Mednick, S.A., W.F. Gabrielli, and B. Hutchings
1984 Genetic influences in criminal convictions: Evidence from an adoption cohort. *Science* 224:891-894.
Michard-Vanhee C.
1988 Aggressive behavior induced in female mice by an early single dose of testosteroned is genotype dependent. *Behavior Genetics* 18:1-12.
Needleman, H.L., A. Schnell, D. Bellinger, A. Leviton, and E. Allred
1990 Long term effects of childhood exposure to lead at low dose: An eleven-year follow-up report. *New England Journal of Medicine* 322:83-88.
O'Callaghan, M.A.J., and D. Carroll
1987 The role of psychosurgical studies in the control of antisocial behavior. In S.A. Mednick, T.E. Moffitt, and S.A. Stack, eds., *The Causes of Crime: New Biological Approaches*. New York: Cambridge University Press.
Plomin, R., J.C. DeFries, and G.E. McClearn
1989 *Behavior Genetics: A Primer*, 2nd ed. San Francisco: W.H. Freeman.
Raine, A., and F. Jones
1987 Attention, autonomic arousal, and personality in behaviorally disordered children. *Journal of Abnormal Child Psychology* 15:583-599.
Raine, A., and P.H. Venables
1988 Enhanced P3 evoked potentials and longer P3 recovery times in psychopaths. *Psychophysiology* 25:30-38.
Raine, A., P.H. Venables, and M. Williams
1990a Relationships between central and autonomic measures of arousal at age 15 and criminality at age 24 years. *Archives of General Psychiatry* 47:1003-1007.
1990b Autonomic orienting responses in 15-year-old male subjects and criminal behavior at age 24. *American Journal of Psychiatry* 147:933-937.
Rumsey, J.M., and J.L. Rapoport
1983 Assessing behavioral and cognitive effects of diet in pediatric populations. Pp. 101-162 in R.J. Wurtman and J.J. Wurtman, eds., *Nutrition and The Brain*, Vol. 6. New York: Raven Press.
Scott, J.P.
1942 Genetic differences in the social behavior of inbred strains of mice. *Journal of Heredity* 33:11-15.
1966 Agonistic behavior of mice and rats: A review. *American Zoologist* 6:683-698.
Selmanoff, M.K., J.E. Jumonville, S.G. Maxson, and B.E. Ginsburg
1975 Evidence for a Y chromosome contribution to an aggressive phenotype in inbred mice. *Nature* 253:529-530.
Siddle, D.A.T., A.R. Nicol, and R.H. Foggit
1973 Habituation and over-extinction of the GSR component of the orienting response in anti-social adolescents. *British Journal of Social and Clinical Psychology* 12:303-308.

Siegel, A., and M. Brutus
  1990   Neural substrates of aggression and rage in the cat. Pp. 135-233 in A.N. Epstein and A.R. Morrison, eds., *Progress in Psychobiology and Physiological Psychology*. San Deigo, Calif.: Academic Press.
Siegel, A., and H. Edinger
  1981   Neural control of aggression and rage. Pp. 203-240 in P. Morgane and J. Panksepp, eds., *Handbook of the Hypothalamus*. New York: Marcel Dekker.
  1983   Role of the limbic system in hypothalmatically elicited attack behavior. *Neuroscience and Biobehavioral Reviews* 7:395-407.
Streissguth, A.P., D.C. Martin. H.M. Barr, B.M. Sandman, G.L. Kirchner, and B.L. Darby
  1984   Intrauterine alcohol and nicotine exposure: Attention and reaction time in 4-year-old children. *Developmental Psychology* 20(4):533-541.
Streissguth, A.P., H.M. Barr, P.D. Sampson, J.C. Parrish-Johnson, G.L. Kirchner, and D.C. Martin
  1986   Attention, distraction and reaction time at age 7 years and prenatal alcohol exposure. *Neurobehavioral Toxicology and Teratology* 8:717-725.
Streissguth, A.P., P.D. Sampson, and H.M. Barr
  1989   Neurobehavioral dose-response effects of prenatal alcohol exposure in humans from infancy to adulthood. *Annals of the New York Academy of Sciences* 562:145-158.
Vale, J.R., D. Ray, and C.A. Value
  1972   The interaction of genotype and exogenous neonatal androgen: Agonistic behavior in female mice. *Behavioral Biology* 7:321-334.
van Oortmerssen, G.A., and C.M. Bakker
  1981   Artificial selection for short and long attack latencies in wild *Mus musculus domesticus*. *Behavior Genetics* 11:115-126.
Venables, P.H.
  1988   Psychophysiology and crime: Theory and data. In T.E. Moffitt and S.A. Mednick, eds., *Biological Contributions to Crime Causation*. Dordrecht, Netherlands: Martinus Nijhoff.
Venables, P.H., and A. Raine
  1987   Biological theory. Pp. 3-28 in B. McGurk, D. Thornton, and M. Williams, eds., *Applying Psychology to Imprisonment: Theory and Practice*. London: Her Majesty's Stationery Office.
Virkkunen, M.
  1986   Reactive hypoglycemia tendency among habitually violent offenders. *Nutrition Reviews* 44:94-103.
Williams, D.
  1969   Neural factors related to habitual aggression: Consideration of those differences between those habitually aggressive and others who have committed crimes of violence. *Brain* 92:503-520.

Genetics and Violence

Gregory Carey

ANIMAL STUDIES

There is no behavioral genetic literature on *violence* in infrahuman species. Rather, the phenotype (i.e., observable behavior) is termed *aggression* or *agonistic behavior*, often occurring as an appropriate, adaptive response to a particular set of environmental circumstances. The extrapolation of such evolutionarily preadapted responses to human homicide or robbery is, of course, tenuous. Nevertheless, the ability to control matings and the intrauterine and postnatal environment dictates that the study of behavioral biology in animals may yield clues to the conditions for onset and cessation of some violent encounters in humans.

The behavioral genetic literature on animal aggression focuses almost exclusively on rats and mice. It has been documented for more than half a century that there are strain differences in the agonistic behavior of male mice (Ginsburg and Allee, 1942; Scott, 1942). The extensive literature on these differences has been reviewed elsewhere (e.g., Brain et al., 1989; Maxson, 1981). Selection studies have also demonstrated significant heritability for murine aggression (e.g., Ebert and Sawyer, 1980; van Oortmerssen and Bakker, 1981). Hence, there is abundant evidence that ge-

Gregory Carey is at the Department of Psychology and the Institute for Behavioral Genetics, University of Colorado, Boulder.

netic polymorphisms influence individual differences in aggression and agonistic behavior in rodents. This genetic liability toward aggression, however, is to a certain degree situation specific (Jones and Brain, 1987). For example, in dyadic male encounters, mice of the BALB/cBy strain are more aggressive than those of the C57BL/6By strain when tested against members of their own strain, but C57BL/6By are more aggressive when tested against mice of other strains. As Maxson (1990) notes, the genetics of aggression in a dyadic encounter depend not only on the individual's own genes but also on the genes of the conspecific partner. Similarly, diurnal variation, season of the year, arena size, test duration, and the operational definition of aggression are but a few of the variables that may change the rank order of aggression in strains (Maxson, 1990).

Similar specificity may occur for sex-appropriate aggression. Early selection studies selected for aggression among males only and did not report differences in aggression among females of the selected lines (Ebert and Sawyer, 1980). However, later researchers argue that females from high aggressive lines demonstrate their aggression in sex-appropriate settings such as postpartum tests (Hood and Cairns, 1988). Similarly, there also appears to be genetic sensitivity to the effects of early neonatal androgens on aggression in mice (e.g., Michard-Vanhee, 1988; Vale et al., 1972).

The recent trend in behavioral genetic research is aimed less at demonstrating the fact of inheritance than at elucidating its genetic correlates and identifying genetic loci that underlie agonistic behavior. Here, polymorphic loci on the Y chromosome may contribute to individual differences in male aggression in the mouse, at least in some strains (Carlier et al., 1990; Maxson et al., 1989; Selmanoff et al., 1976). Given the homology of the mammalian genome, such work may offer insight into the genetics of human behavior.

HUMAN BEHAVIORAL GENETIC STUDIES

INTRODUCTION

Before embarking on a critical overview, the terms of human behavioral genetics require definition. With respect to sibships, phenotypic (i.e., observed) variability is often decomposed into three major components: genetic variance, common environmental variance, and unique environmental variance. The difference between common and unique environmental variance is purely statistical: Common environment includes all environmental factors that contribute to sibling *similarity*, whereas unique environment subsumes all environmental mechanisms that promote sibling *differences*. To express similarity for vertical relationships such as parent and offspring, two major variance components are usually identified: genetic variance and vertical environmental transmission variance. Technically, vertical environmental transmission variance includes *all* environmental mechanisms, even those outside the home, that correlate with parental antisocial behavior and at the same time influence individual differences in offspring antisocial behavior.

Genetic influence is usually quantified by either of two estimates of *heritability*. *Broad sense* heritability is the total genetic variance divided by the phenotypic variance; usually, the only population to permit estimation of broad sense heritability is a large series of identical twins raised apart in random environments. *Narrow sense* heritability is the *additive* genetic variance divided by the phenotypic variance. The difference between additive genetic variance and total genetic variance is a complicated function of allele frequencies, allelic action, and interaction among difference genetic loci. Precise heritability estimates are seldom possible with human behavioral phenotypes. Narrow sense estimates are usually reported, with little or no empirical data to justify the validity of their assumptions.

One does not inherit behavior as one inherits eye color. Hence, when the behavioral phenotype is dichotomized (e.g., criminal offender versus nonoffender), behavioral genetic analysis is aimed at *liability*. Liability is a latent, unobserved variable that is at least ordinally related to risk—the higher an individual's score on the liability scale, the greater is the relative probability that the individual will be an offender. The latent variable of liability is analyzed, not the dichotomized phenotype. Hence, it is appropriate to speak of *heritability of liability* to criminal offending; it is not technically correct to refer to the heritability of criminal offending.

One important specific application of the concept of liability is the multifactorial model. The central assumption of this model is that a large number of factors (many genes, parenting, schooling, peers, etc.) contribute to liability in roughly equal amounts so that some mathematical transformation will be able to scale liability to resemble a multivariate normal distribution within families. In this case, the tetrachoric correlation is the appropriate statistical index used to quantify familial resemblance for liability.

MENDELIAN DISORDERS

Molecular genetic studies of rare, well-defined disorders can elucidate basic physiologic mechanisms. Familial hypercholesterolemia, for example, unraveled aspects of lipoprotein receptor synthesis and transport. Are there any such exploitable models for human violence?

A search through Mendelian Inheritance in Man (MIM), a computerized data base of known or suspected heritable disorders, revealed eight disorders of potential relevance in the sense that the words "aggression," "rage," "violence," or "antisocial" were mentioned in the MIM description. Table 1 lists these disorders, their current MIM numbers, and comments. Except for alcoholism, the genetics of which are unclear, the disorders are only tangentially associated with violence. An appropriate model disorder (e.g., one with presenting symptoms such as inexplicable rage) has yet to be described, although this feature was present in one sibship from a consanguineous mating for Urbach and Wiethe's disease (Newton et al., 1971). Although seizure disorders did not appear in the search, emotional lability is a typical feature of Unverricht and Lundborg myoclonus epilepsy (MIM number 254800), a form of seizure disorder found largely in Finland.

At the other extreme, the genetics of some rare phenotypes associated with violence (e.g., repetitive rapists, pedophiles) have never been studied. It is possible that heritable forms may be uncovered among these rare individuals and that molecular genetics may be used to elucidate basic mechanisms of violence.

A potentially important finding that emerged after this review was completed deserves mention. Brunner et al. (1993a,b) reported on a single, large Dutch kindred in which an unusual number of males were affected with moderate intellectual deficiency and aggression. The pattern of transmission was consistent with X-linkage and perfectly correlated with a deficiency in the gene for the enzyme monoamine oxidase-A. Studies on the frequency of this gene and its association with aggression in the general population have yet to be conducted.

CHROMOSOMAL ANOMALIES

The report by Jacobs et al. (1965) of a high prevalence of men with an extra Y chromosome (karyotypes 47,XYY and 48,XXYY) among incarcerated males sparked research (and debate) into the issue of whether supernumerary Y individuals are at high risk for

TABLE 1 Known or Suspected Genetic Disorders Associated With Aggression

Disorder	MIM Number[a]	Comments
Urocanase deficiency	276880	Vaguely defined disorder; aggression noted in one pedigree.
Gilles de la Tourette symptom	137580	Neurologic disorder with involuntary motoric and vocal tics. Both aggression and self-mutilation have been reported, but neither is symptomatic of the disorder.
Precocious puberty, male limited	176410	Gonadotropin-independent gonadal testosterone secretion; male-limited, autosomal dominant transmission with onset of sexual precocity as early as 1 year; blockage of both androgen and estrogen synthesis was associated with a reduction of aggression in a series of nine boys.
α-Mannosidosis	248500	Aggressive tendency noted in cattle with α-mannosidase deficiency, but not known to be associated with the deficiency in humans.
Alcoholism	103780	Not known to follow strict Mendelian transmission patterns; well-established phenotypic correlation with aggression.
Lipoid proteinosis of Urbach and Wiethe	247100	One pedigree reported in which a sib suffered with attacks of rage.
Deafness-hypogonadism syndrome	304350	Only a single pedigree reported, consistent with X linkage; antisocial and immature behavior noted in males of the pedigree.
Fragile X syndrome	309550	Variable expression with frequent mental retardation; antisocial tendencies noted but not a central part of the syndrome.

[a]Mendelian Inheritance in Man.

violence, crime, and other psychopathology. Early research compared the prevalence of sex chromosome aneuploidy among selected samples (e.g., prisoners, psychiatric patients) with the prevalence among newborn screens or other controls. The initial reviews of 47,XYY and 48,XXYY syndromes by Owen (1972) and Hook (1973) stressed that prospective research on well-defined populations was necessary for accurate assessment of any relationship between XYY and violence.

Several prospective studies have now followed aneuploid children into adolescence. Results on 47,XYY suggest behavioral development well within the normal range, but with minor deficits in intelligence, other cognitive skills, and perhaps, in emotional and social skills (Bender et al., 1987). Study of a Danish birth cohort (Witkin et al., 1976) found a high prevalence of criminal registration among XYY individuals (5 of 12) that is not statistically different from the prevalence among 47,XXY (3 of 16) but is greater than the base rate among normal XY males of the same height (9% of slightly more than 4,000 men). Criminal histories of the five XYY individuals were not characterized by violence and aggression. Convictions were for minor offenses adjudicated by mild penalties, prompting the investigators to suggest that the relationship was likely due to nonspecific factors such as lowered intelligence. Personal follow-up revealed that the 12 XYY individuals had statistically significant but clinically minor differences from controls in sexuality, aggression, and testosterone levels (Schiavi et al., 1984, 1988; Theilgaard, 1984).

The prospective results dispel the myth of the XYY as a "hyperaggressive, supermasculine sociopath" and, in its place, portray a group of individuals within the normal range but with an array of relatively nonspecific behavioral differences in attention and cognition, motoric skills, and personality. For example, the sexuality of XYY individuals is characterized more by insecurity and difficulty in developing and maintaining satisfying relationships with women than by stereotyped hypermasculinity (Schiavi et al., 1988). It is possible that nonspecific behavioral problems may increase risk among these individuals for later criminal offenses.

GENETICS AND PERSONALITY TRAITS

Given that genes do not code directly for crime and violence, it might be reasonable to suspect that genetic diathesis is mediated through personality traits and cognitive styles. Kinship correlations for intelligence have been summarized by Bouchard and

McGue (1981), and the general pattern suggests important contributions from both genes and family environment. A survey of the genetic literature on personality traits is too vast to report here. Hence, this review is limited to two major domains of personality. First, scales purporting to measure *aggression* are reported, with the name of the scale dictating acceptance of a study for review. In some studies, the exact items bear strong content resemblance to the concept of violence used by the Panel on the Understanding and Control of Violent Behavior (e.g., Tellegen et al., 1988); in other cases, the content appears related more to a broader, almost psychoanalytic, notion of intrapunitiveness (e.g., Partanen et al., 1966); and in yet other studies, item content was not fully specified (e.g., Rushton et al., 1986). The second area for review is scales specifically constructed to predict juvenile delinquency. These scales include the Psychopathic-deviate (Pd) scale of the Minnesota Multiphasic Personality Inventory (MMPI; Hathaway and McKinley, 1940); the Socialization (So) scale of the California Psychological Inventory (CPI; Gough, 1969); and the aggression scale of the Missouri Children's Picture Series (MCPS).

Table 2 summarizes the results of this review. The studies are broadly classified into twin and adoption strategies. The twin results are consistent with the overall twin literature on personality: on the average, identical twins correlate higher than fraternal twins. A notable exception is reported by Plomin et al. (1981), the only study using blind ratings of aggression in a test situation (ratings of child's aggression against a Bobo doll). All other studies used self-report inventories or parental ratings. Unfortunately, sample sizes in Plomin et al. (1981) are too small to detect whether this difference is attributable to method of assessment, to aggression in childhood versus adulthood, or to sampling error from a trait with modest heritability.

In the Minnesota series of twins raised apart (Gottesman et al., 1984; Tellegen et al., 1988), the correlations are as great as those for adult twins raised in the same household. Although standard errors for these correlations are large, they suggest that the great similarity of twins raised together is not due exclusively to such processes as imitation or reciprocal interaction (Carey, 1986) that might invalidate the twin design.

There is little relevant adoption data. Different scales were administered to the Texas Adoption Project sample (Loehlin et al., 1985, 1987) in adolescence (the CPI in Loehlin et al., 1985) and in early adulthood (the MMPI in Loehlin et al., 1987). The patterning here is very similar to that of other adoption studies

TABLE 2 Genetics Studies of Personality Measures Related to Delinquency or Aggression

Study	Measure	Males Group	N	R	Females Group	N	R
Genders Analyzed Separately							
Owen and Sines	MCPS aggression	MZ	10	.09	MZ	8	.58
(1970)		DZ	11	−.24	DZ	13	.22
Gottesman	CPI socialization	MZ	34	.32	MZ	45	.52
(1966)		DZ	32	.06	DZ	36	.26
Scarr	ACL n aggression	MZ	24	.35			
(1966)		DZ	28	−.08			
Partanen et al.	Aggression items	MZ	157	.25			
(1966)		DZ	189	.16			
Loehlin and	CPI socialization	MZ	202	.52	MZ	288	.55
Nichols (1976)		DZ	124	.15	DZ	193	.48
	ACL aggressive	MZ	216	.20	MZ	293	.24
		DZ	135	−.05	DZ	195	.06
Rowe (1983)	Number of	MZ	61	.62	MZ	107	.66
	delinquent acts	DZ	38	.52	DZ	59	.46
Rushton et al.	23 aggression items	MZ	90	.33	MZ	106	.43
(1986)		DZ	46	.16	DZ	133	.00
		DZ-OS	98	.12			

Study	Measure	MZ N	R	DZ N	R
Genders Pooled					
Gottesman (1963, 1966); Reznikoff and Honeyman (1967)	MMPI psychopathy	120	.48	132	.27
Plomin et al. (1981)	Median (three objective aggression ratings)	53	.39	31	.42
Ghodsian-Carpey and Baker (1987)	CBC aggression MOCL aggression	21 21	.78 .65	17 17	.31 .35
Pogue-Geile and Rose (1985)	MMPI psychopathy	71 71	.47 .23	62 62	.15 .20
Rose (1988)	MMPI psychopathy	228	.47	182	.23
Tellegen et al. (1988)	MPQ aggression	217	.43	114	.14

TABLE 2 (Continued)

Study	Measure	MZ N	MZ R	DZ N	DZ R
Twins Raised Apart (Minnesota Sample)					
Tellegen et al. (1988)	MPQ aggression	44	.46	27	.06
Gottesman et al. (1984)	MMPI psychopathy	51	.64	25	.34

Study	Measure	Relationship	N	R
Adoption Studies				
Loehlin et al. (1985)	CPI socialization	Adoptive father-child	241	−.03
		Adoptive mother-child	253	−.02
		Biological father-child	52	.16
		Biological mother-child	53	.06
		Adoptive-adoptive sibs	76	.03
		Adoptive-biological sibs	47	.10
		Biological-biological sibs	15	−.01
Loehlin et al. (1987)	MMPI psychopathy	Adoptive father-child	180	.07
		Adoptive mother-child	177	.01
		Biological father-child	81	.12
		Biological mother-child	81	.07
		Birth mother-adopted child	133	.27
		Adoptive-adoptive sibs	44	.02
		Adoptive-biological sibs	69	.06
		Biological-biological sibs	20	−.06
Parker (1989)	CBC aggression items	Adoptive sibs (age 4)	45	.54
		Natural sibs (age 4)	66	.42
		Adoptive sibs (age 7)	17	.28
		Natural sibs (age 7)	19	.55

NOTE: ACL = adjective checklist; CBC = child behavior checklist; CPI = California Psychological Inventory; MCPS = Missouri Children's Picture Series; MMPI = Minnesota Multiphasic Personality Inventory; MOCL = Mothers' Observational Checklist; MPQ = Multidimensional Personality Questionnaire.

(e.g., Scarr et al., 1981)—zero-order correlations with adoptive parents and small positive correlations with biological parents. This is consistent with the twin data in suggesting heritability, although estimates of the genetic effect are lower in the adoption than in the twin studies. The fact that adoptive relatives bear little resemblance to one another suggests that processes such as imitation and common family environment have weak effects on these psychometric predictors of delinquency. Parker's (1989) recent analysis of data from the Colorado Adoption Project (Plomin et al., 1981) gives a very different picture. Based on maternal ratings of aggression, both adoptive and biological siblings show strong, roughly equal, resemblance. These data agree with the Plomin et al. (1981) results in suggesting important common environment effects for childhood aggression, but do not agree with the small twin studies of Owen and Sines (1970) and Ghodsian-Carpey and Baker (1987).

Together, the personality data imply a genetic contribution to individual differences for important *correlates* of violence. This inference is stronger for older adolescents and adults than for children. A glaring lack in this literature (as well as in the genetic literature on criminal offenders) is the absence of data to permit multivariate genetic analysis of personality traits such as aggression and criminal offending.

Since this review was completed, several important studies of parental ratings of childhood aggression and delinquency in child and adolescent twins have been completed and initial results suggest important heritability. Gottesman and Goldsmith (in press) should be consulted for a review.

JUVENILE ANTISOCIAL BEHAVIOR

Here, research on actual crime or antisocial behavior during adolescence is reviewed. Early twin studies of juvenile delinquency, summarized in Table 3, gave strong evidence of important common environment effects and weak evidence for heritability. Many of these early studies did not select samples or define phenotypes with the rigor required by modern research standards (see Slater and Cowie, 1971; Christiansen, 1977, for reviews). More recently, Rowe (1983, 1985, 1986) analyzed the number of self-reported antisocial behaviors from junior and senior high school twins. The twin correlations (given previously in Table 2) demonstrate significant heritability and, agreeing with the concor-

TABLE 3 Pooled Twin Concordance Rates for Juvenile Delinquency in Identical and Same-Sex Fraternal Twins[a]

Gender	Identical		Fraternal	
	Number of Pairs	Percent Concordant	Number of Pairs	Percent Concordant
Female	12	92	9	100
Male	55	89	30	73

[a]Based on the review by Cloninger and Gottesman (1987), eliminating pairs studied by Kranz (1936) in which concordance was not reported separately by gender.

dances in Table 3, implicate common environment, albeit without reaching statistical significance.

Adoption studies support the importance of family environment in early antisocial behavior. Cadoret et al. (1983) reported a significant main effect for an adverse adoptive home and some form of gene-environment interaction in three different adoptee samples. Bohman (1971) and colleagues (Bohman and Sigvardsson, 1985; Bohman et al., 1982) prospectively studied Swedish children from unwanted pregnancies. At age 15, those who remained with their own biological parents or who were placed as foster, but unadopted, children had almost twice the rate of antisocial-like behavior problems (truancy, running away, misuse of alcohol and drugs, repeated thefts) as their classmate controls. Children who were formally adopted, however, showed slightly *lower* rates than their controls and much lower rates than both the foster children and those remaining with their parents, despite a high frequency of criminality and alcohol abuse among their birth parents. Bohman suggested that any adverse genetic liability was neutralized by the benefits provided by secure adoption.

Together, the early twin studies, the nonsignificant trend in Rowe's study, the Cadoret analyses, and the Bohman results provide strong evidence for a family environment effect on juvenile antisocial behavior. Both the Rowe and the later Cadoret studies suggest that genetics cannot be ignored during this period. Perhaps the most important research question for the future is the investigation of genetic and family environmental contributions to adolescent antisocial behavior that persists into adulthood and the detection of reasons why this behavior ceases in many adolescents.

ADULT ANTISOCIAL BEHAVIOR

The bulk of the literature relevant to genetics and violence comprises studies of adult criminality. The results from early twin studies are summarized in Table 4. Compared to the concordances for juvenile delinquency (see Table 3), there appears stronger evidence for the role of genes in adult antisocial behavior than in adolescent delinquency. Although the methodology of most of these twin studies is poor by absolute standards, they are largely the same studies that generated the juvenile data. Strong methodological biases might be expected to affect both the younger and the older twins in a single series, yet the concordance rates appear to differ.

The more modern studies have undergone recent review by Rowe and Osgood (1984), Ellis (1982), and Walters and White (1989). Here, the samples are reviewed and described with an eye on the generalizability of the results.

Norwegian Twin Study

Dalgard and Kringlen (1976) identified male twins born in Norway between 1921 and 1930, of whom at least one member had appeared in a national police register by the end of 1966. The 139 intact and located pairs were interviewed personally, or information was gathered through family history interviews with relatives. Like other selected samples, the registered probands were

TABLE 4 Pooled Twin Concordance Rates for Adult Criminality in Identical and Same-Sex Fraternal Twins (early studies)[a]

Gender	Identical		Fraternal	
	Number of Pairs	Percent Concordant	Number of Pairs	Percent Concordant
Female	7	86	4	25
Male	112	67	101	37

[a]Based on the reviews by Gottesman et al. (1983) and Cloninger and Gottesman (1987), eliminating pairs studied by Lange (1930) in which gender of twins was not specified. Two of 43 DZ probands in Kranz's (1936) sample are female but are included in the male calculations because concordance of the two pairs was not specified in the original article.

characterized by lower education, lower occupational status, and increased rates of alcohol problems.

With a broad definition of criminal involvement (legally punishable behavior reported to the criminal register), there was little difference in concordances—37 and 31 percent for monozygotic (MZ) and dizygotic (DZ) pairs. Slightly better discrimination was found with a stricter definition of criminal involvement (crimes of violence, sexual assault, theft and robbery)—41 and 26 percent, respectively, for MZ and DZ twins—but the difference does not reach significance. Unfortunately, base rates for criminal registration were not provided. Neither were analyses performed jointly on alcohol use and crime. An intriguing but statistically nonsignificant finding that emerged through the interviews was that concordant identical twins tended to collude more often in the same criminal act than fraternal pairs.

Danish Twin Study

Christiansen (1968) identified all twins born on the Danish islands between 1880 and 1910, where both members survived to age 15, and traced the twins through national and local police or penal registers. Updates of the sample were provided by Christiansen (1974, 1977), Cloninger et al. (1978), and after Christiansen's death, by Gottesman et al. (1983) and most recently by Cloninger and Gottesman (1987).

On virtually every type of classification for crime, MZ twins are more concordant than DZ twins. One illuminating finding is that despite base rate differences in criminal registration between males and females, tetrachoric correlations of liability are remarkably similar—.74 for both male and female MZ twins, .47 for male-male DZ and .46 for female-female DZ pairs. In contrast, the correlation for male-female pairs is .23, suggesting either sex-by-genotype interaction (i.e., the loci that contribute to individual differences in males do not have identical effects in females) or common environment-by-sex interaction (i.e., those environmental factors that contribute to twin concordance do not have the same effects in males and females).

One major difficulty in interpreting these twin results is the difference in base rates for the three zygosity groups illustrated in Table 5. In both genders, MZ twins have the highest—and opposite-sex DZ twins the lowest—registration rates. The difference is significant for males and almost reaches significance for females. Such a pattern suggests an "imitation" or collusion effect

TABLE 5 Prevalence of Registered Criminality in Danish Twins by Sex and Zygosity

Zygosity	Males		Females	
	Number[a]	Percent Criminal	Number[a]	Percent Criminal
Identical	730	13.42	694	2.59
SS fraternal	1,400	12.29	1,380	2.17
OS fraternal	2,073	9.55	2,073	1.47
χ^2	11.00[b]		4.70[c]	

[a]Total number of individuals.
[b]$p < .001$.
[c]$.10 > p > .05$.

NOTE: SS = same sex; OS = opposite sex.

SOURCE: Data from Cloninger and Gottesman (1987).

that would increase variance (and hence prevalence) as a function of the magnitude of the imitation and heritability (Carey, 1986; 1992).

Danish Adoption Studies

The Danish adoption studies on registered criminality began with identification of 1,145 male nonfamilial adoptees and a series of matched nonadoptees (hereafter termed "controls") born in the city and county of Copenhagen from 1927 to 1947. Both adoptees and controls were between ages 30 and 44 when they were followed up through national registers for criminality and psychiatric disorder. Detailed analyses on criminal registration in this sample were done by Hutchings (1972) and reported in Hutchings and Mednick (1977); analyses on the diagnosis of sociopathy were performed by Schulsinger (1972). Two major findings are consistent with a hypothesis of genetic effects. First, despite adoption into households offering a representative standard of living in Denmark, adopted males had almost twice the rate of criminal registration as their controls (16.2 versus 8.9%). Because 30.8 percent of biological fathers had a criminal record, compared to 12.6 percent of adoptive fathers and 11.1 percent of control fathers, the increased rate of criminality among adoptees is consistent with their inheritance of a greater genetic liability toward crime than controls. (This finding, however, is not necessarily proof of a genetic etiology because adoption per se may have elevated crime rates.) Second, criminality in the biological fathers was correlated with criminality in the adopted sons. Among adoptees whose biological fathers were registered, 22.7 percent were registered for criminal offenses and another 35.7 percent were on record as having committed only minor offenses. Among adoptees whose fathers were not registered, the respective percentages are 13.6 and 33.7 percent. This pattern suggests a heritable effect for more serious crime, but perhaps a lack of one for minor offenses.

Also evident in the data was a significant family environment effect. Given criminal registration in an adoptive father, 27.1 percent of adoptees had criminal offenses and another 37.5 percent had minor offenses. When the adoptive father was not registered, the respective percentages were 14.5 and 33.9 percent, again suggesting important family environmental effects for more serious crime.

The sample of Danish adoptees was later extended to the female adoptees, and to the entire country of Denmark, and reported on by Mednick et al. (1983, 1984). The expanded sample lacked the controls of the earlier sample and used court convictions instead of criminal registration as the index of crime. Results on biologic parent criminality replicated earlier results. The probability of court conviction among adoptees increased almost linearly with the number of court convictions for biological parents. Of importance here was that the effect was evident only for property crime and not for violent offenses. (The precise definition of a violent offense is not offered by the authors.) Curiously, the significant effect of adoptive parent criminality found in the Copenhagen adoptees did not replicate in the extended sample, and no reason was offered for this lack of consistency.

Further analyses of this sample have been provided by Van Dusen et al. (1983a,b), Baker (1986), Baker et al. (1989), and Moffitt (1987). Significant predictors of adoptee convictions include socioeconomic status of both the adoptive and the biological parents, psychiatric history of the biological parents (especially personality disorder and substance abuse), and late placement and number of placements before final adoption. Again, these relationships are significant for nonviolent convictions. Either violent offenses were not analyzed (Baker, 1986; Baker et al., 1989), or the results were largely insignificant (Van Dusen et al., 1983a,b; Moffitt, 1987). An exception was a single analysis by Moffitt, reported in Mednick (1987). Here, violent offenses in male adoptees were predicted by the number of convictions among biological fathers and antisocial types of psychiatric disorder in biological mothers.

Swedish Adoption Study

In addition to his study of unwanted pregnancies, Bohman (1978) identified 2,324 individuals born in Stockholm, Sweden, between 1930 and 1949 who were given up by birth parents for nonfamilial adoption. These adoptees and their birth parents were tracked through official Swedish records for alcohol abuse (having a fine imposed for intemperance, supervision by a local temperance board, or treatment for alcohol abuse) and criminality (defined as a conviction with more than 60 "day fines," a fine prorated to the defendant's income). At the time of the record search, adoptees' ages ranged from their early twenties to early forties.

Like the biological parents in the Danish sample, biological mothers and fathers both had high rates of registration, but one striking feature distinguishes these two Scandinavian samples— none of the adoptive parents in Sweden appeared on the criminal register (Bohman et al., 1982), whereas the Danish adoptive parents had rates of criminal registration only slightly below population base rates (Mednick et al., 1983). Thus, the Swedish adoptees may be regarded as a special sample selected for higher-than-average genetic liability but lower-than-average family environment liability to crime. What effect did this peculiar selection have? Apparently, the effects cancel each other. The base rate for criminal registration among male adoptees was 12 percent compared to an 11 percent population risk (Bohman et al., 1982). This fact alone argues against misguided views of genetic determinism of behavior that is not amenable to environmental intervention.

The initial results were strikingly negative for any genetic effects on crime: 12.5 percent of the male adoptees with a criminal biological father themselves had a criminal record, compared to 12 percent of the male adoptees whose biological fathers did not have criminal records. Respective prevalences for the adoptees with and without criminal mothers were 12.6 and 12.4 percent. Base rates for criminal conviction among females were too low to permit meaningful analysis.

More detailed analyses on 76 percent of this adoptive sample were reported by Bohman et al. (1982), Cloninger et al. (1982), and Sigvardsson et al. (1982). (The reduction in sample size was due to deletion of adoptees with incomplete information, late placement, or intrafamilial adoption.) For male adoptees, there was a nonsignificant correlation between adoptee conviction and biological parent conviction—13.2 percent of 258 male adoptees with either a convicted father or a convicted mother were themselves convicted, compared to 10.4 percent of 604 adoptees whose biological parents had not been convicted ($\chi^2_1 = 1.37$, $p = NS$). Analogous figures for female adoptees were not reported.

Suspecting heterogeneity for criminality, Bohman et al. (1982) conducted a series of discriminant analyses aimed at distinguishing subgroups of male adoptees. The final classification suggested that alcohol abuse moderated the genetic relationship to criminality. Criminal male adoptees who were also registered for alcohol abuse tended to have committed a larger number of offenses, were more likely to have been convicted of a violent offense, received longer jail sentences, but were less often registered for property crimes than male adoptees registered only for criminality. Moreover, biological parental variables also distinguished the male adoptee subgroups. Most important here are the findings that violent paternal offenses predicted offspring *alcohol abuse* better than they predicted offspring *criminality* and that adoptees registered for criminality but not for alcohol abuse tended to have fathers registered for property crimes and fraud. On the environmental side, male adoptees with criminality only were distinguished by late placement in adoptive homes, the number of placements, and a longer duration spent with the biological mother relative to other adoptee groups.

Others (Walters and White, 1989) have criticized this study for not reporting and controlling for the number of statistical tests, but the original investigators did partly replicate the discriminant function results on female adoptees (Sigvardsson et al., 1982). Of greater concern is what the results mean.

The inability to uncover statistically significant findings on criminality per se is a clear failure to replicate the results in neighboring Denmark. Perhaps this is due to differences in registration practices between the two countries. Was there a tendency in Sweden to report instances of joint alcohol abuse and illegal activity to the temperance board, and not to the police, with perhaps the reverse tendency in Denmark? A second possibility is that the selection of adoptive households in Sweden was so extreme that it overcame potential genetic liability for crime or violence. The pattern might occur if the relationship among genotype, family environment, and crime were nonlinear or if there were strong interaction between genotype and family environment. The pattern of cross-fostering risk in the Stockholm sample was consistent with gene-environment interaction, although the interaction term is not significant (Cloninger et al., 1982).

At the same time, the Swedish results demonstrate that the genetics of violence cannot be divorced from the study of alcohol abuse (and, in U.S. society today, probably other substance abuse as well). There may be important genetic differences among individuals who commit serious offenses as a function of liability toward alcohol and drug abuse.

Iowa Adoption Studies: Crowe

By checking penal institution records in Iowa, Crowe (1972) identified 41 female offenders (90% of whom were felons) who gave up a total of 52 children for adoption. A consecutive series of control adoptees was identified through state records and matched on age, sex, race, and approximate time of adoption. At an average age of 25 (range from 15 through 45), the adoptees were traced through Iowa arrest records. The adopted offspring of the female felons were more likely to have had an arrest record (15%) and conviction (14%) than the control adoptees (4% and 2%, respectively). Five of the 52 probands were incarcerated in Iowa for an offense, compared to none of the controls (Crowe, 1972). Personal follow-up on 70 percent of the sample revealed that 13 percent of probands met diagnostic criteria for antisocial personality, whereas only one control (2%) was diagnosed as a probable antisocial personality (Crowe, 1974). Several placement variables (age at placement and length of time in temporary care before adoption) predicted antisocial outcomes.

Iowa Adoption Studies: Cadoret

Cadoret continued his previous line of adoption research by identifying a series of adopted children from Iowa adoption records whose biological parents had some recorded psychopathology. (This sample did not overlap with that of Crowe.) A second, matched group of adoptees whose record data did not mention psychopathology in the biological parents served as controls. Both groups were personally followed up between the ages of 10 and 37. Sample attrition was high among the adoptive families—more than 30 percent of those contacted refused to participate (Cadoret, 1978).

Sample sizes became small when adoptees were grouped by age and by diagnosis of biological parents—only nine male and three female adult adoptees had a biological parent with a diagnosis of antisocial personality—so most analyses were performed on counts of antisocial symptoms (e.g., lying, truancy, trouble with the law) in the adoptees. Both an antisocial and an alcoholic biologic background predicted the number of antisocial symptoms, although the significance levels of the prediction varied slightly from publication to publication (Cadoret, 1978; Cadoret and Cain, 1980; Cadoret et al., 1983, 1985, 1986). Later publications stress the independence between genetic liability to antisocial personality and alcohol abuse (Cadoret et al., 1985).

On the environmental side, discontinuous mothering of adoptees, adoptive parent psychopathology, and divorce or separation in the adoptive family predicted antisocial outcomes. Male adoptees appeared more likely than females to have high antisocial behavior counts in the presence of these adverse environments (Cadoret and Cain, 1980). Recently, substance abuse in the adoptees could be predicted from biologic parents' antisocial behavior and alcohol abuse (Cadoret et al., 1986).

Since this review was completed, the preliminary reports in three different samples of twins have also pointed to heritability for antisocial behavior. The first sample was a small series in which twins were ascertained in a psychiatric setting and demonstrated both heritability of symptoms of antisocial personality disorder and important genetic correlations between antisocial behavior and alcohol and substance abuse (Carey, 1993; Miles and Carey, 1993). The second sample was a population-based twin sample of children and adolescents and suggested modest heritability for a classification of symptoms of conduct disorder (Eaves et al., 1993). The final sample consisted of male Vietnam-era armed services veteran twins. Analysis of symptoms of antisocial personality disorder suggested moderate heritability for adult symptoms with a stronger influence of common environment for adolescent symptoms (Lyons et al., 1994).

VIOLENCE AND HUMAN GENETICS

Only three modern samples permit meaningful analyses of violent offenders—the Danish adoption sample, the Stockholm adoption sample, and the Danish twins. Table 6 presents a summary of their findings. In the Danish adoption study, the only publicly reported data are the proportions of male adoptees who participated in a violent offense as a function of the number of criminal convictions in their biological parents (Mednick et al., 1983, 1984, 1988). Mednick et al. (1983) also cite other analyses (e.g., violent

TABLE 6 Results of Modern Genetic Studies of Violent Offenders

Sample	Variables	Results
Danish adoptees	Number of court convictions of biological parents with percent of male adoptees who committed a crime of violence	Very slight but insignificant trend toward increased participation rates in violent offenses among male adoptees as a function of number of biological parental convictions.
Stockholm adoptees	Presence of one or more violent crimes in biological fathers, with criminal registration and alcohol abuse registration in their male adoptive offspring	Insignificant, slightly negative relationship between biological father's violent offense and adoptive son's registration for crime; positive and significant relationship for son's registration for alcohol abuse.
Danish twins	Concordance for crimes against person, irrespective of other offenses, in MZ and DZ males	Both MZ and DZ concordance and correlations for crimes against person are greater than 0; MZ correlation is greater than DZ correlation (.77 versus .52), but difference is not significant

crimes in parents and in adoptees) that did not uncover a significant relationship, but no data are presented.

From Bohman et al. (1982:Table 5), it is possible to reconstruct contingency tables relating the probability that biological fathers registered for at least one violent offense will have an adopted son registered for crime or alcohol abuse. (The investigators did not publish data on violent offenses in offspring.) The association between paternal violence and adoptee's criminal registration in the Stockholm adoption study is slightly negative— 7.9 percent of the adopted sons of biological fathers registered for a violent crime are criminal, compared to 10.6 percent of the sons of fathers who were not registered for a violent offense. This negative relationship also appeared for women (Sigvardsson et al., 1982:Table 4).

Paternal violence, however, relates significantly to sons' but not daughters' registration for alcohol abuse. Given a biological father registered for a violent offense, 24.7 percent of the sons exhibited alcohol abuse; if the biological father did not have a violent offense on record, 16.6 percent of the sons were registered for alcohol problems (χ^2 = 4.145, p < .05, tetrachoric r = .145).

Finally, Cloninger and Gottesman (1987) analyzed the male MZ and same-sexed male DZ Danish twin series by crimes against persons, a category previously defined by Christiansen to include crimes of violence and sexual offenses. In contrast to the adoption studies, the twin data demonstrate a strong familial effect for these crimes. Concordances for the 24 MZ and 39 DZ probands were, respectively, 41.7 and 20.5 percent, compared to respective base rates of 3.3 and 2.8 percent among the two types of twins. The tetrachoric correlations for both zygosities are significantly greater than 0.0: r_{MZ} = .77 ± .11 and r_{DZ} = .52 ± .10. Although they suggest a genetic effect, a likelihood ratio test for heritability is not significant (χ^2_1 = 1.775, .20 < p < .10).

Major limitations of all three studies are the reliance on official records and insufficient detail in any of the studies to assess the magnitude of the correlation between recorded crimes and violent acts as defined by the panel. For example, would a significant proportion of crimes that involved an assault be registered under a Scandinavian equivalent of disturbing the peace? Inaccuracies of classification would make it difficult to uncover heritable effects and could be used to explain the adoption results. If violence had the same approximate biological father adopted offspring correlation as property crime (about .15 judging from Baker et al., 1989) and the base rates for violence range from about 6 to 10 percent among biological fathers (Mednick et al., 1983) and 2 to 5 percent among their adopted sons, then the Danish sample has between .55 and .82 power to detect an effect at the .05 level. On the other hand, if inaccuracy in classification lowers the correlation to .10, power is diminished to .25-.45. The fact that twin concordance is significant ensures that the record classification is not totally unreliable but at the same time does not permit actual quantification of reliability.

Together the data do not suggest a strong role for heredity in violence. On the one hand, the positive correlation between violence in biological parents and alcohol abuse in adopted sons and the trend of the twin correlations suggest a genetic effect. On the other hand, the failure in both adoption studies to detect a significant relationship between violent offending and other indices of crime in separated relatives is evidence that any putative genetic factor is weak. Whatever the case for genetics, the strong correlations in MZ and DZ twins suggest that family factors shared by siblings would be a profitable avenue for future research.

SUMMARY AND CRITICAL OVERVIEW OF THE RESULTS

1. *There is a trend in most studies, albeit not always a statistically significant one, consistent with the hypothesis of a genetic effect on adult and perhaps adolescent antisocial behavior.* The consistency of the literature is not to be confused with generalizability of the results to contemporary violence in the United States. The latter issue is discussed in point 10.

2. *The genetics of antisocial behavior do not fit a simple model.* Here, a central comparison involves the estimates of heritability from the Danish twin and adoption studies, given that sample sizes are large and that cross-national variations in definitions of crime and its registration are minimized. By using tetrachoric correlations, the twin data suggest a heritability of around 55 percent. Correlations from the Danish adoption data (Baker et al., 1989) suggest heritabilities about one-half this amount. On a simple additive genetic model, both estimates should be similar. The joint effects of marital assortment, temporal trends over time, nonadditive genetic variance, special twin effects, etc., must be considered in studying the genetics of antisocial behavior.

3. *There is reason to question the application of simple genetic models to the traditional twin design in studying the genetics of adult antisocial behavior.* Both the Danish and the Norwegian twin series provide evidence of imitation, collusion, or reciprocal interaction between twins, especially MZ twins, for criminal participation. Rowe (1985) also reported that adolescent twins often engage in delinquent acts together. Although collusion might invalidate the assumptions of the traditional twin design, the application of statistical models that include imitation effects (Carey, 1986; Eaves, 1976) or direct measurement of potential interactive effects (e.g., the analyses performed by Rowe, 1985) might illuminate important sources of sibling similarity that cannot be isolated without the genetic information provided by twins. Along these lines, data from single-offspring families and data expressed as a function of sibship constellations must complement the twin data (see Carey, 1986; Eaves, 1976).

4. *The adoption studies have consistently reported a correlation between adoptee antisocial behavior and some variable intervening between birth and final placement in the adoptive home.* Some preplacement variable has predicted adoptee antisocial behavior in the Crowe, Cadoret, Danish, and two Swedish adoption studies. The relevant construct, however, eludes identification because the *same* variable (e.g., number of placements

versus disruptive mothering) has not been replicated across studies or within genders of a single study (e.g., compare Bohman et al., 1982, with Sigvardsson et al., 1982).

5. *Evidence for a genetic effect specific to offenses involving physical aggression is weak.* Part of the problem is that the relatively low base rate for violent offenses such as murder, assault, and rape requires samples much larger than those already gathered to give meaningful and replicable results. What is clear is that genetic liability toward participation in criminal activity is not due exclusively to liability toward physical aggression; otherwise, one would find little heritability of liability for crime per se but rather a tendency for violent crime to aggregate in twins and biological relatives of violent adoptees.

6. *The available evidence is inconsistent with a major polymorphism on the Y chromosome associated with antisocial behavior.* A major polymorphic gene on the Y chromosome could theoretically explain some of the large gender differences in criminal participation and also some of the heritable individual differences within males. Although the Y chromosome appears to contribute to aggressive behavior in the mouse (Maxson et al., 1989), a similar polymorphism in humans would predict greater father-son than father-daughter or mother-son resemblance. The tetrachoric correlations for the Danish adoption study (see Baker et al., 1989) do not show this pattern.

7. *The selection of adoptive families may make it difficult, in adoption study designs, to generalize about the effects of family environment that are relevant for criminal participation of offspring.* Registered criminality among adoptive parents was nonexistent in the Swedish adoption study; antisocial behavior was rare among the adoptive parents in Cadoret's study; and a five-year period free of criminal registration was a requirement for adoptive families in Denmark. These findings are consistent with some selection against prospective adoptive parents with a criminal record and raise the possibility that adoptive homes are nonrepresentative with respect to family environmental factors that contribute to offspring antisocial liability. Under these circumstances, it would be unwise to attribute the lack of adoptive parent-offspring correlation for criminality to an absence of vertical environmental transmission. The study of unselected, intact nuclear families in concert with the biological and adoptive families would provide important information about vertical transmission, but such control families have not been utilized in most adoption research.

8. *There is evidence for a genetic association between antisocial behavior and alcohol abuse and possibly other substance abuse.* The Danish adoption and twin studies did not analyze substance abuse, and the Norwegian twin study, although it gathered data on alcohol use, did not report analyses of its association with criminality in twins. Crowe reported no excess of alcohol abuse among the adopted offspring of female felons. The Swedish study reports a genetic association between biological paternal violence and adoptee alcohol abuse, and the investigators suggested genetic heterogeneity—the joint occurrence of criminality and alcohol abuse may have a different genetic liability than that of criminality alone. The association in Cadoret's series varies slightly with the type of data analysis, but points in directions similar to those of the Stockholm study. There is also a strong suspicion that alcohol abuse is itself genetically heterogeneous, with one meaningful subgroup related to antisocial personality (Cloninger, 1987).

9. *There is a strong tendency, especially in the area of juvenile antisocial behavior, to implicate some correlate of home environment as an etiological component in antisocial behavior.* Analyses of four different samples—Cadoret et al. (1975), Cadoret (1978), Crowe (1972) as analyzed by Cadoret et al. (1983), and Bohman (1971)—reveal significant findings in this area. The base rate for criminality in the adult Stockholm adoptees is also congruent with the hypothesis that adoption neutralized genetic liability. Although the expanded Danish sample did not reveal a significant correlation with adoptive parent criminality, urban environment predicted adoptee crime independent of biological background (Gabrielli and Mednick, 1984).

10. *It is difficult to integrate the genetic literature on antisocial behavior into contemporary criminological research on violence.* There are two major reasons for this. The first is extrapolation over space and time; the second is extrapolation over measures of violence. A considerable amount of current research in the United States targets high-crime areas, usually urban, and uses survey instruments. There is no sample in the genetic literature that can readily extrapolate to an urban U.S. population studied with survey instruments.

With regard to extrapolation over time and over cultures, it must be recalled that heritability is a function of the amount of environmental variability for a trait. As environmental variance decreases, heritability increases, and as environmental variance increases, heritability decreases. It is theoretically possible that increased rates of violence associated with drug abuse, for example, might reduce the magnitude of a genetic effect on antisocial behavior or even change the nature of that effect (e.g., loci that contribute to the reinforcing effects of cocaine may become more important). The larger samples in the genetic literature are Scandinavian. One could argue that their relative cultural and socioeconomic homogeneity would provide a ripe medium for maximizing genetic effects. In addition, the Scandinavian samples involve cohorts that went through the major portion of risk for initial antisocial behavior before the upsurge of drug use in the 1960s and 1970s. (All the Danish twins went through their risk period before World War II.)

Another limitation of the large Scandinavian studies is the exclusive reliance on official records for defining phenotypes such as criminality and violence. Although ethics may preclude personal contact with retrospectively identified adoptive families, field research that compares classification via official records with classification using contemporary survey instruments would increase the interpretability of the Scandinavian results. If such research has already been conducted, the results have not been integrated into the primary reports on the adoptees and twins.

The results of the smaller U.S. studies, especially Crowe's, are very consistent with the Scandinavian research. In two of the studies, Cadoret's and Rowe's, the phenotypes include counts of adolescent antisocial behaviors that include many peccadillos (e.g., lying, truancy) when compared to an adult court conviction (used in the Danish study). Taken at face value, the Rowe and Cadoret results suggest heritability of "mild" antisocial behavior. If there were cross-national generalization of these results, one would expect the Scandinavian material to show a similar pattern. Arrests have apparently been recorded in the Danish study (see Mednick et al., 1988) but have not been subjected to genetic analysis. Also, two of the U.S. samples were conducted in Iowa, and the third noted few respondents from inner-city areas (Rowe, 1983).

Can we extrapolate these results to major U.S. metropolitan areas? In the primary genetic literature, there is no attempt on the part of the authors to provide information congruent or discrepant with such extrapolation. Hence, there is no positive evidence to permit this generalization, but at the same time, there is no direct negative evidence against the induction. So, given the consistency of the genetic literature, it is perhaps best to conclude that it would be unwise to overlook the possibility of a genetic contribution to individual differences in antisocial behavior in contemporary criminological research.

PROSPECTS FOR FUTURE HUMAN GENETIC RESEARCH

1. Given the difficulty and expense—and the great importance—of collecting data on twins and adoptees, criminology would immediately benefit from more extensive and more standardized description and data analysis of existing samples. For example, it is curious that a significant genetic effect is reported for court convictions (irrespective of alcohol abuse) in Denmark but not in a similar age cohort in Sweden. Tetrachoric correlations have been reported separately for males and females in the Danish adoptee sample but not for the Stockholm sample. Complete cross-tabulations for violent crime in the Danish sample have not been reported. One inexpensive way to address these problems is to fund a meeting of the original investigators to present preplanned standardized analyses and deal with issues of replicability. Such a meeting could also lead to heuristic hypotheses about the nature of adoption preplacement effects, cross-cultural extrapolation of results, etc.

2. Although the issue of specialization in crime has not been fully clarified, pedigree studies of well-defined, homogeneous groups of offenders should be encouraged. There are very few direct family data on such important phenotypes as pedophiles and repetitive rapists.

3. Although future research might isolate one or more genetic syndromes associated with violence, polygenic influences on a number of different personality, cognitive, and psychopathological traits may be the major source of genetic predisposition to antisocial behavior. Future large-scale, epidemiological research with genetically informative samples should concentrate less on demonstrating the *fact* of a genetic diathesis and more on *uncovering the sources* of familial liability, both genetic and environmental. Consequently, data must be gathered to permit multivariate analysis involving, on the one hand, antisocial behavior and, on the other hand, personality factors such as impulsivity and sensation seeking, cognitive variables such as risk perception, and psychopathology, especially alcohol and drug abuse.

4. Identification of a personality trait or even a genetic locus that contributes to antisocial liability would be a major contribution to pure science, but it is not clear that such a discovery would have immediate practical applications for prevention and intervention. If genetic diathesis is multifactorial, then any single factor will be only a weak predictor. In addition, questions of genetic engineering of antisocial behavior are more appropriate

for science fiction than for contemporary science. This raises quandaries about the best direction of future genetic research. Is it better to study the genetics per se of antisocial behavior, or is it preferable to utilize genetics as a control variable so that less equivocal statements can be made about *environmental* factors? In the area of a child's cognitive growth, several variables hypothesized to be important environmental contributors (e.g., parental educational level) may be more genetic than environmental in origin (Plomin et al., 1985). Although there is an established correlation between several parental traits (e.g., inadequate supervision of offspring) and a child's antisocial behavior, the nature of these correlations is unclear. To what extent is inadequate parenting an environmental contributor to delinquency, and to what extent is it symptomatic of a genetically influenced diathesis that is transmitted to the child? The judicious use of research designs that control for genetics will better elucidate the environmental factors amenable to intervention than will other nonexperimental strategies.

5. The findings on adverse home environment and points 3 and 4 above, indicate that a critical—and unstudied—population is the unrelated sibship residing in the same home. Judging from family distributions in the Texas Adoption Project and the Colorado Adoption Project, a significant proportion of adoptive parents either adopt two children or have a natural child of their own. Although considerable effort would be needed to identify a large number of these sibships, such a study should be feasible and results would permit strong inference about environmental etiology.

SOCIOBIOLOGY, EVOLUTIONARY THEORY, AND VIOLENCE

The behavioral genetic approach presented above seeks to identify sources of individual differences. Compared to other areas of social science research, it is strongly empirical but atheoretical. A different perspective on genetics and human antisocial behavior is offered by some ethological and sociobiological research that begins with strong theoretical assumptions and examines how well empirical data agree with the predictions from theory.

Several attempts (e.g., Ellis, 1987) have been made to explain the diverse correlates of human antisocial behavior in terms of individual differences in reproductive strategies, notably the r/K types systematized as a heuristic for categorizing between-species differences. The r-strategy is associated with relative ecological instability and tends to promote such qualities as small body size, rapid sexual maturation, short life span, and profligate reproduction. The K-strategy is associated with environmental stability and gives advantage to large body size, delayed maturation with parental care, longevity, and economical reproduction. By extrapolating the r/K between-species distinction to a continuum of within-species variability, human aggression is viewed as part of the r end and law abidingness as part of the K end of the dimension.

A second research thrust has been the study of familial homicide by Daly and Wilson (1988a,b). Based largely on the concept of inclusive fitness, they predict that homicide (and presumably other aggression) should vary inversely with the degree of genetic relatedness of assailant and victim and with the reproductive capacity of the victim.

This present round of sociobiological/evolutionary research is more empirical than many previous attempts to extrapolate from other species to human behavior. The research is also relatively new, so a large body of well-controlled studies has yet to be developed to permit assessment of the predictions. For example, Daly and Wilson present Canadian homicide data suggesting that filicide victimization is considerably greater for infants who have yet to reach their first birthday than for older children. To what extent is this due to inhibition because of the reproductive and temporal investment already made in an older child, to what extent is it a consequence of maternal postpartum psychoses, and to what extent is it due to the physical vulnerability of the young infant to a punitive blow that might injure but not kill an older child?

For theories about the r/K-strategies, the relevant multivariate data that would permit one to assess the proportion of variance in violent criminal participation attributable to the latent r/K-variable have not been reported. Presumably, individual differences in these strategies have some heritable component. The requisite twin or adoption data must be gathered and subjected to multivariate analysis.

GENETICS, RACE, AND VIOLENCE

Racial differences in arrests, homicides, etc., have sometimes been interpreted in terms of mean genetic differences among human groups (Rushton, 1988a,b). From a genetic perspective, race differences fall into the general category of *group* differences. There are two distinct questions that may be asked about genetics and violence in different groups. First, is the heritability of violence large *within* each group? Within-group heritabilities may be derived through adoption or twin studies of each group. Second, to what extent are differences in violence *between* groups due to the genetic differences between them? This question addresses *between*-group heritability.

Empirical data suggest the possibility of important within-group heritability in European populations and some American populations of European ancestry. There are no comparable twin or adoption data to document within-group heritability among American minorities.

Even in the presence of such data, it would be incorrect to infer that genetic differences contribute to differences between groups. Within-group heritabilities are insufficient to predict the extent to which mean group differences are genetic (DeFries, 1972; Loehlin et al., 1975). For example, the demonstration of within-group heritability among Danes and among Iowa whites does not imply that differences in crime rates between Denmark and Iowa are due to gene pool variance. Differences in prevalence and incidence of violence between Northern and Southern Ireland might be explained more easily by social and political milieus than by DNA variation.

Wilson and Herrnstein (1985) consider it inconceivable that the rapid historical changes in homicide could be accounted for by changes in allelic frequencies. The two- to threefold increase in homicide rates that occurred between the 1960s and 1970s (Zahn, 1989) is almost certainly due to environmental factors, not all of which have been identified. Hence, it is plausible that large mean group differences could be perpetuated by environmental factors. Despite the fact that there are analytical models in genetics that attempt to account for such types of environmental diffusion (e.g., Boyd and Richerson, 1984; Cavalli-Sforza and Feldman, 1981; Lumsdem and Wilson, 1981), little is known about their empirical validity, let alone their direct application to human violence. It is not clear that it would even be possible to research the genetics of group differences in violence without first identifying these mechanisms for environmental diffusion and their prevalence and impact in different groups.

On the genetic side, it is also improbable that various human populations, reproductively isolated for millennia, will evolve genetic liabilities equal to the last decimal place. What do we know about mean genetic differences? Studies of red cell polymorphisms, proteins, and isozymes have consistently shown considerably more variability *within* a race than *between* races (Hartl, 1980; Nei, 1985; Nei and Roychoudhury, 1974, 1982). To put this in different terms, the "average" U.S. Oriental, "average" U.S. black, and "average" U.S. white are genetically more similar to one another than three randomly selected individuals within, say, the U.S. white population. Hence, if these results can be extrapolated to a polygenic system contributing to liability for violence, then the genetic effects on race differences in violence are probably small and secondary to the genetic differences within races.

Faced with a literature equivocal on whether there is a significant genetic effect on violence in Scandinavian samples, the absence of data on within-group heritability for minorities, abrupt historical trends largely unexplainable by genetic drift or natural selection, and molecular genetic evidence suggesting a low between-to-within ratio of genetic group differences, one might conclude that models attributing racial differences mostly to environmental factors have more plausibility than those that explain them mostly in terms of genetics. At the very least, there is no positive evidence to suggest that heritability plays an important role in group differences in violence within the United States.

REFERENCES

Baker, L.A.
    1986 Estimating genetic correlations among discontinuous phenotypes: An analysis of criminal convictions and psychiatric-hospital diagnoses in Danish adoptees. (Special Issue: Multivariate behavioral genetics and development.) *Behavior Genetics* 16:127-142.
Baker, L.A., W. Mack, T.E. Moffitt, and S. Mednick
    1989 Sex differences in property crime in a Danish adoption cohort. *Behavior Genetics* 19:355-370.
Bender, B.G., M.G. Linden, and A. Robinson
    1987 Environment and developmental risk in children with sex chromosome abnormalities. *Journal of the American Academy of Child and Adolescent Psychiatry* 26:499-503.
Bohman, M.
    1971 A comparative study of adopted children, foster children and children in their biological environment born after undesired pregnancies. *Acta Paediatrica Scandinavica* 60(Suppl 221):5-38.

1978   Some genetic aspects of alcoholism and criminality: A population of adoptees. *Archives of General Psychiatry* 35:269-276.
Bohman, M., and S. Sigvardsson
1985   A prospective longitudinal study of adoption.   Pp. 137-155 in A.R. Nicol, ed., *Longitudinal Studies in Child Psychology and Psychiatry.* Somerset, N.J.: John Wiley & Sons.
Bohman, M., C.R. Cloninger, S. Sigvardsson, and A.-L. von Knorring
1982   Predisposition to petty criminality in Swedish adoptees: I. Genetic and environmental heterogeneity. *Archives of General Psychiatry* 39:1233-1241.
Bouchard, T.J., and M. McGue
1981   Familial studies of intelligence: A review. *Science* 212:1055-1958.
Boyd, R., and P.J. Richerson
1984   *Culture and the Evolutionary Process.* Chicago: University of Chicago Press.
Brain, P.F., D. Mainardi, and S. Parmigiani
1989   *House Mouse Aggression: A Model for Understanding the Evolution of Social Behavior.* London: Harwood Academic Publishers.
Brunner, H.G., M. Nelen, X.O. Breakfield, H.H. Ropers, and B.A. Van Oost
1993a   Abnormal behavior associated with a point mutation in the structural gene for monoamine oxidase A. *Science* 262:578-580.
Brunner, H.G., M.R. Nelen, P. van Zandvoort, N.G.G.M. Abeling, A.H. Van Gennip, E.C. Wolters, M.A. Kuiper, H.H. Ropers, and B.A. Van Oost
1993b   X-linked borderline mental retardation with prominent behavioral disturbance: Phenotype, genetic localization, and evidence for disturbed monoamine metabolism. *American Journal of Human Genetics* 52:1032-1039.
Cadoret, R.J.
1978   Psychopathology in adopted-away offspring of biologic parents with antisocial behavior. *Archives of General Psychiatry* 35:176-184.
Cadoret, R.J., and C.A. Cain
1980   Sex differences in predictors of antisocial behavior in adoptees. *Archives of General Psychiatry* 37:1171-1175.
Cadoret, R.J., L. Cunningham, R. Loftus, and J. Edwards
1975   Studies of adoptees from psychiatrically disturbed biologic parents: II. Temperament, hyperactive, antisocial, and developmental variables. *Journal of Pediatrics* 87:301-306.
Cadoret, R.J., C. Cain, and R.R. Crowe
1983   Evidence for a gene-environment interaction in the development of adolescent antisocial behavior. *Behavior Genetics* 13:301-310.
Cadoret, R.J., T. O'Gorman, E. Troughton, and E. Heywood
1985   Alcoholism and antisocial personality: Interrelationships, ge-

netic and environmental factors. *Archives of General Psychiatry* 42:161-167.

Cadoret, R.J., E. Troughton, T.W. O'Gormon, and E. Heywood
1986   An adoption study of genetic and environmental factors in drug abuse. *Archives of General Psychiatry* 43:1131-1136.

Carey, G.
1986   Sibling imitation and contrast effects. *Behavior Genetics* 16:319-341.
1992   Twin imitation for antisocial behavior: Implications for genetic and family environment research. *Journal of Abnormal Psychology* 101:18-25.
1993   Multivariate genetic relationships among drug abuse, alcohol abuse and antisocial personality. *Psychiatric Genetics* 3:141.

Carlier, M., P.L. Roubertoux, M.L. Kottler, and H. Degrelle
1990   Y chromosome and aggression in strains of laboratory mice. *Behavior Genetics* 20:137-156.

Cavalli-Sforza, L.L., and M.W. Feldman
1981   *Cultural Transmission and Evolution: A Quantitative Approach.* Princeton, N.J.: Princeton University Press.

Christiansen, K.O.
1968   Threshold of tolerance in various population groups illustrated by results from Danish criminological twin study. Pp. 107-116 in A.V.S. de Reuck and R. Porter, eds., *Ciba Foundation Symposium on the Mentally Abnormal Offender.* London: J. & A. Churchill, Ltd.
1974   Seriousness of criminality and concordance among Danish twins. In R. Hood, ed., *Crime, Criminology, and Public Policy.* London: Heinemann.
1977   A review of studies of criminality among twins. In S.A. Mednick and K.O. Christiansen, eds., *Biosocial Bases of Criminal Behavior.* New York: Gardner Press.

Cloninger, C.R.
1987   Neurogenetic adaptive mechanisms in alcoholism. *Science* 236:410-416.

Cloninger, C.R., and I.I. Gottesman
1987   Genetic and environmental factors in antisocial behavior disorders. Pp. 92-109 in S.A. Mednick, T.E. Moffitt, and S.A. Stack, eds., *The Causes of Crime: New Biological Approaches.* New York: Cambridge University Press.

Cloninger, C.R., K.O. Christiansen, T. Reich, and I.I. Gottesman
1978   Implications of sex differences in the prevalences of antisocial personality, alcoholism, and criminality for familial transmission. *Archives of General Psychiatry* 35:941-951.

Cloninger, C.R., S. Sigvardsson, M. Bohman, and A.-L. von Knorring
1982   Predisposition to petty criminality in Swedish adoptees: II. Cross-fostering analysis of gene-environment interaction. *Archives of General Psychiatry* 39:1242-1247.

Crowe, R.R.
1972   The adopted offspring of women criminal offenders: A study of their arrest records. *Archives of General Psychiatry* 27:600-603.
1974   An adoption study of antisocial personality. *Archives of General Psychiatry* 31:785-791.
Dalgard, O.S., and E. Kringlen
1976   A Norwegian twin study of criminality. *British Journal of Criminology* 16:213-232.
Daly, M., and M. Wilson
1988a *Homicide.* Hawthorne, N.Y.: Aldine de Gruyer.
1988b Evolutionary social psychology and family homicide. *Science* 242:519-524.
DeFries, J.C.
1972   Quantitative aspects of genetics and environment in the determination of behavior. Pp. 5-16 in L. Ehrman, G.S. Omenn, and E. Caspari, eds., *Genetics, Environment, and Behavior: Implications for Educational Policy.* New York: Academic Press.
Eaves, L.J.
1976   A model for sibling effects in man. *Heredity* 36:205-214.
Eaves, L.J., J.L. Silberg, J.K. Hewitt, M. Rutter, J.M. Meyer, M.C. Neale, and A. Pickles
1993   Analyzing twin resemblance in multisymptom data: Genetic applications of a latent class model for symptoms of conduct disorder in juvenile boys. *Behavior Genetics* 23:5-19.
Ebert, P.D., and R.G. Sawyer
1980   Selection for agonistic behavior in wild female *Mus musculus.* *Behavior Genetics* 10:349-360.
Ellis, L.
1982   Genetics and criminal behavior: Evidence through the end of the 1970s. *Criminology* 20:43-66.
1987   Criminal behavior and r/K selection: An extension of gene-based evolutionary theory. *Deviant Behavior* 8:149-176.
Gabrielli, W.F., Jr., and S.A. Mednick
1984   Urban environment, genetics, and crime. *Criminology* 22:645-652.
Ghodsian-Carpey, J., and L.A. Baker
1987   Genetic and environmental influences on aggression in 4- to 7-year-old twins. *Aggressive Behavior* 13:173-186.
Ginsburg, B.E., and W.C. Allee
1942   Some effects of conditioning on social dominance and subordination in inbred strains of mice. *Physiological Zoology* 15:485-506.
Gottesman, I.I.
1963   Heritability of personality: A demonstration. *Psychological Monographs: General and Applied* 77(9):1-21.
1966   Genetic variance in adaptive personality traits. *Journal of Child Psychology and Psychiatry* 7:199-208.

Gottesman, I.I., and H.H. Goldsmith
in Developmental psychopathology of antisocial behavior: Insert-
press ing genes into its ontogenesis and epigenesis. In C.A. Nelson,
ed., *Threats to Optimal Development: Integrating Biological,
Psychological and Social Risk Factors.* Hillsdale, N.J.: Lawrence
Erlbaum Associates.
Gottesman, I.I., G. Carey, and D.H. Hanson
1983 Pearls and perils in epigenetic psychopathology. Pp. 287-300 in
S.B. Guze, E.J. Earls, and J.E. Barrett, eds., *Childhood Psychopa-
thology and Development.* New York: Raven Press.
Gottesman, I.I., G. Carey, and T.J. Bouchard
1984 The Minnesota Multiphasic Personality Inventory of Identical
Twins Raised Apart. Paper presented at the 15th annual meet-
ing of the Behavior Genetics Association, Bloomington, Ind.
Gough, H.
1969 *Manual for the California Psychological Inventory,* rev. ed. Palo
Alto, Calif: Consulting Psychologists Press.
Hartl, D.L.
1980 *Principles of Population Genetics.* Sunderland, Mass.: Sinauer
Associates.
Hathaway, S.R., and J.C. McKinley
1940 A multiphasic personality schedule (Minnesota): I. Construc-
tion of the schedule. *Journal of Psychology* 10:249-254.
Hood, K.E., and R.B. Cairns
1988 A developmental genetic analysis of aggressive behavior in mice:
II. Cross-sex inheritance. *Behavior Genetics* 18:605-619.
Hook, E.B.
1973 Behavioral implications of the human XYY genotype. *Science*
179:139-150.
Hutchings, B.
1972 Environmental and Genetic Factors in Psychopathology and Crimi-
nality. Unpublished M. Phil. thesis, University of London.
Hutchings, B., and S.A. Mednick
1977 Criminality in adoptees and their adoptive and biological parents: A pilot study. Pp. 127-141 in S.A. Mednick and K.O.
Christiansen, eds., *Biosocial Bases of Criminal Behavior.* New
York: Gardner Press.
Jacobs, P.A., M. Brunton, M.M. Melville, R.P. Brittain, and W.F. McClermont
1965 Aggressive behavior, mental sub-normality, and the XYY male.
Nature 208:1351-1352.
Jones, S.E., and P.F. Brain
1987 Performances of inbred and outbred laboratory mice in putative
tests of aggression. *Behavior Genetics* 17:87-96.
Kranz, H.
1936 *Lebenschieksale Krimineller zwillinge.* Berlin: Springer-Verlag.
Lange, J.
1930 *Crime and Destiny* (C. Haldane, Translator). New York: Charles
Boni. (Original work published in 1929.)

Loehlin, J.C., and R.C. Nichols
1976 *Heredity, Environment, and Personality: A Study of 850 Sets of Twins.* Austin: University of Texas Press.
Loehlin, J.C., G. Lindzey, and J.N. Spuhler
1975 *Race Differences in Intelligence.* San Francisco: W.H. Freeman.
Loehlin, J.C., L. Willerman, and J.M. Horn
1985 Personality resemblances in adoptive families when the children are late-adolescent or adult. *Journal of Personality and Social Psychology* 48:376-392.
1987 Personality resemblance in adoptive families: A 10-year followup. *Journal of Personality and Social Psychology* 53:961-969.
Lumsdem, C., and E.O. Wilson
1981 *Genes, Mind, and Culture.* Cambridge, Mass.: Harvard University Press.
Lyons, M.J., L. Eaves, M.Y. Tsuang, S.E. Eisen, J. Goldberg, and W.T. True
1993 Differential heritability of adult and juvenile antisocial traits. *Psychiatric Genetics* 3:117.
Maxson, S.C.
1981 The genetics of aggression in vertebrates. In P.F. Brain and D. Benton, eds., *The Biology of Aggression.* Amsterdam: Sijthoff & Noordhoff International Publishers.
1990 Methodological issues in genetic analyses of an agonistic behavior (offense) in male mice. In D. Goldowitz, R.E. Wimer, and D. Wahlsten, eds., *Techniques for the Genetic Analysis of Brain and Behavior.* Amsterdam: Elsevier Science Publishers.
Maxson, S.C., A. Didier-Erickson, and S. Ogawa
1989 The Y chromosome, social signals, and offense in mice. *Behavioral and Neural Biology* 52:251-259.
Mednick, S.A.
1987 Introduction—Biological factors in crime causation: The reactions of social scientists. In S.A. Mednick, T.E. Moffitt, and S.A. Stack, eds., *The Causes of Crime: New Biological Approaches.* New York: Cambridge University Press.
Mednick, S.A., W.F. Gabrielli, Jr., and B. Hutchings
1983 Genetic influences in criminal behavior: Evidence from an adoption cohort. Pp. 39-56 in K.T. Van Dusen and S.A. Mednick, eds., *Prospective Studies of Crime and Delinquency.* Boston: Kluwer-Nijhoff Publishing.
1984 Genetic influences in criminal convictions: Evidence from an adoption cohort. *Science* 224:891-894.
1987 Genetic factors in the etiology of criminal behavior. Pp. 74-91 in S.A. Mednick, T.E. Moffitt, and S.A. Stack, eds., *The Causes of Crime: New Biological Approaches.* New York: Cambridge University Press.

Mednick, S.A., P. Brennan, and E. Kandel
   1988 Predisposition to violence. (Special Issue: Current theoretical perspectives on aggressive and antisocial behavior.) *Aggressive Behavior* 14:25-33.
Michard-Vanhee, C.
   1988 Aggressive behavior induced in female mice by an early single dose of testosterone is genotype dependent. *Behavior Genetics* 18:1-12.
Miles, D., and G. Carey
   1993 The genetics of antisocial personality disorders: A psychiatric sample. *Behavior Genetics* 23:558-559.
Moffitt, T.E.
   1987 Parental mental disorder and offspring criminal behavior: An adoption study. *Psychiatry* 50:346-360.
Nei, M.
   1985 Human evolution at the molecular level. In T. Ohta and K. Aoki, eds., *Population Genetics and Molecular Evolution*. Tokyo: Japan Scientific Societies Press.
Nei, M., and A.K. Roychoudhury
   1974 Genetic variation within and between the three major human races of man, Caucasoids, Negroids, and Mongoloids. *American Journal of Human Genetics* 26:421-443.
   1982 Genetic relationship and evolution of human races. *Evolutionary Biology* 14:1-59.
Newton, F.H., R.N. Rosenberg, P.W. Lempert, and J.S. O'Brien
   1971 Neurological involvement in Urbach-Wiethe's disease (lipoid proteinosis): A clinical, ultrastructural, and chemical study. *Neurology* 21:1205-1213.
Owen, D.R.
   1972 The 47,XYY male: A review. *Psychological Bulletin* 78:209-233.
Owen, D.R., and J.O. Sines
   1970 Heritability of personality in children. *Behavior Genetics* 1:235-247.
Parker, T.
   1989 Television Viewing and Aggression in Four and Seven Year Old Children. Paper presented at Summer Minority Access to Research Training meeting, University of Colorado, Boulder.
Partanen, J., K. Bruun, and T. Markkanen
   1966 *Inheritance of Drinking Behavior: A Study on Intelligence, Personality, and Use of Alcohol of Adult Twins*. Helsinki: Keskuskirjapaino.
Plomin, R., T.T. Foch, and D.C. Rowe
   1981 Bobo clown aggression in childhood: Environment, not genes. *Journal of Research in Personality* 15:331-342.

Plomin, R., J.C. Loehlin, and J.C. DeFries
1985   Genetic and environmental components of "environmental" influences. *Developmental Psychology* 21:391-402.
Pogue-Geile, M.F., and R.J. Rose
1985   Developmental genetic studies of adult personality. *Developmental Psychology* 21:547-557.
Reznikoff, M., and M.S. Honeyman
1967   MMPI profiles of monozygotic and dizygotic twin pairs. *Journal of Consulting Psychology* 31:100.
Rose, R.J.
1988   Genetic and environmental variance in content dimensions of the MMPI. *Journal of Personality and Social Psychology* 55:302-311.
Rowe, D.C.
1983   Biometrical genetic models of self-reported delinquent behavior: A twin study. *Behavior Genetics* 13:473-489.
1985   Sibling interaction and self-reported delinquent behavior: A study of 265 twin pairs. *Criminology* 23:223-240.
1986   Genetic and environmental components of antisocial behavior: A study of 265 twin pairs. *Criminology* 24:513-532.
Rowe, D.C., and D.W. Osgood
1984   Heredity and sociological theories of delinquency: A reconsideration. *American Sociological Review* 49:526-540.
Rushton, J.P.
1988a Race differences in behavior: A review and evolutionary analysis. *Personality and Individual Differences* 9:1009-1024.
1988b The reality of race differences: A rejoinder. *Personality and Individual Differences* 9:1035-1040.
Rushton, J.P., D.W. Fulker, M.C. Neale, D.K. Nias, and H.J. Eysenck
1986   Altruism and aggression: The heritability of individual differences. *Journal of Personality and Social Psychology* 50:1192-1198.
Scarr, S.
1966   The origins of individual differences in adjective check list scores. *Journal of Consulting Psychology* 30:354-357.
Scarr, S., P.L. Webber, R.A. Weinberg, and M.A. Wittig
1981   Personality resemblance among adolescents and their parents in biologically related and adoptive families. *Journal of Personality and Social Psychology* 40:885-898.
Schiavi, R.C., A. Theilgaard, D.R. Owen, and D. White
1984   Sex chromosome anomalies, hormones, and aggressivity. *Archives of General Psychiatry* 41:93-99.
1988   Sex chromosome anomalies, hormones, and sexuality. *Archives of General Psychiatry* 45:19-24.
Schulsinger, F.
1972   Psychopathy: Heredity and environment. *International Journal of Mental Health* 1:190-206.

Scott, J.P.
    1942   Genetic differences in the social behavior of inbred strains of mice. *Journal of Heredity* 33:11-15.

Selmanoff, M.K., S.C. Maxson, and B.E. Ginsburg
    1976   Chromosomal determinants of intermale aggressive behavior in inbred mice. *Behavior Genetics* 6:53-69.

Sigvardsson, S., C.R. Cloninger, M. Bohman, and A.-L. von Knorring
    1982   Predisposition to petty criminality in Swedish adoptees: III. Sex differences and validation of the male typology. *Archives of General Psychiatry* 39:1248-1253.

Slater, E., and V. Cowie
    1971   *The Genetics of Mental Disorder.* London: Oxford University Press.

Tellegen, A., D.T. Lykken, T.J. Bouchard, Jr., K.J. Wilcox, N.L. Segal, and S. Rich
    1988   Personality similarity in twins reared apart and together. *Journal of Personality and Social Psychology* 54:1031-1039.

Theilgaard, A.
    1984   A psychological study of the personalities of XYY and XXY men. *Acta Psychiatrica Scandinavica Supplementum* 69:1-133.

Vale, J.R., D. Ray, and C.A. Vale
    1972   The interaction of genotype and exogenous neonatal androgen: Agonistic behavior in female mice. *Behavioral Biology* 7:321-334.

Van Dusen, K.T., S.A. Mednick, W.F. Gabrielli, Jr., and B. Hutchings
    1983a Social class and crime in an adoption cohort. *Journal of Criminal Law and Criminology* 74:249-269.
    1983b Social class and crime: Genetics and environment. Pp. 57-71 in K.T. Van Dusen and S.A. Mednick, eds., *Prospective Studies of Crime and Delinquency.* Boston: Kluwer-Nijhoff Publishing.

van Oortmerssen, G.A., and T.C. Bakker
    1981   Artificial selection for short and long attack latencies in wild *Mus musculus domesticus. Behavior Genetics* 11:115-126.

Walters, G.D., and T.W. White
    1989   Heredity and crime: Bad genes or bad research? *Criminology* 27:455-485.

Wilson, J.Q., and R.J. Herrnstein
    1985   *Crime and Human Nature.* New York: Simon and Schuster.

Witkin, H.A., S.A. Mednick, F. Schulsinger, E. Bakkerstrom, K.O. Christiansen, D.R. Goodenough, K. Hirschorn, C. Lundsteen, D.R. Owen, J. Philip, D.B. Rubin, and M. Stocking
    1976   Criminality in Xyy and XXy men. *Science* 193:547-555.

Zahn, M.A.
    1989   Homicide in the twentieth century: Trends, types, and causes. In T.R. Gurr, ed., *Violence in America*, Vol. 1. Newbury Park, Calif.: Sage Publications.

The Neurobiology of
Violence and Aggression

Allan F. Mirsky and Allan Siegel

Over the past four decades, there has been an increasing body
of data in the human literature on neuropsychiatric disorders that
raises the question about a possible relationship between the abnormal function of specific regions of the brain and the occurrence of violent and aggressive behavior. The view that violence
and aggression are human behaviors symptomatic of an underlying brain disorder, rather than simply acts to be punished under
law, is relatively new. It is true that the distinction "between the
harmful act that was traceable to fault and that which occurred
without fault" extends back to ancient Hebrew law (American Bar
Association, 1983:7-271). However, the scientific facts that have
been offered as the basis for what Monroe has referred to as "the
episodic dyscontrol syndrome" (Monroe, 1970) or other involun-
tary acts, are of relatively recent origin. The region of the brain
most often linked with this form of behavioral dysfunction is
referred to as the "limbic system." Research on the limbic system (Figure 1) identified an apparently unitary cerebral region or
limbic lobe (Broca, 1878) at the juncture of the forebrain and brain
stem, which Papez (1937) and MacLean (1952) later identified as

Allan F. Mirsky is at the Laboratory of Psychology and Psychopathology, National Institutes
of Health; and Allan Siegel is at the Department of Neurosciences, New Jersey Medical School.

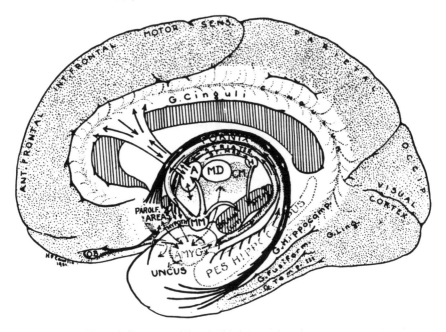

FIGURE 1   Diagrammatic representation of the principal subcortical connections of the limbic system viewed from the mesial surface of one
hemisphere.   Important connections to the brain stem reticular formation have been omitted, and others are only approximately represented.
Some abbreviations: A, anterior nucleus of thalamus; Ant., anterior;
AMYG, amydala; Int., intermediate; CM, center median; Sens., sensory;
Occip., occipital; G., gyrus; Stria Ter., stria terminalis; St. Med., stria
medullaris; MD, medial dorsal nucleus of thalamus; O.B., olfactory bulb;
Ling., lingual; Temp., temporal; Hypoth., hypothalamus; MM, mammillary bodies; Parolf., parolfactory; H., habenular nucleus. SOURCE: Pen-
field and Jasper (1954).

the cerebral substrate of emotional behavior.   The work of Klüver
and Bucy (1939) and of Rosvold et al. (1954) is also relevant here.
These researchers demonstrated that monkeys surgically deprived
of portions of their limbic system had major changes in their
social and affective behavior.   These studies had a significant impact on the thinking about the relation between cerebral structures and/or systems and abnormal behavior, including a number
of neuropsychiatric disorders.   In terms of its overall organization,
the limbic system includes the hippocampal formation, amygdala,
septal area, cingulate gyrus, and prefrontal cortex (according to
some authors).   Several other brain structures have been consid-

ered part of this system because of their neural associations with limbic structures. These include the hypothalamus and midbrain periaqueductal gray matter (PAG). Collectively, these regions comprise a functional unit that is sometimes referred to as the "limbic-hypothalamic-midbrain axis." The importance of this research was that there was a brain system that could be implicated in neuropsychiatric disorders, including those in which violence or aggressiveness was a major symptom. This system could be a focus of research, both clinical and laboratory based, and could provide the target or basis for new treatment possibilities. Some of this research is reviewed in the section of this paper on human studies of aggression and violence.

In 1974, one of the authors reviewed the literature on the relationship between aggressive behavior and brain disease, and concluded that the available evidence did not support the view that aggressive behavior in humans could be attributed to brain disease (Mirsky and Harman, 1974). The question arises as to whether, in the ensuing 16 years since that paper was written, sufficient additional data have been gathered to alter that conclusion. Recent studies related to that question are also reviewed below.

In view of the putative relationship between brain dysfunction and aggressive behavior, it is the purpose of this paper to summarize briefly the neural mechanisms of aggressive behavior as discovered from animal models, and to review human studies on the relationships among brain dysfunction, neuropsychiatric disorders (including abnormal development), cognitive processes, and the symptoms of violence and aggression. The section that deals with animal models focuses on two behaviors that can be readily elicited in the cat: quiet biting "predatory" attack and affective defense. It is our belief that the neural substrates and mechanisms underlying these distinctive forms of aggressive behavior in the cat may also regulate aggressive reactions at the human level or, possibly, provide a framework for understanding human violence and aggression.

In the thesis advocated here, an analogy can be made between the relationship of the limbic system to the hypothalamus and midbrain PAG and that of the motor cortex and reticulospinal fibers to the spinal cord concerning the modulation of "emotional" and "motor" systems, respectively. With respect to motor systems, the final common output pathways for the expression of motor responses such as fine movements of the extremities or walking movements are governed by the activity of cells located in the gray matter of the ventral horn of the spinal cord. In contrast, descending neurons from the reticular formation and motor systems of the cerebral cortex that synapse upon the neurons of the ventral horn serve to modulate their activity and program the sequence of their neuronal discharge patterns, respectively. In a similar manner, the hypothalamus and midbrain PAG constitute the integrating mechanisms whose outputs serve as the "final common pathway" for the expression of aggressive forms of behavior. We propose that the limbic system thus serves the critically important function of modulating the activity of neurons in the hypothalamus and PAG and of programming the sequence of neuronal discharge patterns within these structures. We illustrate this in the examples of research cited below.

ANIMAL MODELS OF AGGRESSION

FELINE MODELS OF AGGRESSIVE BEHAVIOR ELICITED BY BRAIN STIMULATION

Our focus is on two models of aggression that can be elicited by electrical or chemical stimulation of the hypothalamus and midbrain periaqueductal gray matter in the cat—quiet biting "predatory" attack behavior and affective defense behavior. Affective defense behavior may be classified as having aversive properties (such as those associated with fear), whereas predatory attack is associated with more positively reinforcing objectives such as the acquisition of food. The study of these two forms of aggression in the cat may thus provide a more complete picture of the neural substrates of aggression than either model studied alone; moreover, research in the cat has been very systematic and has certain elegant qualities.

Quiet biting attack is predatory in nature and is characterized by stalking of the prey object (usually an anesthetized rat), which is then followed by biting of the back of its neck. The cat may also strike the rat with its forepaw (Flynn et al., 1970; Wasman and Flynn, 1962). This behavior occurs under natural conditions, and includes capturing and killing of a rat or a mouse in the open field (Leyhausen, 1979). In the laboratory, predatory attack behavior can be elicited by electrical stimulation along a region beginning from the rostrolateral and perifornical hypothalamus (Wasman and Flynn, 1962; Siegel and Pott, 1988) and extending caudally through the ventral aspect of the midbrain (Bandler and Flynn, 1972) and ventral aspect of the midbrain PAG (Bandler,

1984; Siegel and Pott, 1988) to the lateral tegmental fields of the pons (Berntson, 1973).

Affective defense behavior was originally described by Hess and Brugger (1943). This form of attack behavicr, in contract with predatory attack, is associated with noticeable affective signs such as piloerection, retraction of the ears, arching of the back, marked pupillary dilatation, vocalization, and unsheathing of the claws. This response can also be evoked under natural environmental conditions. Examples include the affective defense reactions that occur when a cat's territory is invaded by another animal, when a threatening stimulus is introduced into the cat's environment, or when a female cat perceives that its kittens are threatened by another animal. Furthermore, electrical or chemical stimulation applied at the appropriate forebrain or brain stem sites in a cat will elicit affective defense responses with its forepaw that are directed at a moving object such as an awake rat, cat, or experimenter. Although predatory attack requires the presence of a prey object for an attack response to occur, affective defense can be elicited in an impoverished environment.

It should be noted that affective defense behavior is explosive in nature, oftentimes directed at conspecifics; produces a powerful sympathoadrenal response; and thus may share common features with violent "episodic dyscontrol" behavior in the human. Affective defense reactions are generally elicited from sites located throughout the rostrocaudal extent of the medial preoptico-hypothalamus and dorsal aspect of the PAG (Wasman and Flynn, 1962; Fuchs et al., 1985a,b; Shaikh et al., 1987; Siegel and Pott, 1988).

ORGANIZATION AND CONTROL OF AGGRESSIVE BEHAVIOR IN THE CAT

A number of basic research problems have been considered over the past several decades with respect to affective defense behavior and predatory attack in the cat. Due mainly to the pioneering efforts of John Flynn, a number of the basic properties associated with predatory attack have been clearly delineated. For example, one class of studies has provided an analysis of how selective components of the central nervous system linked to the expression of quiet biting attack interact with sensory stimuli to facilitate the occurrence of this response (MacDonnell and Flynn, 1966a,b; Bandler and Flynn, 1972). Another set of studies was directed at identifying the mechanisms that regulate motor control of the attack response. Such studies helped to produce a better understanding of how the motor cortex and trigeminal system regulate such responses as paw striking and jaw closure (Edwards and Flynn, 1972; MacDonnell and Fessock, 1972).

In this paper, we have chosen to summarize or provide references to information that we believe is critical to understanding the neural bases of aggressive behavior as studied in the cat. This includes (1) the anatomic substrates and pathways that underlie the expression and control of each of these forms of attack behavior; (2) the regions along the limbic-midbrain axis that serve to enhance or diminish the likelihood of these responses; this encompasses, as well, the effects of temporal lobe seizures on attack behavior; and (3) the role of the opioid peptide system in the regulation of affective defense behavior.

It should be noted that the structures of the limbic-hypothalamic-PAG axis would appear to constitute the neural substrates for the motivational properties of the response mechanism. In contrast, the outputs of the hypothalamus and PAG to lower brain stem neurons appear to constitute the initial neurons in a system of pathways that descend to the spinal cord or that make synapse with lower motor neurons of the brain stem. This system of fibers, therefore, may comprise part of the motor components of the behavioral response. A second level of motor function may arise from the cerebral cortex. In this context, Flynn et al. (1970) postulated a "patterning mechanism" in which it was hypothesized that sensory and motor regions of the cerebral cortex receive inputs from hypothalamic cell groups linked to the attack response. In turn, the pattern of neuronal responses evoked in the cerebral cortex thus results in a set of output signals to motor and autonomic regions of the lower brain stem and spinal cord that constitute a coordinated attack response.

THE ANATOMY OF AGGRESSIVE BEHAVIOR

PREDATORY ATTACK

Afferent Connections

Predatory attack can be elicited from a variety of regions throughout the forebrain and brain stem, an area that extends from the anterior hypothalamus through the midbrain PAG to the level of the pontine tegmentum (see Figures 2A and 2B).

Concerning the hypothalamic sites from which predatory attack can be elicited, Smith and Flynn (1980a) identified cells in a

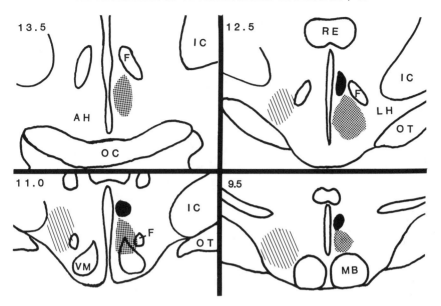

FIGURE 2A   Distribution of regional sites within the preopticohypothal-amus from which affective defense (stippled area), quiet biting attack (striped area), and flight behavior (dark area) can be elicited most fre-quently by electrical stimulation. Data for this figure and for Figure 2B are based on experiments conducted in the laboratory of Allan Siegel. Number in upper left-hand corner of figure indicates the frontal plane of the section. Abbreviations: AH, anterior hypothalamus; F, fornix; IC, internal capsule; LH, lateral hypothalamus; MB, mammillary bodies; OC, optic chiasm; OT, optic tract; RE, nucleus reuniens; VM, ventromedial nucleus.

number of regions that are known or believed to modulate this response. Several of these key structures include the midbrain PAG, locus coeruleus, substantia innominata, bed nucleus of the stria terminalis (BNST), and central nucleus of the amygdala. Other afferent sources of the lateral hypothalamus include the lateral septal nucleus, diagonal band of Broca (Brutus et al., 1984; Krayniak et al., 1980), and midline thalamus (Siegel et al., 1973). A more detailed discussion of the anatomic pathways from limbic nuclei that modulate the attack response is presented below in the sec-tion "Limbic-Midbrain Modulation of Aggressive Behavior in the Cat."

The other major sites from which predatory attack can be elicited include the following brain stem regions: the midbrain PAG, ventral tegmental area, and pontine tegmentum. A major

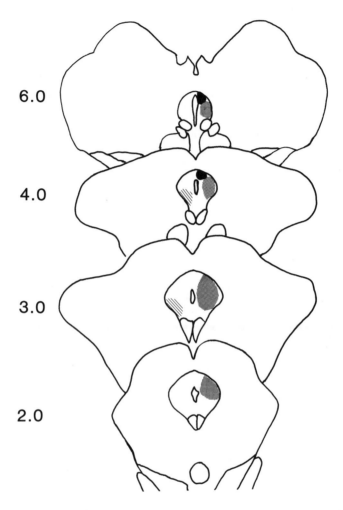

FIGURE 2B   Distribution of regional sites within the midbrain periaqueductal gray matter from which affective defense (stippled area), quiet
biting attack (striped area), and flight (dark area) can be elicited most
frequently by electrical stimulation.   Number on the left side of each
figure indicates the frontal plane of the section.   Note that flight and
affective defense sites are generally situated dorsal to those sites from
which quiet biting attack is elicited.   A recent study by Bandler (1984),
however, has suggested that affective defense reactions characterized by
howling and growling can also be elicited from ventral portions of the
periaqueductal gray, especially when stimulation is applied at caudal aspects of this structure.   SOURCE: Siegel and Pott (1988).

input to the PAG arises from predatory attack sites in the perifornical lateral hypothalamus (Chi and Flynn, 1971; Fuchs et al., 1981). With respect to the ventral tegmental area, the major projections to this structure arise from the gyrus proreus (prefrontal cortex) and perifornical hypothalamus. The key role of the ventral tegmental area in predatory attack is supported by the studies of Bandler and Flynn (1972), Proshansky et al. (1974), and Goldstein and Siegel (1980).

The precise sites at which attack was obtained included mainly the parabrachial region of the tegmentum. Structures that may influence pontine control of predatory attack include the perifornical lateral hypothalamus, PAG, BNST, and central and lateral nuclei of the amygdala (Smith and Flynn, 1979).

Efferent Connections

In an early study, Chi and Flynn (1971) placed lesions at sites in the lateral hypothalamus from which predatory attack was elicited. The procedures enabled these investigators to trace the course of the degenerating axons from the lesion site. The results demonstrated both ascending and descending projections from lateral hypothalamic attack sites. Ascending projections were noted to pass through the preoptic zone into the diagonal band of Broca and septal area, regions known to modulate the attack response. Descending projections could be followed through the hypothalamus into the midbrain ventral tegmental area and PAG.

More recently, Fuchs et al. (1981) examined the projection system from lateral hypothalamic attack sites. The anteriorly directed projections from the attack sites were similar to those described by Chi and Flynn (1971). Fibers were traced through the anterior hypothalamus to the preoptic region, diagonal band of Broca, and lateral septal area. The significance of these projections remains unknown, but these fibers may constitute part of a "feedback" pathway that serves to regulate how these limbic forebrain structures, in turn, control the attack mechanism at the levels of the hypothalamus and brain stem. With regard to the descending projections, fibers were followed caudally through the medial forebrain bundle into the ventral tegmental area and PAG. Of particular interest is the fact that fibers traced from attack sites in the perifornical region were also observed to terminate in the locus coeruleus and motor nucleus of the trigeminal complex as well as the tegmental fields of the pons (Figure 3).

The significance of the latter (trigeminal) projection is that it

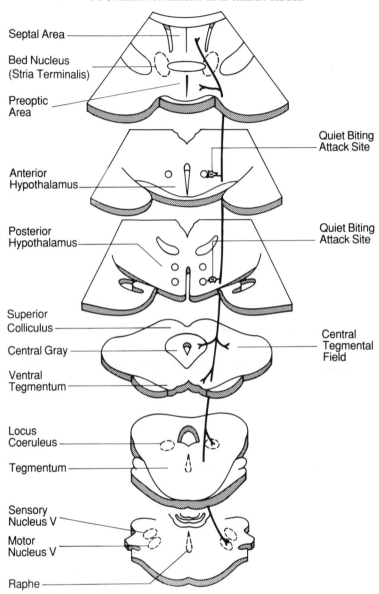

FIGURE 3 Diagram indicating the principal ascending and descending projections of the perifornical lateral hypothalamus associated with quiet biting attack behavior. Of particular interest and presumed importance are the connections from the perifornical region to the periaqueductal gray, tegmental fields, locus coeruleus, and the motor nucleus of the fifth nerve. SOURCE: Siegel and Pott (1988).

establishes the anatomic substrate by which jaw opening and jaw closing are controlled by the perifornical hypothalamus for the biting component of the attack response. Furthermore, the projection to the locus coeruleus provides the possible substrate by which noradrenergic pools from this nucleus can be activated to generate the arousal component of the attack response. Other research (Shaikh et al., 1987) suggests that the predatory attack system at the level of the PAG (see Figure 4) is organized along several possible plans: (1) that the PAG serves as a feedback system to the perifornical hypothalamus from which the primary integrated output to the lower brain stem is organized; and/or (2) that the PAG serves as a feed-forward relay of the perifornical hypothalamus, but does so through the use of short axons passing to the tegmental fields. The precise function of such fibers remains unknown, but they might comprise part of a multisynaptic pathway (presumably via reticulospinal fibers) for the regulation of autonomic and somatic motor components of the predatory attack response. The projection to the raphe nucleus, however, is more likely to be associated with the modulation of the attack response. This judgment is based on the fact that stimulation of the raphe nucleus has been shown to suppress predatory attack (Shaikh et al., 1984) and that administration of para-chlorophenylalanine (a monamine suppressor) can facilitate the occurrence of this response (MacDonnell et al., 1971).

In this discussion, the likely anatomic substrates are described over which the autonomic and several of the somatic motor components of the predatory attack response are expressed. However, little has been said of the anatomic substrates governing how the hypothalamus might regulate visual processes central to the organization of the attack response. Research by Pott and MacDonnell (1986) and by Ogren and Hendrickson (1976) provides suggestions as to how this may be accomplished. The projection from the posterior lateral hypothalamus to the pulvinar represents the initial limb in a disynaptic pathway from the hypothalamus to the visual cortex that may be essential for the integration of visual information as well as for controlling visual pursuit movements during the attack sequence.

AFFECTIVE DEFENSE BEHAVIOR

Afferent Connections

As noted earlier in this chapter, the sites in the brain at which affective defense reactions can be elicited have been identified

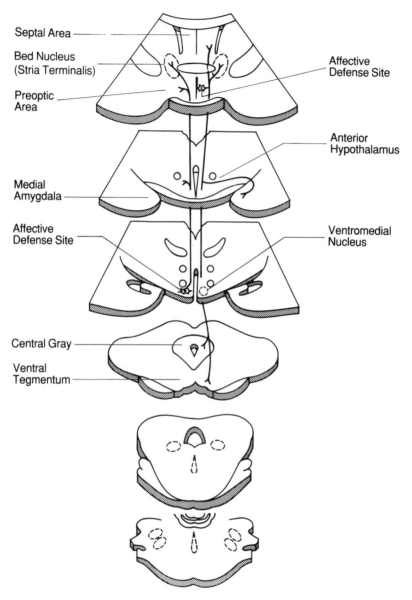

Septal Area

Bed Nucleus
(Stria Terminalis)

Preoptic
Area

Affective
Defense Site

Anterior
Hypothalamus

Medial
Amygdala

Affective
Defense Site

Ventromedial
Nucleus

Central Gray

Ventral
Tegmentum

FIGURE 4   Diagram indicating the principal ascending and descending projections of the medial hypothalamus associated with affective defense behavior. Of significance is the fact that fibers mediating this response arise from the ventromedial nucleus and primarily project rostrally into the anteromedial hypothalamus, medial preoptic region, and bed nucleus of the stria terminalis. Note that the fibers that supply the midbrain periaqueductal gray in association with affective defense arise primarily from the anteromedial hypothalamus. SOURCE: Siegel and Pott (1988).

(Figures 2A and 2B) and include primarily the medial preoptico-hypothalamus and dorsal aspect of the midbrain PAG. The work of Smith and Flynn (1980b), Stoddard-Apter and MacDonnell (1980), and Shaikh et al. (1987) suggests that the cell bodies of origin that project to these regions lie within other areas of the hypothalamus, limbic areas including the medial amygdala and medial septal nuclei, and the midbrain PAG itself. Forebrain structures projecting to the relevant areas of the PAG (i.e., from which affective defense could be elicited) include the lateral hypothalamus, and the ventromedial, dorsomedial, and anterior hypothalamus (Bandler and McColloch, 1984; Bandler et al., 1985). The anteriomedial hypothalamus may provide the most significant behaviorally relevant impact to the PAG (Fuchs et al., 1985a).

Efferent Connections

Space limitations are such that we can only summarize the extensive work on identification of the efferent connections from the structures involved in affective defense. These are depicted in Figures 4 and 5. Included are the anterior medial hypothalamus and the medial preoptic region; in addition, the expression of the entire affective defense response involves the central tegmental fields of the midbrain and pons, locus coeruleus, and motor and sensory nuclei of the trigeminal complex (Fuchs et al., 1985b). This latter pathway is probably important for the vocalization component of the affective defense response, since its axons regulate the jaw-opening reflex. Research continues on the description of this system.

LIMBIC-MIDBRAIN MODULATION OF AGGRESSIVE BEHAVIOR IN THE CAT

As noted in the beginning of this chapter, it has long been recognized that the limbic system plays a major role in the regulation of emotional behavior. Such a view was originally derived from early theoretical papers of Papez (1937) and MacLean (1949), and has been reinforced by more recent authors such as Monroe (1978) who postulated the "episodic dyscontrol" syndrome as a consequence of limbic system dysfunction. In addition, other support comes from clinical investigations that have identified a relationship between limbic system disorders associated with either tumors or epilepsy of this region and "episodic dyscontrol"-like behavior (Malamud, 1967; Bear and Fedio, 1977; Falconer, 1973; Heimburger et al., 1978; Hood et al., 1983; Martinius, 1983;

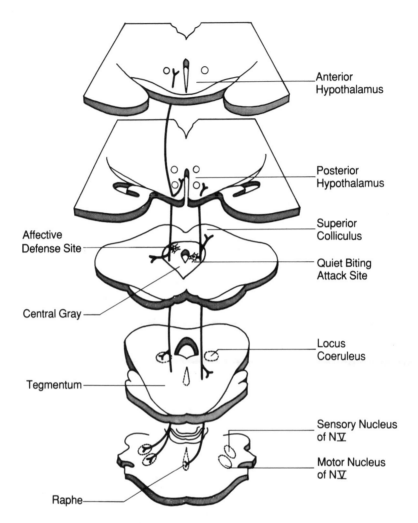

FIGURE 5   Diagrams indicating the principal efferent projections from midbrain periaqueductal gray sites associated with affective defense (left side) and quiet biting attack behavior (right side). Note that fibers associated with affective defense, which arise from the periaqueductal gray, are distributed rostrally to the medial preopticohypothalamus, from which this response can also be elicited, and caudally to the locus ceruleus, tegmental fields, and trigeminal complex. In contrast, the fibers associated with quiet biting attack have a more limited distribution. Ascending fibers synapse within the posterior lateral hypothalamus from which this response can also be elicited, while descending fibers supply the central tegmental fields and median raphe. SOURCE: Siegel and Pott (1988).

McIntyre et al., 1976; Ounsted, 1969; Serafetinides, 1965). (Further discussion of human studies on aggression and the implications of the results of these studies to our understanding of underlying neural regulatory mechanisms appears later in this chapter). On the basis of these clinical reports, we have proposed that the limbic system functions as a major "modulator" of aggressive behavior. Implicit in this assumption is that the limbic system regulates the tendency for aggressive reactions to occur or be suppressed by its direct or indirect actions upon the hypothalamus or midbrain PAG. Findings relevant to these issues have been described fully in other review articles (Siegel and Edinger, 1981, 1983; Siegel and Brutus, 1990), which provide details as to the methods and the results of various limbic system manipulations. It is clear that limbic structures have a major modulatory action, and these are summarized in Table 1.

Thus, the principal finding of these studies is that the limbic system (including ventral and dorsal hippocampus, septal area, amygdala, and portions of the prefrontal cortex and cingulate gyrus) modulates predatory attack and affective defense behavior (Brutus et al., 1986). The anatomic studies detailing the outputs of the limbic structures indicate that the primary sites of interaction from the limbic system most probably include the medial and lateral hypothalamus with respect to modulation of aggressive reactions. While it is also conceivable that limbic structures may modulate PAG neurons intrinsic to the expression of affective defense and predatory attack, little information is presently available that bears upon this possibility. Further studies along these lines would be helpful in clarifying this issue. We have also shown that limbic seizures may produce enduring modulating effects in the post-ictal period on predatory attack and affective defense behavior (Brutus et al., 1986).

NEUROPHARMACOLOGY OF AGGRESSION:
THE OPIOID PEPTIDE SYSTEM

Over the past several decades, a number of studies by various investigators have been directed at examining the possible role of several different putative transmitter systems, such as the monoamines and acetylcholine, in the regulation of affective defense and predatory attack. Unfortunately, much of this work remains incomplete, and a clear-cut understanding of the actions of these transmitters on feline aggression is not yet available. Nevertheless, we have summarized the results of these studies in Table 2.

TABLE 1 Effects of Stimulation of Limbic System Structures on Predatory Attack and Affective Defense Behavior

Structure	Effect		Interneuron	Final Common Pathway
	PA	AD		
Hippocampus:				
Ventral	F	S	Lateral septal n.	MFB
Dorsal	S	F	Lateral septal n. Diagonal band of Broca	MFB
Lateral septal n. (medial aspect)	S	F	Diagonal band of broca	MFB
Lateral septal n. (lateral aspect)	F	?		MFB
Basomedial amygdala and pyriform cortex	S	F	Stria terminalis to BNST	Stria terminalis BNST fibers to hypothalamus
Central/lateral amygdaloid nuclei	F	S	Substantia innominata	MFB
Substantia innominata Lateral aspect	S	?	(Feedback to amygdala?)	Stria terminalis/BNST hypothalamic fibers
Medial aspect	F	?		MFB
Prefrontal cortex	S	S	Mediodorsal n. and midline thalamus	Thalamohypothalamic fibers
Anterior cingulate gyrus	S	?	Mediodorsal n. and midline thalamus	Thalamohypothalamic fibers
BNST	S	F		BNST fibers to hypothalamus

NOTE: AD = affective defense behavior; BNST = bed nucleus of stria terminalis; F = response facilitation; MFB = medial forebrain bundle; n. = nucleus; PA = predatory attack; S = response suppression.

TABLE 2 Effects of Drugs on Aggressive Behavior in Cats

Neurotransmitter System	Route of Administration	Effect on Affective Defense	Effect on Quiet Biting Attack	Effect on Flight	Reference[a]
Cholinergic Muscarinic	Systemic	↑			George et al., 1962
		↑			Leslie, 1965
		↑			Zablocka and Esplin, 1964
		↑			Berntson, 1976
		↑			Berntson and Leibowitz, 1973
		↑			Berntson and Leibowitz, 1973
		↑	↑		Katz and Thomas, 1975
Nicotinic	Systemic	↓	↓		Berntson, 1976
Nicotinic anticholinesterase	Intraventricular	↑		↑	Feldberg and Sherwood, 1954
		↑		↑	Feldberg and Fleischhauer, 1962
Acetylcholine	Intracerebral Hypothalamus, PAG	↑		↑	Allikmets, 1974
		↑		↑	Allikmets, 1974
	Hypothalamus	↑(HD)		↑(LD)	Meyers, 1964
	Hypothalamus	↓(LD)			Kono, 1984
	Hypothalamus	↑(HD)			Kono, 1984
	Hypothalamus		↑		Karmos-Varzegi and Karnos, 1977
	Ventral tegmentum		↑		Karmos-Varzegi and Karnos, 1977
	Hypothalamus			↑	Desci and Karmos-Varszegi, 1969

continued on next page

TABLE 2 (Continued)

Neurotransmitter System	Route of Administration	Effect on Affective Defense	Effect on Quiet Biting Attack	Effect on Flight	Reference[a]
Serotonergic					
P-Chlorophenylalonine	Systemic	↑	↑		Ferguson et al., 1970
		↑	↑		MacDonnell et al., 1971
		↑	↑		MacDonnell and Fessock, 1972
Dopaminergic					
	Systemic	↑			Maeda, 1976
					Maeda et al, 1985
D-1 antagonists	Intraventricular	↑			Beleslin et al., 1985
Adrenergic					
Norepinephrine	Intraventricular	↓			Feldberg and Sherwood, 1954
	Intracerebral hypothalamus	↑			Barrett et al., 1987
Epinephrine	Intracerebral hypothalamus	↓			Meyers, 1964

NOTE: ↑ = response facilitation; ↓ = response suppression; D = dose; HD = high dose; LD = low dose; PAG = midbrain periaqueductal gray.

[a]References are contained in Siegel and Pott (1988) from which this table was adapted.

There are, however, a number of investigations that point to a possible role of the opioid peptide system in the regulation of feline aggression. The choice for these series of experiments was prompted by the following observations: (1) that withdrawal from opiates is associated with heightened aggressiveness, and that opiate addicts appear to replenish "pathologically" diminished stores of endogenous opioids to levels that support noncombative adaptive behaviors (Khantzian, 1974, 1982; Wurmser, 1973); (2) that several regions of the limbic-midbrain from which attack can be elicited or modulated are richly endowed with opiate receptors, enkephalin-positive cells, and axon terminals (Atweh and Kuhar, 1977; Moss et al., 1983); and (3) that several studies, conducted in rodents and in the monkey, have shown that peripheral administration of the opioid antagonist naloxone results in a heightened aggressiveness and defensive behavior (Kalin and Shelton, 1989; Puglisi-Allegra and Oliverio, 1981; Tazi et al., 1983).

The results of these studies, using the feline model of affective defense, have in fact shown that the opioid peptide system can powerfully modulate affective defense behavior elicited from the hypothalamus or midbrain PAG (Brutus et al., 1988; Brutus and Siegel, 1989; Shaikh and Siegel, 1989). In contrast, the predatory attack system seems not to be opioid dependent. These studies suggest that several of the primary structures involved in opioid modulation of this response include the nucleus accumbens, BNST, and midbrain PAG. The next goal in this line of research should be a determination of the cell bodies of origin of the opioidergic fiber systems that supply each of these important nuclear modulatory groups. Such information would represent an important new step in our understanding of the nature of the inhibitory regulatory system for affective defense behavior.

SUMMARY AND CONCLUSIONS BASED ON DATA FROM STUDY
OF FELINE MODELS OF AGGRESSION

In reviewing the large numbers of studies conducted over the past two decades with respect to the neural bases of feline aggressive behavior, the following conclusions appear warranted. The sites in the brain at which predatory attack and affective defense can be elicited or modulated are now well established. The basic anatomic pathways utilized for the expression or modulation of the attack responses have also been identified, as well as the basic physiologic properties of these response systems. Moreover, we are now beginning to understand how the basic putative transmitters may function to regulate these responses. However, it is this area of investigation in which our knowledge is most limited. For example, we have little understanding of where the key synapses for monoaminergic regulation of aggressive reactions may be situated; nor do we understand their actions on the attack mechanisms at each of these synapses. Furthermore, with regard to the opioid peptide system, as noted above, we have no knowledge of the nuclear groups whose axons project to such key regions for the expression and modulation of affective defense as the BNST, nucleus accumbens, and PAG. Nor do we fully understand the cellular bases for opioid modulation at each of these synapses. Accordingly, it would appear that the most promising lines of research in the study of the neurobiology of aggression lie in attempting to obtain answers to these critical questions. Certainly, a thorough understanding of the neuropharmacology of aggressive behavior and the substrates at which transmitters act along the limbic-midbrain axis will be required before any attempts at rational pharmacologic intervention strategies for the control of human aggression based on this work can be considered.

HUMAN STUDIES OF AGGRESSION AND VIOLENCE

INTRODUCTORY AND DEFINITIONAL STATEMENTS

Adaptive Versus Maladaptive Violence and Aggression: Animal and Human Models

It would be well, in a consideration of the relationship between human violence and brain function, if a clear and defensible distinction could be drawn between adaptive (and socially acceptable) forms of violence and those that are maladaptive and symptomatic of an overt or presumed disorder of the brain. The research on animal models of aggression reviewed in the first section of this paper is related to adaptive aggression or "violence."

It seems inappropriate to refer to predatory attack (quiet biting attack) and affective defense behaviors as violent ("intentional infliction of physical harm") since they are part of the survival mechanisms of the species. This would require us to conclude that the cat is intentionally inflicting physical harm on a mouse when it kills and eats it for food, or to label as "violent" a female cat who may kill or severely injure an intruder in the course of an affective defense of her litter. This seems incorrect. In some sense, the goals or methods of the two (i.e., animal and human) investigations appear incompatible and/or inconsistent. That is to say, we study attack and defense mechanisms in normal cats (and other animals) and violent behavior in abnormal (presumably) human subjects. Are there really any links between the two bodies of research? We must assume that the mechanisms and structures involved in the final common neural pathway for the expression of attack and defense are basically similar in all higher mammals. A further assumption is that the mechanisms for activating or suppressing the aggression/violence pathway(s) are vastly more complex in humans because of the increased size of the forebrain. Nevertheless, it seems reasonable to hypothesize that humans who are engaging in maladaptive, violent behavior have some brain abnormality that somehow temporarily (and, possibly, repeatedly) short-circuits the aggression/violence pathways leading to truly violent behavior. We consider the evidence bearing on this hypothesis in the review and discussion below.

To return to the issue of defining adaptive versus maladaptive aggressive behavior in humans, there is on the one hand likely to be general agreement that serial killers or sexual offenders who maim their victims are displaying disordered, maladaptive violent behavior. On the other hand, the label of "adaptive" would be applied to the behavior of soldiers who kill enemy personnel in the course of battle and to the actions of skilled boxers who inflict physical punishment on (and occasionally kill) an adversary. Unfortunately, most of the examples of human violence and aggression considered in this section appear to fall somewhere between these two extremes. Violent or aggressive behavior may be quite adaptive in young male persons raised and living in impoverished environments; it may also be one of the few means of expression available to persons with impaired cognitive or language capacities. Surveys of violence and aggression have implicated age, gender, socioenvironmental, and/or cognitive variables; it may be virtually impossible to identify the portion of the variance that is attributable to adaptive, as opposed to nonadaptive, factors. Nevertheless, to the extent possible we will try to bear this distinction in mind in the course of reviewing and discussing the areas of human violence.

Measuring the Dependent Variable

The end points in measuring aggression/violence in animal models (see above) are simple, highly reliable, and almost stereotypic motor behaviors. By contrast, the measurement of human aggressiveness/violence is beset with confusing, highly controversial, and often contradictory legal, psychologic, medical, and sociologic issues. The acts that involve the criminal justice system range from robbery and aggravated assault through rape and homicide, whether these acts are actually committed, merely threatened, or attempted. The data on which research has been based are manifold. They consist of records obtained from official sources (prison, hospital, and outpatient clinical records), which may be biased in significant ways because of the nature of the clients who are involved with that institution. Some data have been gathered by questioning the subject about his or her behavior, or by questioning those who are familiar with the subject, such as teachers, nurses, peers, parents, and other relatives. These two sources cover the bulk of investigations in this review. In only a relatively small proportion of studies are there actual observations of the commission of aggressive or "violent" acts (which would be formally similar to the animal model studies) during the course of the investigation; these consist of a small number of treatment studies of hospitalized psychiatric patients and one study of epileptic patients monitored on closed-circuit television.

Since violent behavior is a relatively low-frequency act, and since it is likely to occur in circumstances where direct observation is unlikely, any conclusions we may draw from our review of research in human violence are based on second- or third-hand reports of behavior of varying (or unknown) degrees of reliability and validity.

Most likely, these criticisms lose their force in the case of documented repeat offenders in the criminal justice system. However, in other instances, including poorly documented studies of the effects of therapeutic brain lesions on "aggression, hyperkinesis, destructive tendencies and wandering tendency" (Ramamurthi, 1988), it is difficult to evaluate the merit of the contribution. This must temper our conclusions and will also influence the recommendations we make for future work in this area.

VIOLENCE IN PERSONS WITH NEUROPSYCHIATRIC DISORDERS

In the introduction to this paper we made reference to the concept that a person who commits crimes, of either a violent or a nonviolent nature, may need to be treated as one suffering from a disorder, rather than as a criminal. The work of Papez (1937) and MacLean (1952) identifying and promulgating the limbic system as the cerebral region necessary for the support and expression of emotion provided the impetus and rationale for a considerable body of research attempting to discover disturbed brain function in persons with neuropsychiatric disorders, especially disorders of emotion. Structures comprising the limbic system (amygdala, hippocampus, cingulate gyrus, portions of the thalamus and hypothalamus, and their connections) have been special targets of this research (Figure 1). However, the limbic system is ordinarily difficult to observe, measure, or study directly (except in certain neurosurgical cases); therefore, most studies of neuropsychiatric patients have attempted to measure brain abnormalities as best they could by examining electroencephalogram (EEG) records or searching for "soft" or hard neurologic signs, such as the presence of seizures. The advent of newer, sophisticated, and relatively noninvasive imaging techniques such as computerized tomography (CT) scans, regional cerebral blood flow, magnetic resonance imaging, and positron emission tomography (PET) has made the study of limbic and limbic-related structures in humans more feasible. Some imaging research with neuropsychiatric patients exhibiting violent behavior has been done and is reviewed later in this paper.

Schizophrenia

Although there may be a connection in the view of lay persons between the co-occurrence of major psychiatric disorders such as schizophrenia and violent behavior, the evidence is in fact not very compelling. We report here six recent investigations involving schizophrenia.

There are a few reports of increased aggressiveness or assaultiveness in patients with the diagnosis of schizophrenia (Tardiff and Sweillam, 1980; Krakowski et al., 1986; Karson and Bigelow, 1987) (Table 3).* Tardiff and Sweillam (1980) attribute the violence to either the sex (male) or the socioeconomic status of the patient (low). Krakowski et al. (1986) focus on the role of violence in the disorder of the patient and remind us that the carrying out of assaults is incompatible with the disorganized psychotic states seen in severe psychiatric illness. Karson and Bigelow (1987) report an increased incidence of assaultive behavior in schizophrenic patients in comparison with those carrying other diagnoses. However, the assaultive patients were also younger and tended to have a prior history of violence.

Robertson (1988) makes much the same point as Krakowski et

---

*Tables 3 through 17 appear at the end of this paper, beginning on page 112.

al. (1989): Schizophrenic men are socially incompetent and there-fore more easily detected in the commission of a criminal act, including violence, and more easily detained. As a result, it is difficult to compare the rate of criminal offenses among the men-tally ill with that of the general population. The last two studies in this group (Convit et al., 1988; Herrera et al., 1988) point to the fact that both illegal drugs (PCP) and legal drugs (haloperidol—an antipsychotic medication) can increase the frequency of violent acts in schizophrenic men. These findings cast further doubt on the specificity of any association between schizophrenia and the tendency to commit violent acts.

Alcoholism

Only two studies (Table 4) from a vast literature are reported here; the relationships among alcohol use, brain damage, psychopathy, and violence are very complex and beyond the scope of this re-view. One study (Buydens-Branchey et al., 1989) deals with the intervening variable of deficit of serotonin metabolism as a pre-cursor of antisocial behavior. The other study by Coid (1982) reviews the literature and concludes that although violence is an alcohol-related problem, the relationship is not a direct one. Fur-ther research into alcohol-related damage to the brain is the most fruitful area of study.

Epilepsy

A number of studies on violence in neuropsychiatric popula-tions have focused on epilepsy. Before discussing this research, some definitions are needed. An important distinction concerns the division between acts of violence or aggression that allegedly occur in the course of an actual clinical seizure and are properly to be considered part of the seizure or ictus itself (i.e., ictal mani-festations) and violent or aggressive behavior that occurs between seizures (i.e., in the interictal period). Research on interictal aggression is summarized in Table 5. Before discussing these studies, however, it is necessary to review the ictal data. With respect to ictal manifestations, an earlier study by Ajmone Marsan and Ralston (1957) reviewed several hundred seizures in epileptic persons, induced in the laboratory by injec-tion of metrazol for diagnostic purposes. The results of their review indicated that aggressive behavior or angry feelings were extremely rare. Much more commonly reported were feelings of fear. Similar observations about the rarity of ictal rage or aggression were made by Gloor (1967) and Rodin (1973) and were reviewed by Mirsky and Harman (1974). Epileptic patients may thrash about in the course of a generalized seizure, or may appear to strike out if attempts are made to restrain them; however, the impaired cognitive state associated with a generalized seizure is scarcely compatible with a directed attack upon another person.

Although it seems unlikely that ictal violence or aggression is a significant human problem, there has nevertheless been a scale devised to rate aggressive behavior during seizures and to certify whether such behavior was truly ictal (Delgado-Escueta et al., 1981). This scale (Table 6) was derived from videotaped seizures suggestive of violent behaviors; it is of interest that it was based on a population of 13 patients out of a pooled sample of more than 5,000 cases. Approximately one-quarter of 1 percent of epileptic persons thus show ictal violence or aggressiveness; this minuscule figure is not in disagreement with the earlier findings with metrazol-induced convulsions by Ajmone Marsan and Ralston (1957) and others.

Concerning *interictal* violence and aggression, in contrast, the picture is considerably more murky. One of the models for conceptualizing this behavior is to assume that it is *state*-related. That is to say, the violence represents an occult or hidden seizure equivalent for some persons with epilepsy; if one were fortunate enough to have recording electrodes deep within the appropriate region of the limbic system of the brain of the patient, one would be able to record evidence of seizure (i.e., EEG spike) activity. This is notwithstanding that there may be no other overt indications of convulsive activity. Enormous controversy has raged over whether or not there has ever been a convincing demonstration of this phenomenon, since it provides a model and rationale for surgical treatment (i.e., ablation) of the offending seizure focus (Mark and Ervin, 1970; Valenstein, 1980).

An alternative view of interictal aggression or violence in persons with epilepsy is that it is a *trait* associated with this type of cerebral disorder rather than an ictal occurrence. According to this view, the emotional mechanisms within the brain (usually referring to the limbic system) of such a person are damaged or modified in some way by the seizure disorder such that aggressive or violent outbursts are more likely to occur in them than in nonepileptic persons. A number of questions are raised by these purported findings. Are there actually patients who conform to this description? To what extent are they representative of the entire population of persons with epilepsy? To what extent are those behaviors, if they occur, a characteristic of the age, sex, and socioeconomic characteristics of the patients, as opposed to their epilepsy? If the behaviors occur at rates greater than those seen in appropriately selected control subjects, are there other equally plausible interpretations of the behavior rather than what might be called temporolimbic irritability?

Another question pertaining to the design of research in the area concerns whether one starts with patients suffering from seizure disorders and examines the prevalence of violence/aggressiveness in these cases. Such a study might be done using admissions to a state hospital or a clinic (e.g., Tardiff and Sweillam, 1980). The alternative is to start with persons who have been involved with the criminal justice system (or other medical-legal authority) because of dangerous or assaultive behavior and to ascertain the prevalence of seizure disorders. An example of this approach is the study by Bach-y-Rita et al. (1971) (summarized in Table 5). The first approach (i.e., to start with seizures) suffers from the potential problem of sampling bias. To what extent is a state hospital or large clinic dealing with a truly representative sample of seizure cases? What proportion of patients, unknown to the system, seek and receive treatment from private practitioners (or other sources of care) and never enter the statistical pool? The second approach (i.e., to start with the aggressive behavior) may also suffer from a sampling bias. Those cases most likely to be referred and to be the objects of clinical attention are those in whom some brain abnormality is suspected or demonstrated. It is also more likely that persons who have engaged in violent/aggressive/assaultive behavior during their lives would have suffered from head injuries leading to seizure disorders. The association between the abuse received during childhood (including possible head injuries) and later abusive behavior toward others appears to be a clinical truism (e.g., Tarter et al., 1984; Dodge et al., 1990). What is sorely lacking in studies of this type are (1) a national register of persons with seizure disorders and/or (2) an epidemiologic sample of a large and varied catchment area. The availability of such sources of information would enable us to draw unambiguous conclusions about associations and, possibly, about the direction of causality. We return to this issue later.

Some of these questions are addressed in the studies reviewed here; definitive answers are not available to most of them. At issue, particularly with respect to interictal violence/aggression, is whether or not surgical treatment (i.e., resections of brain tissue) of patients who exhibit violent behavior can be rationalized as the equivalent of removal of a seizure focus. Surgical treatment of intractable seizures has been a successful therapy for at least 40 years (Penfield and Jasper, 1954).

Aggression as an Epileptic Equivalent

A review of the literature published during 1974-1989 indicates that there has been a substantial emphasis on the possible relation between seizure disorders and aggressive behavior. Twelve studies or literature reviews have examined this question; of these, seven concluded, on the basis of the review either of clinical case material or of the literature, that there is an association in patients between the occurrence of seizure disorders and the likelihood of aggressive or violent behavior (Weiger and Bear, 1988; Lewis et al., 1983; Bear and Fedio, 1977; Bear et al., 1981; Devinsky and Bear, 1984; Perini, 1986; Engel et al., 1986). It should be noted that four of these studies are based on the work or views of Bear (Weiger and Bear, 1988; Bear and Fedio, 1977; Bear et al., 1981; Devinsky and Bear, 1984), for whom the greater aggressiveness of epileptic patients (specifically those with temporal lobe foci of abnormality) is an example of the enhanced emotional responsivity that characterizes behavioral transactions in such persons. This, in turn, is attributed by Bear to the sensitization of the limbic system and greater autonomic responsivity resulting from the epileptogenic focus (Bear et al., 1981).

Three additional references concluded that although there may indeed be an association between epilepsy and violence, this is a spurious connection, related to other factors (Hermann, 1982; Virkkunen, 1983; Stone, 1984). Among these other factors are impaired cognition (Hermann, 1982) and an angry reaction borne out of despair over the incurable nature of the disorder (Stone, 1984). Further, Virkkunen concludes that aggression in epileptics is a "multifactorially determined . . . phenomenon" (Virkkunen, 1983:647).

Two studies concluded that there is no relation between epilepsy and violence that cannot be accounted for in terms of the socioeconomic class of the patients studied (i.e., more assaultive behavior is associated with lower socioeconomic class; Tardiff and Sweillam, 1980:Table 3; Treiman, 1986:Table 5). The study by Tardiff and Sweillam (1980) is particularly impressive since it is based on a sample of more than 9,000 patients admitted to psychiatric hospitals in a one-year period.

Violence as the Independent Variable—Is There a Brain Disorder in Such Patients?

In the work reviewed in this section, the primary focus of the study was on the symptom of violence per se, rather than as a component or additional symptom of some other disorder (i.e., schizophrenia, alcoholism, epilepsy). Two subgroups of investigations are presented: in one, violent sexual offenders were the focus of the study (Table 8); the other dealt with non-sex offenders (Table 7). In all of those studies, however, the intent of the investigators was to determine whether available neurodiagnostic tests could detect a consistent pattern of brain disorder or disorders that could account for the symptom(s). A particular interest has to do with whether or not limbic system pathology or dysfunction can be identified.

Violent Offenders—Nonsex Crimes

There were seven studies or literature reviews related to this topic during our selected reporting period (ca. 1974-1989) (Table 7). In six of these, it was concluded that there was an association between aggressive/violent behavior and brain abnormalities (Bach-Y-Rita et al., 1971; Lewis et al., 1986; Volkow and Tancredi, 1987; Hendricks et al., 1988; Andrew, 1980; Brickman et al., 1984). The studies ranged from comparisons of left-handed juvenile delinquents (less violent) with right-handed juvenile delinquents (more violent) (Andrew, 1980), through neuropsychologic assessment of felonious delinquents (Brickman et al., 1984), to studies of episodically violent patients (Bach-Y-Rita et al., 1971; Volkow and Tancredi, 1987) and persons sentenced to death for violent crimes (Lewis et al., 1986). The studies by Volkow and Tancredi (1987) and by Hendricks et al. (1988) are of special interest because they employed, respectively, positron emission tomography and regional cerebral blood flow measurements to assess brain abnormalities. Both of these "high-tech" biomedical studies reported positive findings. Langevin et al. (1987) found impairment on the Reitan neuropsychologic battery (presumably reflecting brain damage or dysfunction) in one-third of 18 males facing murder or manslaughter charges. However, as noted above in the discussion of violent acts among men with a diagnosis of schizophrenia (Table 3), the high prevalence of alcohol and drug abuse in these accused murderers prevents an unequivocal interpretation of the results. Only the study by Mungas (1983) (Table 7), on the basis of comparisons among various subgroups of patients differing in the degree or intensity of violent behavior, concluded that there was no association between violence and brain dysfunction. The remaining citation, a literature review by Eichelman (1983), concluded that there is no necessary link between limbic system pathology and the induction of aggressive behavior; the two phenomena are independent. Moreover, other brain regions may be contributory as well.

Violent Offenders—Sex Crimes

In Table 8 are summarized the results of six recent studies or reviews of violent sex offenders; five of these were conducted by the same group of researchers (Langevin et al., 1985, 1988, 1989; Hucker et al., 1988; Garnett et al., 1988). They involve a variety of innovative research techniques such as the measurement of sexual arousal as elicited by standard audiotaped erotic stimuli, and the measurement of brain pathology with CT scans and PET scans. In addition, hormone profiles, as well as careful and detailed neuropsychologic and personality examinations, were conducted. Hendricks et al. (1988) measured regional cerebral blood flow (xenon inhalation technique) in a group of child molesters.

Although the results of these innovative studies are provocative, there are no uniform findings that can be used to implicate any particular brain region in the pathophysiology of the violent sexual offender. Some results (Hendricks et al., 1988) suggest a nonspecific lower brain metabolism in such persons. Some of the findings of the Langevin group (Langevin et al., 1985; Hendricks et al., 1988) are compatible with the presence of abnormal functioning in the right temporal lobe, particularly among sexual sadists. The results of later studies (Langevin et al., 1989), however, seem to implicate left hemisphere problems, related to language, particularly among pedophilic offenders. The authors conclude that the findings are very provocative but not conclusive; the brain abnormalities are likely to be subtle in such cases and will require extremely subtle and sensitive measurement techniques. Moreover, as some of their work suggests, there may be specific abnormalities associated with different types of sex offenders (i.e., sadism, pedophilia, etc.).

TREATMENT OF VIOLENCE

Electroconvulsive Therapy (ECT) and Pharmacotherapy

It must follow as the night the day that if violent offenders are considered to be suffering from a brain disorder, efforts to treat them would be instituted. Moreover, if violence is an epileptic equivalent, would not violent patients be helped with antiepileptic drugs? Although there is undoubtedly a large literature associated with this problem, we have presented only a small number of representative recent studies. ECT (Schnur et al., 1989) may be efficacious in patients with episodic dyscontrol (or Intermittent Explosive Disorder; American Psychiatric Association (1987) (Table 9).

Monroe (1975) reviews studies of the use of anticonvulsants in the treatment of episodic dyscontrol, concluding that these compounds may be a useful adjunctive treatment. Finally, two studies employ pharmacotherapy in treating the symptoms of aggressive behavior in persons with geriatric problems (De Cuyper et al., 1985) or Huntington's disease (Stewart et al., 1987).

Psychosurgical Treatment of Violent Offenders

Despite the reservations or at least the caution implied or stated by most authors concerning brain-behavior connections in this controversial field, there were three reports of surgical ablations of limbic or limbic-related structures as treatment of violent or aggressive behaviors (Sano and Mayanagi, 1988; Ramamurthi, 1988; Dieckmann, 1988; see Table 10). All involve stereotaxic lesions in the medial hypothalamus (or amygdala, in the case of Ramamurthi). The persons on whom the surgery is performed are violent/aggressive cases (Sano and Mayanagi, 1988; Ramamurthi, 1988) or "sexual delinquents" (Dieckmann, 1988). In addition to the forthright surgical approach, these studies are notable in that they often target young subjects (many of Sano and Mayanagi's cases were under 15 as were most of Ramamurthi's). The numbers of cases reported varied from 14 (Dieckmann et al.) through more than 600 (Ramamurthi); the reports were all favorable, although the follow-up varied in quality. Independent assessments of outcome, as well as untreated control groups, were absent from these reports.

It is difficult to assess these reports, not only for the reasons stated, but because the criteria for assessing or diagnosing violent or aggressive behavior may differ in Japan (Sano and Mayanagi, 1988) and India (Ramamurthi, 1988) from those used in the United States. It is not clear that violence and aggression have unambiguous cross-cultural meaning. What for instance, is the Western equivalent of "wandering tendency," one of the criteria used by Ramamurthi in selecting candidates for surgical therapy? The use of neurosurgical treatment for violence thus leads to a set of complex and controversial issues. Some of these issues are discussed in the volume edited by Valenstein (1980). The last study in this category is the report of a single psychosurgery case seen by one of the authors over a nearly 40-year follow-up period (Mirsky and Rosvold, 1990). Immediately premorbid, the patient was in her second year of medical school. After 11 years of hospitalization and treatment, the patient was considered dangerous (she was a biter) to all health care personnel. She could not be approached safely before her psychosurgical procedure(s); postoperatively, she was extremely docile. Although the violent behavior disappeared after surgery, the apparently intractable schizophrenia was the reason for the operation, not the violence. The latter had developed as a symptom sometime during the course of the progressive deterioration accompanying her 11-year schizophrenic illness. The case is of some heuristic value in that it emphasizes that violent behavior may be part of a more complex syndrome of disorder, in this case schizophrenia. The progressive deterioration was the primary reason for the radical operation(s), and the violence presumably stemmed from that circumstance. It may be that the aggressive biting was one of the few forceful means of communication left to the patient in her deteriorated state. In this sense, this case study is similar to the reports summarized in Table 1 (suggesting no close relation between schizophrenia and violence) and points to a possible interpretation of violence as a primitive, aggressive method of communication in a person with poor or absent communication skills. This theme is developed further in subsequent sections.

Treatment of Violence—A Commentary

The view that violence and aggressiveness are disorders is not one that has been accepted universally. Some authors have criticized medical, pharmacological, and surgical approaches to violence, not only as based on incomplete, if not poor, scientific evidence, but also as part of a conspiracy by those in power to suppress individual liberties, differences, or dissent. The roots of the dissent are said to stem in part from frustration over economic inequalities in society (Breggin, 1980; Chorover, 1979, 1980). Although most of the criticisms of what might be called the neurobiologic approach to human violence have centered around neurosurgical treatment of psychiatric disorders (i.e., "psychosurgery"), some have argued that the conspiracy extends to pharmacotherapy of mental disorders as well (Breggin, 1980; Chorover, 1980). Chorover has cited stimulant drug treatment of hyperactive children as another example of the way in which "in both families and societies, practical and political interests may be served by attributing blame, by identifying symptoms as causes, and by controlling individuals whose behavior is defined to be dangerous or disturbing" (Chorover, 1980:133). Although it is true that these authors have been primarily polemicists rather than gatherers of data, they are representative of those who, for a variety of reasons, oppose neurobiologic approaches to complex medical-social problems. This may be a minority view, but it is clear that there is no unanimity of opinion that seeking the neurobiologic causes of violent and aggressive behavior (and basing treatments on these data) is an unalloyed good.

SOCIAL CLASS, AGE, AND GENDER RELATED TO VIOLENCE

The fact that violent and aggressive behavior is more prevalent among young male members of lower socioeconomic groups has been mentioned in conjunction with a number of studies we have reviewed (e.g., Tardiff and Sweillam, 1980; Kindlon et al., 1988; see Table 11); this variable or group of variables is often a confounding factor in the interpretation of the results of research on violence. A recent statistical summary cited by Roberts (1990) shows a precipitous rise in male homicide rates (for nonrelatives of the same sex) at ages 15-19, which does not moderate until age 40-44. Even then, the curve does not approach the rates for females until ages 50-54 or later. The rates for females remain relatively stable throughout life, except for a slight bump occurring at age 20-24. This gender difference may be rooted in evolution and is undoubtedly heavily influenced by cultural and sociologic factors (Daly and Wilson, 1989, cited by Roberts, 1990); however, the question arises as to whether there are differences between male and female brains that could account for, or at least be related to, the differential propensity to violence. The work of two investigators is possibly relevant here. Lansdell has shown, in a series of studies beginning in 1962 (Lansdell, 1962), that male patients with seizure disorders who undergo surgical resection of the temporal lobe for the relief of epilepsy tend to show greater behavioral deficits postsurgery than female patients. "These unique results with these operations show a resiliency of the female brain, compared with the male brain, with regard to these two types of surgery . . ." (Lansdell, 1989).

The studies of Lansdell and a confirmatory finding by Sundet (1986) are summarized in Table 11. One additional study is included in this table although it does not pertain directly to the effects of temporal lobe surgery or epilepsy. Instead, this study by Hampson and Kimura (1988) shows that women exhibit large and significant "reciprocal performance fluctuations over the menstrual cycle in two types of skills." The possibility exists that these hormonal effects in the female may have some bearing on the apparently greater resilience of the female brain; further, it is conceivable that these sex differences may have some bearing on the greater propensity for male aggressiveness/violence. If it is true that violence is a symptom due to a lesion somewhere in the brain, then the increased prevalence of the symptom in the male may be due to the greater vulnerability of the male brain to injury and/or to the greater apparent plasticity of the female brain. Clearly, much additional research would be needed before the merit of this hypothesis could be evaluated.

PSYCHOPHYSIOLOGIC STUDIES RELATED TO VIOLENCE

There are numerous psychophysiologic techniques that have been employed in the study of disordered populations. However, even a cursory review of these would go beyond the limits of this paper. We will refer here only to some recent studies of Event-Related Brain Potentials (ERPs) in populations relevant to the current discussion: prison inmates rated high on a psychopathy scale, although not necessarily selected for unusual violence or aggressiveness. Other research, however (Hare and McPherson, 1984), has shown that psychopaths are more likely than other criminals to commit violent and aggressive crimes. In this series of studies (Jutai et al., 1987; Hare and Jutai, 1988) summarized in Table 12, the findings suggest that high psychopathy is associated with poor information processing capacities. Specifically, those inmates rated high in psychopathy showed both ERP and neuropsychologic evidence of left hemisphere dysfunction and/or weak or unusual lateralization of language functions. The results are compatible with the view that psychopaths have fewer language processing resources than do normal individuals.

The results of these studies are important for two reasons: (1) they highlight a possible cognitive deficit in persons who may commit violent crimes; and (2) they exemplify a powerful noninvasive technique (ERPs) for studying brain functions and information processing in violence-prone populations. Such research, as Duncan (1990) and colleagues (1985, 1989) have shown, can help illuminate specific information processing deficits in a variety of clinical populations, including schizophrenics and persons with eating disorders or seasonal affective disorders. These methods add to the armamentarium of noninvasive techniques that may be especially appropriate for research on child populations.

DEVELOPMENTAL RESEARCH RELATED TO VIOLENCE

Neurodevelopmental Issues

In numerous papers over the past 15 years there has been attention to developmental issues that are potentially related to the etiology of violent/aggressive behavior. We have already alluded to some of these studies in previous sections (i.e., Tarter et al., 1984). Two major themes emerge from this work: the first, and most often discussed, is that the tendency to display aggressiveness and violence represents a neurodevelopmental maturational deficit. The specific form of the hypothesized relationship differs somewhat from author to author, but the basic message is similar for all: in children who exhibit violent and aggressive behaviors there is evidence of substantial neuropsychologic deficit in such functions as memory, attention, and language or verbal skills. The aggressive behavioral manifestations may thus represent: (1) maladaptive communication skills in persons who are impaired in their capacity to communicate and/or (2) a frustration-elicited response based on inability to compete in the cognitive arena with peers (Mungas, 1988; Miller, 1987; Andrew, 1981; Yeudall et al., 1982; Brickman et al., 1984; Piacentini, 1987; Woods and Eby, 1982; Lewis et al., 1988; Hermann, 1982) (Table 13). Some unpublished data from a recent study of attention and other cognitive skills in second-grade children judged by teachers to be unusually aggressive, provide additional support for these hypotheses. Boys rated as aggressive showed poorer attention skills in tasks requiring the encoding of information and in sustained concentration. Aggressive girls were poorer than controls on sustained concentration tasks only (Anthony et al., in press; Mirsky, 1989). Zahn et al. (1991) recently reported similar deficits in externalizing (acting-out) boys (i.e., their performance on sustained attention tasks under conditions of stimulus uncertainty was inferior to that of control boys).

A somewhat unusual finding—an exception to the general trend of these studies—is reported by Kindlon et al. (1988). Although they reported the usual greater incidence of aggressiveness in 12- to 16-year-old boys of lower socioeconomic status, they failed to find that children with hard neurologic findings (cerebral palsy, seizure disorders, head injury accompanied by significant coma) or definite EEG or imaging evidence of damage showed greater aggressivity and delinquency than controls. It would be of considerable interest to have had neuropsychologic-cognitive data on these children. It may be that aggressive disorders in brain-injured children are seen only in the presence of cognitive impairments (i.e., leading to communication difficulties). In any event, the data are of interest and must temper any conclusions about obligatory relationship between head injury and externalizing behaviors. It would also be well to conduct follow-up studies of this cohort.

The second theme in these studies, which is not necessarily independent of the first, is that violence, abuse, and deprivation of a loving relationship in childhood predispose the individual to later delinquency, including homicide (Lewis et al., 1983; Tarter et al., 1984; Walsh and Beyer, 1987). The first theme is, in fact, consonant with that developed in earlier sections, namely that aggressive/violent behavior may stem from involvement of the limbic system (and related regions of the brain); however, the mediating variable is neuropsychologic deficit in communication, attention, and other cognitive skills. It is thus not the case, according to this view, that there is a direct relationship between limbic system abnormality and aggression. Rather, the link appears to be through the effects of altered cognitive capacities (pursuant to limbic system damage) of the individual. The point that perhaps should be made in this context is that while the limbic system may subserve emotional behavior, the structures of which it is comprised also serve a major role in the support of cognitive behaviors. For example, the role of the hippocampus in the support of memory is well known (Milner, 1969). However, questions remain as to how such injury to the developing brain could occur and whether there are social-environmental conditions that could foster such developmental defects. In fact, there appears to be a host of pernicious influences on the developing nervous system that could, in the perinatal period (or later), account for serious compromise of the developing brain; these influences are more likely to be present in lower socioeconomic circumstances.

Possible Etiological Variables

Genetics   We discuss only two recent contributions to this area. These are summarized in Table 14. In the study by Mednick and Kandel (1988) of 173 recidivistically violent offenders (from the Danish health records) it was found that early commission of violent crimes predicted later commissions (a point also made strongly by Farrington, 1989). Moreover, when the offenses of adopted-away children were compared with those of biologic and adoptive parents, it was found that biologic parent-child relationships predicted property convictions, but not violent offenses. This finding supports the view that the circumstances leading to the commission of violent acts are more likely due to nongenetic (i.e., environmental) than genetic determinants.

The other study, by Walker et al. (1989), explored the relation between schizophrenic diathesis and the stress of parental maltreatment in a group of acting-out children with significant aggressive and delinquency problems. The design (summarized in Table 14) allowed the authors to conclude that the combination of parental schizophrenia and maltreatment at home was associated with a progressive increase in delinquency in both sexes over time.

Lead The possible etiologic role of small amounts of lead in the environment, in contributing to cognitive and social failure, has long been the primary research interest of Needleman and collaborators (Table 15). Not only does environmental lead (in air contaminated by leaded fuels, from paint in older houses, from plumbing systems) contribute to poorer cognitive capacities (including attention) but in follow-up studies in a Boston cohort recently reported (Needleman et al., 1989), it is also associated with significant academic failure: a sevenfold increase in the failure to graduate high school—"a serious impairment in life success." The study by Thomson et al. (1989) confirms Needleman, with an independent sample from Edinburgh. This work also implicated lead levels more directly in deviant antisocial and hyperactive behavior in school children. The remaining paper in this group (Lansdown, 1986) suggests that the relationship between lead levels and behavior may not exist in children from higher socioeconomic strata. If this were to be the case, then cofactors (some of which are discussed below) may be involved.

Alcohol Although considerable data have been reported on the later effects on the child of maternal ingestion of alcohol during pregnancy, the most complete and systematic data are probably provided by the work of Streissguth and colleagues (Table 16). This is a continuing investigation of a group of 500 Seattle children that began with a group of pregnant women. Significant correlations between maternal alcohol ingestion during pregnancy and the attention performance of the child at age 4 have been reported (Streissguth et al., 1984). This finding was replicated and extended to other cognitive behaviors at age 7 1/2 (Streissguth et al., 1986, 1989). This study provides strong presumptive evidence that the roots of academic failure (and later delinquency) may lie in maternal drinking (even of moderate amounts) during pregnancy.

Cocaine, Opiates, and Tobacco Data now exist (Table 17) showing that maternal use of cocaine has a significant depressant effect on reflexes and state control in infants and a general impairing effect on neonatal neurobehavioral capacities (Chasnoff and Griffith, 1989; Chasnoff et al., 1985). Although the children in such cohorts may be too young to allow reliable assessment of cognitive capacities, it seems reasonable to assume that many of them may present significant academic and social problems as they develop and enter the school system. In more or less the same vein, Table 17 summarizes some recent reports of reviews of later effects in the child of maternal use during pregnancy of opiates (Olofsson et al., 1983; Wilson et al., 1979) and tobacco (Rush and Callahan, 1989). It should also be noted that Streissguth found nicotine effects on performance in her maternal alcohol studies independent of alcohol effects (Table 16).

Some Areas Not Addressed—Some Roads Not Traveled

In the course of this review of neurobiologic factors of possible relevance to the development of aggression and violence, it became evident early that the literature on related factors or related clinical populations was staggeringly large and could not be addressed in a single paper of this size. The areas chosen for discussion are obviously in part a reflection of the interests and biases of the authors; other authors would probably have chosen to emphasize different aspects of the problem and would have made different selections from the voluminous related literature. The following list of undoubtedly relevant problems, factors, or clinical groups either has not been discussed in this review or has been mentioned so fleetingly as to have been essentially ignored:

Attention deficit disorders, with or without hyperactivity
Social learning of violence

Autonomic nervous system influences
Neurochemical factors
Neural specificity/plasticity
Kindling of seizure foci

The issue of whether or not human aggression and violence can be linked unequivocally to disordered brain mechanisms remains very much in doubt. The literature surveyed in this report is suggestive of some relationship between pathophysiologic processes in the limbic system and the tendency to engage in assaultive and violent behavior; however, the issues of sample size, sampling bias, and the co-occurrence of other variables that could account for the excess violence in the patient or offender groups under study remain for the most part unresolved. Although it is true that many studies of violent persons have found evidence of brain abnormalities, the possible role of a violent lifestyle (leading to head injuries) is generally difficult or impossible to control. Violent lifestyles, moreover, for many young male residents of impoverished areas, may be adaptive or at least difficult to avoid. The role of parental mistreatment leading to injury, not only to the brain but to self-esteem, also needs to be considered: violence begets violence. Possibly, there is one issue that can or should be laid to rest: violence occurring in the course of a seizure (ictal violence) is an almost nonexistent phenomenon.

In the course of this review, a number of suggestions were made that could simplify the process of doing research in this area and help to clarify the relationship of brain disorders to aggression, violence, and other types of antisocial behaviors. One would involve the creation of a national health registry of the type that exists in a number of Scandinavian countries. This resource has, in fact, been exploited in a most profitable manner by many American and Scandinavian researchers to illuminate and clarify the role of nature and nurture in the development of schizophrenia spectrum disorders (Rosenthal and Kety, 1968). Data from the police or court systems would automatically be entered into the registry. This would allow us to ascertain, with some assurance, the actual prevalence and incidence of antisocial, as well as other, behaviors in various populations at risk. The necessary precautions would have to be observed to restrict access to the data to qualified researchers, and to prohibit the use of the information for any purpose other than research. This is virtually identical to the recommendation made by one of the authors in 1974 (Mirsky and Harman, 1974:204):

Only large scale survey data with competent, standardized reporting of neurologic, psychiatric and behavioral information will permit us to describe adequately the relation between epilepsy, temporal lobe disease and disordered social behavior. . . . And the potential additional information would insure the value of the project. For example, if an association were found between aggressive, violent or other socially unacceptable behavior and a series of neurologic, demographic and environmental variables, then a prospective study could be done in which a group of infants or young children at risk for such complications in later life could be extended every medical, environmental and social benefit. A matched control group without such advantages would be constituted that would not receive such comprehensive ministrations, and the outcome of the two groups could be evaluated at 5, 10 and 15 years. If the treatment were successful in preventing or reducing the development of unacceptable behavior, it could then be extended . . . to all children at risk.

It may be possible to implement a less grandiose and utopian scheme by performing population-based surveys in one or more cities. Such surveys have been conducted by the National Institute of Mental Health (NIMH) for the purpose of identifying persons with psychiatric disorders and have been useful for mental health planning. Subjects at risk could be identified in this way, and controlled studies involving various types of interventions could be launched. Interventions could include training targeted at improving general classroom competence, as implemented by Kellam and Rebok (1992) in the Hopkins-Baltimore preventive intervention project (see below) or aimed at specific cognitive weaknesses such as attention and concentration. While it is true that violent acts may be relatively rare, their occurrence may reflect a much more prevalent and widespread problem associated with disordered cognition: failed academic opportunities, underachievement of occupational level, increased nonviolent (as well as violent) crimes with the attendant staggering costs of courts and jails.

We believe that this review has highlighted the fact that violent behavior is more likely the consequence of impaired cognitive processes than of altered emotional states. Our efforts should therefore be directed toward understanding, controlling, ameliorating, and reducing the effects of brain-damaging environments.

Some modest efforts along roughly these lines have been undertaken by NIMH in the creation of prevention intervention research centers. One of the authors (AFM) is collaborating with the staff at the Department of Mental Hygiene at Johns Hopkins University and is involved in the planning, assessment, and evaluation

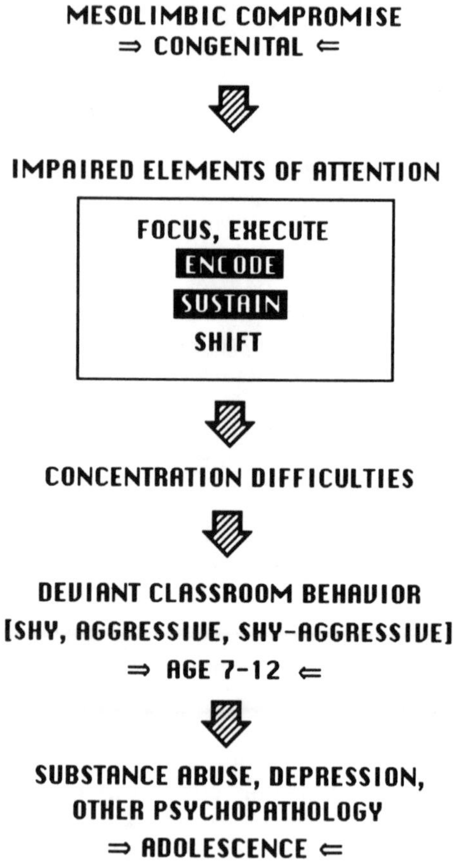

MESOLIMBIC COMPROMISE
⇒ CONGENITAL ⇐

IMPAIRED ELEMENTS OF ATTENTION

FOCUS, EXECUTE
ENCODE
SUSTAIN
SHIFT

CONCENTRATION DIFFICULTIES

DEVIANT CLASSROOM BEHAVIOR
[SHY, AGGRESSIVE, SHY-AGGRESSIVE]
⇒ AGE 7-12 ⇐

SUBSTANCE ABUSE, DEPRESSION,
OTHER PSYCHOPATHOLOGY
⇒ ADOLESCENCE ⇐

FIGURE 6   This model is based on data gathered in the course of the collaborative Johns Hopkins University—NIMH prevention intervention research project referred to in the text, as well as the results of the studies by Kellam and colleagues (1975) in the Woodlawn district of Chicago.

of cognitive-based interventions. A tentative model, based on some of the preliminary results from this collaboration, is included in Figure 6. The model starts with the assumption of early congenital damage to limbic or related mesencephalic structures of the brain. This in turn leads to difficulty with certain "elements of attention" (Mirsky, 1989) in elementary school, and thence to teacher-identified concentration difficulties and early academic failures. These may result in deviant classroom behaviors (shyness, aggressiveness, shyness and aggressiveness in combination) and to adolescent behaviors related to aggressiveness ("other psy-chopathology"). This model is also based, in part, on the results of the studies of Kellam and colleagues in the Woodlawn district of Chicago and in Baltimore (Kellam et al., 1975; Kellam and Rebok, 1992).

We need to increase funding for prevention-intervention centers such as the one that led to the model presented in Figure 5. The operation of centers of this type could in fact be coordinated with case finding from the proposed national health registry. In this manner, children at particular risk for the development of later antisocial behavior could be identified early and provided with ameliorative cognitive and/or medical treatments. The costs of such programs would undoubtedly be high, but in time their benefits would outweigh the costs and would reduce substantially expenses associated with the operation of the criminal justice system.

ACKNOWLEDGMENTS

Support for this work was obtained, in part, from National Institutes of Health Grant NS 07941-20 and a grant from the Foundation of the University of Medicine and Dentistry of New Jersey (AS). Thanks are due to the following persons for their devoted assistance in the preparation of this manuscript: Sandra M. Wilkniss, Mary La Padula, Susann M. Nourizadeh, and Susan J. Lochhead.

REFERENCES

Ajmone Marsan, C., and B.L. Ralston
1957   *The Epileptic Seizure.* Chicago: Chas. C. Thomas.
American Bar Association
1983   *First Tentative Draft Criminal Justice Mental Health Standards.* Washington, D.C.: Second Committee on Association Standards for Criminal Justice, American Bar Association.
American Psychiatric Association
1987   *Diagnostic and Statistical Manual of Mental Disorders,* 3rd ed., revised. Washington, D.C.: American Psychiatric Association.
Andrew, J.M.
1980   Are left-handers less violent? *Journal of Youth and Adolescence* 9(1):1-9.
1981   Imbalance on the weights test and violence among delinquents. *International Journal of Neuroscience* 14:35-40.

Anthony, B., G. Rebok, D. Pascualvaca, P. Jensen, M.B. Ahearn, S.G. Kellam, and A.F. Mirsky
in   Epidemiological investigation of attention performance in chil-
press dren. II. Relationships to classroom behavior and achievement. *Journal of Child Psychology, Psychiatry and Allied Disciplines.*

Atweh, S.F., and M.J. Kuhar
1977 Autoradiographic localization of opiate receptors in the rat brain. 111. The telencephalon. *Brain Research* 134:393-405.

Bach-Y-Rita, G., J.R. Lion, C.E. Climent, and F.R. Ervin
1971 Episodic dyscontrol: A study of 130 violent patients. *American Journal of Psychiatry* 127(11):49-54.

Bandler, R.
1984 Identification of hypothalamic and midbrain neurons mediating aggressive and defensive behavior by intracerebral microinjections of excitatory amino acids. Pp. 369-392 in R. Bandler, ed., *Modulation of Sensorimotor Activity During Alterations in Behavioral States.* New York: Alan R. Liss.

Bandler, R., and J.P. Flynn
1972 Control of somatosensory fields for striking during hypothalamically elicited attack. *Brain Research* 38:197-201.

Bandler, R., and T. McCulloch
1984 Afferents to midbrain periaqueductal gray region in the "defense reaction" in the cat as revealed by horseradish peroxidase: 11. The diencephalon. *Behavioural Brain Research* 13:279-285.

Bandler, R., T. McCulloch, and B. Dreher
1985 Afferents to midbrain periaqueductal gray regions involved in the "defense reaction" in the cat as revealed by horseradish peroxidase: 1. The telencephalon. *Brain Research* 330:109-119.

Bear, D.M., and P. Fedio
1977 Quantitative analysis of interictal behavior in temporal lobe epilepsy. *Archives of Neurology* (Chic.) 34:454-467.

Bear, D.M., L. Schenk, and H. Benson
1981 Increased autonomic responses to neutral and emotional stimuli in patients with temporal lobe epilepsy. *American Journal of Psychiatry* 138(6):843-845.

Bernston, G.G.
1973 Attack, grooming and threat elicited by stimulation of the pontine tegmentum in cats. *Physiology and Behavior* 11:81-87.

Breggin, P.R.
1980 Brain-disabling therapies. Pp. 467-493 in E.S. Valenstein, ed., *The Psychosurgery Debate: Scientific, Legal and Ethical Perspectives.* San Francisco: W.H. Freeman and Company.

Brickman, A.S., M. McManus, W.L. Grapentine, and N. Alessi
1984 Neuropsychological assessment of seriously delinquent adoles-

cents. *Journal of the American Academy of Child Psychiatry* 23(4):453-457.

Broca, P.
1878 Anatomie comparee des circonvolutions cerebrales. Le grand lobe limbique et la scissure limbique dans la serie des mammiferes. *Revue Anthropologique* 1:385-498.

Brutus, M., and A. Siegel
1989 Effects of the opiate antagonist naloxone upon hypothalamically elicited affective defense behavior in the cat. *Behavioural Brain Research* 33:23-32.

Brutus, M., M.B. Shaikh, H.E. Siegel, and A. Siegel
1984 An analysis of the mechanisms underlying septal area control of hypothalamically elicited aggression in the cat. *Brain Research* 310:235-248.
1986 Effects of experimental temporal lobe seizures upon hypothalamically elicited aggressive behavior in the cat. *Brain Research* 366:53-63.

Brutus, M., S. Zuabi, and A. Siegel
1988 Effects of D-Ala(2)-MET(5)-enkephalinamide microinjections placed into the bed nucleus of the stria terminalis upon affective defense behavior in the cat. *Brain Research* 473:147-152.

Buydens-Branchey, L., M.H. Branchey, D. Noumair, and C.S. Lieber
1989 Age of alcoholism onset: Relationship to susceptibility to serotonin precursor availability. *Archives of General Psychiatry* 46:231-236.

Chasnoff, I.J., and D.R. Griffith
1989 Cocaine: Clinical studies of pregnancy and the newborn. *Annals of the New York Academy of Sciences* 562:260-266.

Chasnoff, I.J., W.J. Burns, S.H. Schnoll, and K.A. Burns
1985 Cocaine use in pregnancy. *New England Journal of Medicine* 313(11):666-669.

Chi, C.C., and J.P. Flynn
1971 Neuroanatomic projections related to biting attack elicited from the hypothalamus in cats. *Brain Research* 35:49-66.

Chorover, S.L.
1979 *From Genesis to Genocide.* Cambridge, Mass.: The MIT Press.
1980 The psychosurgery evaluation studies and their impact on the commission's report. Pp. 245-264 in E.S. Valenstein, ed., *The Psychosurgery Debate: Scientific, Legal, and Ethical Perspectives.* San Francisco: W.H. Freeman and Company.

Coid, J.
1982 Alcoholism and violence. *Drug and Alcohol Dependence* 9:1-13.

Convit, A., Z.C. Nemes, and J. Volavka
1988 History of phencyclidine use and repeated assaults in newly admitted young schizophrenic men. *American Journal of Psychiatry* 145(9):1176.

Daly, M., and M. Wilson
1989   Homicide and cultural evolution. *Ethological Sociobiology* 10:99.
De Cuyper, H., H.M. van Praag, and D. Verstraeten
1985   The effect of milenperone on the aggressive behavior of psychogeriatric patients. *Neuropsychobiology* 13:1-6.
Delgado-Escueta, A.V., R.H. Mattson, L. King, E.S. Goldensohn, H. Spiegel, J. Madsen, P. Crandall, F. Dreifuss, and R.J. Porter
1981   The nature of aggression during epileptic seizures. *New England Journal of Medicine* 305(12):711-716.
Devinsky, O., and D. Bear
1984   Varieties of aggressive behavior in temporal lobe epilepsy. *American Journal of Psychiatry* 141(5):651-656.
Dieckmann, G., B. Schneider-Jonietz, and H. Schneider
1988   Psychiatric and neuropsychological findings after stereotactic hypothalamotomy, in cases of extreme sexual aggressivity. *Acta Neurochirurgica Supplement* 44:163-166.
Dodge, K.A., J.E. Bates, and G.S. Pettit
1990   Mechanisms in the cycle of violence. *Science* 250:1678-1683.
Duncan, C.C.
1990   Current issues in the application of P300 to research in schizophrenia. Pp. 117-134 in E.R. Straube and K. Hahlweg, eds., *Schizophrenia—Concepts, Vulnerability, and Intervention.* New York: Springer-Verlag.
Duncan, C.C., W.H. Kaye, W.M. Perlstein, D.C. Jimerson, and A.F. Mirsky
1985   Cognitive processing in eating disorders: An ERP analysis. *Psychophysiology* 22:588.
Duncan, C.C., P.J. Deldin, A.F. Mirsky, R.G. Skwerer, F.M. Jacobsen, and N.E. Rosenthal
1989   Phototherapy enhances visual P300 in patients with seasonal affective disorder. *Psychophysiology* 26:S 21.
Edwards, S., and J.P. Flynn
1972   Corticospinal control of striking in centrally elicited attack behavior. *Research* 41:51-65.
Eichelman, B.
1983   The limbic system and aggression in humans. *Neuroscience and Biobehavioral Reviews* 7:391-394.
Engel, J.J., S. Caldecott-Hazard, and R. Bandler
1986   Neurobiology of behavior: Anatomic and physiological implications related to epilepsy. *Epilepsia* 27(Suppl. 2):S3-Sl3.
Falconer, M.A.
1973   Reversibility by temporal lobe resection of the behavioral abnormalities of temporal lobe epilepsy. *New England Journal of Medicine* 289:454-455.
Farrington, D.P.
1989   Childhood aggression and adult violence: Early precursors and later life outcomes. Pp. 5-29 in K.H. Rubin and D. Pepler, eds., *The Development and Treatment of Childhood Aggression.* Hillsdale, N.J.: Lawrence Erlbaum.

Flynn, J.P., H. Vanegas, W. Foote, and S. Edwards
1970 Neural mechanisms involved in a cat's attack on a rat. Pp. 135-173 in R. Whalen, ed., *The Neural Control of Behavior*. New York: Academic Press.
Fuchs, S.A.G., M. Dalsass, H.E. Siegel, and A. Siegel
1981 The neural pathways mediating quiet biting attack behavior from the hypothalamus in the cat: A functional autoradiographic study. *Aggressive Behavior* 7:51-68.
Fuchs, S.A.G., H.M. Edinger, and A. Siegel
1985a The organization of the hypothalamic pathways mediating affective defense behavior in the cat. *Brain Research* 330:77-92.
1985b The role of the anterior hypothalamus in affective defense behavior elicited from the ventromedial hypothalamus of the cat. *Brain Research* 330:93-107.
Garnett, S., C. Nahmias, G. Wortzman, R. Langevin, and R. Dickey
1988 Positron emission tomography and sexual arousal in a sadist and two controls. *Annals of Sex Research* 1:387-399.
Gloor, P.
1967 Discussion. Pp. 116-124 in C.D. Clemente and D.B. Lindsley, eds., *Aggression and Defense*. Los Angeles: University of California Press.
Goldstein, J.M., and J. Siegel
1980 Suppression of attack behavior in cats by stimulation of ventral legmental area and nucleus accumbens. *Brain Research* 183:181-192.
Hampson, E., and D. Kimura
1988 Reciprocal effects of hormonal fluctuations on human motor and perceptual-spatial skills. *Behavioral Neuroscience* 102(3):456-459.
Hare, R.D., and J.W. Jutai
1988 Psychopathy and cerebral asymmetry in semantic processing. *Personality and Individual Differences* 2:329-337.
Hare, R.D., and L.M. McPherson
1984 Violent and aggressive behavior by criminal psychopaths. *International Journal of Law and Psychiatry* 7:35-50.
Heimberger, R.F., I.F. Small, J.G. Small, V. Milstein, and D. Moore
1978 Stereotactic amygdalotomy for convulsive and behavioral disorders. Long-term follow-up study. *Applied Neurophysiology* 41:43-51.
Hendricks, S.E., D.F. Fitzpatrick, K. Hartmann, M.A. Quaife, R.A. Stratbucker, and B. Graber
1988 Brain structure and function in sexual molesters of children and adolescents. *Journal of Clinical Psychiatry* 49(3):108-112.
Hermann, B.P.
1982 Neuropsychological functioning and psychopathology in children with epilepsy. *Epilepsa* 23:545-554.

Herrera, J.N., J.J. Sramek, J.F. Costa, S. Roy, C.W. Heh, and B.N. Nguyen
1988   High potency neuroleptics and violence in schizophrenics. *Journal of Nervous and Mental Disease* 176(9):558-561.

Hess, W.R., and M. Brugger
1943   Das subkorticale zentrum der affektinen abwehrreaktion. *Helvetica Physiolooica Pharmacologics Acta* 1:35-52.

Hood, T.W., J. Siegfried, and H.G. Wieser
1983   The role of stereotactic amygdalotomy in the treatment of temporal lobe epilepsy associated with behavioral disorders. *Applied Neurophysiology* 46:19-25.

Hucker, S., R. Langevin, R. Dickey, L. Handy, J. Chambers, and S. Wright
1988   Cerebral damage and dysfunction in sexually aggressive men. *Annals of Sex Research* 1:33-47.

Jutai, J.W., R.D. Hare, and J.F. Connolly
1987   Psychopathy and event-related brain potentials (ERPS) associated with attention to speech stimuli. *Personality and Individual Differences* 8(2):175-184.

Kalin, N.H., and S.E. Shelton
1989   Defensive behaviors in infant rhesus monkey: Environmental cause and neurochemical regulation. *Science* 243:1718-1721.

Karson, C., and L.B. Bigelow
1987   Violent behavior in schizophrenic inpatients. *Journal of Nervous and Mental Disease* 175(3):161-164.

Kellam, S.G., and G.W. Rebok
1992   Building developmental and etiological theory through epidemiologically based preventive intervention trials. Pp. 162-195 in J. McCord and R.E. Tremblay, eds., *Preventing Antisocial Behavior: Interventions from Birth Through Adolescence.* New York: Guilford Press.

Kellam, S.G., J.D. Branch, K.C. Agrawal, and M.E. Ensminger
1975   *Mental Health and Going to School.* Chicago: University of Chicago Press.

Khantzian, E.J.
1974   Opiate addiction: A critique of theory and some implications for treatment. *American Journal of Psychotherapy* 28:59-70.
1982   Psychological (structural) vulnerabilities and the specific appeal of narcotics. *Annals of the New York Academy of Sciences* 398:24-30.

Kindlon, D., N. Sollee, and R. Yando
1988   Specificity of behavior problems among children with neurological dysfunction. *Journal of Pediatric Psychology* 13(1):39-47.

Klüver, H., and C.P. Bucy
1939   Preliminary analysis of functions of the temporal lobe. *Archives of Neurological Psychiatry* 42:979-1000.

Krakowski, M., J. Volavka, and D. Brizer
1986   Psychopathology and violence: A review of literature. *Comprehensive Psychiatry* 27(2):131-148.
Krakowski, M.I., A. Convit, J. Jaeger, S. Lin, and J. Volavka
1989   Neurological impairment in violent schizophrenic inpatients. *American Journal of Psychiatry* 146(7):849-853.
Krayniak, P.F., S. Weiner, and A. Siegel
1980   An analysis of the efferent connections of the septal area in the cat. *Brain Research* 189:15-29.
Langevin, R., M.H. Ben-Aron, R. Coulthard, G. Heasman, J.E. Purins, L. Handy, S.J. Hucker, A.E. Russon, D. Day, V. Roper, J. Bain, G. Wortzman, and C.D. Webster
1985   Sexual aggression: Constructing a predictive equation, a controlled pilot study. Pp. 39-76 in R. Langevin, ed., *Erotic Preference, Gender Identity, and Aggression in Men: New Research Studies*. Hillsdale, N.J.: Lawrence Erlbaum Associates.
Langevin, R., M. Ben-Aron, G. Wortzman, R. Dickey, and L. Handy
1987   Brain damage, diagnosis, and substance abuse among violent offenders. *Behavioral Sciences and the Law* 5(1):77-94.
Langevin, R., J. Bain, G. Wortzman, S. Hucker, R. Dickey, and P. Wright
1988   Sexual sadism: Brain, blood, and behavior. *Annals of the New York Academy of Science* 528:163-171.
Langevin, R., G. Wortzman, P. Wright, and L. Handy
1989   Studies of brain damage and dysfunction in sex offenders. *Annals of Sex Research* 2:163-179.
Lansdell, H.
1962   A sex difference in effect of temporal-lobe neurosurgery on design preference. *Nature* 194(4831):852-854.
1989   Sex differences in brain and personality correlates of the ability to identify popular word associations. *Behavioral Neuroscience* 103(4):893-897.
Landsdown, R.
1986   Lead, intelligence, attainment and behaviour. *Lead Toxicity: History Environmental Impact* 235-267.
Lewis, D.O., R. Lovely, C. Yeager, and D. Della Femina
1988   Toward a theory of the genesis of violence: A follow-up study of delinquents. *Journal of the American Academy of Child and Adolescent Psychiatry* 28(3):431-436.
Lewis, D.O., S.S. Shanok, M. Grant, and E. Ritvo
1983   Homicidally aggressive young children: Neuropsychiatric and experiential correlates. *American Journal of Psychiatry* 140(2):148-153.
Lewis, D.O., J.H. Pincus, M. Feldman, L. Jackson, and B. Bard
1986   Psychiatric, neurological, and psychoeducational characteristics of 15 death row inmates in the United States. *American Journal of Psychiatry* 143(7):838-845.

Leyhausen, P.
1979   The Predatory and Social Behavior of Domestic and Wild Cats. New York: Garland STPM Press.
MacDonnell, M.F., and L. Fessock
1972   Some effects of ethanol, amphetamine, disulfiram and PCPA on seizing of prey in feline predatory attack and on associated motor pathways. Quarterly Journal of Studies on Alcohol 33:437-450.
MacDonnell, M.F., and J.P. Flynn
1966a Control of sensory fields by stimulation of hypothalamus. Science 152:406-1408.
1966b Sensory control of hypothalamic attack. Animal Behavior 14:339-405.
MacDonnell, M.F., L. Fessock, and S.H. Brown
1971   Aggression and associated neural events in cats and effect of chlorophenylalanine compared with alcohol. Quarterly Journal of Studies on Alcohol 32:748-763.
MacLean, P.
1949   Psychosomatic disease and the visceral brain: Recent developments bearing on the Papez theory of emotions. Psychosomatic Medicine 11:338-353.
1952   Some psychiatric implications of physiological studies on frontotemporal portion of limbic system (visceral brain). Electroencephalography and Clinical Neurophysiology 4:407-418.
Malamud, N.
1967   Psychiatric disorders with intracranial tumors of limbic system. Archives of Neurology 17:113-123.
Mark, V.H., and F.R. Ervin
1970   Violence and the Brain. New York: Harper & Row.
Martinius, J.
1983   Homicide of an aggressive adolescent boy with right temporal lesion: A case report. Neuroscience and Biobehavioral Reviews 7:419-422.
McIntyre, M., P.B. Pritchard, and C.T. Lombroso
1976   Left and right temporal lobe epileptics: A controlled investigation of some psychological differences. Epilepsia 17:377-386.
Mednick, S.A., and E.S. Kandel
1988   Congenital determinants of violence. Bulletin of the American Academy of Psychiatry Law 16(2):101-109.
Miller, L.
1987   Neuropsychology of the aggressive psychopath: An integrative review. Aggressive Behavior 13:119-140.
Milner, B.
1969   The memory defect in bilateral hippocampal lesions. Research Reports 11:43-52.
Mirsky, A.F.
1989   The Neuropsychology of Attention: Implications for the Devel-

opment of Neuropsychiatric Disorders. Paper presented at a meeting of the American Psychological Association, New Orleans, La.

Mirsky, A.F., and N. Harman
1974 On aggressive behavior and brain disease—Some questions and possible relationships derived from the study of men and monkeys. Pp. 185-210 in R.E. Whalen, ed., *Advances in Behavioral Biology*. New York: Plenum Press.

Mirsky, A.F., and H.E. Rosvold
1990 The case of Carolyn Wilson—A 38-year follow-up of a schizophrenic patient with two prefrontal lobotomies. Pp. 51-75 in E. Goldberg, ed., *Contemporary Neuropsychology and the Legacy*. Hillsdale, N.J.: Lawrence Erlbaum Associates.

Monroe, R.R.
1970 *Behavioral Disorders*. Cambridge, Mass.: Harvard University Press.
1975 Anticonvulsants in the treatment of aggression. *Journal of Nervous and Mental Disease* 160(2):119-126.
1978 *Brain Dysfunction in Aggressive Criminals*. Lexington, Mass.: Lexington Books.

Moss, M.B., E.J. Glazer, and A.I. Basbaum
1983 The peptidergic organization of the cat periaqueductal gray: The distribution of immunoreactive enkephalin neurons and terminals. *Journal of Neuroscience* 13:803-816.

Mungas, D.
1983 An empirical analysis of specific syndromes of violent behavior. *Journal of Nervous and Mental Disease* 171(6):354-361.
1988 Psychometric correlates of episodic violent behaviour: A multidimensional neuropsychological approach. *British Journal of Psychiatry* 152:180-187.

Needleman, H.L., A. Schell, D. Bellinger, A. Leviton, and E. Allred
1989 Long term effects of childhood exposure to lead at low dose: An eleven-year follow-up report. *New England Journal of Medicine* 322:83-88.

Ogren, M.P., and A.E. Hendrickson
1976 Pathways between striate cortex and subcortical regions in *Macaca mulatta* and *Saimiri sciureus*. Evidence for a reciprocal pulvinar connection. *Experimental Neurology* 780-800.

Olofsson, M., W. Buckley, G.E. Andersen, and B. Fries-Hansen
1983 Investigation of 89 children born by drug-dependent mothers 11. Follow-up—10 years after birth. *Acta Paediatric Scandinavica* 72:407-410.

Ounsted, C.
1969 Aggression and epilepsy. Rage in children with temporal lobe epilepsy. *Journal of Psychosomatic Research* 13:237-242.

Papez, J.W.
1937 A proposed mechanism of emotion. *AMA Archives of Neurology and Psychiatry* 38:725-743.

Penfield, W., and H.H. Jasper
1954   Epilepsy and the Functional Anatomy of the Human Brain. Boston: Little Brown and Company.
Perini, G.I.
1986   Emotions and personality in complex partial seizures. Psychotherapy and Psychosomatics 45:141-148.
Piacentini, J.C.
1987   Language dysfunction and childhood behavior disorders. Advances in Clinical Child Psychology 10:259-285.
Pott, C.B., and M.F. MacDonnell
1986   Enhancement of optokinetic responses by lateral hypothalamic areas associated with attack in cats. Physiology and Behavior 38:415-422.
Proshansky, E., R.J. Bandler, and J.P. Flynn
1974   Elimination of hypothalamically elicited biting attack by unilateral lesion of the ventral midbrain tegmentum of cats. Brain Research 77:309-313.
Puglisi-Allegra, S., and A. Oliverio
1981   Naloxone potentiates shock-induced aggressive behavior in mice. Pharmacology Biochemistry and Behavior 15:513-514.
Ramamurthi, B.
1988   Stereotactic operations in behavior disorders: Amygdalotomy, hypothalamotomy. Acta Neurochirurgica Supplement 44:152-155.
Roberts, S.S.
1990   Murder, mayhem, and other joys of youth. Journal of NIH Research 2:67-72.
Robertson,G.
1988   Arrest patterns among mentally disordered offenders. British Journal of Psychiatry 153:313-316.
Rodin, E.A.
1973   Psychomotor epilepsy and aggressive behavior. Archives of General Psychiatry 28:210-213.
Rosenthal, D., and S. Kety
1968   The Transmission of Schizophrenia. Oxford: Pergamon.
Rosvold, H.E., A.F. Mirsky, and K.H. Pribram
1954   Influence of amygdalectomy on social behavior in monkeys. Journal of Comparative Physiological Psychology 47:173-178.
Rush, D., and K.R. Callahan
1989   Exposure to passive cigarette smoking and child development. Annals of the New York Academy of Sciences 562:74-100.
Sano, K., and Y. Mayanagi
1988   Posteromedial hypothalamotomy in the treatment of violent aggressive behavior. Acta Neurochiruraica Supplement 44:145-151.
Schnur, D.B., S. Mukherjee, J. Silver, G. Degreer, and C. Lee

1989   Electroconvulsive therapy in the treatment of episodic aggressive dyscontrol in psychotic patients. *Convulsive Therapy* 5(4):353-361.

Serafetinides, E.A.
1965   Aggressiveness in temporal lobe epileptics and its relation to cerebral dysfunction and environmental factors. *Epilepsia* 6:33-42.

Shaikh, M.B., and A. Siegel
1989   Naloxone induced modulation of feline aggression elicited from midbrain periqueductal gray. *Pharmacology Biochemistry and Behavior* 31:791-796.

Shaikh, M.B., M. Brutus, H.E. Siegel, and A. Siegel
1984   Differential control of aggression by the midbrain. *Experimental Neurology* 83:436-442.

Shaikh, M.B., J. Barrett, and A. Siegel
1987   The pathways mediating affective defense and quiet biting attack from the midbrain central gray. *Brain Research* 437:9-25.

Siegel, A., and M. Brutus
1990   Neural substrates of aggression and rage in the cat. Pp. 135-233 in A.N. Epstein and A.R. Morrison, eds., *Progress in Psychobiology and Physiological Psychology.* San Diego, Calif.: Academic Press.

Siegel, A., and H. Edinger
1981   Neural control of aggression and rage. Pp. 203-240 in P. Morgane and J. Panksepp, eds., *Handbook of the Hypothalamus.* New York: Marcel Dekker.
1983   Role of the limbic system in hypothalamically elicited attack behavior. *Neuroscience and Biobehavioral Reviews* 7:395-407.

Siegel, A., and C.B. Pott
1988   Neural substrates of aggression and flight in the cat. *Progress in Neurobiology* 31:262-283.

Siegel, A., R. Troiano, and H. Edinger
1973   The pathway from the mediodorsal nucleus to the hypothalamus in the cat. *Experimental Neurology* 38:202-217.

Smith, D.A., and J.P. Flynn
1979   Afferent projections to attack sites in the pontine tegmentum. *Brain Research* 164:103-119.
1980a Afferent projections to quiet attack sites in cat hypothalamus. *Brain Research* 194:29-40.
1980b Afferent projections to affective attack sites in the cat hypothalamus. *Brain Research* 194:41-51.

Stewart, J.T., M.L. Mounts, and R.L. Clark
1987   Aggressive behavior in Huntington's disease: Treatment with propranolol. *Journal of Clinical Psychiatry* 48(3):106-108.

Stoddard-Apter, S., and M.F. MacDonnell
1980   Septal and amygdalar efferents to the hypothalamus which fa-

cilitate hypothalamically-elicited intraspecific aggression and associated hissing in the cat. An autoradiographic study. *Brain Research* 193:19-32.

Stone, A.A.
1984 Violence and temporal lobe epilepsy. *American Journal of Psychiatry* 141(12):1641.

Streissguth, A.P., D.C. Martin, H.M. Barr, B.M. Sandman, G.L. Kirchner, and B.L. Darby
1984 Intrauterine alcohol and nicotine exposure: Attention and reaction time in 4-year-old children. *Developmental Psychology* 20(4):533-541.

Streissguth, A.P., H.M. Barr, P.D. Sampson, J.C. Parrish-Johnson, G.L. Kirchner, and D.C. Martin
1986 Attention, distraction and reaction time at age 7 years and prenatal alcohol exposure. *Neurobehavioral Toxicology and Teratology* 8:717-725.

Streissguth, A.P., P.D. Sampson, and H.M. Barr
1989 Neurobehavioral dose-response effects of prenatal alcohol exposure in humans from infancy to adulthood. *Annals of the New York Academy of Sciences* 562:145-158.

Sundet, K.
1986 Sex differences in cognitive impairment following unilateral brain damage. *Journal of Clinical and Experimental Neuropsychology* 8(1):51-61.

Tardiff, K., and A. Sweillam
1980 Assault, suicide, and mental illness. *Archives of General Psychiatry* 37:164-169.

Tarter, R.E., A.M. Hegedus, N.E. Winsten, and A.I. Alterman
1984 Neuropsychological, personality, and familial characteristics of physically abused delinquents. *Journal of the American Academy of Child Psychiatry* 23(6):668-674.

Tazi, A., R. Danizer, P. Mormede, and M. Le Moal
1983 Effects of post-trial administration of naloxone and ß-endorphin on shock-induced fighting in rats. *Behavioral and Neural Biology* 39:192-202.

Thomson, G.O.B., G.M. Raab, W.S. Hepburn, R. Hunter, M. Fulton, and D.P.H. Laxen
1989 Blood-lead levels and children's behaviour—Results from the Edinburgh lead study. *Journal of Child Psychiatry* 30(4):515-528.

Treiman, D.M.
1986 Epilepsy and violence: Medical and legal issues. *Epilepsia* 27 (Suppl. 2):S77-Sl04.

Valenstein, E.S., ed
1980 *The Psychosurgery Debate: Scientific, Legal, and Ethical Perspectives.* San Francisco, Calif.: W.H. Freeman and Company.

Virkkunen, M.
1983   Psychomotor epilepsy and violence. *American Journal of Psychiatry* 140(5):646-647.

Volkow, N.D., and L. Tancredi
1987   Neural substrates of violent behavior: A preliminary study with positron emission tomography. *British Journal of Psychiatry* 151:668-673.

Walker, E., G. Downey, and A. Bergman
1989   The effects of parental psychopathology and maltreatment on child behavior: A test of the diathesis-stress model. *Child Development* 60:15-24.

Walsh, A., and J.A. Beyer
1987   Violent crime, sociopathy and love deprivation among adolescent delinquents. *Adolescence* 22(87):705-717.

Wasman, M., and J.P. Flynn
1962   Directed attack elicited from hypothalamus. *Archives of Neurology* 61:220-227.

Weiger, W.A., and D.M. Bear
1988   An approach to the neurology of aggression. *Journal of Psychiatric Research* 22(2):85-98.

Wilson, G.S., R. McCreary, J. Kean, and J.C. Baxter
1979   The development of preschool children of heroin-addicted mothers: A controlled study. *Pediatrics* 63(1):135-141.

Woods, B.T., and M.D. Eby
1982   Excessive mirror movements and aggression. *Biological Psychiatry* 17(1):23-32.

Wurmser, J.
1973   Psychoanalytic considerations of the etiology of compulsive drug use. *Journal of the American Psychoanalytic Association* 22:820-843.

Yeudall, L.T., D. Fromm-Auch, and P. Davies
1982   Neuropsychological impairment of persistent delinquency. *Journal of Nervous and Mental Disease* 170(5):257-265.

Zahn, T.P., M. Kruesi, and J.L. Rapoport
1991   Reaction time indices of attention deficits in boys with disruptive behavior. *Journal of Abnormal Child Psychology* 19(2)233-252.

Table 3 Studies Relating Violence to Psychiatric Disorders

Author(s) Title/Date	Population Studied	Procedure	Results/Conclusions
Tardiff, K. and A. Sweillam Assault, Suicide, and Mental Illness 1980	9,365 patients admitted to public psychiatric hospitals in a one year period.	Data taken from information from the New York State Dept. of Mental Hygiene. Type of data-demographics, DSM II diagnosis, and prior use of psychiatric or other human services.	21% of patients had some type of violent, assaultive, or suicidal problems, isolated or in combination, prior to admission. Found increased assaultive behavior for male patients (females of lower socioeconomic class more assaultive behavior). Evidence that assaultive behavior is associated with lower socioeconomic class. Did not confirm that patients with assaultive problems are more likely to have a history of seizure activity.
Krakowski, M., J. Volavka, and D. Brizer Psychopathology and Violence: A Review of the Literature. 1986	Clinical lit. Review paper	The authors examine the role of psychopathology in the occurrence of violence. They review several studies in order to determine what variables play a crucial role in this relationship.	Violence is indicative of different psychiatric impairments dependent upon the nature of the psychiatric illness of which it is a manifestation. In order to be understood, violence cannot be isolated from various other clinical characteristics of the disorder of which it is a part. Carrying out of assaultive acts involves a certain amount of "intact functioning which is not compatible with severe impairment found in disorganized psychotic states," it is not solely based on specific dysfunction.
Karson, C., and L. B. Bigelow Violent Behavior in Schizophrenic Inpatients 1987	140 inpatients 97 of the patients were diagnosed schizophrenic. Mean age = 29 yrs.(71 men and 26 women). The remaining 43 (34 men and 9 women) had other diagnoses. Mean age= 31. None had responded well to neuroleptic treatment.	Chart review. Should be aware that patients with a history of violent attacks on others were, as best could be determined, excluded for admission.	41 of the 97 patients with schizophrenia became assaultive during hospitalization. Only four of the 43 patients with other diagnoses became assaultive ($p<.0001$). These assaultive patients were significantly younger and had a greater proportion of previous history of violence than nonassaultive patients.

112

Table 3 (continued)

Author(s) Title/Date	Population Studied	Procedure	Results/Conclusions
Robertson, G. Arrest Patterns among Mentally Disordered Offenders. 1988	Four groups of prisoners. One group consisted of 61 schizophrenic men, one of 30 men who were suffering from an affective illness, usually of psychotic intensity, one of 35 men with no record of a psychotic or other serious mental illness, but with a criminal history of violence, and a fourth group of 41 normal men with no history of mental illness and no record of serious violence. All groups were matched for age.	The experimenters examined the variables involved in the offense (circumstances, living arrangements at the time of the offense), the arrest, and the detention of the subjects. Information was obtained through an interview with the subject prior to cognitive testing.	The author clearly states that the normal groups in the present study are quite atypical of the offender population in general. Crimes of violence are greatly overrepresented. However, the comparisons made concerning the personal circumstances of the subjects and the offenses indicate real differences between the normal and mentally ill offenders and the difference would probably have been even larger if the population had been less atypical. The study showed that the schizophrenic men were more isolated, usually without a home. "Many of the offenses committed by the schizophrenic men involved petty larceny or criminal damage, and it is argued that the social incompetence and debilitated state of these men made them vulnerable to detection and detention. Furthermore, it is proposed that this increased vulnerability is large enough to make it very difficult indeed to compare the rate of criminal offending of the mentally ill with that of the general population."

113

Table 3 (continued)

Author(s) Title/Date	Population Studied	Procedure	Results/Conclusions
Convit, A., Nemes, Z. C., Volavka, J. History of Phencyclidine Use and Repeated Assaults in Newly Admitted Young Schizophrenic Men. 1988 Letter	79 schizophrenic men 36 years or younger.	The men were administered a test battery which comprised a quantified neurological evaluation, drug and alcohol use questionnaires and a psychosocial assessment. Through chart reviews, ward journals and "as-needed" medication records behavior of each was monitored for 6 months or until discharge. Assaults were defined as physical contact made with another patient or a staff member. The intraclass correlation coefficient between those rating assaultive behavior was 0.95. The purpose was to compare patients with a history of phencyclidine (PCP) use to those with no history of PCP use. Use of PCP was determined through self-report mainly because patients were not identified until several days following admission and PCP cannot be tested for in urine or blood beyond 6 hours following ingestion.	24 of the patients had reported using PCP at least once. No significant differences were found with respect to time of follow-up, age, race, age at first psychiatric hospitalization, history of violent crime, neurological abnormality score, and other drug or alcohol use between those who had used PCP and those who had not. PCP users were, however, significantly more likely to repeatedly assault during the 6 months of follow-up. They also had more assaults than nonusers. The authors state that although their findings are tentative they should stimulate further research in the role of PCP in assaultive behavior of psychiatric inpatients.

114

Table 3 (continued)

Author(s) Title/Date	Population Studied	Procedure	Results/Conclusions
Herrera, J. N, Sramek, J. J., Costa, J. F., Roy, S., Heh, C. W., and Nguyen, B. N. High Potency Neuroleptics and Violence in Schizophrenics. 1988	16 males, mean age of 33.3 years (range 25 - 40 years), with a DSM-III diagnosis of schizo-phrenia. They were shown to be resistant to previous attempts at neuroleptic treatment. These patients did not have unusual histories of violence as de-termined by a retro-spective chart review. Patients with organic cerebral brain disease, mental retardation or those who did not ex-hibit active symptoms were excluded from the present study. Also, those younger than 18 or older than 55, those with a physical illness, or those who had abused illicit drugs or alcohol in the 2 weeks before the study were excluded.	The study began with a preliminary 14-day washout period using placebo capsules. Following this period, a 6 week clinical trial of fixed-dose haloperidol was started. This was followed by a second placebo period and another 6-week double-blind medication trial. Patients were administered fixed dosages of either clozapine or chlorpromazine. Those given haloperidol and chlorpromazine were given benztropine as well while on these drugs. Clinical changes were evaluated at entrance and weekly during the placebo washout period using Brief Psychiatric Rating Scale, Simpson-Angus Rating Scale for Extra-pyramidal Effects, and the Abnormal Involuntary Movement Scale. The Lion Scale of Inpatient Violence was completed on a daily basis.	Patients were significantly more violent during haloperidol treatment than during treatment with low-potency neuroleptic drugs or during the placebo period. Through examining Lion data taken during the haloperidol period, two distinct patient groups were established. A violent group and a nonviolent group were formed. The violent subgroup showed a deterioration on BPRS scores during the haloperidol treatment. From these findings it seems that some patients display increased violent behavior when given moderately high-dose haloperidol. It is not evident whether this effect is specific to haloperidol or specific to the relatively high dosage.

115

Table 4 Aggression in Persons with Alcoholism

Author(s) Title/Date	Population Studied	Procedure	Results/Conclusions
Buydens-Branchey, L., M. H. Branchey, D. Noumair, and C. S. Lieber Age of Alcoholism Onset 1989	112 male alcoholic patients, ages 25 to 60, who had been admitted consecutively to a detoxification unit for one week and then to a rehabilitation unit for four to six weeks.	Each received a Structured Clinical Interview, SADS, a questionnaire and the Buss-Durkee Inventory for aggressive tendencies. Blood was taken one to two days after admission, then weekly until discharge. Plasma free tryptophan levels were determined.	Each received a Structured Clinical Interview, SADS, a questionnaire and the Buss-Durkee Inventory for aggressive tendencies. Blood was taken one to two days after admission, then weekly until discharge. Plasma free tryptophan levels were determined.
Coid, J. Alcoholism and Violence 1982	Coid, J. Alcoholism and Violence 1982	The author examines five different hypotheses of association between violent behavior and alcoholism.	"No direct relationship is apparent." It would, however, still be one of the alcoholic's various alcohol-related problems. He found a strong association between previous personality abnormalities and violence by alcoholics. Perhaps there is a common origin. There remains a tremendous need for research into behavioral disorders. He suggests research into alcohol-induced brain damage and how it can contribute to violent behavior as an area with more potential.

Table 5 Studies in Seizure Disorders and Aggressive Behavior

Author(s) Title/Date	Population Studied	Procedure	Results/Conclusions
Bear, D. M., and P. Fedio Quantitative Analysis of Interictal Behavior in Temporal Lobe Epilepsy. 1977	Fifteen patients with right temporal and twelve with left temporal foci were compared with twelve normal adults comparable in age, education, socioeconomic class, and geographic distribution. Nine of the controls suffered from debilitating neurological disease.	Eighteen traits were assessed. Both patients and one observer for each patient completed the same true-false questionnaire. Clinical records were used to obtain epidemiological variables for the patients.	Patients with right temporal foci exhibit "polishing" behavior, or exaggeration of valued qualities. They deny anti-social behavior. Left temporal patients emphasize negative qualities and minimize outstanding conscientious behavior. The psychological features found in these two groups were not found in patients with socially debilitating neurological disorders. Continual behavior, thought, and affect changes, then, appear to be specific to patients with temporal epileptic focus. The present results support the theory that man possesses a hemispheric asymmetry in expression of affect.
Bear, D., L. Schenk, and H. Benson Increased Autonomic Responses to Neutral and Emotional Stimuli in Patients with Temporal Lobe Epilepsy. 1981	Three male and two female patients with an unambiguous history of complex partial seizures and EEG localization of epileptic spike foci to temporal lobes. Ages 25 - 54 Two male and five female control subjects. Ages 21 - 30.	Seventeen 35-mm color slides served as the stimuli. Subjects were asked to rate each photo on a 5-point scale (1 - most pleasant and 5 - most unpleasant). Silver-silver chloride cup electrodes on the nondominant hand (which was restrained) measured the Galvanic skin conductance changes.	The patients with temporal lobe epilepsy exhibited significantly greater palmar conductance in response to emotionally neutral or charged visual stimuli. This result shows consistency with "increased emotional responsivity." This increase in autonomic response was seen in interictal periods with no obvious association to clinical seizures. Further investigation into different populations is necessary to study the specificity of the enhanced autonomic responding. If these findings are confirmed they may serve as implications for treatment strategies

Table 5 (continued)

Author(s) Title/Date	Population Studied	Procedure	Results/Conclusions
Devinsky, O., and D. Bear Varieties of Aggressive Behavior in Temporal Lobe Epilepsy. 1984	Five patients with temporal lobe epilepsy	Case reports	Aggressive behavior in each case occurred subsequent to the development of an "epileptic focus clinically localized to the limbic system." Aggressive behavior occurs in different patterns for various cases of temporal lobe epilepsy. Most temporal lobe epileptics do not display overt violent behavior. The authors do, however, consider aggression an important behavioral disturbance related to this illness and suggest different forms of treatment including psychotropic medication, psychotherapy and "in select cases" neurosurgery.
Engel, J. S. Caldecott-Hazard, and R. Bandler Neurobiology of Behavior: Anatomic and Physiological Implications Related to Epilepsy. 1986	Clinical Lit. Review	The authors address the controversial issue of whether enduring alterations in neuronal function that result in interictal behavior disturbances can be produced by epileptic seizures.	Some disorders in interictal behavior may reflect unrecognized ictal events. The evidence they reviewed implicates that not all behavioral problems suffered by epileptic patients can be attributed to psychosocial factors. Behavioral disturbances attributed to antiepileptic drugs or certain structural lesions could also stem partially from epileptiogenic mechanisms.
Lewis, D. O., S. S. Shanok, M. Grant, and E. Ritvo Homicidally Aggressive Young Children: Neuropsychiatric and Experiential Correlates. 1983	55 children admitted to a psychiatric service. Primarily a diagnostic service where patients remain 90 days on the average. 21 were homicidally aggressive.	Data were obtained from hospital records, which included family, medical, developmental histories, psychiatric evaluations, physical examinations, neurological assessments, educational assessments, psychological testing, and EEGs (for most). Also reviewed the types of medications that had been prescribed. Four raters determined from information provided by the chart whether the individual had ever been homicidally aggressive.	Homicidally aggressive children were significantly more apt to have had a seizure, have a father who displayed violent, often homicidal, behavior, have a mother with a history of hospitalization for a psychiatric disorder, and have attempted suicide. The homicidal and nonhomicidal children were not distinguished by psychiatric symptoms and diagnoses. The most significant variable appeared to be whether the child had been exposed to a violent father. Witnessing irrational violence engenders rage that can be expressed through suicidal behavior when turned inward or aggressive (homicidal) behavior when directed outward.

Table 5 (continued)

Author(s) Title/Date	Population Studied	Procedure	Results/Conclusions
Perini, G. I. Emotions and Personality in Complex Partial Seizures 1986	Eleven patients with left temporal epileptogenic foci and thirteen patients with right temporal foci were included. They did not differ in seizure frequency, age of onset, or type of aura.	Eight personality traits were measured using the emotion profile index (EPI). The personality inventory (PI) (Bear & Fedio, 1977) was used to assess specific behavioral changes.	Result showed that left patients gave a negative image of themselves and displayed a paranoid, depressed personality. they were more guilt ridden and aggressive than right patients. On the other hand, right patients rated themselves positively. Both groups manifest epileptic behavioral syndrome according to the PI.
Weiger, W. A., and D. M. Bear An approach to the Neurology of Aggression. 1988	Clinical lit. Review paper highlighting hierarchical neural controls over aggression and characteristic syndromes of human aggression due to lesions in various areas	The authors compare and contrast the functions of the hypothalamus, amygdaloid complex, and orbital prefrontal cortex. The examine sensory inputs, effector channels and integration principles drawn from observations in human and animal studies. The authors then speculate on the application of this approach to research on criminal violence.	Abnormalities (or lesions) at various levels of the brain yield distinctive forms of aggressive behavior. The authors note that although aggressive behavior may result from neurological lesions, it is not necessarily of organic origin. Neurological abnormalities may be the result of, rather than the cause of aggressive behavior. Further research in this area is suggested. Three specific brain structures are central to the discussion of the functional anatomy of aggression (the hypothalamus, the amygdala and temporolimbic cortex, and the frontal neocortex. The authors do mention, however, that other brain structures and certain neurotransmitters are factors involved in the control of aggressive behavior.

119

Table 5 (continued)

Author(s) Title/Date	Population Studied	Procedure	Results/Conclusions
Hermann, B. P. Neuropsychological Functioning and Psychopathology in Children with Epilepsy. 1982	Fifty children with epilepsy ages 8 - 12 yrs. Two groups were created based on neuropsychological function. 25 had good neuropsychological function and 25 had poor neuropsychological function. The subjects were also classified according to seizure type.	Each child received a complete neurological evaluation including an EEG. The two groups were compared in several domains. Social competence and behavioral functioning were measured using the Child Behavior Checklist. For neuropsychological data each child completed the Luria-Nebraska Neuropsychological Battery - Children's Version.	A highly significant relationship between adequacy of neuropsychological function and the level of aggressive behavior in children with epilepsy was determined. Those subjects with a poor performance on the neuropsychological battery also retained significantly higher scores on the Aggression scale of the CBCL as compared with children who performed better on the neuropsychological tests. Neuropsychological functioning was also associated with increased scores on the Total Behavioral Problems measure and decreased scores on the Total Social Competence scale.
Virkkunen, M. Psychomotor Epilepsy and Violence. 1983	Letter to the Editor	Letter discussing "Psychomotor Epilepsy and Violence in a Group of Incarcerated Adolescent Boys." (Lewis, 1982)	The author criticizes Lewis's definition of psychomotor epilepsy citing that Livingston et al. (1980) defined the characteristics that should not be considered symptoms of psychomotor epilepsy and that the two are conflicting. Evidence exists that aggression in epileptics is "a multifactorially determined interictal, not ictal, phenomenon." The author provides several clinically supported examples of aggressive behavior, with the specific symptoms described by Lewis, originating from disorders other than epilepsy.

120

Table 5 (continued)

Author(s) Title/Date	Population Studied	Procedure	Results/Conclusions
Stone, A. A. Violence and Temporal Lobe Epilepsy. 1984	Letter to the Editor	The letter is in regard to "Varieties of Aggressive Behavior in Temporal Lobe Epilepsy" (Devinsky & Bear, 1984).	The author criticizes Devinsky's and Bear's presentation of the types of treatment potentially helpful in suppressing aggression in temporal lobe epileptics. Their mention of neurosurgery, he feels, did not indicate the risk involved and "perpetuate[s] the misleading impression about the positive benefits of neurosurgical intervention to control violence" in this population. The author states that increase in aggression may merely be a very human reaction to the knowledge of the limiting nature of the possibly incurable disease that the epileptic endures. He claims that they can draw no new conclusions from the evidence gathered in their study.
Treiman, D. M. Epilepsy and Violence: Medical and Legal Issues 1986	Clinical Lit. Review	Review of literature focused on the relationship between epilepsy and violent crime.	A two-to fourfold greater prevalence of epilepsy was found in a prison population as compared with a control population. This prevalence is, however, similar to that found in other lower socioeconomic populations from which a majority of prisoners come. The prevalence of epilepsy in persons convicted of violent crimes is no greater than that in other prisoners used as normal controls. No evidence was found in support of the notion that violence is more common among epileptics than others or that temporal lobe epileptics are more violent than persons with other forms of epilepsy. Ictal violence rarely occurs and when it does is considered "resistive." Finally, they list five criteria that should be used to determine whether an epileptic seizure was the cause of a particular violent act (Delgado-Escueta et al., 1981).

121

Table 6 The Ictal Aggression Scale

Author(s) Title/Date	Population Studied	Procedure	Results/Conclusions
Delgado, A. V., R. H. Mattson, L. King, E. S. Goldenshohn, H. spiegel, J. Madsen, P. Crandall, F. Dreifuss, R. J. Porter The Nature of Aggression During Epileptic Seizures. 1981	19 patients believed to display aggressive behavior during epileptic seizures. These 19 were taken from a group of approximately 5400 epileptic patients. On closed-circuit television, 13 showed aggressive motions during their seizures. These 13 were then rated.	Aggressive behavior was analyzed by closed-circuit television. The panel used its own rating scale. 1 - nondirected aggressive motion 2 - violence to property 3 - threatening violence to a person 4 - mild violence to a person 5 - moderate violence to a person 6 - severe violence to a person	The panel compiled five relevant criteria to be used in determining whether a certain violent crime resulted from an epileptic seizure. 1. The diagnosis of epilepsy should be established by one or more neurologists with "special competence" in epilepsy. 2. The presence of epileptic automatisms should be recorded by closed-circuit television, EEG biotelemetry, and the history. 3. Aggression during epileptic automatism should be confirmed in a videotape-recorded seizure. Ictal epileptiform patterns must be recorded by EEG 4. The act (aggressive or violent) should be typical of the patient's seizures. 5. Clinical judgement should be provided by a neurologist regarding the credibility that such an act was part of a seizure.

Table 7 Studies of Violence as the Independent Variable--Is There a Brain Disorder in Violent Non-Sex Offenders?

Author(s) Title/Date	Population Studied	Procedure	Results/Conclusions
Andrew, J. M. Are Left-Handers Less Violent? 1980	139 probationed juvenile delinquents mean age = 15.5. The group was made up of offenders from the intensive super-vision probation program (with the exception of the left-handed Anglo females who were taken from a non-intensive unit.	The Violence Scale (Andrew, 1974) was used as the measure of violent crime. The scale gives a numerical value for each type of crime (from 1.0 to 56.0). Handedness was determined solely by the hand used in writing (though different degrees of hand dominance exist).	Violence Scale scores were higher for right-handed juvenile offenders than they were for left-handed juvenile offenders. The most violent subgroup was right-handed boys and the least violent was left-handed girls. Left-handers are overrepresented among male offenders, but not among female offenders or violent male offenders. The authors reexamined previous studies and were not able to explain these results. They suggest further research into types of left-handedness, cerebral organization and considerations of sex-handedness interaction.
Bach-Y-Rita, G., J. R. Lion, C. E. Climent, and F. R. Ervin Episodic Dyscontrol: A Study of 130 Violent Patients. 1971	130 Patients - chief complaint - explosive violent behavior	A two-year study. "When pertinent and possible," EEGs, neurological and psychological tests, pneumoencephalograms, and other tests.	Found abnormalities in EEGs, histories of seizure-like impairment. Feel interaction with environment is a crucial factor leading to episodes. See patients as having problems with coping with demands place on their insufficient egos. Both the physiological/neuroanatomical and the psychological areas require further research.

123

Table 7 (continued)

Author(s) Title/Date	Population Studied	Procedure	Results/Conclusions
Brickman, A. S., M. McManus, W. L. Grapentine, and N. Alessi Neuropsychological Assessment of Seriously Delinquent Adolescents. 1984	71 subjects (40 male; 31 female) were chosen based on four criteria: 1. commission of violent felonies 2. commission of multiple non-violent felonies 3. multiple placements within the training school system; and 4. assaultive in-program behavior requiring medical attention for the victim. Mean age=16.3 yrs. 39 of the subjects were white, 26 - black, 3 - Hispanic 3 - mixed race.	The Luria-Nebraska Neuropsychological Battery (LNNB) was administered to all subjects. This examination consists of 269 items organized into 11 categories: Motor, Rhythm, Tactile, Visual, Receptive Language, Expressive Language, Reading, Writing, Arithmetic, Memory, and Intelligence. Educational levels for the subjects were assessed (as accurately as possible).	Violent and recidivist delinquents display a distinctly abnormal pattern of functioning. This pattern is not limited to higher "intellectual" functions, but rather encompasses a wide range of functions. The only significant differences between the subgroups, however, are in expressive speech and memory scales. This could be attributable to education or concentration problems. Upon reviewing the specific neuropsychological dysfunction, the authors suggest that the temporal lobe may be implicated.
Eichelman, B. The Limbic System and Aggression in Humans. 1983	Clinical lit. Review paper	Review includes the following areas. 1. naturally occurring and iatrogenic brain lesions. 2. electrical disturbances 3. pharmacologic intervention. 4. central neurochemical concentrations - possible limbic lobe involvement	"Neuroanatomic and EEG data tend to link the limbic system with human aggression." However, the existence of limbic pathology does not necessitate the induction of aggressive behavior, and aggressive behavior can occur without clear demonstration of specific limbic pathology. The author feels that concentration should not focus solely on limbic regions as a generating source.

Table 7 (continued)

Author(s) Title/Date	Population Studied	Procedure	Results/Conclusions
Mungas, D. An Empirical Analysis of Specific Syndromes of Violent Behavior. 1983	138 neuropsychiatric out-patients. Five homogeneous subgroups: two - closely resembled episodic dyscontrol syndrome (differed in severity); one infrequent but severe violence; one-infrequent, less severe violence; one - no history of violence.	Violent behavior was rated in four parameters. The violent behavior could be directed either toward others or toward property. The four parameters are: frequency; severity; appropriateness to environmental context; organization of the act/directedness. Seven historical variables were rated as well: behavior disorder as a child; developmental abnormalities; independent functioning level; home environment as a child; drug abuse; neurodiagnostic evaluation; head trauma. Ratings were made retrospectively based on an in-depth chart review.	Results showed that violent behavior is not necessarily a result of brain dysfunction in general. Perhaps more specific kinds of abnormalities predispose one to this behavior. Evidence is provided that distinct sub-groups of patients displaying violent behavior exist that can be differentiated according to behavior patterns and clinical correlates. Results did not support a specific etiology of violence. A relationship between temporal lobe abnormalities and violence is not supported by these results. The author notes that this is probably not a representative sample of violent persons in the population as a whole due to the nature of the clinical population. Suggests further research in delineating relationships between specific etiologies and behavioral symptoms to aid in creating guidelines for differential diagnoses and treatments.
Volkow, N. D. and L. Tancredi Neural Substrates of Violent Behavior: A Preliminary Study with Positron Emission Tomography. 1987	four psychiatric patients (inpatients) with a history of repetitive purposeless violent behavior.	All subjects received full psychiatric, physical, and neurological examinations. All had a CT scan (with and without contrast), an EEG, and PET using oxygen-labelled water for cerebral blood flow and fluorodeoxy-glucose measurement. Left vs. right temporal lobes, and frontal vs. occipital cortices were compared using PET images of three slices at the thalamic level.	All of the patients displayed temporal cortex asymmetry. The left showed lower metabolism and blood flow than seen in the right. These patients displayed defects in cerebral function that were widespread, not originating from one focal brain lesion. All subjects showed signs of temporal dysfunction. The authors emphasize that violent behavior in these cases is not purely organic in origin. They support the idea that violent behavior springs from complex interactions involving environmental stimuli, hormones, neurotransmitters, and various neural systems. PET seems to have much potential in studying cerebral function and dysfunction in individuals exhibiting violent behavior.

125

Table 7 (continued)

Author(s) Title/Date	Population Studied	Procedure	Results/Conclusions
Langevin, R., Pen-Aron, M., Wortzman, G., Dickey, R., and Handy, L. Brain Damage, Diagnosis, and Substance Abuse among Violent Offenders. 1987	18 males who faced charges of murder or manslaughter. 21 males facing non-homicidal violence charges (assaulters). All of these men had physically attacked another person. 16 males charged with nonviolent sex offenses (controls). 43% of the killers' victims and 33% of the assaulters' victims were male. The controls faced charges of property offenses.	Each subject received the Halstead-Reitan, the Luria-Nebraska batteries and the WAIS-R. CT scans were done. EEGs of baseline, hyper-ventilation, and photic stimulation were done (in some cases sleep and alcohol EEGs were done). ICD9 diagnoses were assigned, the MMPI and Assertiveness Inventory were administered. The MAST and Clarke Drug Use Survey were administered to assess substance abuse.	Killers were found to be more like nonhomicidal violent offenders than nonviolent offenders. Neuropsychological variables were significant in one-fifth to one-quarter of violent offenders. No epileptics were present in any group. Based on present results, almost one-third of killers can be expected to show clinically significant neuropsychological impairment on the Reitan. The neurological impairment was not significantly associated with diagnosis, drug and alcohol abuse, age, intelligence or education. "Overrepresentation of alcohol and drug abusers in all groups, however, may be masking differences between violent and nonviolent groups." Killers and assaulters were more often considered alcoholic, they abused alcohol with a higher frequency and experienced mood dysphoria, congruent with their behavior (i.e. hostility, paranoia) rather than feelings of relaxation and pleasure normally experienced. "The best predictor of assaultive behavior has been the use of alcohol and drugs, but its role in violence is far from established." It may be the interaction of neuropathology and substance abuse along with other factors which underlie the commission of homicide. Perception of the violent offenders themselves may be distorted according to this study. This study was carried out in a medium security setting; the authors suggest that it should be replicated in a maximum security setting as well as others.
Lewis, D.O., Pincus, J.H., Feldman, M., Jackson, L., and Bard, B. Psychiatric, Neurological and Psychoeducational Characteristics of 15 Death Row Inmates in the United States. 1986	15 death row inmates chosen because of the imminence of their deaths rather than for evidence of neurological damage.	Detailed family, medical, social, educational and neurological histories on all subjects were recorded. All received psychiatric evaluations. Eleven subjects were run through batteries of psychological tests (Wechsler Adult Intelligence Scale-Revised, the Bender-Gestalt test, and the Rorschach test). Psycho-educational assessments with selected subjects were obtained for eight of the subject.	Many condemned individuals in the U.S. are probably victims of, until now, unrecognized psychiatric illnesses or neurological disorders. This information should be pertinent in considerations of reducing the severity of sentences. The authors suggest that, given that this group is a sufficient representation of the entire population of death row inmates, these individuals may be less capable of obtaining services or presenting the information essential for purposes of mitigation. Comparisons of with violent inmates not sentenced to death would be beneficial in further investigating this possibility.

126

Table 8 Studies of Violence as the Independent Variable--Is there a Brain Disorder in Violent Sex Offenders?

Author(s) Title/Date	Population Studied	Procedure	Results/Conclusions
Langevin, R., Ben-Aron, M., Coulthard, R., Heasman, G., Purins, J., Handy, L., Hucker, S., Russon, A., Day, D. Roper, V., Bain, J., and Webster, C. Sexual Aggression: Constructing a Predictive Equation A Controlled Pilot Study 1985	Twenty sexual aggressives (SAs) charged with rape, attempted rape, or indecent assault. Twenty nonviolent non-sex offenders served as controls and were matched with SAs on age and education. Controls faced charges of theft, fraud, and drug possession. SAs mean age =26.75 and controls mean age =24.5. SAs were also broken down into sadists (n=9) and nonsadists (n=11). Assignment to the sadist group was based on erotic preference for control of victims, their fear, terror, destruction, torture and/or unconsciousness.	The phallometric study consisted of four classes of audiotaped stimuli: 1. normal consenting intercourse 2. intercourse plus aggression (rape) 3. aggression to a female with no sex contact, and 4. neutral statements. The twenty stimuli (5 in each category) were recorded by a female. This method was devised by Abel et al, 1977). Subjects received a 10 min. relaxation tape because it increases overall responsiveness. Penile volume change in the first 10 seconds and a max volume change in the 100 second interval were the dependent variables. Abel's rape index was also used (response to rape stimuli/total response to intercourse stimuli). The Derogatis Sexual Functioning Inventory and the Clarke SHQ were used to compare sex history of the two groups. The Bem Sex Role Inventory and the Feminine Gender Identity Scale were used in examination of androgyny and gender identity. A CT-Scan and the Reitan Battery were used to assess brain pathology. Subjects had blood drawn between 8:00 am and 10:30 am (in 15 min intervals) for hormone analysis. Medical histories were examined. The MAST scale and penile reactions to erotic stimuli were used to compare the groups. Then each subject received enough alcohol to produce a BAL of 50 mg.% and then shown erotic slides. Max	[Extensive lit review in introduction] A summary of findings is provided on p. 6.: Rape Index: does not discriminate rapists and nonrapists. Sex History: sadists show more toucherism and frottage, exhibiting and sadomasochism. Some SAs crossdress orgasmically. Gender Identity: sadists are femine identified or undifferentiated. Alcohol & Drugs: abuse common in all groups, but least in sadists. Sex Hormones: LH and FSH may be abnormal in some sadists. DHAS, cortisol and prolactin may be elevated in rapists in general. Brain Pathology: Right temporal lobe abnormalities in sadists. Aggression: not discriminating. SAs generally reacted more to audiotaped stimuli than did controls, but not differentially to rape and consenting intercourse. Both SAs and controls reacted significantly more to rape and consenting intercourse than to violence or neutral stimuli. Nonsadistic sexual aggressives experienced double the penile output of controls and sadists responded the least. Sadists were more common in the present sample of SAs than in many others (45%). As far as the nature of their acts is concerned, "we know that control, fear and terror, injury, and unconsciousness play a role but we are ignorant about the relative importance of each of these components and what other factors may be important." "That orgasmic crossdressing was consistently found does not fit the authors' stereotype of SAs. Sex hormone results showed that alcohol and drug abuse were associated with normal testosterone levels. The "incongruous levels of both testosterone and LH/FSH" were informative. A complete hormone profile on a substantial number of rapists would be beneficial. 45% of all subjects showed some brain pathology, but no significant differences were found in the two groups. Sadists alone, however, showed significant temporal horn dilation and atrophy. The CT scan detected the temporal lobe anomaly but WAIS and Reitan

127

Table 8 (continued)

Author(s) Title/Date	Population Studied	Procedure	Results/Conclusions
Langevin et al. 1985 (continued)		penile volume change within 30 and 60 seconds after stimulus onset were the dependent variables. A drug survey was given that asked about current frequency, max frequency, and accompanying affect. The Buss-Durkee Hostility Inventory, MMPI and Clarke History of Aggression Test were administered to look at aggressive tendencies. Three psychiatrists (independently) examined SAs on dangerousness using a Dangerousness Rating Scale and assessment of the amount of force used in the offence, likelihood of recidivism, the extent of force likely to be used in foreseeable future offenses, and global rating of dangerousness.	findings only suggested it. The authors suggest possible exclusion of the latter two in future investigations. Ascribing sexual aggression to substance abuse is difficult in this study because abuse was prominent in both SAs and controls. "We can surmise that alcohol increases sexual reactivity and elicits erotic reactions that are indiscriminate." The study of aggression in this particular sample was difficult. The authors indicate that psychometrically acceptable measures of aggression/violence proneness are almost nonexistent. The question arose, "Are measures of general aggression/violence inadequate or are sexual aggressives not really generally aggressive after all?"
Hucker, S., Langevin, R., Dickey, R., Handy, L., Chanbers, J., and Wright, S. Cerebral Damage and Dysfunction in Sexually Aggressive Men 1988	51 sexually aggressive men charged with, or convicted of, rape, attempted rape, indecent assault or sexual assault. (20 taken from Langevin et al., 1985). 22 sadists and 21 nonsadistic sexual aggressors and eight were unclassified. All victims of the offenses were females 16 years of age or older. 36 nonviolent nonsex	The Clarke Sex History Questionnaire, criminal history, medical records and interview by experienced sexological investigators were used to determine whether a diagnosis of sexual sadism was appropriate according to DSM III criteria. Reliability of the diagnosis was checked using two raters to interview and examine records of the same 10 sex offenders and two controls - 92% agreement overall was found. CT scans with no contrast material were carried out. 14 slices were taken	Reliability of both CT scan interpretations and DSM III diagnoses of sexual sadism were examined. There was 90% interrater agreement. CT scans showed a larger proportion of right temporal lobe abnormalities in sadistic sexual assaulters than in both nonsadists and controls. Results of the Luria-Nebraska Neuropsychological Battery showed that non-sadistic sexual assaulters were most impaired and that this impairment tended to be global, not lateralized to one or the other hemisphere. "The ability of the Luria battery to localize brain pathology is poor and hampered by lack of established normative data." The authors suggest that tests (such as EEG or PET) focusing on brain activity during sexual arousal would be very helpful in studying brain pathology in these groups. The fact that the CT scan, Luria battery and Reitan battery results do not

128

Table 8 (continued)

Author(s) Title/Date	Population Studied	Procedure	Results/Conclusions
Hucker, S. et al. 1988 (continued)	offenders served as controls. They had charges of theft, fraud, or other property of-fences.	from skull base to vertex. To further focus on the temporal horns overlapping cuts were taken in the temporal area and primary coronal cuts were taken through the middle fossa. Temporal lobes were special-ly examined because of an expected "association of damage to that area and sexually anomalous behavior." The Luria-Nebraska Neuropsycho-logical Test battery and Hartford Shipley Aptitude Test were given to 31 sexual assaulters and 12 controls seen after the 1985 study. WAIS was administered to the original sample. The Hartford Shipley and the WAIS scores were combined and all converted to WAIS Full Scale IQ equivalents. Alcohol and drug abuse variables were examined The Clarke Sex History Question-naires were examined to delineate erotic preferences and better classify groups.	overlap needs to be considered. Although the temporal lobe has been implicated in sexual behavior, the structures of the limbic system and temporal regions that are related to sexual arousal in humans have not been clearly identified. When these temporal lobe sub-structures have been clarified, MRI and PET techniques may be very useful for more detailed research. Studying a larger number of sexual assaulters who have no history of drugs and alcohol abuse (i.e., adolescent sex offenders) may reduce the "problem of diffuse brain damage due to such habits which may have influenced the present results by obscuring small localized abnormalities.

Table 8 (continued)

Author(s) Title/Date	Population Studied	Procedure	Results/Conclusions
Garnett, E.S., Nahmias, C., Wortzman, G. Langevin, R. and Dickey, R. Positron Emission Tomography and Sexual Arousal in a Sadist and Two Controls. 1988	Two males, ages 23 and 26 were screened for physical and mental health, using the MMPI, Clarke Sex History Questionnaire, Gender Identity Scale, MAST, Halstead-Reitan Neuro-psychological test battery, WAIS-R, Clarke Drug Use Survey, and Clarke Medical History They were considered normal. One 30 year old male sadist with a history of antisocial behavior and admitted interests in bondage, crossdressing, and sadomasochistic acts. His crimes focused on ritualized rape with torture.	The study comprised three sessions. The first was a pretest for habituation to the PT (this session was the same as the others except for the injection of radioactive substance). In subsequent sessions the subject was injected with FDG. Then he listened to a 40-minute erotic or sexually neutral tape while laying still with closed eyes. Both tapes were recorded by same male voice. Penile circumference changes were monitored.	The controls showed significant (p<.0001) differences in penile circumference between the erotic and the neutral stimuli, with few or no penile reactions during the neutral stimuli. They were aware of sexual arousal while listening to the erotic tape but not the neutral tape. Arousal was greatest at 15 min. after stimulus onset. The men did show reduction of arousal during periods of blood sampling. For the sadist, penile reactions were smaller and more erratic than for the controls. He was also aware of mild sexual arousal to erotic stimuli and no arousal to neutral stimuli. He showed peaks of arousal during blood sampling periods. In the PT scans for the controls, both temporal lobe areas were most activated (as should be expected during sustained auditory stimulation). There was, however, an unexpected lack of differential activation of the limbic area during erotic stimulation. Greater levels of erotic stimulation may be required to excite specific limbic centers to prepare for actual sexual behavior rather than merely fantasy. "The predominant accumulation of radioactivity occurred in the right hemisphere and spread across the whole cortex in the right side of the brain." Both types of stimuli showed similar patterns of activation. More energy was apparent in the erotic as compared with the neutral stimuli. These observations also held true for the sadist but with one interesting difference. The metabolic activation was in the cortex of both hemispheres, unlike predominant right hemisphere activation in controls.

Table 8 (continued)

Author(s) Title/Date	Population Studied	Procedure	Results/Conclusions
Langevin, R., Worzman, G., Wright, P., and Handy, L. Studies of Brain Damage and Dysfunction in Sex Offenders. 1989	160 extrafamilial child sexual abusers, 123 incest perpetrators, 108 sexual aggressors against adult females. The sexual offenders were either facing charges of sexual assault, had been convicted of this type of offence, or were involved in a post-prison treatment program for sex offenders. 36 nonviolent, nonsex offenders with charges of theft and/or fraud served as the control group. "This group controls for both patient and offender status as well as, in general, being better matched on age, education and social class."	Erotic preferences were checked when possible using a phallometric test of sexual preference (Freund et al., 1972). Each offender received the Halstead-Reitan Neuropsychological Test Battery including WAIS-R. CT scans were carried out with no contrast material. Scans were interpreted by a neuroradiologist blind to the nature of participants and test results. They also received the Wechsler Memory Scale and the Differential Aptitude Test Space Relations Test. Information on age, education, admitter status, substance abuse, and history of violence was gathered. The MAST DAST, Clarke Drug Use Survey, and Clarke Violence Scale were given.	Intelligence scores of sex offenders were basically average but the distribution was skewed to below normal. Especially in pedophiles, verbal abilities tended to be lower than nonverbal abilities. This was not a consistent effect, however. Homosexual and heterosexual pedophiles displayed lower abilities than controls in overall IQ. Bisexual pedophiles displayed lower abilities than controls in performance but not in verbal. The difference in scores (Verbal - Performance) reached significance which, the authors feel, suggests right hemispheric brain impairment. HR battery results showed greater impairment in pedophilic offenders than in controls. "Results for heterosexual and homosexual men suggested left hemisphereic language-mediated problems but also frontal lobe rigidity and perhaps impulsiveness, compared to the other groups." The CT results were inconsistent with earlier work carried out by the authors. They suggest that it is important to determine which techniques (EEG, CT, MR, etc.) will best identify the structures in and functions of the brain associated with sexual aggression. They feel that CT scans may miss important features that MR scans may target. The Wechsler Memory Scale did not differentiate subgroups of the offenders "Brain damage and dysfunction among sex offenders, if critical in the etiology of their sexual anomalies, is likely to be subtle and may well be specific to sexual behavior." The authors provide suggestions for further research.

131

Table 8 (continued)

Author(s) Title/Date	Population Studied	Procedure	Results/Conclusions
Hendricks, S. E., Fitzpatrick, D. F., Hartmann, K. Quaife, M. A., Stratbucker, R. A. and Graber, B. Brain Structure and Function in Sexual Molesters of Children and Adolescents. 1988	Sixteen male patients charged specifically with sexual assault against individuals 14 years of age or younger. Mean age = 34 (20 to 63 years). A control group comprised of two women and fourteen men. Mean age=30.1 (23 to 47 years).	Hemispheric rCBF was measured using the Xenon inhalation technique of Orbist as modified by Meyer. CT images were taken (11 or 12 axial slices for each subject).	Child molesters exhibited relatively low rCBF values. This finding replicates the finding in Graber et al. (1982). Skull thickness and density seemed to be a major difference between the control and child molester groups. Two potential explanations for this are variation in ventricular size and variation in overall brain size. In this study significant negative correlations between skull thickness and ventricular area for the two higher slices were found. No evidence was found suggesting a relationship between skull thickness and amount of total brain tissue. The authors do not draw any conclusions about the variations in cerebral structure and function and their potential causative relationship to sexual molestation of children. They suggest further research in this area.
Langevin, R., Bain, J., Wortzman, G. Hucker, S., Dickey, R., and Wright, P. Sexual Sadism: Brain, Blood and Behavior. 1988	Lit review	Literature review on sadism. The authors focus on behavior patterns, endocrine abnormalities, and brain abnormalities.	"The behavior of sadists is bizarre and poorly understood. There are gross endocrine and brain abnormalities in a small number of these men. Approximately two-fifths show subtle temporal lobe brain abnormalities that are logically linked to sexual behavior and require further exploration. It would be interesting to explore the interface of the endocrine system and the brain - that is, to determine if there are interactive processes that may be related to the development of sexual anomalies, perhaps early in life as suggested by Kolarsky et al. Certainly, biological factors cannot determine whether an individual will act on his sexual impulses. Many psychological factors, such as family background and substance abuse, play a significant role in the dangerousness of the individual. However, it appears that biological factors are noteworthy in sexual sadism. Brain pathology, especially, shows some correlation with force used in offenses and likelihood of recidivism, and for this reason alone it merits further study" (p. 170).

Table 9 Treatment of Violent Offenders with ECT or Pharmacotherapy

Author(s) Title/Date	Population Studied	Procedure	Results/Conclusions
Schnur, D. B., Mukherjee, S., Silver, J., Degreef, G. and Lee, C. Electroconvulsive Therapy in the Treatment of Episodic Aggressive Dyscontrol in Psychotic Patients. 1989	Five cases of episodic aggressive dyscontrol. All of them met the DSM-IIIR criteria for Organic Delusional Disorder. They would also meet DSM-IIIR criteria for Intermittent Explosive Disorder except for their chronic schizophrenia-like psychosis. Their aggressive behavior was severe and they displayed actual violence directed at and sometimes injuring staff members and other patients. Four patients exhibited recurrent spontaneous seizures which a counseling neurologist diagnosed as epilepsy.	ECT was administered on an alternate day, three per week schedule. ECT was given with a constant current, brief pulse, bi-directional square wave stimulus with the use of a MECTA device. Seizures were monitored by the cuff method and EEG. The standard bifrontotemporal electrode placement was used. A retrospective review of clinical records was used to determine the number of spontaneous seizures and episodes of aggression during the month prior to ECT. They were also observed during ECT treatment and one month thereafter. Only violent acts that resulted in physical danger to the victim or the perpetrator were considered aggressive episodes. Medication alterations during ECT included only reductions in neuroleptic and anticonvulsant medications. Actual case reports are provided.	Each patient experienced a reduction in frequency of episodic aggressive behavior associated with ECT. The improvement tended to occur early in the treatment schedule. A modest reduction in psychosis was also observed. ECT was associated with total remission of seizures in all patients but one in which the postictal state was modified. Gross clinical evidence of an organic brain syndrome was not associated with ECT. Clinical improvement in aggressive behavior often remained the same for long periods following the discontinuation of ECT. Whether this is directly related to ECT or a reflection of post-ECT pharmacotherapy is not clear. Only one patient showed improvement in psychotic symptoms. "In cases of episodic aggressive dyscontrol, the antiaggressive effects of ECT may be independent of its antipsychotic properties. The possible specificity may not be generalizable to other variants of aggressive behavior." The authors do not wish to claim that ECT is efficacious in the treatment of aggressive behavior in general.

Table 9 (continued)

Author(s) Title/Date	Population Studied	Procedure	Results/Conclusions
De Cuyper, H., H. M. van Praag, D. Verstraeten The Effect of Milenperone on the Aggressive Behavior of Psychogeriatric Patients 1985	20 female in-patients (of a psychogeriatric ward) with chronic aggressive behavior toward others seen as negativism, dysphoria, verbal and/or motor excitation	A double-blind study where the therapeutic activity of milenperone is compared with a placebo. The patients were divided into two random groups. The second phases involved a doubling of the dosage in both groups. Two scales were used as measures. The Paranoid Belligerence Scale and a scale developed from the Visual Analogue Line	Addition of milenperone to the psychotropic medication already in use significantly decreased the aggressiveness scores. Both groups did show a proportional improvement at the start. In the last phase of the study, however, aggressive behavior in the placebo group increased again.
Stewart, J. T., M. L. Mounts, and R. L. Clark Aggressive Behavior in Huntington's Disease: Treatment with Propranolol 1987	Three white males ranging in age from 44 to 50 and ranging in history of diagnosed Huntington's disease from 5 to 7 years.	Three case reports. Treatment with Propranolol.	Propranolol therapy was effective in all three patients. Careful titration on an individual basis is stressed.

134

Table 9 (continued)

Author(s) Title/Date	Population Studied	Procedure	Results/Conclusions
Monroe, R. R. Anticonvulsants in the Treatment of Aggression. 1975	Literature Review.		"A significant number of violent acts are committed by individuals in whom central nervous system instability can be demonstrated by special EEG activation procedures utilizing alpha-chloralose as the activating agent. Furthermore, subcortical electrograms suggest that this instability is related to a circumscribed ictal phenomenon in the limbic system. The abruptness of the aggressive act, the fact that the behavior is so often out of character for the individual and inappropriate for the situation, as well as the confusion and partial amnesia which accompany these episodes lend clinical support for the ictal hypothesis. Some anticonvulsants not only block the activated abnormalities on the EEG but also lead to dramatic clinical improvement in those individuals showing repeated and frequent aggressive behavior. For instance, in one study 46.7 per cent and 53.3 per cent of the patients demonstrated activated abnormalities on no drug and placebo, respectively. When these same patients were receiving chlorpromazine or trifluoperazine, the activation rates were 60.0 per cent and 73.3 per cent, respectively. On the other hand, when these same patients were placed on a regimen of chlordiazepoxide the activation rate was reduced to 20 per cent (p<.01). Another study involved severely disturbed chronically hospitalized psychotic patients whose aggressive uncontrolled outbursts relegated them not only to a locked ward, but often to isolation rooms despite high doses of phenothiazines. A regiment of chlordiazepoxide and/or primidone added to their current medication led to dramatic improvement in 23 patients and some improvement in 17 others. Only 15 subjects showed no response to this regimen."

135

Table 10 Treatment of Violent Offenders with Psychosurgery

Author(s) title/Date	Population Studied	Procedure	Results/Conclusions
Sano, K., and Y. Mayanagi Posteromedial Hypothalamotomy in the Treatment of Violent, Aggressive Behavior. 1988	60 cases operated on between 1962 and 1977. 29 were under the age of 15. There were 44 males and 16 females. Most patients had a history of epileptic seizure connected with a certain amount of mental retardation preceding the development of the behavior disorder.	Stereotactic lesions in the ergotropic portion of the posterior hypothalamus (posteromedial hypothalamotomy) The procedure was used only when aggression became un-maintainable by drugs and the isolation of the patient became necessary.	For 37 cases precise information of their postoperative daily life was gleaned from interviews with family members and professional people who had been involved with the patients for an extended period. 18 cases (49%) showed excellent results. They displayed no violent or aggressive behavior post-operatively. 11 cases (30%) showed good results with operatively. 11 cases (30%) showed no violent or aggressive behavior postoperatively but remained easily excitable. As a whole these 29 cases (78%) showed satisfactory results. 5 cases showed fair results and 3 showed poor results. These assess-ments were made in a 10-25 year postoperative period. In the meantime 13 had died from seemingly unrelated causes. The author feels that the experiences of the present group support early surgical intervention.

excellent	4	9		1	4	0	0
good	2	5		0	1	1	2
fair	0	1		0	0	3	1
poor	0	0		1	0	0	2

Table 10 (continued)

Author(s) title/Date	Population Studied	Procedure	Results/Conclusions
Ramamurthi, B. Stereotactic Operations in Behavior Disorders: Amygdalotomy, Hypothalamotomy. 1988	603 operations for control of conservatively untreatable aggression. 481 cases of bilateral amygdalotomies and 122 cases of mostly secondary posteromedian hypothalamotomy. Most of the patients were children below the age of 15. Some had epilepsy.	The types of behavior problems that required this treatment were: aggression, hyperkinesis, destructive and self-destructive tendencies, and wandering tendency. The stereotactic operations were decided upon following two years of treatment with psychotropic drugs yielding no appreciable relief. Pre- and post-operative psychological assessments were available for only 60 of the patients.	The stereotactic operations were beneficial for approximately 60% of the patients. When the disorder was accompanied by epilepsy the chances of improvement increased. Following the bilateral amygdalotomies 39% of the patients showed good to excellent results, and 37% showed moderate improvement. When the amygdalotomy failed after the first attempt the patient underwent a second one. Half of the patients who required this showed improvement following the second operation. Results of a 3 year follow-up showed that 55% maintained good condition, 15% maintained moderate condition, and 30% failed to respond to the treatment. Benefits from the operation were assessed from 3 points of view. (1) improvement in restlessness or violence -good/excellent when the subject remained calm and quiet despite provocation and moderate when aggression is absent or diminished when not provoked. (2) the beneficial effect of a quiet patient on the siblings and relatives of the patient: measured by responses of the parents and relatives whose quality of life has rapidly improved, and also by the increase in demand for these operations. (3) better possibilities of child educating measured by improvement in restlessness. The authors suggest that gastric acid levels could potentially be factors significant to prognosis.

137

Table 10 (continued)

Author(s) title/Date	Population Studied	Procedure	Results/Conclusions
Dieckmann, G., B. Schneider-Jonietz, and H. Schneider Psychiatric and Neuropsychological Findings after Stereotactic Hypothalamotomy, in Cases of Extreme Sexual Aggressivity. 1988	14 cases; 8 had thorough psychiatric and psychological examinations during the follow-up.	1970 -1972. Unilateral ventromedial hypothalamotomy in the non-dominant hemisphere for aggressive sexual delinquency. The 8 with thorough examinations each underwent 1. a physical examination 2. physiological recordings (EEG, EMG). 3. external anamnesis and behavior observation (psychiatric exploration, semistructured interview) 4. questionnaires (Frieburg personality inventory) 5. psychological tests (HAWIE, LPS, BENTON, d2, GIESE, TAT, FPT)	Results showed a decrease in domination by sexual drive, an increase in rapidity of visual image formation, and increase in fluency in semantic contexts, an increase in coordinative perception processes, positive changes in some personality dimensions (i.e. openness, self-criticism, poise), an increase in appetite,and a decrease in color perception. Of greatest importance in this study is that the structure of the individual's sexuality did not change, but a decrease in the probability of specific aggressive behavior was observed. The authors stress that a complete psychic and organic investigation is of great importance to the success of this type of treatment. They also state that it has been shown that intervention by psychosurgery does not hinder an individuals self-development and does not prevent him from making decisions about how to live.

138

Table 10 (continued)

Author(s) title/Date	Population Studied	Procedure	Results/Conclusions
Mirsky, A. F. and Rosvold, H. E. The Case of Carolyn Wilson -- A 38-year Follow-Up of a Schizophrenic Patient with 2 Prefrontal Lobotomies. 1990	Case history of a woman who suffered from schizophrenia for at least 11 years. She received a prefrontal lobotomy in December 1946 at 37 years of age and another, more posterior one, in July of 1947.	This case report looks at Ms. Wilson's life since surgery, reconstructs the events leading up to the surgery, and summarizes the neuropsychological studies conducted over the past 40 years. The authors also conducted a psychophysiological evaluation of P300 in a go/no go task in which the target stimulus is a rare tone. She was also given a full-scale neuropsychological evaluation and a neurological evaluation which included a full battery of neurological tests, a CT scan and a PET scan.	It appears that the first lobotomy (December 1946) had not substantial effect on Carolyn's behavior. For this reason, the authors decided to consider both procedures as a single, radical prefrontal lobotomy. Preoperatively she displayed florid psychotic behavior which included mania, both verbal and physical aggression, incontinence of urine and feces. Along with these characteristics, she refused to eat, drink or care for herself. She was considered a dangerous patient. When examined several months following the July 1947 procedure and just prior to discharge, Carolyn's assaultive, destructive, and delusional behavior had largely disappeared. She was approachable and tractable. In the postoperative examinations, her demeanor was described as placed and unruffled. Strong affect or emotion was rarely if ever seen. Most IQ assessments over the last 40 years have shown that Carolyn is in the bright-normal to superior range. When tested at age 75 an across-the-board reduction in intellectual capacity was observed (not necessarily unusual for this age). Carolyn had a pervasive remote memory deficit. Next to this the most prominent cognitive loses are in tasks requiring a flexible approach to problem solving and ability to modify behavior to suit the situation. This was most apparent in performance on the Wisconsin Card Sorting Test. The scalp distribution of the P300 was not typical of a 72 year old. Carolyn showed an atypical distribution (in elderly subjects P300 is usually of equal amplitude at Fz, Cz and Pz) which may be due to the extensive prefrontal damage. The PET scan revealed that "the anterior 30-40% of the entire cerebrum, comprising primarily the prefrontal regions, was severely, if not totally, hypometabolic. This suggests that the prefrontal areas of Carolyn's brain are largely nonfunctional."

Table 11 Gender Differences in the Effects of Temporal Lobe Neurosurgery

Author(s) Title/Date	Population Studied	Procedure	Results/Conclusions
Lansdell, H. A Sex Difference in Effect of Temporal-Lobe Neurosurgery on Design Preference. 1962	22 patients with temporal-lobe epilepsy were tested before and after surgery. One man and two women had operations on the left because test showed that the dominant hemisphere for speech in them was on the right.	Subjects were tested on the Graves design judgment test. This test measures "certain components of aptitude for the appreciation or production of art structure." In the 22 cases, the amount of tissue removed was less than usual, 3-51/2 cm from the tip of the lobe on the right and 21/2-41/2 cm on the left.	The men having had resections in the non-dominant hemisphere showed a drop in mean aptitude score post-operatively. Women with the same operation showed a rise in mean score. For those groups of men and women with resections in the dominant hemisphere, the opposite effect was observed. Men showed a rise in score while women showed a drop. There was, however, a significant interaction of sex and side only post-operatively. Three patients had score changes that were opposite to the general trend, and two others showed no change. Three of these five did have atypical operations which may help explain the inconsistency in behavior. The changes in scores on the Graves test were not found to be related to changes in scores on the Wechsler-Bellevue intelligence tests and, therefore, do not appear to be "intellectual" changes. "Although the post-operative changes on the Graves test are clear, they appear to be transitory. No significant relation to the surgery was discernible for either sex or side of operation in the group of 42 patients tested a year or more after their operations." The results of the women tested do not support the hypothesis that an "impairment in some aspect of human visual perception occurs after removals from the temporal lobe of the non-dominant hemisphere." "Effects of the operations suggest that some physiological mechanisms underlying artistic judgment and verbal ability may overlap in the female brain but are in opposite hemispheres in the male."

Table 11 (continued)

Author(s) Title/Date	Population Studied	Procedure	Results/Conclusions
Lansdell, H. Sex Differences in Hemispheric Asymmetries of the Human Brain. 1964	Letter		Review of a study done by Conel (1963) who investigated eight brains of 4-year-old children for hemispheric differences. Only when the sexes were separated were two noteworthy differences found. Myelination was greater in the left FAy-hand area than in the same area in the right hemisphere in 4 of the 5 female brains. In the 3 male brains this was reversed. Secondly, in the 4 female brains the number of exogenous fibers in layer 1 of areas FAy and PB is greater on the right. In 2 of the 3 male brains this number was greater on the left. Though the results provided did not reach statistical significance, the author speculates whether "these anatomical differences could be related to the finding that side differences in the tactual thresholds on the thumbs of young children are not the same for the two sexes." An earlier study by Matsubara, cited by Lansdell, suggests that the right vein of Troland is larger than the left in girls. This was not true for boys. This is the major vein in the hemisphere opposite to that used in speech so differences in venous drainage could possibly be related to the superiority of girls over boys in certain verbal skills. He suggests that "the sex of the patients is a factor which should be heeded in investigation of the laterality of cerebral function."

141

Table 11 (continued)

Author(s) Title/Date	Population Studied	Procedure	Results/Conclusions
Lansdell, H. and Urbach, N. Sex Differences in Personality Measures Related to Size and Side of Temporal Lobe Ablations. 1967	46 epileptic patients who had had unilateral temporal lobe removals. Others excluded had IQs below 90, invalid MMPIs, and possible complicating factors in the lateralization of their damage and its effects. 18 of the 46 had left temporal removals (13 men (ML), 5 women (WL)). 28 had right temporal removals (12 men (MR), 16 women (WR)). Mean age of all = 33.6 years and an average of 3.4 years had elapsed since surgery at the time of the questionnaire.	Subjects received the card form of the MMPI. Only MMPI records with <60 and L<9 were included in the analysis. The Wechsler-Bellevue Intelligence Scale was also frequently administered at the same time, but for some a year or more earlier. Subjects also received the Graves Design Judgment Test.	"The mean extent of right temporal removals, 5.1, was greater than the mean of the left removals, 4.5 (p<.01) this difference was slightly greater for women than for men, and the range of the extent of the removals was greater in men, 1.0 to 7.6, than in women, 2.5 to 7.3. Men showed a larger correlation between size of ablation and WB verbal to nonverbal ratio. The difference in range of extent of removals may account for this. It would not, however, account for differences found in the signs of the correlations on the DJT and MMPI. DJT results suggest that the larger the left temporal removal in males, the stronger their preference for simple forms. The correlation between extent of removal and scores on the "Schizophrenia" scale of the MMPI was negative for the MR group, distinguishing it from the other three groups. The ML group showed a relatively lower verbal component of the WB ratio. This is consistent with previous documentation of selective effect (in males) of left temporal removals on the interpretation of proverbs. The same analyses were performed and the same pattern of results found in a less selected group of 76 Ss. "As suggested by the results of this investigation, research on the lateralization of cerebral function may often benefit from checking for the presence of sex differences."

142

Table 11 (continued)

Author(s) Title/Date	Population Studied	Procedure	Results/Conclusions
Lansdell, H. Sex Differences in Brain and Personality Correlates of the Ability to Identify Popular Word Associations. 1989	117 neurosurgical subjects. 89 had previously had temporal lobe surgery and 28 had subcortical surgery (which involved coagulations in the thalamus). Mean age = 51.1 yrs. Neurosurgical subjects were tested three times - with the third test occurring about 17.6 months post-surgery with temporal lobe patients and 16.4 months post-surgery for subcortical patients. A group of 55 (of the 89) temporal lobe patients were only tested once - a year or more post-surgery. Mean = 60.2 months post-surgery. 306 nonsurgical subjects served as comparisons. 164 had epilepsy, 80 were classified as "other neurological patients," and 62 visitors or patients who appeared to be normal neurologically. Mean age=36.3 yrs.	Subjects were administered the multiple choice Word Association Test (WAT) and the Minnesota Multiphasic Personality Inventory (MMPI), box form. The MMPI was only administered to those subjects whose IQs were average or above. The two factor scores A (anxiety) and R (repression) were used for this study. Form I of the Wechsler-Bellevue Intelligence Scale (WB) was used prior to surgery and again one year or longer later. Form II was used in the period immediately following surgery.	The mean WAT error score was no different for the neurological subjects as compared to a college and a university sample in a previous study. Somewhat of an affect on the WAT scores was observed after neurosurgery to the left hemisphere. The most obvious impairment was observed a year or so later in the males. The female patients showed improvement on the third test (one year or more post-surgery). Results on the WAT for the 55 temporal lobe patients, who were first tested years after surgery, also showed that the scores for males were more affected than those for females when intelligence was controlled for. "These unique results with these operations show a resiliency of the female brain, compared with the male brain, with regard to these two types of surgery and the ability assessed by the WAT." Scores of the MMPI and WB did not show comparable changes post-surgery. In the males tested after temporal lobe surgery, a higher number of errors on the WAT tended to be associated with higher A and lower R scores (MMPI). High A and low R scores usually indicate that the person is "introspective, ruminative" and "lacking in common sense" among other characteristics. "These results with the WAT are unique in implying that most men undergoing these types of neurosurgery can suffers some permanent selective impairment that affects this aspect of their understanding of normal human thought processes." The authors note marked controversy in reports of sex differences in the effect of lateralized brain damage.

143

Table 11 (continued)

Author(s) Title/Date	Population Studied	Procedure	Results/Conclusions
Sundet, K. Sex Differences in Cognitive Impairment Following Unilateral Brain Damage. 1986	Patients were taken from a neuropsychological register of brain-injured patients. 232 of them were tested with WAIS and of these 83 met the criteria for inclusion in the present study. The criteria were: (1) lesion diagnosed as positively unilateral either through operation or autopsy, or through use of CT, angiography, or pneumoencephalogram; (2) nonaphasic; and (3) functionally right-handed. Of the 83, 19 were males with left-sided brain lesions (LBL), 32 were males with right-sided brain lesions (RBL), 15 females with LBL, and 17 females with RBL.	Patients were administered the authorized Norwegian version of the WAIS. Information on age, time since onset, etiology, and intra-hemispheric localization was collected from the records.	The author states that "sex differences in the pattern of WAIS impairment following unilateral brain damage may be regarded as a cross-cultural phenomenon." Results showed that the WAIS discrepancy score did diagnose laterality of lesion in the males but did not in the females. The author believes that this indicates a sex-related difference in cognitive functioning. Some references to previous studies are made in support of these findings. The author suggests that characteristics of cognitive strategies and how they "compensate for chronic deficits" be the focus of further research on sex differences.

144

Table 11 (continued)

Author(s) Title/Date	Population Studied	Procedure	Results/Conclusions
Hampson, E. and Kimura, D. Reciprocal Effects of Hormonal Fluctuations on Human Motor and Perceptual-Spatial Skills. 1988	34 spontaneously cycling women ranging from 20-39 years (mean= 24.65 yrs.). They were predominantly university students. Their menstrual cycles were regular (25-25 days). 32 were right-handed.	The portable version of the Rod-and-Frame test was administered. The battery includes a finger-tapping test, the Purdue Pegboard, and the Manual Sequence Box. Each was tested twice, once during menstruation when levels of estrogen and progesterone are low (day 3 - day 5) and once during the midluteal estrogen and progesterone high (7 days before expected menstrual onset). Days 1 and 2 were avoided because of possible confounders (i.e., physical discomfort, etc.).	A number of investigators have previously reported that males are significantly more accurate than females (to 2 and even 3 degrees) on both the original and the portable versions of this test. Subjects showed significantly less accuracy at aligning the rod to true vertical during the midluteal phase test than during menstruation. On the other hand, performance improved during the midluteal phase on most tests targeting manual skill. Subjects showed greater speed and accuracy on these tests. A mood inventory was given prior to each test. No significant phase-related mood differences were observed. "We have thus demonstrated reciprocal performance fluctuations over the menstrual cycle in two types of skills, namely speeded manual coordination and a perceptual-spatial skill. The dissociation in the pattern of change is particularly interesting in light of the sex differences usually reported for these tests: High levels of female hormones enhanced performance on tests at which females excel but were detrimental to performance on a task at which males excel. The size of the hormone effect in relation to the size of the sex difference exceeded 75% for the Rod-and-Frame test and 65% for the Manual Sequence Box task. Together these suggest that the sex differences on these, and perhaps other cognitive tests, may have a substantial hormonal basis."

145

Table 12 Psychophysiological Studies of Psychopaths

Author(s) Title/Date	Population Studied	Procedure	Results/Conclusions
Jutai, J. W. and Hare, R. D. Psychopathy and Selective Attention During Performance of a Complex Perceptual-Motor Task. 1983	39 white male inmates of a medium security prison. Based on Cleckley's conception of psychopathy (1976), each subject was rated by two investigators on a 7-point scale of psychopathy and a 22-item psychopathy checklist. Two groups were established. A high psychopathy group (Group H) comprised 11 inmates with a combined rating of 12-14 and a mean checklist score of 35.36. The low psychopathy group (Group L) comprised 10 inmates with a combined rating of 3-8 and a mean checklist score of 20.35.	Electrodes were attached to record ERPs. Subjects went through some pregame tasks which included alternating periods of just listening to a series of tone pips through headphones and 5-min eyes open 5-min eyes closed periods. These were carried out to determine whether auditory ERP differences between groups existed. The main experiment consisted of a set of tank games and a set of jet games. During the trials, subjects were instructed to score as many hits as possible on the enemy target. Tone pips were presented during each trial, but the subjects were told that the pips were task-irrelevant. At the end of testing, subjects used 4-point scales to rate how difficult and how exciting each type of game was.	No group differences in heart rate or electrodermal activity were found. The authors believe that this lends support to the view that most autonomic differences between psychopaths and other comparison groups are more prone to occur when tasks involved are monotonous or threatening than when they art the notion that psychopathy is not associated with abnormal electrocortical responsivity during tasks involving passive attention. No group differences were found in amplitude or latency of N100 and P200 of the auditory ERPs during the passive attending periods. The question remains whether or not the similarity in electrocortical responses during passive attention in psychopaths and others is due to the use of similar behavioral and cognitive strategies. In the main part of the experiment where attention was divided between tone pips and the primary task, groups H and L differed in attention allocation. Specifically, Group H showed much smaller N100 responses in the first trial than Group L. It appears that Group L was less successful initially at ignoring the task-irrelevant stimuli in order to focus attention on the primary task than Group H. This was concluded from the fact that the responses of Group H were less than half as large in the first trial as during the passive attention to the tones; those of Group L were slightly larger. As the task progressed Group L became better at shifting attention toward relevant features and away from irrelevant ones, N100 responses got smaller and performance improved. Although the N100 responses were small from the onset for Group H, they were not associated with especially good performance. Also, during later trials performance deteriorated. "While psychopaths may have found it easy to ignore stimuli they had been told were irrelevant, they may have failed to develop an efficient strategy for the distribution of their attentional resources among the various elements of the primary task."

Table 12 (continued)

Author(s) Title/Date	Population Studied	Procedure	Results/Conclusions
Jutai, J. W., Hare, R. D., and Connolly, J. F. Psychopathy and Event-Related Brain Potentials (ERPs) Associated with Attention to Speech Stimuli. 1987	33 white male inmates selected from a medium security institution. The subjects were consistently right-handed, had normal (or corrected to normal) eye sight and normal hearing. A 7-point psychopathy scale was used to assess psychopathy in attempts to maximize separation between the two groups. Also, a 22-item psychopathy checklist was used to aid in this assessment. Two groups were established. Psychopaths (group P) comprised 11 inmates with a global rating of 7 and a score on the psychopathy checklist of at least 34. Nonpsychopaths (group NP) comprised 13 inmates with a global rating of 1-4 and a score on the psychopathy checklist of at most 24. There was no significant difference between groups in age and years of formal education.	Two phonemes, /v/ and /ts/, with onset and offset characteristics served as the speech stimuli. They were presented binaurally through stereo headphones. Subjects received 64 presentations of each phoneme. One block of 32 simply required passive listening and the other block of 32 required pressing of a microswitch on each presentation. This was a preliminary experiment to determine whether group differences in the nature of the ERPs elicited by phonemes or in the effect of motor activity on the ERPs existed. No significant differences were found. The experiment itself consisted of a Single-Task, in which one phoneme (the target stimulus) was presented less often than the other, a Game-only condition, which served as a baseline for the Dual-Task, and a Dual-Task, where the subject played the video game as best he could. In the Dual-Task the subject performed the phonemic oddball paradigm while simultaneously playing some sort of video game (essentially a combination of the Single-Task and Game-only tasks). Subjects received $.05 for each point scored and lost $.05 for each target stimulus missed. ERPs were recorded throughout.	Results of this study and a previous study (Jutai and Hare, 1983) both failed to find inter-hemispheric differences in N100 responses in psychopaths attending to tones in a passive listening task. Thus the hypothesis that "psychopaths are characterized by asymmetric low left-hemisphere arousal'" is not supported by these ERP data. In the Single-Task, ERPs of the psychopaths were normal. In the Dual-Task, however, the ERPs of the psychopaths showed a prominent slow wave which was most apparent at Cz and T3. "Because there were no differences between Groups P and NP in the amplitude of N100 responses, and because the N100 responses of both groups were the same in the Dual-Task as they were in the Single-Task, Group P's slow wave in the Dual-Task is not readily explainable in terms of the group differences in central arousal. It is unlikely that Group P was uncertain (or equivocal) about the outcome of target discrimination, because target detection rates were high and false alarms were rare. Rather, Group P may have failed to learn as much as did Group NP about the likelihood that a target would appear on a subsequent trial, and consequently, equivocated about stimulus probability, producing a large slow wave." Slow wave activity was much more prominent in T3 than in T4 for the psychopaths. This could indicate some type of left hemisphere dysfunction, or, perhaps, these subjects differ in cerebral organization of language functions. The authors are cautious about speculation because of "relative lack" of empirical data on cognitive functioning and perception in psychopaths.

147

Table 12 (continued)

Author(s) Title/Date	Population Studied	Procedure	Results/Conclusions
Hare, R. D. and Jutai, J. W. Psychopathy and Cerebral Asymmetry in Semantic Processing. 1988	39 white male inmates of a medium security prison. Each subject was right-handed. To establish groups, each was evaluated with a 22-item Psychopathy checklist. The inmates were divided into high, medium and low (H,M, L) psychopathy groups. Each group consisted of 13 inmates. The mean checklist scores were: H>32; L<23 and M=23-32. 13 right-handed men recruited from a federal employment center served as a noncriminal comparison group (NC). Groups were similar on demographic and socioeconomic variables.	The examiners were blind to the group each subject was assigned to. A divided visual field procedure was used to research cerebral organization of language processes in psychopaths. A 2-field Cambridge tachistoscope was used to present the 4-letter concrete noun printed on a white card that served as the stimulus. The stimuli were presented in either the left or the right visual hemifield. Subjects participated in three tasks: the SR task where the stimulus had to match a pretrial word, the SC task where the stimulus was an exemplar of a specific category and the AC task where the stimulus was an exemplar of an abstract category.	Reaction times did not differ significantly among Groups H, L and NC. Responses in each of these groups were somewhat faster to stimuli that appeared in the right visual hemisphere than those that appeared in the left visual hemisphere. In general, reaction times were faster in the SR task than in the AC task as expected. Of most importance was the finding that psychopathic criminals differed from other criminals and noncriminals in the asymmetry of errors on an abstract categorization task. Group H showed normal asymmetry on the SR task, but showed a left visual field advantage (reversed asymmetry) on the AC task. In actuality the psychopaths made a large number of errors in the right visual field and thus a right visual field deficit may be more descriptive than a left visual field advantage. Group H did not show improvement in right visual field performance during the AC task. "The results, along with those obtained in a recent dichotic listening study, lead us to speculate that psychopathy may be associated with weak or unusual lateralization of language function, and that psychopaths may have fewer left hemisphere resources for processing language than do normal individuals."

148

Table 13 Studies Related to Neurodevelopmental Issues in Aggression and Violence

Author(s) Title/Date	Population Studied	Procedure	Results/Conclusions
Andrew, J. M. Imbalance on the Weights Test and Violence among Delinquents. 1981	41 adjudicated Caucasian male legal offenders within four age groups: 7 age 12-14 yrs 13 age 15-16 yrs 10 age 17-20 yrs 11 age 21-29 yrs. Only right-handed subjects were included. These males were consecutively referred for psychological evaluation for treatment planning.	Each subject received the Weights Test (a psychological test of proprioceptive motor/cognitive function) and was rated for violence on the Violence Scale. Each offender's most serious crime was rated according to the Violence Scale and used in the study.	Results from the Weights Test, as hypothesized, divided the sample of Caucasian male offenders into more versus less violent groups. Scores from the Violence Scale were highest for subjects considered mildly impaired, according to the Weights Test, and lowest for those subjects considered less impaired and those considered more impaired ("moderately impaired"). The authors speculate on the applicability of the deficit theory and the imbalance theory to these results. They do not provide any concrete conclusions.

149

Table 13 (continued)

Author(s) Title/Date	Population Studied	Procedure	Results/Conclusions
Brickman, A. S., M. McManus, W. L. Grapentine, and N. Alessi Neuropsychological Assessment of Seriously Delinquent Adolescents. 1984	71 subjects (40 male; 31 female) were chosen based on four criteria: 1. commission of violent felonies 2. commission of multiple non-violent felonies 3. multiple placements within the training school system; and 4. assaultive in-program behavior requiring medical attention for the victim. Mean age=16.3 yrs. 39 of the subjects were white, 26 - black, 3 - Hispanic 3 - mixed race.	The Luria-Nebraska Neuropsych-ological Battery (LNNB) was administered to all subjects. This examination consists of 269 items organized into 11 categories: Motor, Rhythm, Tactile, Visual, Receptive Language, Expressive Language, Reading, Writing, Arithmetic, Memory, and Intelligence. Educational levels for the subjects were assessed (as accurately as possible).	Violent and recidivist delinquents display a distinctly abnormal pattern of functioning. This pattern is not limited to higher "intellectual" functions, but, rather, encompasses a wide range of functions. The only significant differences between the subgroups, however, are in expressive speech and memory scales. This could be attributable to education or concentration problems. Upon reviewing the specific neuropsychological dysfunction, the authors suggest that the temporal lobe may be implicated.

150

Table 13 (continued)

Author(s) Title/Date	Population Studied	Procedure	Results/Conclusions
Hermann, B. P. Neuropsychological Functioning and Psychopathology in Children with Epilepsy. 1982	Fifty children with epilepsy ages 8 - 12 yrs. Two groups were created based on neuropsychological function. 25 had good neuropsychological function and 25 had poor neuropsychological function. The subjects were also classified according to seizure type.	Each child received a complete neurological evaluation including an EEG. The two groups were compared in several domains. Social competence and behavioral functioning were measured using the Child Behavior Checklist. For neuropsychological data each child completed the Luria-Nebraska Neuropsychological Battery - Children's Version.	A highly significant relationship between adequacy of neuropsychological function and the level of aggressive behavior in children with epilepsy was determined. Those subjects with a poor performance on the neuropsychological battery also retained significantly higher scores on the Aggression scale of the CBCL as compared with children who performed better on the neuropsychological tests. Neuropsychological functioning was also associated with increased scores on the Total Behavioral Problems measure and a decreased scores on the Total Social Competence scale.
Lewis, D. O., Lovely, C. Yeager, and D. Della Femina Toward a Theory of the Genesis of Violence: A follow-up Study of Delinquents 1988	77 more violent and 18 less violent males incarcerated at a correctional school. The mean age was 15.3 yrs. (12.4-17.4). 41% of the subjects were black, 37% white, 21% Hispanic, and 1% Oriental.	A follow-up study (Lewis et al., 1979). Data were taken from adult F.B.I. and state police records.	All subjects, with the exception of six, had an adult criminal record. The average number of adult offenses was 11.58. Seventy-seven percent of the more violent juveniles along with sixty-one percent of the less violent juveniles went on to commit adult aggressive offenses. It seems, then, that juvenile violence, as the only variable, did not predict which individual would go on to exhibit violent criminal behavior as an adult and which would not. The authors suggest studying neuropsychological, cognitive and psychiatric variables; history of abuse, and familial vulnerabilities may also play a role. Studying the interaction of these variables may lead to a more accurate model from which to predict adult violent crime.

151

Table 13 (continued)

Author(s) Title/Date	Population Studied	Procedure	Results/Conclusions
Lewis, D. O., S. S. Shanok, M. Grant, and E. Ritvo Homicidally Aggressive Young Children: Neuropsychiatric and Experiential Correlates. 1983	55 children admitted to a psychiatric service. Primarily a diagnostic service where patients remain 90 days on the average. 21 were homicidally aggressive.	Data were obtained from hospital records, which included family, medical, developmental histories, psychiatric evaluations, physical examinations, neurological assessments, educational assessments, psychological testing, and EEGs (for most). Also re-viewed the types of medications that had been prescribed. Four raters determined from information provided by the chart whether the individual had ever been homicidally aggressive.	Homicidally aggressive children were significantly more apt to have had a seizure, have a father who displayed violent, often homicidal, behavior, have a mother with a history of hospitalization for a psychiatric disorder, and have attempted suicide. The homicidal and nonhomicidal children were not distinguished by psychiatric symptoms and diagnoses. The most significant variable appeared to be whether the child had been exposed to a violent father. Witnessing irrational violence engenders rage that can be expressed through suicidal behavior when turned inward or aggressive (homicidal) behavior when directed outward.
Miller, L. Neuropsychology of the Aggressive Psychopath: An Integrative Review. 1987	Clinical lit. Review paper on aggressive psychopaths	The review has been compiled according to the hypothesis that neuropsychological approaches may be useful on this domain because they allow for elaboration on the relationship between cognitive and affective dimensions, and generate new hypotheses about brain-behavior correlations.	It is hypothesized that the "disinhibited psychopath" suffers from a neurodevelopmental maturational deficit that accounts for decrease in, and practically lack of ability to regulate attention, affect, thought, and behavior through inner speech. In situations involving social frustration or confusion, behavior regresses and more primitive aggressive reaction strategies are employed to cause changes in the social milieu. The author argues that psychopathy is not a "frontal lobe disease," but that derangements in development of the frontal system foundations for the control of attention, cognition, affect, and volition may lead to subsequent disorders in this area that allow for antisocial behavior in a specific group of aggressively disinhibited persons. The author feels that a neuropsychological approach would be complimentary to the psychodynamic and cognitive approaches already in use in studying this area.

152

Table 13 (continued)

Author(s) Title/Date	Population Studied	Procedure	Results/Conclusions
Mungas, D. Psychometric Correlates of Episodic Violent Behaviour. 1988	Three groups of neuropsychiatric out-patients 1. (n=35); frequent, impulsive violence 2. (n=57) non-violent group 3. (n=31) much less frequent violence and more provoked	Subjects were given the Minnesota Multiphasic Personality Inventory (MMPI), the Holtzman Inkblot Technique (HIT), and neuropsychological tests battery measuring cognitive ability. Specifically, four categories of cognitive functioning were tested: attention and concentration, visual-perceptual abilities, memory and new learning, and language-related abilities (these tests included WAIS subtests among others).	Results of this study apply mainly to impulsive, poorly provoked violent behavior. Violent individuals often showed abnormalities in language and visual-perceptual skills and in ability to perceive and evaluate complex environmental situations effectively, but not in other cognitive areas. The groups were not significantly discriminated by IQ measure or by the cognitive index. Violent behavior and poor language skills could be related through lack of adaptive verbal mediation of behavior in environmental situations that are emotionally charged. It may also be that the relationship of language deficits to violent behavior is not a causal one, and that dominant hemisphere dysfunction itself could be the cause. Results of this study support the first hypothesis. The authors strongly support a neuropsychological model in brain abnormality research.
Piacentini, J. C. Language Dysfunction and Childhood Behavior Disorders. 1987	Chapter/ Clinical Lit. Review	The chapter describes the nature of the relationship between language dysfunction and behavior disorders in children. Literature examining the relationship from the standpoint of either domain is reviewed.	Children with language dysfunctions are at higher risk to develop behavior disorders than are children who show no language disorders. Further investigation into the chronology of development of language and speech and behavior disorders would provide necessary etiological information that is beneficial to development of appropriate treatments. The author suggest that population studies should follow in order to determine how environmental, familial, genetic, and biological variables affect this type of relationship.

153

Table 13 (continued)

Author(s) Title/Date	Population Studied	Procedure	Results/Conclusions
Tarter, R. E., A. M. Hegedus, N. E. Winsten, and A. I. Alterman Neuropsychological, Personality, and Familial Characteristics of Physically Abused Delinquents. 1984	101 delinquent adolescents consecutively referred to Western Psychiatric Inst., by a judge, for a comprehensive neuro-psychiatric assessment. 82% were male, 18% female, 35% black 65% Caucasian, 27% physically abused, remaining 74% - control group.	Developmental and familial information was obtained through psychiatric evaluation, social worker's report, probation officers and past records. From this the environmental correlates of child abuse were assessed. A battery of cognitive, behavioral, and personality tests were also given. Cognitive Measures: Wechsler Intelligence Scale for Children-Revised (for 16 and under), Wechsler Adult Intelligence Scale, Peabody Individual Achievement Test, Detroit Tests of Learning Aptitude, Pittsburgh Initial Neuropsychological Test System. Behavioral and Personality Measures: Matching Familiar Figures Test, Minnesota Multiphasic Personality Inventory, Devereux Adolescent Behavior Scale. Familial and Developmental Indices: Family Environment Scale, Family History, Developmental Measures.	Abused delinquents performed relatively more poorly than nonabused delinquents on specific intellectual, educational, and neuropsychological measures. Abused delinquents are more likely to commit assaultive crimes. They present themselves as less domineering but, also, are less likely to exhibit feelings of inferiority. They generally come from more disrupted families, often involving parental alcoholism, criminality, and separation. Cognitive impairments in abused children were primarily focused in verbal or linguistic areas. The authors support the proposed notion (Luria, 1966; Lewis, 1979) that this may "indicate an underlying inability to self-regulate behavior or acquire rule governed behavior." The authors suggest that the distinguishing characteristics found in an abusive family seem to place these abused delinquents at high risk for a poor subsequent adulthood.

154

Table 13 (continued)

Author(s) Title/Date	Population Studied	Procedure	Results/Conclusions
Walsh, A., and J. A. Beyer Violent Crime, Sociopathy and Love Deprivation among Adolescent Delinquents. 1987	131 male delinquents previously on probation with the Ada County Juvenile Probation Dept., Idaho. All were white. Age is a constant since all cases were drawn from inactive files - each contained the subjects delinquency history up to the age of eighteen.	Based on the observation by Wechsler that adolescent sociopaths tend to score higher on performance than on verbal sections of the Wechsler IQ scale. Examined P-V discrepancy, love deprivation, and delinquency and their interaction. Asked social workers and probation officers to rate items in probation files on how well they indicate love deprivation (1-10). Used all items above 6 as indicators and subsequently rated subject's family file. All crimes on each record were scored using the Violence Scale (Andrew, 1978)	Results showed that the a good deal of the explanatory power of characteristically higher scoring on the performance than on the verbal sections of the Wechsler IQ scale among juvenile delinquents with regard to violent behavior is mediated by love deprivation. Low verbal scores, lower full-scale scores, or social class were not significantly related to performance > verbal, only love deprivation was. The authors feel this evidence supports the hypothesis that early emotional stresses affect development of autonomic nervous system function.
Woods, B. T., and M. D. Eby Excessive Mirror Movements and Aggression. 1981	170 inpatients in a child psychiatric unit	170 patients (113 males; 57 females) ages 10 - 15. Received neuropsychological testing which included standardized testing for mirror movements. Patient assigned a DSM II diagnosis of the following were classified as aggressive. 1. Unsocialized Aggressive Reaction. 2. Impulsive, Explosive, Sociopathic or Antisocial or Dissocial Relevancy or Character Reaction. 3. GAP diagnosis of Tension Discharge Disorder	Aggressive patients, especially the subgroup of aggressive males, had significantly more mirror movements than the nonaggressive patients. The authors feel that this may reflect an underlying effect in brain function: "a lack of inhibition of inappropriate activity."

155

Table 13 (continued)

Author(s) Title/Date	Population Studied	Procedure	Results/Conclusions
Yeudall, L. T., D. Fromm-Auch, and P. Davies Neuropsychological Impairment of Persistent Delinquency. 1982	99 adolescents (64 males; 35 females) were consecutively admitted to the Youth Development Centre. Mean age = 14.8 years (13 - 17) for delinquent group. Eight were on medication at the time of testing. High percentage of nonprescription drug usage. The nondelinquent group: 47 adolescents from regular classrooms (29 males; 18 females). Mean age = 14.5 yrs.	Subjects received the Halstead-Reitan Battery along with 12 other neuropsychological tests (total = 40 tests in standard order) Ages 16 and under were required to complete the Wechsler Intelligence Scale for Children-Revised. Subjects ages 17 and up were given the Wechsler Intelligence Scale.	84 per cent of the profiles were abnormal in the delinquent; 11 per cent were abnormal in the non-delinquent group. In other words, a high percentage of delinquents showed a neuropsychological deficit. There was a high degree of statistical difference between the control and delinquent groups based on neuropsychological and psychological test scores. Following further analysis of these deficits, it appeared that these deficits implicated anterior dysfunction that is greater in the nondominant than in the dominant hemisphere. Using these results the authors suggest that delinquents may be deficient in ability to plan their actions and perceive consequences of these actions. These conclusions are discussed with regard to a low number of violent adolescents in the sample as compared with a high percentage of delinquents showing signs of depression.

156

Table 13 (continued)

Author(s) Title/Date	Population Studied	Procedure	Results/Conclusions
Kindlon, D., Sollee, N. and Yando, R. Specificity of Behavior Problems Among Children with Neurological Dysfunction. 1988	248 children, 172 boys and 76 girls, aged 4 - 16 years. They were broken down into three groups: the neurological dysfunction (ND) group comprised 81 children who had to have either (1) a diagnosis of neurological dysfunction such as cerebral palsy, seizure disorder, or head injury with resultant coma lasting at least 1 week or (2) findings definitively indicative of brain damage on the EEG, CT scan, or BEAM. The specific developmental disorders group (SDD) - based on DSM-II designation - comprised 167 children. They showed no hard evidence of neurological dysfunction, or history of serious head injury, or abnormalities on EEG, CT scan, or BEAM. The psychiatric clinic populations (PC) comprised 856 children described by Achenbach (1978). All groups were broken down by age and sex: ages 6-11 and 12-16.	Parents or guardians of the children in the ND and SDD groups completed the CBCL (either at the time of evaluation or three years later - retrospectively). This same data were obtained for the PC group from Achenbach (1979). Children in each group were divided into three levels of socioeconomic status.	Children in the ND and SDD groups had a higher rate of disturbance on many factors, but they exhibited less aggressivity and delinquency than did the PC group. Although children with neurological dysfunction (as a group) show various psychological problems, these problems are less likely to be characterized as externally directed aggressive types of behavior. SDD and ND samples are more alike than PC. This indication of lower Aggressiveness and Delinquency for the nonpsychiatric groups is particularly apparent in the younger ages. For both sexes in the 6-11 age group, differences were not explainable on the basis of socioeconomic status. On the contrary, in boys in the 12-16 age group, lower socioeconomic status was associated with higher scores on the Aggressiveness and Delinquency

157

Table 14 Genetic Determinants of Violence

Author(s) Title/Date	Population Studied	Procedure	Results/Conclusions
Mednick, S. A., and Kandel, E. S. Congenital Determinants of Violence. 1988	173 recidivistically violent offenders who had committed two or more violent offenses (potential "specialists") were isolated from a Danish birth cohort consisting of 31,436 men born in Copenhagen between Jan. 1944 and Dec. 1947. 28,879 were still alive at the time of the study. At 27 years of age 37.8% (10,918) had had at least one police contact for a criminal law offense; 2.5% (721) had committed at least one violent offense and 173 had committed two or more violent offenses These 173 account for 0.6% of the 28,879 and were held accountable for 43.4% of the violent offenses. For the adoption study, also in Denmark, a birth cohort of 14,427 nonfamilial adoptions from 1924-1947 were used.	The authors examine various types of studies designed to determine whether congenital factors contribute to a predisposition to repeated violent behavior. These include family, twin and adoption studies. The authors include both inherited characteristics and perinatal experiences in their idea of contributing congenital factors. They feel adoption studies "provide the most fertile ground for study." For each adoption psychiatric hospital diagnoses and court conviction histories were recorded for the adoptee, the biological mother and father, and the adoptive mother and father. Occupation was included as a record of socioeconomic status.	Calculations showed that a first-time violent offender was 1.94 times more likely to commit a violent act in the future than was a first-time property offender. When analyses were limited to offenses committed prior to 18 years of age the effect was still prominent, thus, the data indicate that "specialization" for violence can also be found in juvenile offenders. The authors define specialization as such: "Specialization has been observed if an individual who commits a violent offense is more likely to commit a subsequent violent offense than an individual who commits a property offense." Results of the adoption study showed a "definite relationship between biological parent and adopted-away son for property convictions, but there is no significant relationship for violent offenses." A study by Cloninger et al. (1987) found similar results. However a study by Moffit (1984) did show a genetic effect for violence. As far as perinatal factors are concerned, the authors briefly describe two studies (Mednick, 1983 and Drillie, 1964) which found that perinatal problem indices related to later violent crime than property offenses. They also showed, however, that stable family rearing appeared to compensate for perinatal damage.

158

Table 14 (continued)

Author(s) Title/Date	Population Studied	Procedure	Results/Conclusions
Walker, E., Downey, G., and Bergman, A. The Effects of Parental Psychopathology and Maltreatment on Child Behavior: A Test of the Diathesis-Stress Model. 1989	Of a total subject pool of 144 children included in a longitudinal study of risk factors in development, 53 boys and 49 girls were selected for the present study. Mean ages were 9.75 and 9.24 yrs. respectively. 44 of the boys and 40 of the girls were located for a second behavioral evaluation. Families were distributed across three groups (schizophrenia, psychiatric control, and normal control). Number of children in each parental diagnosis group were as follows.Schizophrenia/ maltreatment: 5 boys and 4 girls Schizophrenia/no maltreatment: 9 boys and 6 girls; psychiatric comparison/maltreatment: 9 boys and 5 girls; psychiatric comparison/ no maltreatment: 5 boys and 4 girls; normal comparison/maltreatment: 8 boys and 16 girls; normal comparison/no maltreatment: 17 boys and 14 girls.	This study was designed to investigate "the main and interactive effects of parental psychopathology and maltreatment on child behavior." The Achenbach Child Behavior Checklist was used. Parents (of the 84 for whom a second evaluation was obtain) completed a the CBCL a second time one year later. It was often necessary to rely on the disturbed parent to complete the CBCL since many households were single-parent homes. Despite this possible weakness, the ratings did show "adequate internal consistency and test-retest reliability." Two variables were used as measures of quality of rearing environment: maternal educational level and household composition (single- vs. two-parent households).	The emphasis in this study was on the importance of investigating multiple risk factors simultaneously. The authors were interested in testing the assumptions of the diathesis-stress model. Boys from maltreating homes displayed significantly greater "externalized behavior problems". These included aggression and delinquency (first assessment). Girls from maltreating homes displayed increased aggression as well (first assessment). The combination of parental schizophrenia and maltreatment was associated with a progressive increase in delinquency in both sexes over time. "These results support a diathesis-stress model of psychopathology."

159

Table 15 Lead as a Cause of Behavioral Pathology

Author(s) Title/Date	Population Studied	Procedure	Results/Conclusions
Lansdown, R. Lead, Intelligence, Attainment and Behaviour. 1986	Lit Review	Lit Review	The body of large-scale surveys suggests at least a possibility that lead is causally related to deficits in cognitive functioning. Even the most recent epidemiological studies have failed to produce convincing consistent evidence for an association between moderate levels of lead and behavioural patterns in general; even less is the evidence on lead and hyperactivity. There have been several suggestions and the work on chelation means that the topic should not be dismissed. More than this cannot be said." The author claims that many of the early investigations in this field were based on a population of socially disadvantaged children. He points out that the more recent studies in Germany and Britain have indicated that there may be "virtually no causal relationship among children from more advantaged homes." If this is proven to be true, perhaps the contradictions in the earlier studies can be explained.

Table 15 (continued)

Author(s) Title/Date	Population Studied	Procedure	Results/Conclusions
Needleman, H. L., and Gatsonis, C. A. Low Level Lead Exposure and the IQ of Children: A meta-analysis of modern studies. 1988	12 studies of childhood exposure to lead and intellectual performance. Each study employed multiple regressions analyses with IQ as the dependent variable, lead as the main effect, and controlled for non-lead covariates. 12 other studies were excluded from the analysis because of inadequate analysis, over-control or under-control of co-variates or clinical toxicity. Of these 5 showed a lead effect and 7 did not. Of those used for the meta-analysis 7 used blood lead levels and 5 used tooth lead levels - they were classified as such.	A meta-analysis of 12 studies drawn from all studies on lead exposure and children's neurobehavioral development published since 1972. 8 studies measured IQ by the WISC-R scales, two used the Stanford Binet IQ scales, one used the British Ability Scales, and one used the McCarthy Scales.	Of four previous reviews on the studies of lead at low dose one came to a qualified negative conclusion, one to a positive conclusion and two found the evidence inconclusive. These reviews were inconsistent partly due to a "limitation inherent in the method of narrative reviewing." By using the meta-analysis method, 11 of the 12 studies reported a negative coefficient for lead (the range of the effect size measured by the t-value for the lead term was -3.86 to -.03). Sample sizes ranged from 75 to 724. Joint p-value for the 7 blood lead studies was <.0001 (for all three methods of analysis). This did not change when any of the 7 studies was excluded. Joint p-value for the 5 tooth studies fell between .004 and .0003. When any one study was excluded the p-value for the remaining 4 ranged from .025 to <.0005. "The hypothesis that lead impairs children's IQ at low dose is strongly supported by this quantitative review. The effect is robust to the impact of any single study."

161

Table 15 (continued)

Author(s) Title/Date	Population Studied	Procedure	Results/Conclusions
Needleman, H. L., Schell, A., Bellinger, D., Leviton, A. and Allred, E. Long Term Effects of Childhood Exposure to Lead at Low Dose; An Eleven-Year Follow-Up Report 1989	132, of an original 270 children tested in 1979, 18 to 19 year olds.	Dentine levels measured in 1976-1977 for each subject were used to for computing mean concentrations. Venous blood lead levels were obtained in the present study as well. After the first 48 subjects, in which lead levels exceeding 7 μg/dl did not exist, venous blood withdrawal was discontinued. One examiner, blind to lead status, rated subjects on behavior. Subjects were administered the CPT, Symbol digit substitution, hand-eye coordination, simple visual reaction time, finger tapping, pattern memory, pattern comparison, serial digit learning, vocabulary, switching attention, mood scales, California Verbal Learning Test, Boston Naming Test, Rey-Osterreith Complex Figures Test, Word Identification Test, Self Report of Drug Use, Self Report of Delinquency, and review of school records was carried out.	The present study showed that the effects of lead on academic progress and cognitive function found in earlier studies continue to be apparent in this population as young adults. A seven-fold increase in failure to graduate from high school was found along with lower class standing, greater absenteeism, impaired reading skills (scores 2 grades below expected - which qualify as reading disability), deficits in vocabulary, fine motor skills, reaction time and hand-eye coordination. All of these indicate "a serious impairment in life success." The estimates of cognitive and academic difficulties made on the basis of these 132 subjects taken from the original sample are probably conservative. Those not tested in the present study tended to have more lead, lower IQ scores, lower teachers' ratings and were generally of lower socioeconomic status than those tested. "The association between lead and outcome reported here meet six observational criteria that support the validity of causal inference: proper temporal sequence, strength of association, presence of a biological gradient, non spuriousness, consistency and biological plausibility."

Table 15 (continued)

Author(s) Title/Date	Population Studied	Procedure	Results/Conclusions
Thomson, G. O. B., Raab, G. M., Hepburn, W. S., Hunter, R., Fulton, M., and Laxen, D. P. H. Blood-Lead Levels and Children's Behaviour--Results from the Edinburgh Lead Study 1989.	501 boys and girls ages 6-9 years in ta defined central Edinburgh area. 43% of the families of these children were in social class I or II and 85% owned their own homes.	These 501 children were tested in school by a trained psychologist. The British Ability Scales and tests of mental speed were administered. The children's teachers completed the Rutter behavior questionnaire. The family of each child was interviewed and one parent (usually the mother) received an ability test. Shed deciduous teeth were collected and the child's exposure to environmental lead was assessed.	Results showed a significant relationship between blood-lead and measures of deviant behavior taking into account any confounding variables. The 501 children tested had a mean blood-lead level value of 10.4 μg/dl. The measures of deviant behavior were shown to be influenced by sex, mother's performance on a matrices test, history of family disruption and total number of cigarettes smoked in the household. There was a stronger effect for boys than for girls in a lead by sex interaction, but evidence for such an interaction is limited. Data from the Edinburgh study suggest that a small tendency exists for an association between blood-lead and deviant behavior even after controlling for confounders. "This relationship may reflect a causal association whereby low level lead exposure acts to influence deviant anti-social and hyperactive behaviour in pupils." Lead and behavior could possibly be associated by reverse causality which means the way children behave may lead to variations in body lead burden. "The hyperactive, acting out aggressive child may well behave in ways which increase his/her lead levels. Despite this caveat the results reported here add to the growing evidence that lead at low levels of exposure probably has a small but harmful effect on children's behavior."

Table 16 Alcohol Ingestion in Pregnancy--Cognitive Effects

Author(s) Title/Date	Population Studied	Procedure	Results/Conclusions
Streissguth, A. P., Martin, D. C., Barr, H. M., and MacGregor Sandman, B., Kirchner, G. L., Darby, B. L. Intrauterine Alcohol and Nicotine Exposure: Attention and Reaction Time in 4-Year-Old Children. 1984	452 singleton-born children - a 4-year-old follow-up cohort. The present study is part of a longitudinal prospective study which started when the mothers were pregnant. Those included in the longitudinal study were children of heavier drinkers, more moderate drinkers, infrequent and nondrinkers. For the present evaluation, subjects were 4 years and 3 months of age. The mothers were predominantly white (87%); married (86%); and middle-class (80%). For the 4-year olds in the present study, follow-up rate was 86% from those seen at 8 months.	Independent variables measured were maternal alcohol use, maternal nicotine use, maternal caffeine, maternal drug use, and maternal diet. A psychometrist, blind to the drinking history of the children's mothers and prior assessments of development, administered a vigilance task at the end of a 1.5 hr test battery. The child was required to press a button each time a kitten appeared on the window of a Victorian house. The psychometrist recorded the total amount of time (in the 13 minute task) that the child was nonoriented to the stimulus board and whether the child was oriented to the apparatus at each stimulus presentation.	Both alcohol and nicotine were related significantly to errors of omission and commission, and to the ratio of correct responses to total responses. "These findings held up even when the test for alcohol was adjusted for nicotine, when the test for nicotine was adjusted for alcohol, and when both were also adjusted for maternal caffeine use, nutrition and education and child's birth order." Alcohol effects were still present upon removal of the children of the smoking mothers from the analysis leaving 248 subjects. Upon deletion of the heavier drinkers from the analysis, leaving 285 subjects who were children of lighter drinkers, a significant nicotine main effect remained. That increased attentional errors and longer reaction time occurs in 4-year-old children exposed to heavier doses of alcohol in utero is a new finding. These children pressed to significantly fewer of the target stimuli and made significantly more extraneous button presses. They did not differ significantly in their orientation to the target stimuli. Speed of responding, in these children, became significantly slower as the session went on. On the contrary, nicotine exposed children showed significantly less frequent orientation to the target stimuli. Their errors of omission could be explained by this. Alterations in reaction time were no associated with nicotine exposure. Decreased orientation and attention in children exposed to nicotine is consistent with studies carried out by Denson et al. (1975) and Nichols and Chen (1981) that showed a relationship between hyperkinesis in young children and maternal smoking during pregnancy. "Attentional errors assessed from a vigilance paradigm such as the one used in the present study have been previously associated with poor academic achievement (Kirchner and Knopf, 1974) and hyperactivity (e.g., Porrino et al., 1983) in school-age children." These attentional errors are considered to be the basis of learning disabilities found in school-age children (Doyle, 1976).

164

Table 16 (continued)

Author(s) Title/Date	Population Studied	Procedure	Results/Conclusions
Streissguth, A. P., Barr, H. M., Sampson, P. D., Parrish-Johnson, J. C., Kirchner, G. L., and Martin, D. C. Attention, Distraction and Reaction Time at Age 7 Years and Prenatal Alcohol Exposure. 1986	486 subjects of the original 500 that began the longitudinal study (Streissguth, 1981) were brought in for the 7.5 year examination. Of these, 475 had a least partial valid data for this study. This sample included 255 boys and 220 girls. Age range was from 6.5 to 8.5 yrs.	Again the independent variables were maternal alcohol use, maternal smoking, maternal caffeine use, maternal drug use and maternal diet. The subjects were administered the CPT vigilance task. They participated in both the X and the AX tasks. Subjects were examined, after a 2.5 hour psychological battery by one of eight examiners blind to prior developmental assessments and mother's scores on independent variables.	Following statistical adjustment for a variety of variables that may be potentially confounding, results showed that prenatal alcohol exposure was significantly related to attentional deficiencies and reaction time. The results remained "essentially unchanged" even in the presence of various potential confounders. Levels of alcohol exposure where effects on attention can be observed appear to depend on the task and the type of alcohol score used. "Among the tasks derived from this CPT paradigm, MRT and EC-AX (EC = errors of commission), by virtue of their highly significant partial correlations with terms linear in alcohol exposure, were the most sensitive attentional outcomes for assessing the long term effects of prenatal alcohol exposure." It seemed for these two measures that the magnitude of the effect increased with increased exposure or increased number of drinks per occasion. The authors suggest that strong conclusion can only be drawn from these two measures and the vigilance summary score. The fact that EC-AX seemed to be a more sensitive result than EC-X may be due to the subjects' inability to withhold a response (i.e. responding to A in anticipation of X). This would be consistent with the clinical observation (made by the authors) that children suffering from FAS are frequently uninhibited and impulsive (1985). An alternative explanation may be that EC-AX was the last task in the battery and attention was probably waning. Observations made by examiners on distractibility also showed that the most highly exposed children were most easily distracted. The authors state that an important consideration in choosing outcome variables for such studies is that they are suitable for all subjects in the population. In the present study, one child out of 475 was not able to take the CPT task and only 3% showed some resistance to it. They conclude that the CPT is a "suitable endpoint" for attentional studies in these children.

165

Table 16 (continued)

Author(s) Title/Date	Population Studied	Procedure	Results/Conclusions
Streissguth, A. P., Sampson, P. D., and Barr, H. M. Neurobehavioral Dose-Response Effects of Prenatal Alcohol Exposure in Humans from Infancy to Adulthood. 1989	Study 1 included 92 subjects with FAS or FAE (Fetal Alcohol Effects) who were 12 yrs. of age or older. Age range was 12 - 42 with a mean of 18.4 years. 58 were diagnosed as having FAS and 34 as having FAE. 61% were male, 55% were reservation Indians, 22% nonreservation Indians, and 23% non-Indians. Study 2 included a cohort of about 500 children whose mothers were interviewed in the fifth month of pregnancy Of the 1,500 women interviewed all heavier drinkers and a proportion of moderate, light, and infrequent drinkers and nondrinkers were included. The women in this sample were predominantly white (87%) married (86%), middle class (80%), and well educated (58% - some college)	Study 1. Each subject was administered the following tests (standardized conditions were maintained): an IQ test appropriate for the age of the subject, WISC-R or WAIS-R; an auditory receptive ability test; the PPVT; an adaptive and maladaptive behavior test; the VABS; and the Symptom Checklist (SC) developed for this study. Data are not available for all subjects on all tests. Study 2. To measure maternal alcohol, a quantity-frequency-variability interview was conducted and scored according to 25 alcohol scores. Alcohol scores were taken twice, once "during pregnancy" and "prior to pregnancy" (a month or two prior). Both were self reports. The use of cigarettes, caffeine, and other drugs was also taken into consideration. Information on other variables such as major life changes mother-child interaction, age of siblings, injuries and illnesses, etc. were obtained to assess influence on development. A full list of the 150 covariates examined here can be gotten from Streissguth et al. (1986). Dependent variables were assessed on the first and second day, at 8 and 18 months, and at 4 and 7 years.	"Prenatal alcohol exposure produces a wide variety of effects on offspring including intellectual decrements, learning problems, attentional and memory problems, fine and gross motor problems, and difficulty with organization and problem solving. In patients with FAS/FAE, psychosocial problems are observable in adolescence and adulthood that may have their roots in early cognitive deficits. Psychosocial problems associated with moderate exposure levels have not yet been evaluated." In general, prenatal alcohol exposure effects on neurobehavioral variables show a dose response relationship where "high levels of exposure are associated with large magnitude effects, while moderate levels of exposure are associated with more subtle effects." Self-reported binge drinking (5 drinks or more at a given time) and self-reported drinking in the period prior to pregnancy recognition are two of the strongest predictors of later neurobehavioral deficits. No evidence was found that these effects are due to confounding with other drugs or can be accounted for by a small group of outliers. "The comparable findings from the clinical study, the epidemiologic study and the animal literature present convincing evidence of the neuroteratogenicity of alcohol and the long-lasting effects on prenatally-exposed offspring."

Table 17 Cocaine, Opiates, and Tobacco: Effects on Cognitive Development

Author(s) Title/Date	Population Studied	Procedure	Results/Conclusions
Chasnoff, I. J., and Griffith, D. R. Cocaine: Clinical Studies of Pregnancy and the Newborn. 1989	Two groups of cocaine using women. Group 1 comprised 23 women who reached abstinence by the end of the first trimester and did not use cocaine for the remainder of the pregnancy. Group 2 comprised 52 women who used cocaine throughout the pregnancy.	The neonates were all examined at birth by a physician who did not know of the prenatal history. Weight, crown to heel length, and fronto-occipital head circumference were measured. When the infants reached 12 to 72 hours of age the Neonatal Behavioral Assessment Scale was administered.	Performances on the NBAS showed that the children of both groups of women demonstrated impairment in motor ability, orientation, state regulation, and number of abnormal reflexes. Group 1 showed significantly poorer performance on motor cluster than Group 2. 7 of 16 in Group1 and 8 of 36 in Group 2 could not reach alert states at all during testing and thus were unable to engage in any orientation. Those infants in Group 1 were significantly more fragile and "less robust" in capability to complete the testing. This study did show that reaching abstinence from cocaine use at the first trimester increased the number of pregnancies carried to term and improved obstetric outcome. Results of this study support findings that "exposure to cocaine during the prenatal period leads to significant impairment in neonatal neurobehavioral capabilities." The greater impairment in Group 1 infants is difficult to explain. The authors suggest replicating the study with larger samples.

167

Table 17 (continued)

Author(s) Title/Date	Population Studied	Procedure	Results/Conclusions
Chasnoff, I. J., Burns, W. J., Schnoll, S. H., and Burns, K. A. Cocaine Use in Pregnancy. 1985	23 infants born to cocaine-using women. The women were divided into groups according to use or nonuse of narcotics along with the cocaine. These groups were compared to two control groups. Group 1 was made up of 12 women who had conceived while using cocaine, had not history of opiate use, 4 used alcohol at least twice monthly, 6 used marijuana at least 3 times monthly in the first two trimesters of pregnancy. 7 smoked cigarettes throughout pregnancy. Group 2 comprised 11 women who had conceived while using both cocaine and heroin. 2 used alcohol at least twice monthly, 5 used mari-juana at least 3 times monthly through the first two trimesters, and 8 smoked cigarettes throughout pregnancy. Group 3 comprised 15 women who conceived while using heroin. 3 used alcohol at least twice monthly, 7 used	marijuana at least three times monthly in the first two trimesters, and 11 smoked throughout pregnancy. Group 4 comprised 15 women who did not abuse drugs. However, 3 used alcohol at least twice monthly, 7 used marijuana at least three times monthly in the first two trimesters, and 10 smoked cigarettes throughout pregnancy. Group 2 and Group 3 women were started on methadone treatment upon admission to the Perinatal Addiction Project. About 60% of the women in Groups 1 and 2 used cocaine throughout pregnancy. All women except those in Group 4 were enrolled in the Perinatal Addiction Project. Experimenters reviewed the reproductive history of all of the women in the study. Groups 1 and 2 had used cocaine in all prior pregnancies, and Group 3 women had used opiates. At birth all neonates were weighed, measured from crown-to-heel, and fronto-occipital head circumference was recorded. At three days of age, the infants were administered the Brazelton Neonatal Behavioral Assessment Scale by examiners blind to the child's prenatal history.	Results showed that infants exposed to cocaine in utero showed and increase in depressed interactive abilities and significant impairment in organizational abilities as compared with infants whose mothers used methadone and control infants. "Cocaine exposure in utero interferes with an infant's ability to maintain adequate state control in the neonatal period. This puts the infant at high risk. Group 2 infants displayed weaker reflexes and decreased state control than did control infants (Group 4), but they showed no significant deficits in orientation, neither auditory nor visual. The authors feel that this may be the result of a Type II error. The interaction of cocaine and methadone in these infants may have been antagonistic since one is a CNS depressant and one a CNS stimulant. Alcohol, marijuana, and nicotine use in all four groups was similar and thus could not account for discrepancies in the findings.

168

Table 17 (continued)

Author(s) Title/Date	Population Studied	Procedure	Results/Conclusions
Olofsson, M., Buckley, W., Andersen, G. E., and Friis-Hansen, B. Investigation of 89 Children Born by Drug-Dependent Mothers. 1983	A reinvestigation of 72 of 89 previously examined children 11 months to 10 years of age (mean = 3.5 yrs.). 62 mothers (10 of which had 2 children among the 89). None of the mothers had been drug-free for 5 years or more. 19% had been drug-free for 14 days to 5 years. 45% were taking mainly i.v. opiates, 26% were taking methadone, and 10% were using minor tranquilizers. 66% had no job, 16% were working, 18% were in prison on sick leave or participating in an educational program.	Psychomotor development of the children was determined by the Denver Development Screening Test (DDST). Data on physical, social and behavioral history were obtained from interviews with guardians and/or mothers, and other professionals involved in the child's life (i.e. private practitioners visiting health nurses, social welfare authorities and school- and day-care personnel.	Fifteen children (21%) showed an impaired psycho-motor-development. Two of the fifteen retarded children showed major organic abnormality and in another four minor brain damage was suspected. 56% were considered behaviorally abnormal. Lack of concentration, hyperactivity, aggressiveness and lack of social inhibition were the predominant signs.

169

Table 17 (continued)

Author(s) Title/Date	Population Studied	Procedure	Results/Conclusions
Rush, D. and Callahan, K. R. Exposure to Passive Cigarette Smoking in Child Development. 1989	Extensive Literature Review.	The authors looked at the designs of and highlight strengths and weaknesses of all of the studies on maternal smoking and fetal development known to them. They then summarized relationships with somatic, cognitive and behavioral development across all studies. Detailed results are provided in tables within the text.	There is a consistent pattern of depressed cognitive development and tests of school achievement associated with maternal smoking during pregnancy. It still remains beyond current knowledge that these are causally related though a trend is apparent. A consistent pattern of behavioral abnormalities are also reported. Several studies are cited. Again, a causal relationship was not supported based on the available data. A strong and highly significant relationship between amount of cigarette smoking prior to pregnancy and the psychomotor developmental index was observed.

Table 17 (continued)

Author(s) Title/Date	Population Studied	Procedure	Results/Conclusions
Wilson, G. S., McCreary, R., Kean, J. and Baxter, J. C. The Development of Preschool Children of Heroin-Addicted Mothers: A Controlled Study. 1979	77 children, 40 boys and 37 girls, 3 years 1 month to 6 years 4 months of age (mean = 4 years 7 months). 30 had Latin American surnames, 30 were black, 17 were Anglo-American. Three groups were compared with the heroin exposed group. Heroin exposed group: 22 children of mothers who used heroin (as the predominant drug) continuously throughout pregnancy. In one case use of all drugs was terminated at one month of gestation, in another methadone was substituted at six months, and another two substituted other drugs in the last trimester. 7 abused other along with the heroin. Drug environment group: 20 children of mothers who did not use heroin during pregnancy but were involved in the "drug culture". The high-risk comparison group: 15 children labelled as such because of medical factors such	Hospital records were examined for maternal drug use, obstetric complications, amount of prenatal care, complications of labor or delivery, one-minute Apgar score, gestational age, birth weight, type and severity of nursery morbidity, and duration of hospitalization. They received a general physical exam and a neurological evaluation. A structured social service interview was carried out at each child's home. A 19-item scale (Kasmar and Altman) was used by a social worker to rate physical environment of each home. Parental Attitude Research Inventory was given. Standardized tests used: Illinois Test of Psycholinguistic Abilities, the Columbia Mental Maturity Scale, the McCarthy Scales of Children's Abilities, and the Minnesota Child Development Inventory. A perceptual battery designed by Deutsch and Schumer was modified and administered. Parents completed the three subtest of Child Behavior Rating Scales. Pediatricians rated the subjects on alertness, cooperation, attention, activity level, and intensity. During a 5-minute free-play period children were videotaped and rated on attention, activity level, cooperation, independence and confidence by a psychologist. Speech was assessed by a speech pathologist during the	No significant difference was found in the educational attainment, occupational level, and Hollingshead index of social position of the families. Heroin exposed children lived with a substitute mother more commonly than children in any of the three comparison groups. No group differences were found in physical environment. IQ did not differ between groups. Heroin-exposed children showed poorer performance than comparison groups on the General Cognitive Index, on three out of five subtests of the McCarthy Scales of Children's Abilities (perceptual performance, quantitative and memory), on measures of visual, tactile, and auditory perception. The McCarthy tests of skills are "considered organizational processes in the ITPA categorization. They require attention, concentration, short-term memory and the internal manipulation of symbols. Groups did not differ significantly on verbal and motor scales. They were rated as more active, and by parents as having increased difficulty in self-adjustment, social-adjustment and physical-adjustment areas. Group differences could not be based on age, sex, ethnic group, socioeconomic status, or participation in school readiness programs. These were controlled for initially. "Behaviorally, the problems of the heroin-exposed group were related to impulsiveness, aggressiveness, and peer relations. These behavior problems may also be manifestations of impaired attention and organizational abilities."

171

Table 17 (continued)

Author(s) Title/Date	Population Studied	Procedure	Results/Conclusions
(continued) Wilson, G. S., McCreary, R., Kean, J. and Baxter, J. C. The Development of Preschool Children of Heroin-Addicted Mothers: A Controlled Study.	as dysmaturity, intra-uterine growth retard-ation, fetal distress, and disturbed transition. Mothers claim to have abstained from psycho-tropic drug use at any time. Finally, the socio-economic comparison group: 20 children who were born without com-plications. There were no significant group dif-ferences in age, sex or socioeconomic status.	same doll play task.	

172

Hormonal Aspects of Aggression and Violence

Paul Fredric Brain

Relationships between aggressive behavior and the endocrine system have been studied intensively in recent years. This interest has occurred presumably because hormones are naturally occurring secretions of the bodies' endocrine or ductless glands, and are perceived as providing possibly reversible (certainly when compared to psychosurgery) therapies for some clinical conditions that include hyperaggressiveness as a symptom. Hormones are transported throughout the body by the blood stream and represent the slow and chronic component of the neuroendocrine coordinating system that regulates physiological and behavioral activities.

AGGRESSION AND VIOLENCE

TERMINOLOGIES OF AGGRESSION

To provide a critical evaluation of developments within this area, bringing together material from animal and human studies, it is essential initially to clarify some terminologies. Brain (1990b), as well as others (Buss, 1971; Kutash et al., 1978; Goldstein,

Paul Fredric Brain is at the School of Biomedical Sciences, University College of Swansea, United Kingdom.

1986; Huntingford and Turner, 1987; Archer, 1988; Klama, 1988; Browne and Archer, 1989), have recently reexamined the nature of aggression for a general audience. In both animal and human sciences, terms such as "aggression" and "violence" are used with enormous flexibility, making it difficult to tie down firm associations with biologic factors (such as hormones).

Potential for Harm or Damage

The one attribute of aggression about which everyone agrees is that the action must, at least, have the potential for harm or damage. Yet what do we mean by harm? Does harm include only physical harm, or can it include emotional damage or reduced breeding potential? There are behavioral responses that clearly involve harm or potential harm and receive labels other than aggression. For example, harm is definitely involved in predation—an activity that is generally distinguished from aggression by ethologists (students who emphasize behavior's role within the organism's natural environment). Predation is often, but not exclusively, an activity involving members of different species and generally does not involved marked arousal (see below). Harm is also a potential consequence of defensive responses by animals. Consequently, the potential for harm is insufficient cause for an action to be labeled aggression.

Having said this, one could make a convincing case for examining the possibility that behaviors more associated with predation deserve consideration in accounts of human violence. Humans are clearly designed to be omnivorous and do (in most cultures) obtain at least part of their food by predatory behaviors. Indeed, a surprisingly large number of infrahuman primate species are also not averse to taking the occasional prey item. Although accounts of human cannibalism are often rendered rather lurid in popular writing and the activity clearly has a semireligious component in many of the cultures that practice (or used to practice) it, this kind of "predation" has been described in a range of cultures. One can also add that the behaviors of certain psychopaths (efficient killing without many visible signs of emotional arousal) seem to fit the ethological description of predation rather than that of social aggression. Perhaps the detailed studies of the biologic factors involved in activities such as mouse killing by rats (e.g., Karli, 1981) are of relevance to some types of human violence even if they are not aggression.

Intentionality

Intentionality is another feature necessary, according to some authorities, to identify aggression, but it is generally difficult to establish whether responses are deliberate or not. Some authorities maintain that the motives of the "aggressor" are actually unimportant—what matters is whether the "victim" regards the action as intentional or not. Others go one stage further and maintain that a "dispassionate observer" (i.e., an individual outside the encounter) is a better judge of aggression. Although the best way of distinguishing between intentional and accidental acts is to consider the probability of the particular event, one must note that different individuals vary in their willingness to see particular responses as intentional.

Arousal

Biologists generally maintain that aggression has to involve arousal. Arousal is a psychological term applied to evidence of internal changes including alterations in heart rate, respiration, and the distribution of blood in the tissues. Charles Darwin, as early as 1872, advocated that one could deduce something about the arousal state of animals by looking for evidence in postures; the position of hairs, feathers, or combs; and the production of sounds (e.g., spitting, snarling, and crowing). Indeed, one can see animals (e.g. "cornered," subordinate dogs) that are simultaneously fearful and likely to attack. Some authorities (e.g., Scott, 1981; Huntingford and Turner, 1987) have expressed a preference for the term *agonistic behavior* to cover the range of activities (from overt attack to fleeing, submission, and defense) evident in social conflict. There certainly is some merit in this position since organisms in social conflict encounters generally fluctuate between a range of activities.

One should not rely exclusively on this "body language" because one may misinterpret postures and "facial" expressions in animals. In studies of humans, individuals may simulate emotional expressions.

Aversiveness

A final proviso needed, before some authorities will accept that an act is aggressive, is that the "victim" must regard the action as something to be avoided. This requirement is intended to get around the difficulties of sadomasochism in humans and the use of "love darts" by snails, which cause slight tissue damage but appear to facilitate courtship in these hermaphrodite animals.

AGGRESSION AS A CONCEPT

A basic problem with the everyday use of the term aggression is that people generally think they are discussing an entity ("thing") rather than using a concept. We humans essentially have to deal with a complex world in which a vast array of so-called independent variables (potential causes) may be related to an equally large collection of dependent variables (potential consequences). Humans are not computers, and they attempt to make sense of the world by creating intervening variables that link together groups of independent and dependent variables. The concept of aggression is one of these intervening constructs. The trouble with concepts is that they are theoretically definable in many ways—one does not assess a concept by its accuracy but by its usefulness (as an explanatory device).

Aggression and Communication

Aggression in animals involves communication with any or all of the sensory modalities (as we shall see later, hormones can influence the cues used in such communication and the sensitivities of the sensory systems that respond to them). It has also become apparent that most species have a range of threatening and attack-related activities that can be used in different contexts or for different purposes.

Utilities of Aggression

It is recognized that animals fight and threaten for a wide range of reasons, such as selection of mates, obtaining exclusive access to an area (territory) that is a prerequisite for breeding, gaining status within a social hierarchy, or defending themselves from conspecifics and predators. Status determines the animal's ease of access to a mate, food, water, or nest sites. One misconception is the view that, because particular animals may employ aggression to obtain a mate, territory, or elevated social status, behaviors receiving the same label in humans necessarily serve one or more of these functions. There is little evidence that humans are intrinsically territorial, always obtain their mates by crude physical competition, or attain high social status by attacking other individuals. The serious dangers of simplistic extrapolations from animals to humans have been well explored.

Different Tests for Animal Aggression

Another complicating feature of dealing with animal aggression is the striking diversity of tests said to measure this attribute in particular species (see Brain, 1981, 1989b). In, for example, the laboratory mouse, aggression is said to be generated by pairing preisolated males (intermale aggression), by exposure of paired males or females to unavoidable foot or tail shock (shock-elicited aggression), by arranging for an unfamiliar intruder to enter the nest area of a lactating female with her offspring (maternal aggression), by placing a lactating female (or an animal marked with her urine) into an established group of females or castrated males, by giving the subject the opportunity to kill a locust or a cricket (predatory aggression), and by confining subjects in a narrow tube where they may bite a target suspended in front of them thus activating a telegraph key (instrumental aggression). Thus even in the "simple" mouse, the tests used to generate aggression are so varied (and the responses generated so qualitatively different), it is highly improbable that all measure the same motivation. Certainly, housing conditions (Brain and Benton, 1983), genes (Jones and Brain, 1987), hormones (Brain et al., 1983), and drugs do not have consistent influences across these different tests. It has been argued (Brain, 1984a) to be highly probable that these diverse harm-directed activities variously tap offensive, defensive, or even predatory motivations. In some cases, mixtures of motivations appear to be involved. Support for this view is provided by the use of video analysis, which reveals that, in some "ritualized" responses, vulnerable areas (i.e., the head and ventral surface) of the opponent's body are rarely bitten (in so-called offensive intermale aggression); in other tests, vulnerable areas are frequently bitten (e.g., "defensive" maternal attack on a potentially cannibalistic male intruder), and a third category involves directed killing strategies (e.g., predatory aggression). Perhaps one should limit the term aggression to offensive displays, and thus clearly separate these utilities of attack and threat from defensive and predatory functions? Having said this, one can make a strong case for the detailed investigation of offense, defense and predation in animals being of great relevance to understanding the possible roles of biological factors in human behavior. One should note that the terms "offensive" and "defensive" are essentially based on functional explanations of particular events; it is not easy to operationalize them. For example, the action of biting can be used in rodents to carry offspring, to eat, to kill prey, to defend a nest site, or to attack a conspecific. Brain (1984a) has suggested that one can define a number of categories of behavior that all employ fighting and/or threat. These include the following:

• Social conflict: generally intraspecific phenomena involving competition for a substrate (e.g., a mate, territory, social status, or food), the possession of which increases the organism's relative fitness. These are generally ritualized responses in which the potential for serious damage is limited.

• Parental defense: behaviors in both inter- and intraspecific contexts that serve to protect the attacker's young or nest sites from potentially destructive intruders.

• Self-defense: behaviors in both inter- and intraspecific contexts that normally serve to protect the organism per se from potential predators or attacking conspecifics. Such behaviors are generally limited to situations where flight is difficult or precluded, and do not involve ritualization.

• Infanticide: an intraspecific phenomenon involving the killing of young. In males, this may be a method of increasing the individual's reproductive fitness, whereas in females, it is commonly a response to stress or disturbance (recycling of resources?).

• Predation: an inter- or intraspecific response that involves efficient killing and is often followed by feeding activity.

One should note that it is extremely rare (in animal studies) to find purely offensively or purely defensively motivated behavior. The "ethoexperimental" approach to the analysis of animal behavior (Blanchard et al., 1989) seems to offer advantages in studying animal conflict. It basically attempts to fuse the positive features of ethology and experimental psychology by

(1) creating laboratory environments that reflect the natural requirements of the feral ancestors of laboratory animals. For example, when dealing with a socially living primate species from an arboreal habitat, it seems appropriate to study mixed-sex groups in complex environments, offering a range of tactile experience. When dealing with animals such as the laboratory rat, it seems appropriate to offer the species the opportunity to construct burrow and nest systems (or to provide an equivalent) and to investi-

gate the organism under seminocturnal (its major activity occurs at night) conditions.

(2) carrying out detailed, inclusive analyses of behavior that simultaneously record other categories of activity in addition to threat and attack (these can be very revealing in determining how a biological manipulation changes aggressiveness).

HETEROGENEOUS NATURE OF HUMAN AGGRESSION AND VIOLENCE

Aggressive behavior is certainly no less heterogeneous in our own species. The classification of Buss (1971), based on three dichotomies, provides a clear indication of the diversity of human aggression as viewed through the eyes of a social psychologist. Aggression, according to Buss, may be physical or verbal, active or passive, and direct or indirect. Although it is easy to think of animal analogies for punching, stabbing, or shooting (physical/active/direct aggression), it is much harder to think of animal analogies for "failing to carry out a necessary task" (physical/passive/indirect aggression) or "refusing consent" (verbal/passive/indirect aggression). Obviously, the social psychologist includes a much wider range of activities under the heading aggression than does the biologist.

This diversity of human aggression, has led to an enormous range of methods for assessing the attribute in our species. These broadly fall into two categories. In the first, behavior is assessed in situ by seeking the opinions of peers or by questionnaires. These include examining interactions in preschool play groups, determining the reactions (verbally or physiologically) to films or written material, creating experimental conflict situations (such as use of the "hostility" machine), studying individuals in natural high-stress situations, looking at participation in group activities in which hostile outcomes are probable (certain sports and committees), responding as observers in sporting situations (e.g. "football hooliganism"), and even investigating participation in riots. The second category involves relating physiological events to behavioral characteristics largely determined on the basis of past events. These include studies of convicted criminals (here the material is often divided into largely sexual and nonsexual, and a distinction is made between "impulsive" and "premeditated" crime). One should note here that studying the "same" crime does not always mean that one is dealing with the same phenomenon (e.g., rape is said by several authorities such as Groth, 1979, to have several etiologies and is generated by a plethora of influences).

Similar critiques could be advanced for homicide, assault, etc., before adding the complications introduced by procedures such as plea bargaining. Other studies involve investigations of persons resisting attack on themselves or on property, analysis of individuals with a variety of psychiatric and clinical disorders, and comparisons of male and female individuals (e.g., variations in "rough-and-tumble play" and correlating premenstrual tension with crime).

The focus of this review is human violence. Violence is used almost interchangeably with aggression in most reviews. One might add that violence is a term often applied to aggressive actions that attract greater than normal social disapproval. In this respect, the term clearly fulfills Felson's role of being used to label behavior that transgresses normative values. Obviously, since judgments of sections of society may be involved in determining which behaviors receive the labels aggression and violence, relating hormones to such human activities is not easy. It is doubtful whether one can always differentiate adaptive forms of aggression from maladaptive violent and aggressive acts, because this implies a very accurate knowledge of the motivations of all participants at all times. The definition of the Panel on the Understanding and Control of Violent Behavior of violent human behavior as "threatened, attempted, or completed intentional infliction of physical harm by persons against persons" is eminently reasonable and is broadly the type of behavior referred to throughout this account. One must note, however, that even within such a framework, "appropriate" vigor grades into violence. Obviously, areas of contention include physical punishment of children and activities in a range of contact sports.

HORMONES AND AGGRESSION

It is necessary to note initially that what we call aggression is (like any other behavioral concept) influenced by diverse factors that are difficult (impossible?) to disentangle. These include

(1) biological factors (i.e., genes, neural systems, neurotransmitters, and hormones);

(2) situational determinants (i.e., the environmental or social context); and

(3) the accumulated experiences of individuals.

Figure 1 is a schema of the relationship between biology and behavior. If one looks at interindividual forms of aggression, one is

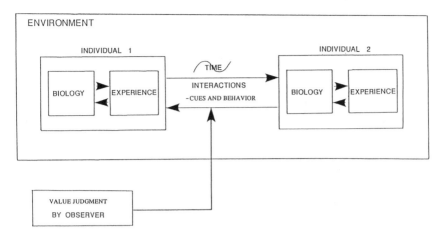

FIGURE 1   Schema showing the relationship(s) between biology and aggression: factors to be considered in rating "simple" interactions between individuals.   SOURCE: Brain (1989a).

really dealing with some quite complex interactions between biology and experience. Some of these effects are mediated by changes in aggressive motivation, some by influencing other behaviors that compete for expression with the aggression, others by changing the social signals that organisms direct toward each other, and yet others by the way in which individuals perceive those social signals.   There are also changes over time and the impact of the particular environment to consider.   One has to add to this complex mix that whether one chooses to call a behavior aggression or not, is based the observer's value judgment.   It is consequently highly improbable that one will find simple relationships between any one biological factor and expressed behavior.

NEUROENDOCRINOLOGY

Although the neural and endocrine systems have tended to be regarded separately, it is currently thought by a variety of authorities that they are best considered components with differing characteristics (in terms of the speed, duration, and diffuseness of their actions) of a single neuroendocrine coordinating system.   Certainly there is an intimate relationship between neural and endocrine factors.   Indeed, hormones, neurotransmitters, and neuromodulators can all be defined as *information-transferring molecules.*   Cues received and integrated by the central nervous system (CNS) can be passed by the neural elements to specialized neurosecretory cells that essentially convert nerve impulses into hormonal output. These cells (in evolutionary terms, the oldest glands) are modified neurons (with many of the elements of such cells) that secrete protein-derived (peptidergic) or amino-derived (aminergic) material. This material may be transported along the modified axonal elements that end in close association with blood vessels (some of these are called neurohemal organs: e.g., the posterior lobe of the pituitary). It is now apparent that many endocrine glands are innervated by conventional neurons, which suggests that direct neural input can modify their secretory activity. Some neurosecretory cells may have direct effects on muscles and other effector organs, such as exocrine ("ducted") glands.

Although some endocrine glands are primarily controlled by direct neural input (the adrenal medulla), others are controlled by tropic hormones from the pituitary (the adrenal cortex, the gonads, and the thyroid), and a third category largely responds to blood-borne metabolites (the pancreas and the parathyroids). Many endocrine systems maintain homeostasis (a balance vis-à-vis the internal environment) by employing negative feedback mechanisms (see Figure 2). In some cases, positive feedback mechanisms may also operate (e.g., involvement of luteinizing hormone (LH) in ovulation in mammals).

DEFINING "HORMONE"

As mentioned earlier, hormones are secretions of endocrine glands that are passed into the bloodstream and are accumulated by target tissues (including the CNS), where they induce particular physiological or behavioral responses. The use of the term hormone originally implied (1) a natural chemical structure that had been extracted from a recognized endocrine gland and (2) the use of the blood system as the transport mechanism employed to reach the target tissue. The term hormone has, however, recently become less precise. Synthetic hormones, fragments of peptide factors, analogues of hormones, and parahormones (e.g., prostaglandins and opioids) have been included within this heading. Is the neurally located material that is immunoreactive to an antibody to adrenocorticotropic hormone (ACTH) truly a hormone, even if the substance is chemically identical to ACTH? This peptide may never get near the bloodstream. Are the bodies' own pain killers, enkephalins and endorphins, hormones? They may be derived from the peptide hormone ß-lipotropin, but are they

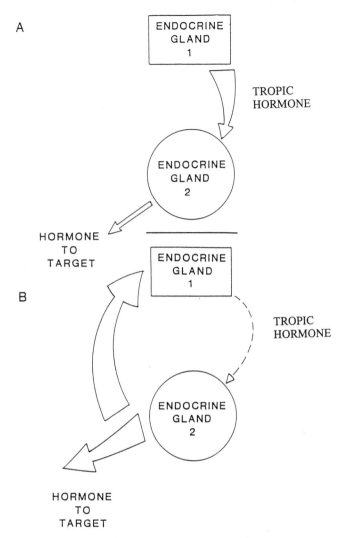

FIGURE 2 Negative feedback arrangement evident between an endocrine gland and a tropic gland (e.g., the adrenal cortex and the anterior pituitary). SOURCE: Brain (1989c). Reprinted by permission of Kluwer Academic Publishers.

best regarded as degradation products? A certain degree of flexibility seems consequently appropriate in this area at the present time. The main systems discussed in this review are the HPG (hypothalamus-pituitary-gonadal axis, which includes hypothalamic luteinizing hormone (LH) releasing factor (LHRF), gonadotropins from the pituitary, and sex steroids from the gonads) and the HPA (hypothalamus-pituitary-adrenocortical axis, which includes corticopin releasing factor (CRF), ACTH, and adrenal steroids) axes. One should recognize, however, that other hormones (e.g., those of the thyroid gland and adrenal medulla; see Valenti and Mainardi, 1989, and Hucklebridge et al., 1981) can also be implicated in changes in violence and aggression (see Figure 3 for basic biochemistry).

A brief note on the respective significance of the different levels in the brain-pituitary-gonadal axis as they relate to violent and aggressive behavior might be appropriate here. Associations between hypothalamic LHRF and violence could be interpreted as the most neural component of the HPA axis directly influencing the CNS, or they could be regarded as suggesting that the axis is activated or deactivated in such responses. If LH is correlated with violence or aggression, it could suggest that this pituitary gonadotropin controls behavior before the major effects of the HPA system on gonadal function are established (this has been the interpretation favored in some seasonally breeding animals), or it might indicate that the feedback mechanisms of steroids are accentuated or suppressed. Correlations between androgens or estrogens and behavior are also capable of being interpreted in several ways (e.g., "direct" effects of hormones on violence via neural actions of steroids, or reflections of changes in the hypothalamus and/or pituitary).

STUDIES INVOLVING HORMONES AND "AGGRESSION"

The literature relating hormones to aggressive behavior has been reviewed on several occasions (see Brain, 1977-1981). One striking feature of this enormous body of data is that the sheer diversity of approaches and methodologies used makes extracting generalities from the data exceedingly difficult. These earlier reviews confirm that the topic of hormones and aggression in vertebrates can be effectively subdivided into

(1) studies on the effects of early hormonal "programming" of adult aggressiveness;

(2) direct effects (presumably via the CNS) of endocrine manipulations on fighting and threat;

(3) indirect effects (presumably via changed social signals, etc.) of endocrine manipulations on fighting and threat;

(4) hormone-aggression correlations; and

(5) influences of fighting on endocrine function.

The types of manipulation involved in items (2) and (3) can be referred to as exogenous modifications. These are produced by

(a) surgical removal of the endocrine glands;
(b) injection of hormone in a solution at a variety of sites (e.g., subcutaneous, intramuscular, intracerebral);
(c) implantation of hormone as a solid phase (as crystalline hormone or with a material such as Silastic) at a variety of sites (see above);
(d) use of antihormones or blocking agents; or
(e) transplantation of hormone-secreting materials (these may be functional endocrine glands or hormone-secreting tumors).

Sometimes treatments are used in combination, as in classical replacement therapy (surgical removal of the gland followed by injection of the hormone associated with the structure). One should note that all manipulations alter more than one component of the endocrine system. Hormones do not function *in vacuo* but operate as part of a complex integrated physiological system.

Endogenous changes are likely implicated in type (4) studies. Here one may use the following indicators of endocrine activity

(a) organ weight,
(b) histology,
(c) morphometry,
(d) histochemistry,
(e) bioassay,
(f) fluorometric assay,
(g) competitive protein binding assay,
(h) radioimmunoassay, or
(i) enzyme immunoassay.

Measurements may be applied to tissues, plasma, or even saliva. Assays are generally judged on the basis of three criteria, namely,

(1) sensitivity (the amount that can be detected);
(2) reproducibility (the ability to get the same result twice running); and
(3) specificity (the ability to measure only the particular hormone).

On these criteria, radioimmunoassay, enzyme immunoassay, and some of the new flourometric assays may be judged the most accurate current means of measuring hormones, but it must be stressed that all methodologies seem to have utility under par-

Source	Name	Abbreviation	Structure
Hypothalamus	Luteinizing Hormone Releasing Factor	LHRF	(pyro)-Glu-His-Trp-Ser-Tyr Gly-Leu-Arg-Pro-Gly-NH_2
	Corticotropin Releasing Factor (human)	CRF	Ser-Gln-Glu-Pro-Pro-Ile-Ser-Leu-Asp-Leu-Thr-Phe-His-Leu-Leu-Arg-Glu-Val-Leu-Glu-Met-Thr-Lys-Ala-Asp-Gln-Leu-Ala-Gln-Gln-Ala-His-Ser-Asn-Arg-Lys-Leu-Leu-Asp-Ile-Ala-NH_2
Anterior Pituitary	Luteinizing Hormone	LH	A glycoprotein consisting of two subunits with a molecular weight of 32,000
	Adrenocorticotropic Hormone	ACTH	Ser-Tyr-Ser-Met-Glu-His-Phe-Arg-Trp-Gly-Lys-Pro-Val-Gly-Lys-Lys-Arg-Arg-Pro-Val-Lys-Val-Tyr-Pro-Asp-Ala-Gly-Glu-Asp-Gln-Ser-Ala-Glu-Ala-Phe-Pro-Leu-Glu-Phe

Gonads	Testosterone	T
	Estradiol	E$_2$
Adrenal Cortex	Cortisol	–
	Corticosterone	B
Adrenal Medulla	Epinephrine	E
	Norepinephrine	NE
Thyroid	Thyroxine	T$_4$
	Triiodothyronine	T$_3$

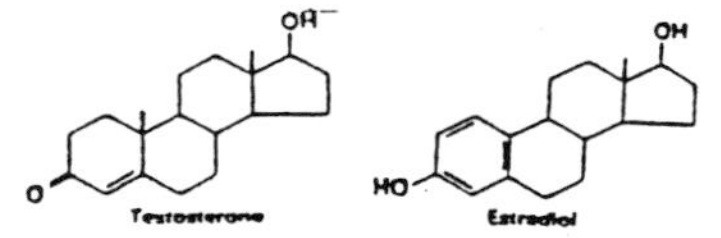

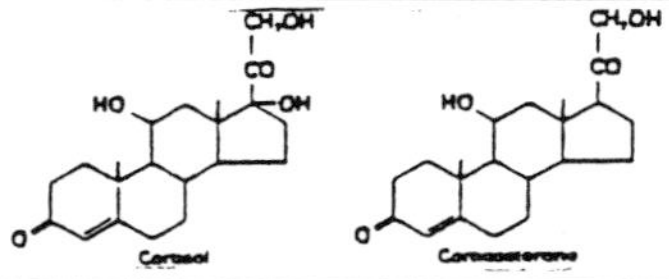

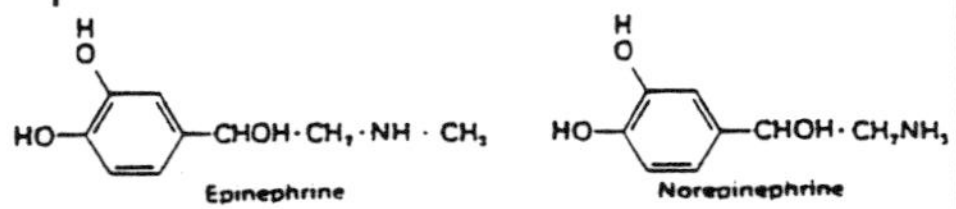

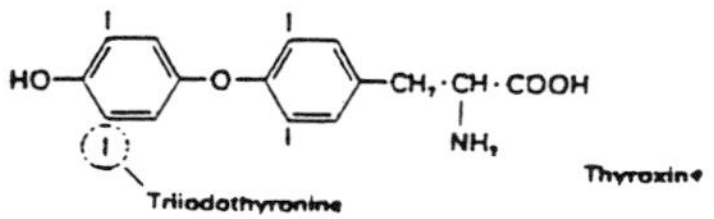

FIGURE 3 Structures of some hormones that appear to be important in aggression research. NOTE: The structures of LHRF, CRF, and ACTH are the sequences of amino acids; ACTH is actually 39 residues long, but only 1-24 are essential for the actions on the adrenal cortex (sequence 4-10 is shared by MSH and may have extraadrenal actions on the brain); T$_3$ has one iodine missing at the position indicated.

ticular circumstances. For example, organ weights or histology may prove useful when the stress of obtaining blood samples is likely to confound the measurements or when the amount of material is very small. It is obvious that the concentrations of hormones at receptors are more likely to be of relevance to ongoing behaviors (including aggression) than changes in serum or plasma concentrations. This having been said, there are a variety of technical and ethical problems that make the obtaining of such data in clinical studies inherently improbable (at present). The best that can currently be achieved in human studies is to examine a range of values and to look at the time courses of these variations in detail.

One must note that a wide variation of sophistication is evident in the work on hormone-aggression correlations emanating from different specialties. A personal view is that modern plasma and saliva (these are especially useful because they reflect unbound hormone and the samples are obtained noninvasively with little associated stress) assays of multiple circulating hormones in humans are most appropriate, whereas attempts to estimate receptor dynamics in the brains of animals are of the greatest utility (these enable one to see how the hormones act at the level of the receptor).

It now seems unlikely that one will be able to show clear relationships between aggression and a single (even if immensely accurate) determination of the plasma value of a single hormone. One may argue that it is important to know

(a) whether the titer is increasing or decreasing and whether the factor is elevated or depressed with relation to a fluctuating baseline;

(b) the distribution of the hormone in the different body compartments (e.g., plasma and target tissues);

(c) the extent to which the receptor (hormone-receiving points on the cell's membrane or within the cell) is occupied by bound hormone and the amount of such receptor (receptor populations can be altered by hormones); and

(d) interactions with other hormones (many hormones change other components of the endocrine system); and

(e) whether hormone production shows estrous, seasonal, or circadian rhythms.

Brain (1989c) has recently (in advocating an ethoexperimental approach to the study of relationships between hormones and behavior) detailed some of the complications that have become apparent to workers in this area. Initially, hormones were assumed to directly influence the expression of a specific behavior (e.g., aggression). In this schema, the endocrine gland is viewed as functioning in isolation, and measures of its output are correlated with a specific behavioral measure. There has, in such cases, been a tendency to assume that the hormone produces the behavior by acting on the CNS.

As the vascular system carries hormones throughout the body, they may be picked up and change processes in a variety of structures that modify behavior. Although motivational changes (mediated via the CNS) are recognized as being one means of expressing hormonal action, these chemicals may modify the production of social cues and/or the sensory systems that detect such factors. Such effects may be expressed in a variety of sensory systems including the somatosensory, visual, olfactory, and auditory modes. There is good evidence that hormones can change the responsiveness of female mammals to male odors and that such sensory input has a powerful impact on social behaviors. One must also stress that other endocrine glands can be targets for the actions of hormones. The situation may be complicated by emphasizing this "cross-talk" between endocrine glands (e.g., the effects of the gonadal system in mammals on the adrenal cortex and vice versa) and by recognizing that the altered target tissues may express their actions on behavior in rather different ways. It is rare for a single modality or mechanism to be employed. Rather than specifying a need to examine phenomena such as "cross-talk" in terms of their impact on aggression, it is simply advocated that one remember that a complex interactive system is involved. The actuality strongly suggests that one should look for multiple endocrine correlates of behavior, recognizing that some will be secondary consequences of other changes.

Another complication is the fact that metabolic conversions may transform initially secreted or applied hormones (in a variety of locations, including the blood and neural sites) to a range of compounds. Testosterone, for example, can be aromatized (chemically converted) to 17 ß-estradiol in certain neural locations and reduced to 5α-dihydrotestosterone in other neural locations and peripheral androgen-dependent tissues (e.g., the seminal vesicles; see Figure 4). Adrenal steroids (because of their chemical similarities) are also often interconverted.

A further complication is that one rarely sees single behavioral responses in isolation. Behavioral elements (which may be individual actions or broad categories of behavior) not only can

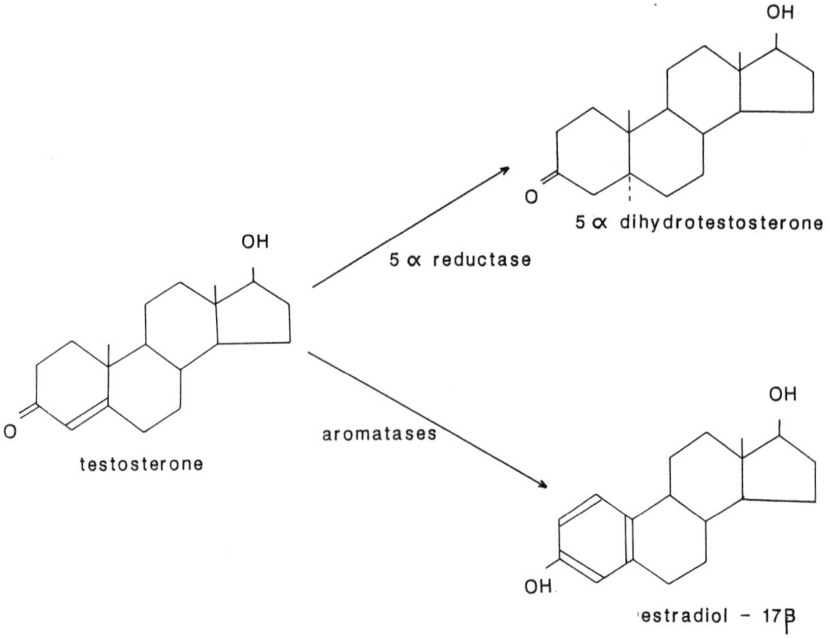

FIGURE 4 Conversion by enzyme systems in different targets of testosterone to 5α-dihydrotestosterone and estradiol-17ß.

interact (e.g., if an animal is showing fearful behavior, it is unlikely to explore) but can be influenced by a variety of hormone-target relationships, some of which can be common to several behaviors, and others exclusive.

The final complication introduced into the scheme is shown in Figure 5, reflecting the obvious fact that there are feedback relationships between many of the subprocesses linking endocrine gland activities to behavioral elements. It is well documented that behavioral experiences can have profound physiological repercussions (see Brain and Benton, 1983). Further, it is difficult to generalize across species and test situations or to infer common underlying mechanisms (Brain, 1979a; Miczek and Krsiak, 1981). Brain et al. (1983) have shown that the radically different laboratory tests used to assess murine aggression produce very different pictures when common manipulations of the gonadal system are attempted. It seems likely that these tests tap different mixtures of motivations (see earlier). Brain (1981) has emphasized the complex interplay between endocrine glands and the variety of target tissues that must be considered in any investiga-

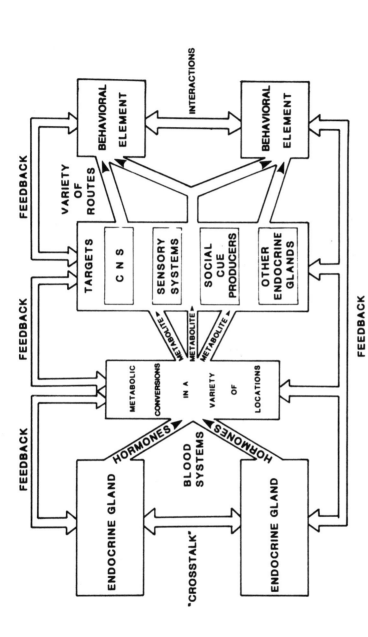

FIGURE 5   Associations between hormones and behavior stressing "feedback" relationships at many points in the scheme. Experiences obtained via behavioral expression can modify targets of hormonal action (especially the CNS), change metabolic conversion rates, and alter secretion of endocrine glands. SOURCE: Brain (1989c). Reprinted by permission of Kluwer Academic Publishers.

tion of hormonal involvement in aggression. It is certainly hard to assess the roles of hormones in aggression because this area of research involves a great number of "strong" and "weak" correlations between a difficult behavioral concept (aggression) and the complex and subtly integrated endocrine system (Brain, 1977).

One of the current difficulties of implicating hormones in agonistic behavior is that much of the accumulated literature in behavioral endocrinology is based on the traditional approaches of physiological psychology (see Carlson, 1977). Here, relatively crude (generally pharmacologic) manipulations of the endocrine system are usually attempted via surgery, implantation, or injection, and the consequences of the modification related to aggression are determined in restricted situations. For example, large doses of testosterone have been given to male chicks soon after hatching, and the effects on their aggressive responding to a hand have been assessed (Andrew, 1975). Sigg et al. (1966) were unable to demonstrate any action of repeated injections of 2 International Units per kilogram of thyroid stimulating hormone (TSH) on social conflict in male hypophysectomized (with the pituitary gland removed) mice. Suchowsky et al. (1969) claimed that estradiol, when given repeatedly at high doses to pairs of male animals, inhibited social conflict in intact and castrated albino mice subjected to a series of tests. The estrogen clearly modifies odor characteristics in such subjects. There are, in direct contrast, fertile developments in ethopharmacology, where a tradition of inclusive measurement of behavior is undertaken in animals who generally have options that are less restricted than those seen in traditional laboratory tests for aggression. Further, a current trend is to look for alternative explanations of hormone actions on behavior to those interpreted as "simple" changes in aggressiveness. A more ethoexperimental approach to behavioral description should enable us to assess the important or fundamental behavioral correlates of hormonal changes.

It would also make sense to use the recently developed, less invasive, sensitive assay techniques to measure as many different aspects of hormonal function as are feasible (including changes in binding site populations as well as titers of hormones). A second gain would follow from assessing behavior in more sophisticated situations. It is, naturally, impossible to show the behavioral impact of a particular hormonal manipulation in an inappropriate environment—many of the effects of hormones important to animals in the wild may only be apparent in more complex, familiar environments where the animal has the possibility of expressing behavior in context. Progress may also depend on a willingness to be more concerned about the temporal relationships between endocrine changes and behavior. Much of what has already been described in the physiological psychology relates to the acute correlation of hormonal change with behavior, whereas it might be more fruitful to look at the organism more chronically and attempt to relate patterns of endocrine fluctuation to life events (e.g., onset of mating or assumption of a territorial habit). It should be possible to blend a reasonable degree of scientific rigor with richer and more flexible arenas for studying behavioral expression.

When commenting on the available literature, it seems worthwhile to stress an earlier claim (Brain, 1981) that it would not be especially surprising if all hormones alter some aspect of aggression in a particular species or a specific situation. As mentioned earlier, hormones can have very diverse actions and can change behavior in a variety of ways. Consequently, any modification of the endocrine environment can have consequences (e.g., altered brain architecture, increased body size, altered odor characteristics, or varied availability of energy substrates) that can change the probability of recording activities capable of being judged violent or aggressive. This having been said, certain hormone-behavior correlations are stronger than others. There is little doubt that the hormones associated with the HPA and HPG axes are more likely to have lasting impacts on behavioral development and concurrent behavioral expression than most other endocrine factors. At this state of our knowledge, perhaps only the hypothalamic-pituitary-thyroid axis, adrenomedullary hormones, and hormones that produce large changes in blood sugar levels (e.g., insulin and glucagon) are worthy of additional consideration. It is my belief that new studies in this area will tap a wider range of endocrine measurements and that they will be especially directed to the so-called metabolites of peptide and steroid hormones. One will also have to pay considerably more attention to changes in binding site populations in receptors, and to the rates of metabolic conversion and the clearance rates of hormones and their products.

The basic subdivisions of material referred to earlier are used in the following sections, with material on nonprimate vertebrates, infrahuman primates, and humans being presented separately.

EARLY PROGRAMMING EFFECTS OF HORMONES

STUDIES ON NONPRIMATES

The frequently recorded sex differences in aggressiveness (see Brain, 1979b) in a variety of situations may reflect variations in the early patterns of endogenous sex steroid secretion and/or adult production of hormones. Androgens have developmental effects in early life when they alter the capacity of the animal to react aggressively in adulthood (Leshner, 1981). The presence of androgens in the mature animal is also a necessary prerequisite for the display of this form of behavior. Dixson (1980) stated that the neural mechanisms that mediate patterns of sexual and aggressive behavior in rodents are profoundly masculinized and defeminized by the influences of androgens on the developing brain (note that masculinization and defeminization are different processes). Male or androgenized but genetically female rats (*Rattus norvegicus*) not only need higher doses of estrogen to produce female-typical sexual responses (e.g., lordosis) than nonmasculinized females (van de Poll and van Dis, 1977), but also fail to respond to testosterone by showing this behavior (van de Poll et al., 1978). The vast majority of studies relating neonatal hormone applications to adult aggressive behavior have employed traditional "aromatizable" androgens (i.e., those that can be metabolized to estrogens in target tissues including the CNS). Male rats and house mice (*Mus musculus domesticus*) show a greater potential for aggressiveness than female counterparts because endogenous hormone secretion by the testis occurs earlier than that by the ovary. The early surge of testosterone was said to "create" the neural circuitry of male rodents from the "undifferentiated" (female) condition. Recently revealed complications in this process are considered in Aron et al. (1990), Brain and Haug (1990), and Goy and Roy (1990).

Many authors have found that testosterone propionate treatment of neonatally castrated male (vom Saal et al., 1976) and female (Mugford, 1974) mice results in these animals showing much higher incidences of fighting in adulthood than counterparts treated with control injections. It seem reasonable to conclude that (in rodents at least) testosterone plays an important role in the genesis and maintenance of some forms of aggressive behavior. One should note, however, that in some species and situations, no sex differences in testosterone-mediated aggression are apparent. This is certainly and pertinently true of same-sex encounters in certain strains of rats (van de Poll et al., 1981).

TABLE 1 Synopsis of the Effects of Early Treatments with Aromatizable Androgens (e.g., testosterone (T) and androstenedione) on Social Conflict

Class	Species	References	Effects
Pisces	*Xiphophorus variatus* (platyfish)	Laskowski, 1954	Augments in female
Aves	*Gallus domesticus* (chicken)	Andrew, 1975	Augments in male
	Anas platyrhynchos (domestic duck)	Balthazart and Stevens, 1975	Augments in male
Mammalia	Mouse	Edwards, 1968, 1969, 1971	Augments in ovariectomized, T-treated female
	Mesocricetus auratus (golden hamster)	Payne, 1976	Augments in gonadectomized, T-treated males and females
	Macaca mulatta (rhesus monkey)	Joslyn, 1973	Augments in female

The fact that early exposure to testosterone (or other aromatizable androgens) increases aggressiveness in a wide range of nonhuman species is emphasized in Table 1. There are considerable species differences in the roles of hormones influencing prepubertal aggressive behavior (Brain, 1978), with the data suggesting that lower vertebrates (such as fish, reptiles, and rodents) are more likely to show hormonal dependence than subjects such as dogs and primates. Although this may be interpreted as a phylogenetic change, one should caution that the longer life span of the less hormonally dependent species may play a role. Certainly primates are more likely to have acquired a range of social experiences before castration or steroid hormone treatment than are rats and mice.

There have been some truly excellent studies on the intrauterine location phenomenon that have added greatly to our knowledge (see review by vom Saal, 1990). The positioning of the fetus during intrauterine development is an important source of variation in the hormonal titers to which both developing male and female rats and mice are exposed. Vom Saal and Bronson (1980) found that fetuses of male mice on day 17 of gestation have three times the amount of circulating testosterone seen in their female counterparts. Intrauterine position in relation to male and female peers may, by changing early sex steroid exposure, consequently influence the potential for different types of behavior in adult mice (vom Saal, 1983). This has been confirmed in that 2M (developing between two males) female mice are aggressive toward and establish dominance over 0M (developing between two females) counterparts (vom Saal and Bronson, 1978). Further, after treatment in adulthood with testosterone, 2M female rats exhibited more mounting of receptive females than did 0M counterparts (Clemens et al., 1978).

It is well established (Brain et al., 1983) in rodents that the impact of castration on aggression becomes less evident in experienced fighters, which suggests that learning overrides hormonal influences. This having been said, there is also some evidence (in mice) that stimulation of gonadal function (e.g., by breeding activity), without the opportunity to fight, reduces the ease of demonstrating suppressed fighting postcastration. Perhaps some of the androgen-induced changes in aggressiveness are relatively persistent once generated?

There is also evidence that early exposure to stress (perhaps including the hormones of the pituitary-adrenocortical axis) modify the adult behavioral potentials (in terms of aggressive and emotional activities) of rats and mice (reviewed in Chevins, 1989).

INFRAHUMAN PRIMATE STUDIES

Goy (1968) studied the effects of early exposure of nonhuman primates to testosterone. Prenatal exposure to testosterone propionate masculinized the external genitalia of female rhesus monkeys and increased their rough-and-tumble and chasing play, features that are usually more typical of the male. Eaton et al. (1973) demonstrated that augmented aggression was still evident in ovariectomized and androgen-treated female adults who had been exposed neonatally to testosterone. Dixson (1980) reviewed the available data on such studies and concluded that "androgen administered prenatally has important consequences for behavior, including aggressive responses, in female rhesus monkeys." He suggested that testosterone (a major androgen in the circulation of fetal rhesus monkeys), which is present in much higher concentrations in males than females (Resko, 1974), influences brain development. Since progesterone is present in higher concentrations in female rhesus monkeys than in males (Resko, 1974), it has been suggested that it protects the female's developing brain from masculinization. Dixson (1980) suggests (not unreasonably) that it is important (given the variability within the order) to assess whether these relationships hold in other species of primate. Marmosets and tamarins (*Callitricidae*) might prove useful because of their tendency to produce twins and the possibility that sexual differentiation is delayed until after birth in such species.

HUMAN DATA

Meyer-Bahlburg (1980) and Meyer-Bahlburg and Ehrhardt (1982) comprehensively reviewed studies on the lasting impact of variations in early hormone exposure on human aggressiveness. Tables 2 through 4 provide synopses of data involving endocrine syndromes (partial androgen insensitivity and congenital adrenal hyperplasia) and treatment with hormones (generally to reduce the probability of miscarriage) created by Meyer-Bahlburg and Ehrhardt (1982). There are few reliable data on the incidences of those syndromes in particular populations (these are anyhow quite variable in their degree of severity), but their importance is not as potential causes of problems in society but as indicators of the normal biological mechanisms that influence behavioral differences between the sexes. The 5α-reductase deficiency syndrome (which is cited later) seems very rare, being associated with 38 hermaphroditic individuals born to 23 interrelated families in two inbred mountain villages in the Dominican Republic.

Although many of the data are characterized by small sample sizes, Meyer-Bahlburg (1980) suggests that exogenous sex hormones that slightly increase aggressiveness in females produce some degree of genital masculinization. He felt that the data from girls and boys resulting from treating toxemic pregnancies with progesterone were inconsistent. There was, however, some evidence of decreased aggressiveness in boys from diabetic pregnancies exposed to progestogen-estrogen combinations and from boys and girls treated with medroxyprogesterone acetate (MPA). It is, of course, uncertain precisely how such behavioral effects are generated since features such as parental rearing styles, degree of exercise, and changes in the musculoskeletal system are involved in such phenomena and are likely to be influenced (directly or indirectly) by early hormonal factors. Early exposure to androgens is said to modify temperament, increasing it in the direction of "impetuous and active." Olweus (1984) has suggested that this factor

TABLE 2 Relationships Between Aggression and Prenatal Endocrine Syndromes

Authors	Medical Syndrome	N	Age (years) at Study	Results of Assessment by Interview	
Money and Ogunro (1974)	Partial androgen insensitivity	8 Males 2 Females	13.5-39	Athletic-competitive energy expenditure	↓?
				Dominance assertion—fighting	↓?
	Controls	None		Dominance assertion—social	↓?
Ehrhardt (1969)	Congenital adrenal hyperplasia	15	5.2-16.1	Fighting in childhood	NS
Money and Ehrhardt (1972)	Female controls (pair-matched)	15	5.8-15.2		
Ehrhardt and Baker (1974)	Congenital adrenal hyperplasia	17	4.3-19.9	Initiation of fighting	↑NS
	Female controls (siblings)	11	6.8-24.7		
	Female controls (mothers)	10	82-49		
Money and Schwartz (1976)	Congenital adrenal hyperplasia	15	15-23	Participation in contact sports	↑?
	Female controls	None		Leadership among friends	NS ?
				Express dominance over others	NS ?
				Temper	NS ?
				History of fights	NS ?
				Most common anger manifestation (physical, verbal, other)	NS ?
				Physical destruction of property	NS ?
Ehrhardt and Baker (1974)	Congenital adrenal hyperplasia	9	4.8-26.3	Initiation of fighting	↑NS
	male controls (siblings)	11	6.8-23.7		

NOTE: ↑ = increased in patients; ↓ = decreased in patients; NS = not statistically significant; ? = conclusion without statistical comparison data.

SOURCE: Modified from Meyer-Bahlburg and Ehrhardt (1982).

TABLE 3 Associations Between Aggression and Prenatal Hormone Treatment in Females

Authors	Hormone Treatment	N	Age (years) at Study	Results of Assessment (by interview, unless stated otherwise)	
Ehrhardt and Money (1967)	Various masculinizing progestins	10	3.8–14.3	Interest in organized team sports	↑?
				Liked to compete with boys in sports	↑?
	Controls	None	14.3	Self-assertive and independent	↑?
				Self-defending and belligerent	↑?
				Aggressive attack in the pecking order	NS?
Ehrhardt (1969)	Various masculinizing progestins	10	3.8	Fighting in childhood	NS
Money and Ehrhardt (1972)	(same sample as above, reanalyzed with control group)		14.3		
	Controls (pair-matched)	10	3.7 14.3		
Zussman et al. (1975, 1977)	Progesterone	12	16–19	Reports of discipline in school	→
				Influence over peers	→
	Controls	12	16–19	Frequency and intensity of anger	→
Reinisch (1981)	Various synthetic progestins	17	6–17	Leifer-Roberts Response Hierarchy: Potential for physical aggression	←
	Sibling control (at least one per hormone-exposed subject)	17		Potential for verbal aggression	NS
Meyer-Bahlburg and Ehrhardt (1982)	Medroxyprogesterone	15	9–14	Self- or mother-reported aggression	→
	Pair-matched controls	15	9–14		

NOTE: See Table 2 for abbreviations.

SOURCE: Modified from Meyer-Bahlburg and Ehrhardt (1982).

TABLE 4 Associations Between Aggression After Prenatal Hormone Treatment in Males

Authors	Hormone Treatment	N	Age (years) at study	Results of Assessment (by interview, unless stated otherwise)	
Zussman et al. (1975, 1977)	Progesterone	18	16-19	Reports of discipline in school	↑
				Influence over peers	→
	Controls	17	16-19	Aggression in childhood	←
				Frequency and intensity of anger	→
Yalom et al. (1973)	Diethylstilbestrol + progesterone	20	16-17	Self-rating of physical aggressiveness	→
	Controls	22	16-17	Clinician's ranking (aggressive-assertive)	→
				Moos Aggression Questionnaire	
				In past week, anger at male	→
				In past week, expression anger	→
				Usually wins fights	→
				Likes to fight	→
				Number of fights	→
				Aggression, aggregate score	→
	Estradiol valerate + hydroxyprogesterone acetate	20	6	Aggressive activity (mother's report)	NS
				Clinician's ranking (aggressivity)	NS
	Controls	17	6	Teacher's ratings:	
				Assertiveness	→
				Toughness	NS
				Disruptiveness	NS
Reinisch (1981)	Various synthetic progestins	8	6-18	Leifer-Roberts Response Hierarchary:	
				Potential for physical aggression	←
	Sibling controls (at least one per hormone-exposed subject)	≥8		Potential for verbal aggression	NS

Study	N	Age	Measure	Result
Kester et al. (1980)				
Diethylstilbestrol	17	18-30	Childhood:	
Controls (pair-matched)	17	18-30	(a) Fighting	→
			(b) Preference for stories with more aggressive themes	NS
			(c) Participation in individual competitive noncontact sports	NS
			Adolescence:	
			(d) Participation in team competitive contact sports	NS
			(e) Participation in team competitive noncontact sports	NS
			(f) Participation in individual competitive noncontact sports	NS
			Adulthood:	
			(g) Preference for TV shows with more aggressive themes	←
			(h) Participation in team competitive contact sports	NS
			(i) Watching individual competitive contact sports	NS
			(j) Watching team competitive contact sports	NS
Diethylstilbestrol, natural progesterone	22	24-29	Childhood:	
			(a)	NS
			(b)	NS
			(c)	←

continued on next page

TABLE 4 (Continued)

Authors	Hormone Treatment	N	Age (years) at study	Results of Assessment (by interview, unless stated otherwise)	
	Controls (pair-matched)	22	24-29	Adolescence:	
				(d)	NS
				(e)	←
				(f)	NS
				Adulthood:	
				(g)	NS
				(h)	NS
				(i)	NS
				(j)	NS
	Natural progesterone	10	10-24	Childhood:	
				(a)	NS
				(b)	←
				(c)	NS
	Controls (pair-matched)	10	10-24	Adolescence:	
				(d)	NS
				(e)	NS
				(f)	NS
				Adulthood:	
				(g)	NS
				(h)	→
				(i)	→
				(k)	NS

Synthetic progesterone	13	19-24	Childhood:	
			(a)	NS
			(b)	NS
			(c)	NS
Controls (pair-matched)	13	19-24	Adolescence:	
			(d)	←
			(e)	NS
			(f)	NS
			Adulthood:	
			(g)	NS
			(h)	NS
			(i)	NS
			(k)	←
Meyer-Bahlburg and Ehrhardt (1982)	Medroxyprogesterone acetate	13	9-14	Self- or mother-reported aggression	→
	Controls (pair-matched)	13	9-14		→

NOTE: ↑/↓ = statistically significant increase/decrease in hormone-exposed subjects; (↑)/(↓) = increase/decrease of statistically borderline significance in hormone-exposed subjects; NS = statistically not significant; ? = conclusion without statistical comparison data.

SOURCE: Modified from Meyer-Bahlburg and Ehrhardt (1982).

has a weak direct action on the aggression of boys and, indeed, may have a stronger indirect action on this potential by increasing the mother's permissiveness for aggression (by exhaustion?).

HORMONES AND PUBERTY

STUDIES ON NONPRIMATES

In most species investigated, the prepubertal male is markedly less aggressive than his mature counterpart in a variety of situations. Maturation of social aggressiveness in the male mouse has been described (Williams and Scott, 1953-1954; Brain and Nowell, 1969; Cairns, 1972; Bernard et al., 1975; Barkley and Goldman, 1977). Indeed, the increased aggressiveness and the surge of androgens that precede puberty have been correlated in mice (McKinney and Desjardins, 1973). Similar maturation effects have been claimed in the domestic cockerel (*Gallus domesticus*; Sharp et al., 1977); the golden hamster (*Mesocricetus auratus*; Goldman and Swanson, 1975), the Mongolian gerbil (*Meriones unguicalatus*; Kaplan and Hyland, 1972), and guinea pigs (*Cavia porcellus*; Willis et al., 1977). There are also puberty-related increases in pituitary and plasma luteinizing hormone in many organisms. In some seasonally breeding species, this gonadotropin may exert a direct effect on fighting propensity (reviewed in Brain, 1977; see below).

Puberty is also associated with changing body signals (visual and olfactory) in many species, which may be related to hormones and can partially account for the changed fighting behavior. In many species, the anabolic (body size-increasing) effects of male sex hormone account for a surge in growth around the time of puberty. Threat and attack behaviors in lower vertebrates may be related directly to body size. Archer (1988) has also suggested that androgens influence the "distractibility" of animals, rendering males more repetitive and "single-minded" in their activities than females. This change may also have a role in attack behavior.

INFRAHUMAN PRIMATE STUDIES

Dixson (1980) reviewed the data concerning puberty and aggression in a range of primate species. Field studies indicate behavioral changes during adolescence in male primates, but further research is needed on this topic (especially since the available evidence does not support any simple conclusion about the effects of androgen on aggressive behavior). Prepubertal castration does not preclude a male from occupying a high-ranking position in a social hierarchy, and androgen therapy does not inevitably enhance aggressiveness or dominance in captive rhesus monkeys. Perhaps androgens influence patterns of "aggressive play" (Symons, 1973; Owens, 1975), which enable the monkeys to practice patterns that will be useful in adult life?

HUMAN DATA

Roberts (1990) has recently reviewed the general parallels between violence in young primates and those seen in our own species. Some of the parallels are extremely striking. Predictors of adolescent, teenage, and adult violence in humans are complex (Farrington, 1989). The best predictors appear to be measures of economic deprivation, family criminality, poor child rearing, school failure, hyperactivity-impulsivity-attention deficient, and antisocial child behavior. In looking for biological predictors, however, Hamburg (1971) failed to find a clear relationship between testosterone levels and aggressive behavior in postpubertal boys. In spite of this, Benson and Migeon (1975) reviewed the physiological and psychological changes occurring around adolescence and puberty in human males. They noted marked changes in serum levels of LH and follicle stimulating hormone (FSH), as well as sex steroids from the testes and adrenal cortices, and tentatively implicated these in the development of a "rebellious attitude" around this time. Hays (1978) provided a review of strategies for studying psychoendocrine aspects of puberty. She speculated that changes in hypothalamic LHRF, which occur around puberty, may (by altering mood) induce sexuality and hostility in our species. She also felt that the involvement of androgens in behavioral changes could be studied by comparing pubertal status of highly aggressive and nonaggressive boys of the same ages. Hays noted that developmental changes are evident with respect to thyroid releasing factor, TSH, prolactin, and somatotropic hormone and that these factors may alter mood and hence aggression. She pointed out four important conclusions from the (then) available data:

(1) Mood changes induced by hormones may be consequences of the instigation of "drives" that have no socially acceptable outlet in young people.

(2) Development may involve changes in behavioral sensitivity to hormones as well as changes in the hormones per se.

(3)   All studies should consider the effects of circadian rhythms on hormonal secretion.

(4)   Interactions between hormones may prove more important than titers of single hormone.

These points are equally valid today.

Archer (1990) has reviewed recent developments in this area. He notes that the only studies that measure hormonal levels and aggressiveness at or soon after puberty in human males are those of Olweus et al. (1980), Mattsson et al. (1980), Olweus (1986), and Susman et al. (1987). The basic data from these nonlongitudinal studies are extracted in Table 5. Olweus (1986) noted particularly that the items involving responses to provocation from his physical and verbal scales correlated best with plasma testosterone. In contrast, Eccles et al. (1988) studied hormones and affect in early adolescence, and recorded that increased levels of androgens were not correlated with increased levels of anger/impatience or aggression.

There has been great interest in the claim that pseudohermaphroditism with changed gender identity and role at puberty may be a consequence of a deficit in 5α-reductase activity (an enzyme that converts testosterone to dihydrotestosterone). There are conflicting claims about the relevance of these data to the debate about the biological versus environmental determination of gender in our species (Aron et al., 1990; Gotz et al., 1990; Money, 1990).

What of other hormones?   Magnusson (1987), in a Swedish study of 82 boys at 13 years of age, found negative correlations between aggressiveness/restlessness and urinary secretion of adrenomedullary epinephrine under both active and passive conditions.   This supported the more general finding of a positive correlation between good social and personal adjustments and elevated epinephrine excretion. Even more intriguingly, there was a strong inverse relationship between epinephrine excretion by these 13 year olds and their adult delinquency at age 18 to 26. More detailed statistical analysis showed, however, that the changes in epinephrine levels were more closely related to motor restlessness than to aggression.   Highly aggressive individuals who are not highly restless have about the same epinephrine levels as subjects who are neither restless nor aggressive.   Magnusson (1988) reviewed these data and warns that the traditional mechanistic model tending to view epinephrine excretion as the cause of both conduct in a current perspective and adult delinquency in a longitudinal perspective is probably inadequate. He suggests consider-

TABLE 5 Recent Data Attempting to Relate Androgens to Puberty in Human Boys

Authors	Sample Size	Age of Subjects	Endocrine Measures	Behavioral Measure	Correlations
Olweus et al. (1980)	58	15-17	Plasma testosterone	Olweus aggression inventory	Positive correlations with verbal and physical scales and combined scores
Mattsson et al. (1980)	40	14-19	Plasma testosterone	Olweus aggression inventory	Positive correlation with verbal but *not* physical scales
Susman et al. (1987)	56	10-14	Plasma testosterone, dehydro-epiandrosterone	Self-ratings and child behavior	Positive relationships between mother's ratings of delinquency or rebelliousness or androstenedione or dehydroepiandrosterone

ing individual subsystems (e.g., cognitions, emotions, physiological factors, and conduct) that are in constant reciprocal interaction and influence each other in current processes with developmental consequences. Simultaneous appearance of low epinephrine level, elevated aggressiveness, and substantial motor restlessness at an early age in subjects later involved in crimes does not necessarily show that these factors are causative or that they have a common etiology. Similar considerations apply to many of the other hormone-behavior associations mentioned here.

Further, Levander et al. (1987) studied 40 recidivists in a Swedish state institution for delinquent males and compared these individuals with 58 "normal" school boys. In spite of the stressful backgrounds of the former, they showed patterns of stress hormone production suggestive of very low psychophysiological arousal. In this group, the common deviant psychoendrocrine pattern consisted of low epinephrine, norepinephrine, and cortisol, with high thyroid hormone levels. Different behavioral subtypes showed different hormonal profiles. Half the subjects had a history of hyperactivity associated with low epinephrine. Levander et al. speculate that elevated thyroid levels represent a compensatory mechanism attempting to correct the deviance in norepinephrine and, to some extent, epinephrine turnover.

Thus, although the available data are sparse, it does seem likely that longitudinal developmental studies involving multiple measures of hormones and varied ratings of aggression are yielding fruitful material.

INFLUENCES OF SEX HORMONES ON ADULT AGGRESSIVE MOTIVATION

The repeatedly demonstrated effects of hormones on behavioral motivation imply that these chemicals have direct actions on the CNS. Brain (1977) has reviewed the lines of circumstantial evidence for such actions on the brain. The following have been suggested:

(a) Some steroid treatments cause morphological and/or receptor population changes in neural structures.

(b) Hormones are sometimes more behaviorally effective when placed in particular neural loci.

(c) Neural regions commonly accumulate specific behaviorally active hormones, a process that can be detected by injecting radioactively labeled hormone and then carrying out autoradiographic examinations of brain sections of treated animals. Pfaff

(1971) has shown that major concentrations of neurally located sex steroids are found in the hypothalamus, the preoptic area, and the septum. Specifically, Naess and Attramadal (1974) documented the medial preoptic nucleus, periventricular nucleus, paraventricular nucleus, septal region, medial amygdaloid nucleus, and ventral premammillary body as accumulating androgens in rodents. The binding characteristics of neural regions that concentrate these compounds can also be altered by administering the hormones in early life, which changes the organism's behavioral potential (McEwen et al., 1970).

(d) Compounds alter neuronal activity in particular regions of the CNS.

(e) Hormonal changes may be correlated with alterations in localized neurotransmitters. There is ample evidence (see Brain, 1977-1979a) that hormones (especially steroids) alter local concentrations of neurotransmitters in the central nervous system and that some of these changes can induce further modifications of endocrine activity (i.e., they may, in some cases, be part of the feedback mechanisms). Hormones not only can change concentrations of neurotransmitters in adult animals but can produce biogenic amine changes in the brains of neonates and even alter neural enzyme systems in immature animals.

One should note that attempts to link hormonal actions to changes in biogenic amines and other factors are much rarer in aggression research than in studies involving sexual behavior (e.g., Meyerson, 1983). Changes in different patterns of these biogenic amines have (in some cases) been related to forms of aggressive behavior in rats and mice (reviewed in Daruna, 1978).

In line with the ethoexperimental approach, one should note that a number of motivational systems may respond to a single manipulations of an animal's endogenous hormones (Adams, 1980). Thus, patrolling-marking, male and female sexual motivational systems and aggressiveness may all be concomitantly changed by exposure to androgens and estrogens. One should also note that other "sex" hormones (e.g., hypothalamic LHRF and pituitary gonadotropins are being implicated in aggression and violence.

STUDIES ON NONPRIMATES

The many studies that have investigated the abilities of sex steroids to maintain motivation for social aggression in mice (e.g., Brain, 1979b; Brain and Bowden, 1979) are of relevance here. These hormones can be further implicated in the modulation of aggressiveness by citing studies correlating endogenous titers with fighting and/or dominance. The motivational states underlying such responses are not, however, identical to spontaneous motivational states (e.g., those that serve locomotor activity; Schallert, 1977). The behavioral influences of androgens in male rodents may depend on their metabolic conversion (Naftolin and Ryan, 1975). There are obvious species and strain differences, but a compelling body of evidence has been accumulated (Larsson et al., 1973; Parrott, 1975, 1976), suggesting that testicular androgens are neurally converted into estrogenic metabolites before motivating ejaculatory behavior in the rat. Brain and Bowden (1979) and Brain (1983) have provided some support for the idea that androgens are similarly neurally aromatized before having their motivational effects on fighting in "TO" strain mice. Brain et al. (1983) have reported that natural (e.g., estradiol) or (especially) synthetic estrogens (e.g., diethylstilbestrol) and aromatizable androgens (e.g., testosterone) are the most effective compounds in terms of their abilities to maintain motivation for social conflict in castrated mice. Since this study, Simon and Whalen (1986) have suggested that the subject's genotype strongly influences which sex steroids are implicated in the control of male aggression in mice. The use of intact male stimulus animals also strongly indicates that estrogen directly increases aggressiveness in mice (all rodent studies in which estrogens suppress aggression depend on treating both intact subjects with these steroids, altering their stimulus effectiveness as targets for attack). Estrogens are implicated in the control of aggressive motivation in other species or sexes of animals. Payne and Swanson (1972), for example, recorded that attacks directed by castrated male golden hamsters toward intact male opponents are restored by injection of the former animals with estradiol benzoate. It is also of interest to note that Harding (1989) has shown that androgens and estrogens interact to modulate social behaviors including aggression in songbirds such as zebra finches (Poephila guttata) and red-winged blackbirds, suggesting that the metabolic products of androgens are important in such species also.

One should reiterate that other aggression models using rodents (especially mice) give different associations between hormones and behavior. Conner et al. (1983) have shown that shock-induced fighting has hormonal correlates that are not too dissimilar from social aggression (e.g., sex differences are evident and androgens have generally stimulatory actions). Svare, in an excellent series of studies (reviewed in Svare and Mann, 1983; Svare, 1989), showed that the major impact of hormones on maternal aggression is via their effects predisposing the animals receiving suckling stimulation. Haug and Brain (1989) have compared and contrasted the attack behavior by group-housed mice on lactating intruders with the more utilized social aggression. They find that this form of attack produces a mirror-image hormonal picture, with castration stimulating attack by males and replacement with androgens or estrogens suppressing this "female" form of attack. One can argue (with considerable justification) that we should attempt to learn as much as possible about the endocrine correlates of many forms of rodent behavior. Not only will there be a gain in theoretical knowledge, but because different kinds of aggression in many species (including our own) are motivated differently (see introductory material), we need such information.

Since this account considers agonistic behavior, it is worth mentioning that avoidance of attack (which can be regarded as the other end of the spectrum of activities that makes up agonistic behavior) seems much more influenced by ACTH and the adrenocortical hormones (see Leshner and Roche, 1977; Leshner, 1980). These studies should be extended because it is clear that "stress-related" hormones are commonly released in social encounters, and it appears that they may influence the progress and eventual outcome of interactions.

The hormonal bases of learning to be submissive or aggressive are of great relevance but far from fully evaluated. Archer (1977) suggested that the testosterone-induced increases in persistence in castrated mice may underpin the effect of this hormone on intermale fighting behavior. Brain (1979a) has reviewed some of the evidence in rodents that ß-lipotropic hormone, ACTH (more especially the 4-10 peptide sequence), melanocyte stimulating hormones, and a variety of related peptides (e.g., enkephalins and endorphins) can alter the acquisition and retention of a variety of reward- and aversion-mediated responses. There is also some evidence that adrenal glucocorticoids influence ongoing or subsequent avoidance reactions. One may consequently suggest that one of the ways in which hormones influence aggression and violence is by mediating learned responses associated with persistence, submission, and avoidance. Consequently, one should look for interactions between hormonal variations and subsequent responding. For example, markedly stressful situations (even in our own species) could predispose individuals to show avoidance, leading to social isolation and subsequent behavioral problems. Anabolic steroids (androgens) might very easily increase persistence in body builders.

Because of the need to examine the effects of antihormones on ethologically assesssed behavior, several reinvestigations have recently been conducted with rats and mice, in which the resultant behavior in a variety of pairings has been analyzed in detail using videotaped records. These studies on the antiandrogen cyproterone acetate, antiestrogens (tamoxifen and CI 680), and progesterone are listed in Table 6. The basic conclusions are that cyproterone acetate suppresses "hostility" in rodents only by reducing the production of androgen-dependent "pheromones" (odor cues). Tamoxifen and CI 680, on the other hand, seem to have real potential as antihostility agents in some forms of attack as assessed in laboratory rodents. Progesterone (and progesterone derivatives such as MPA) produce antiandrogenic effects, blocking the production of gonadal and adrenal androgens in male animals. They also possibly exert calming effects via an anesthetic action (P'an and Laubach, 1964), a property that has been said to account for the hormone's induction of lordosis in rats (Meyerson, 1967).

It is interesting to note that Poshivalov (1982) reported that acute injections of LHRF enhanced intermale social aggression in laboratory mice. Lincoln (1987) similarly found that the LHRF agonist buserelin increased both testosterone levels and aggressive behavior in male red deer (*Cervus elaphus*).

One should comment that the consequences of winning or losing encounters on the HPA and HPG axes, as well as on the secretion of medullary hormones in male lower vertebrates, generate patterns of changes that seem likely to intensify dominance-subordination polarities and/or facilitate social group living. The "winner" tends to show relatively augmented testosterone (increasing male dominance and masculine cues), relatively reduced adrenocortical activity (facilitating body weight increases and gonadal function), and increased norepinephrine (again related to increased active aggressiveness). The "loser" shows the opposite responses and generally becomes more passive, "learns" submissive responses, and elicits less attack.

INFRAHUMAN PRIMATE STUDIES

Dixson (1980) reviewed the data suggesting that intermale aggression increases during the mating season in a number of primate species including *Lemur catta*, *Saimiri sciureus*, and *Macaca mulatta*. It is established that testosterone levels increase at this time. Dixson stressed that these correlations do not demonstrate a causal relationship between changes in testosterone and aggres-

TABLE 6 Recent Ethoexperimental Studies with Antihormones in Rodents

Authors	Material Studied	Male Subjects	Impact on Subject's Sex Accessory Weights	Impact on Behavior
Simon et al. (1987)	Cyproterone acetate	Paired individually housed mice	Suppresses weights	Reductions in offense only when both animals received drug. The antiandrogen augments threat and attack in encounters with non-drug-treated opponents and influences sexual activity, social investigation, and immobility.
Brain et al. (1988)	Tamoxifen	Lister hooded rats	No great influence	Markedly reduces time allocated to offense.
Hasan et al. (1988)	Tamoxifen	TO strain mice	No great influence	Dose dependent effects—but generally reduces offense—changing investigatory behavior at lower doses.
Brain et al. (1988)	CI 680	OFI strain mice	No great influence	Reduces time allocated to offense.
Brain and Hasan (1989)	Progesterone	Lister hooded rats	Suppresses weights	Progesterone reduces offense but also produces pronounced immobility and suppresses sexual behavior.

sive behavior. These associations rely on castration and androgen replacement studies that have been carried out in very diverse ways (see Table 7). The results show that although androgens exert some influences on aggressiveness in some of these primates, the effects are variable and are often influenced by social factors (especially in females). For example, Mirsky (1955) studied the effects of implanting gonadectomized *M. mulatta* with pellets of testosterone or estrogen. None of the treatments produced major effects on position in the hierarchy or on dominant/subordinate behavior in unisexed groups of males or females.

Somewhat in contrast to the data for nonprimates and humans, Doering et al. (1980) reported that LHRF did not influence agonistic behavior in male chimpanzees (*Pan troglodytes*).

Human Data

Brain (1984b) has critically reviewed the use of endocrine manipulations in controlling human aggression (accepting earlier statements that this epithet in humans is applied to a range of phenomena that are even more diverse than those seen in rodents). Castration has been applied to curb sexual aggression in Scandinavian and American populations. In spite of the considerable ethical problems associated with its use and the fact that it changes many aspects of physiology and behavior, such surgery has been claimed to produce impressively low rates of recidivism (one should, of course, examine the impacts of aging and perceptions of one's body here).

As noted earlier, therapies with hormones or antihormones are generally more ethically acceptable than castration because they seem (in theory at least) reversible. Estrogens have been used to control aggressive tendencies in intact men. The synthetic estrogen stilbestrol has been given orally to treat hyperirritable aggression and "excessive libido," but it has many unfortunate side effects including gynecomastia (development of breasts), fluid retention, and phlebothrombosis (production of blood clots), making its use problematic (Dunn, 1941). Chatz (1972) and Field and Williams (1970) advocated intramuscular or subcutaneous injections of long-acting estradiol BPC or estradiol valerate, which allowed the release of otherwise highly dangerous individuals. Both aggressive and sexual drive were essentially eliminated by such treatment.

Antiandrogens (e.g., cyproterone acetate) largely replaced castration (see later) and estrogen therapies in the treatment of Euro-

TABLE 7 Effects of Castration and Androgen Replacement in Adult Male Infrahuman Primates

Authors	Species	Type of Group	Behavioral Measures	Hormone Replacement	Findings
Green et al. (1972)	*Saimiri sciureus*	Adults: 3 males, 3 females	Genital display and feeding order	10 mg testosterone enanthate to dominant female and later to male castrate for several weeks	No effects on penile displays but an indication that feeding order is reduced by castration and restored by androgen.
Epple (1978)	*Saguinus fuscicollis*	Heterosexual pairs with strange intruders	Attack and threat on stimulus animal	Castration only	No declines in male aggressiveness towards males or females, but female residents often inhibited males from making "friendly" contacts.
Dixson and Herbert (1977)	*Miopithecus talapoin*	Heterosexual groups	Attack on lower-ranking males	25 or 75-100 mg testosterone pellets	No changes in hierarchies but testosterone increased aggressiveness towards lower-ranking males.

pean (aggressive?) sexual offenders. Berner et al. (1983) recorded treating 21 inmates of a Vienna prison with a combination of cyproterone acetate (100 milligrams (mg) per day orally for 1-10 years) and supportive psychotherapy. Taking the drug had no effect on release from prison, and the rearrest rate for sexual offenses was 28 percent in individuals that were followed up. There was no comparison to inmates treated with psychotherapy only.

Progesterone derivatives (e.g., A-norprogesterone) and MPA have also been used (especially in the United States) in clinical therapy of human hostility. These compounds all reduce endogenous testosterone production or action and are said to produce variable ameliorative effects on behavior. They certainly have rather complex actions, are not without associated problems, and seem currently to be used with less enthusiasm for a variety of technical, legal, and ethical reasons.

In terms of actions, it is well established that cyproterone acetate blocks endogenous testosterone efficacy by competing with 5α-dihydrotestosterone for receptor sites (Mainwaring, 1975). In contrast, the antiestrogen tamoxifen binds to intracellular estrogen receptors, prevents estrogen uptake, reduces the estrogen surge characteristic of early pregnancy, and alters gonadotropin production (Watson et al., 1975); it also produces dose-related declines in cytosol high-affinity estrogen receptors in a variety of tissues including the hypothalamus (Bowman et al., 1982). Consequently, antihormones can have quite wide repercussions on the endocrine system. One has to add to this the rider that categories of antihormones are not homogeneous. For example, "antiandrogens" may be subdivided into "pure antiandrogens," "antiandrogens with antigonadotropic effects," and "progestins." Administration of a pure antiandrogen (e.g., flutamide) to an intact male increases LH production and consequently augments plasma testosterone. Cyproterone acetate is an antiandrogen with antigonadotropic effects that acts directly on the testis and results in a decline in plasma testosterone. Progestins alter liver steroid metabolism, augmenting the metabolic clearance rate of testosterone (Albin et al., 1973). Some progestins are without apparent actions on testosterone uptake and binding in target tissues, whereas others (e.g., MPA) have a minor inhibitory action in this respect (Suffrin and Coffey, 1973).

Although there are strong indications that neural androgen receptors are implicated in some forms of aggressive behavior, not all forms of "violence" depend on such actions, and the details have been less systematically investigated than in the case of sexual behavior (e.g., Massa et al., 1983). Certainly, we do not know which neural androgen receptor populations are implicated in which aspects of behavior, and we know relatively little about the enzyme changes and transformations involved in androgen-mediated violent behavior. Sheard (1987) maintains that treatment with medroxyprogesterone is "the most common pharmacological approach [to the treatment of aggression] in the USA." The material has also been much used in Canada. This compound has been used to treat "aggression" in temporal lobe epilepsy. For example, O'Connor and Baker (1983) used MPA as an adjuvant in the treatment of three males (22-40 years of age) diagnosed as having chronic schizophrenia. In a double-blind study involving staff evaluations of behavior with the Brief Psychiatric Rating Scale, two of the patients who were assaultive showed significant dose-related (25-75 mg per week) drug improvements.

Many of the early attempts to correlate levels of testosterone with aggression in hostile and nonhostile prisoners (reviewed in Brain, 1984b), for example, have proved difficult to replicate. This seems related to the facts that the behavioral measures (e.g., rating by courts or the individuals per se) were often vague and divorced in time from the endocrine measurements (generally a single plasma determination of testosterone, a hormone that is secreted in a highly fluctuating manner, using samples that were taken in a "stressful" fashion, potentially reducing testosterone secretion). Rather obvious complications such as the incidence of homosexual activity in the populations and alcohol consumption were not controlled.

What then of the current position? Archer (1990) performed a limited meta-analysis on the five available studies that attempted to associate aggression as measured on the Buss-Durkee Hostility Inventory (Buss and Durkee, 1957) with plasma testosterone. The analysis suggested a very low but positive relationship between testosterone levels and overall Buss-Durkee Inventory score for the 230 males tested over the five studies. Social environment was more highly correlated with testosterone level than this score, and there was a closer association between aggression and the hormone when external assessments of the subject's behavior (rather than self-assessments) were made.

One should note that the Buss-Durkee scale is intended to measure aggressive feelings rather than aggressive actions. Indeed, Buss-Durkee factor II (the item correlated with testosterone in the above studies) is a composite of several measures, and there is little evidence of its relevance to violent or dominant behavior. Unfortunately, violent individuals rarely fill out questionnaires at the time of their violent actions! However, Langevin (personal communication, 1990) suggests that the perceptions of aggressive offenders are often very different from their actual behavior. This is an obvious area for future study.

Langevin et al. (1985) performed a comprehensive pilot study on predictive factors of sexual aggression in which hormones were examined as factors (total testosterone, LH, FSH, estradiol, dehydroepiandrosterone sulfate (DHAS), androsterone, cortisol, and prolactin). Adrenal production of sex hormones (notably DHAS) seemed important in sexually aggressive males, and it was felt that sex hormones other than testosterone may prove of relevance to sexual aggression (see also the data on lower vertebrates). It may be possible to distinguish sadists (abnormal LH and FSH) from rapists (elevated DHAS, cortisol, and prolactin). Bain et al. (1987) failed to find significant hormonal differences among murderers, assaulters, and controls but did suggest that further study of the complex interactions of these factors is necessary. There were indications that changes in LH and LHRF might be implicated in some forms of violent behavior. Bain et al. (1988) studied baseline values of eight hormones in sexually aggressive males and found no significant group differences. In an ACTH stimulation test, however, sexual aggressives had lower baseline values of DHAS than controls. These results appear more clear-cut than most, probably because they focus on sexual aggression, distinguish subcategories of this behavior, and seem prepared to measure a range of hormonal factors. One would still like to establish whether these approaches extrapolate to other populations and situations.

The so-called challenge tests of hormonal function were clinically developed initially to assess the integrity of the endocrine system (primarily with a view to detecting pathologies). Consequently, the function of the HPA axis could be challenged by a stimulatory dose of ACTH or by suppression with a synthetic glucocorticoid such as dexamethasone. It is thought that measurement of hormones after such treatments gives one an indication of the reactiveness of the particular endocrine system and may pick up differences between individuals that are not apparent in the "basal" (unchallenged) condition. One should comment that infrahuman animal studies suggest that maximal information is extracted when one has both basal and challenge test data. It is possible that varied endocrine reactivities are unrelated to ongoing associations between hormones and behavior.

McEwen and Pfaff (1985) have emphasized that the effects of

hormones on hypothalamic neurons can involve neurotransmitter effects and neuromodulator actions including LHRF and prolactin. They speculate that such interactions can be involved in processes such as the regulation of aggression. So far as humans are concerned, Tiwary (1974) speculated that LHRF was involved in behavioral change in a young child after testosterone treatment. Such studies are, however, contentious because there is considerable debate concerning the ethics of giving a synthetic analogue of LHRF (goserelin, Zoladex, ICI) to a pedophile in England (Brahams, 1988). The treatment was said to suppress sexual urges in a way not evident with cyproterone acetate or MPA.

Some more recent studies in our species have made use of competitive sporting situations. Although there are problems with these data, they are suggestive. Mazur and Lamb (1980) studied testosterone responses 1 to 2 hours after performance in a tennis doubles match for a cash prize, after obtaining a similar prize by lottery, and after success in medical degree examinations. Testosterone levels were elevated in the successful tennis players (compared to the losers) and the recipients of an M.D. degree but not the lottery winners. They suggest that when a male achieves enhanced status via his own efforts, both mood and testosterone levels are elevated. Elias (1981) measured cortisol, testosterone, and testosterone binding globulin in 15 males at three times in relation to wrestling bouts. Concentrations of both hormones increased during the bouts, but the binding globulin decreased in concentration. Winners of these competitive matches showed greater increases of both hormones than losers of the bouts. Salvador et al. (1987) carried out a pilot study on young male judo competitors in which plasma testosterone and cortisol were also measured. Winning or losing per se did not change the levels of these hormones. These authors did find, however, that previous personal success altered the response. Members of a regional team showed increases in testosterone levels postfighting that were not seen in individuals who had not been selected to represent their locality. The authors also emphasized that physical exercise alone could increase plasma testosterone levels.

Gladue et al. (1989) studied changes in testosterone and cortisol (assessed in saliva by immunoassays) in 40 young male U.S. subjects (18-34 years of age) in response to a nonathletic laboratory reaction time task. Subjects were randomly assigned to "winning" or "losing" categories by varying the computer feedback they received. Within the winning and losing categories, contests could be "close" or "decisive." Postcompetition mood was also assessed.

Winners had higher overall testosterone levels than losers, there being no significant difference between close and decisive wins. Winning and losing had no measurable effect on saliva cortisol measures. Mood was depressed in decisive losers compared to all other categories. The data suggest that the perception of winning or losing differentially influences testosterone output as a consequence of changing mood and apparent status.

Julian and McKenry (1989) suggested (on the basis of sociobiologic theory) that lower levels of aggression are most adaptive for men, particularly at midlife. They consequently studied 37 middle-aged (39-50 years of age) professional males from the American Midwest and used radioimmunoassay-determined serum measures of testosterone as the dependent variable. It was found (by using stepwise multiple regression analysis) that low testosterone titers related to enhanced marital and parental relationships and androgynous behavior. High levels of testosterone were significantly related to emotional expressiveness.

One should perhaps add that there have been repeated (rather anecdotal) suggestions that the anabolic steroids used illegally by some athletes may have a much more profound effect on aggression and hostility than they do on muscle development. There have even been claims from Sweden that the self-administration of these substances played a role in the homicidal behavior of weight lifters or body builders (especially when combined with alcohol ingestion). It seems appropriate to suggest that this claim should be properly evaluated in controlled studies. Any clear association would be a powerful incentive to regulate these substances more widely than at present.

Jeffcoate et al. (1986) studied relationships between dominant behavior (assessed by attendant females) of four males sharing a boating holiday of 14-day duration. This very preliminary study suggested a positive relationship between testosterone level on the boat and the rating of dominant behavior. Hellhammer et al. (1985) measured salivary testosterone levels in young males before, during, and after films containing erotic, sexual, stressful, aggressive, or neutral material. Short-term increases were found 15 minutes after exposure to erotic or sexual films. A decrease was found after exposure to stressful material, but the aggressive film produced no change.

Archer (1990) has emphasized the essentially correlational nature of the existing evidence linking androgenic hormones and measures of aggression in humans. He suggests that future research might involve more extensive longitudinal studies (as in

Olweus et al., 1988) or the manipulation of hormone levels as in investigations of human sexual behavior (e.g., Sherwin et al., 1985). Archer (1990) also suggests that the current methods of measuring aggression are confused and inadequate, often being based on rating scales that measure traits rather than states. Archer advocates (where appropriate) using direct assessments of aggression. Rating by peers, teachers, and staff may be useful in some cases, but one could also employ direct responses to provocation (e.g., Olweus, 1986), diary accounts of anger (Averill, 1982), anger inventories (Siegel, 1986), or the Conflict Tactics Scale (Straus, 1979), which involves asking the subject to rate how often they use particular strategies to solve conflicts. It is not unreasonable to suggest that a broad approach should be taken before attempting to standardize techniques.

It seems unlikely that androgens have a simple causal effect on human aggression and violence, but the patterns of production of sex steroids do appear to alter several factors (e.g., "aggressive feelings," self-image, and social signaling) that predispose individuals toward carrying out actions that can receive this label. Because environmental and experimental factors can profoundly influence androgen production in a wide range of organisms (including man), the impact of such variables on the incidence of violence should be assessed.

The basic methodological problem with the majority of data claiming to examine the relationship between androgens and violence has been the assumption (even if this is refuted) that there will be a simple causal relationship between these "male hormones" and this "masculine" behavior. Consequently, attempts have been made to correlate relatively crude measures of hormones (often without considering the time course of changes and the possibility of metabolic transformations) with proposed indices of aggression (ranging from actual behavior in animals to court records, peer evaluations, self-evaluations, questionnaire techniques, and responses to staged situations). The numbers of individuals employed and the homogeneity of the categories are also often open to question. There seems to be an urgent need for a much more sophisticated approach.

BODILY RHYTHMS AND VIOLENT BEHAVIOR

At several times in this account, we have touched on the fact that bodily rhythms (circadian, sex-cycle related, or seasonal) can have powerful effects on endocrine functioning and consequently on behavior. There is ample evidence from studies with lower vertebrates and infrahuman primates (see Brain, 1977-1979a) that associations between hormones and changes in aggressiveness can be demonstrated in both male and female organisms. It seems especially relevant to comment on the repeated suggestion (see reviews by Dalton, 1964; Lloyd and Weisz, 1975; Bardwick, 1976; Steiner and Carroll, 1977) that the increased hostility and irritability of some females evident in the premenstrual tension (PMT) syndrome has a hormonal component. Various authorities have implicated prolactin, progesterone, testosterone, or even aldosterone (an adrenal hormone that influences water and electrolyte balance). Although hormones may play a role in sensations of discomfort, it seems unlikely that "raging hormones" alone account for the violence sometimes associated with the female reproductive cycle—factors such as membership in religious groups and expectancy of PMT symptoms have a bearing on how individuals react to their physiological changes.

What of the male? It is certainly true that some lower vertebrates and infrahuman primate males show seasonal changes in HPG function that can be related to changes in aggression associated with reproductive activity. Although the cycles of male reproductive hormones are not as obvious as those of the female, there is some evidence of cyclicity (with acute variations) in some animals (e.g., cattle). Sex may certainly influence gonadal function in men. A study of 101 college volunteers has suggested that testosterone titer is positively associated with the number of orgasms achieved through masturbation (Monti et al., 1977) but that there is little relationship among the hormonal change, the Buss-Durkee Hostility Inventory, and a rating for "attributed aggression." Frodi (1977), in a study on 80 college freshmen, found that sexual arousal inhibited aggressive behavior mediated by deliberate angering, possibly as a consequence of an elevation of "anger-induced self-consciousness."

ADRENOMEDULLARY FUNCTION AND HUMAN AGGRESSION

In line with the studies on boys by Magnusson (1987, 1988), Woodman (1983) has reviewed the predictive power (in terms of assessing dangerousness) of examining the ratio of norepinephrine to epinephrine in response to a period of anticipation in a variety of incarcerated 18- to 45-year-old males (with no evidence of brain damage, renal dysfunction, or sensory defects, and a verbal I.Q.

greater than 80 on the Wechsler Adult Intelligence Scale (WAIS)). It was found that subjects with convictions for only violent crimes have a higher ratio of these adrenomedullary hormones than either subjects with a mixed violence and property crime background or those with convictions for sexual offenses. Woodman (1983) suggests that this finding supported the view that increased norepinephrine production (relative to epinephrine) "is found in more aggressive personalities."

"INDIRECT" EFFECTS ON AGGRESSION AND HOSTILITY

It is obvious, when reading the literature on associations between aggression and hormones in lower vertebrates, primates (see Dixson, 1980), and humans, that many of the associations are mediated via changes in the social signals employed between conspecifics (Brain, 1977-1979a). Barnett (1967) defined a "social signal" as "a small amount of energy or matter which induces a large change in the rate of energy release in a system, and it is produced by an animal and acts on another of the same species." Somewhat in contrast, Poole (1985) suggested that a "social signal is behavior which has evolved to convey information to a conspecific with the object of modifying its behavior for the benefit of the signaller." The latter author felt that there were basically two types of social signals, namely, discrete and graded. The alarm call of a ground squirrel (*Spermophilus parryi*) is a discrete signal. Aggressive vocalizations and threat displays, however, generally consist of a series of graded signals. Indeed, threat has been defined as a signal that potentiates withdrawal of a conspecific (Barnett, 1975), and one would expect the use of different intensities of display for different purposes. A comparative approach to social communication has revealed the great variety of signaling methods used by different species.

The endocrine system can serve as a "go-between" in such social communication. Kelley (1981) has described three ways in which endocrine mediation may be involved in social signaling. These are described separately below.

EFFECTS OF HORMONES ON PERCEPTION

The term "perception" refers to the processing of sensory input by the CNS (Gandelman, 1981). It is important to distinguish between the processing of information that is involved in the perception of an individual relationship in a given situation and the mechanisms involved in the elaboration and execution of a behavioral "project."

Perception is frequently assessed in routine clinical examinations. This is not because of an identified need to collect specific information on this important area of human performance, but is due rather to the clinically outmoded notion that the perceptual apparatus is particularly liable to brain damage (reviewed by Thomas et al., 1981). Hormones also alter perception in humans, possibly accounting for some behavioral changes. An animal's hormonal status certainly may affect its perception of stimuli that might act as social signals. Hormones can be regarded as acting on situational factors by altering the perception of signaling between conspecifics (Brain, 1983). Evidence for hormonal involvement in perception has been obtained for all the major sensory systems.

EFFECTS OF HORMONES ON SIGNAL GENERATION

Hormones may also alter the production of signals that serve social functions. The most frequently modified signals are somatosensory, olfactory, visual, or auditory. In many species, such signals have a profound effect on aggression (e.g., anosmic rodents do not fight, and conflict depends on the receiving of appropriate olfactory cues). Since it is certainly true that hormones modify both the perception of cues and the generation of potential signals in humans, it seems well worth examining the possibility that hormones can exert such indirect effects on aggression in our species also. It is worth adding that there is evidence that certain drug actions certainly are expressed in this manner. One of the ways in which alcohol influences human aggression is by interfering with rational social communication, leading to effects such as the "battered alcoholic syndrome." There have been virtually no attempts in clinical studies involving aggression to assess the impact of hormones in this way. This seems to be a rather obvious omission.

EXPERIENCE-INDUCED CHANGES IN
HORMONAL STATUS OF THE RECEIVER

There is good evidence that signals expressed by the behavior of conspecifics can alter the functioning of their recipient's endocrine system. Workers (e.g., Lehrman, 1965; Silver, 1983) have found that behavior produces endocrine changes in a variety of bird species. Harding (1981) also presented several examples of this kind of interaction from her work with hamsters. She observed, for example, increases in plasma prolactin and LH levels after mating in this species. Brain (1989b) has recently reviewed the evidence that fighting and (more specifically) subjection to defeat can produce wide-ranging repercussions in the endocrine systems of rodents (see below).

HORMONE-AGGRESSION CORRELATIONS IN MAMMALS

Although we have already touched on the topic of correlations between measures of aggression and titers of hormones, there are some studies in which the relationship can be regarded as more remote than in injection-behavioral analysis studies. The data presented are remote because aggression and the particular endocrine factor are presumed (rather than measured), and there is little hope of establishing whether direct or indirect hormonal influences are involved. Such studies provide, however, useful further pointers (reviewed in Brain, 1977-1979a).

These wide-ranging investigations (reviewed in Brain, 1977-1979a) involve species differences (e.g., von Euler, 1956); domestication (Popova et al., 1980; Hammer et al., 1990); sex differences (Archer, 1976); variations in genetic constitution (Selmanoff et al., 1975); maturation (Bernard et al., 1975); body size (Barr et al., 1976); housing condition (Brain and Benton, 1983); reproductive status, including seasonal changes (Rose et al., 1978), sexual cycles (Floody and Pfaff, 1977), pregnancy (Svare, 1977) and lactation (Haug and Brain, 1989); photoperiod (Balthazart and Hendrick, 1977); diet (Schultz and Lore, 1987); endocrine dysfunction (Tonks, 1977), and behavioral stereotyping (Wehle et al., 1978). The implied associations are complex but involve many species and varied endocrine factors.

Influences of Fighting on Endocrine Function in Mammals

Brain (1990a) has reviewed the hormonal impact of threat and fighting in rodents. He pointed out that the stress of fighting or defeat in rats and mice can produce temporally complex changes in hypothalamic releasing factors (LHRF and thyrotropin releasing factor); anterior pituitary hormones (TSH, ACTH, LH, and FSH); thyroid hormones; adrenomedullary catecholamines (e.g., epinephrine); adrenal glucocorticoids (e.g., corticosterone); and sex steroids. In general, adrenocortical hormones are increased and sex steroids reduced by such exposure. Dixson (1980) has also reviewed the available data, suggesting that in some infrahuman primates the stress of defeat produces a marked reduction in circulating androgen levels. Archer (1990) points out that stressful experiences (including aggression) can lead to reduced testosterone levels in humans. It is not at all unlikely that such stressors influence as wide a range of hormonal factors as those evident in rats and mice.

HORMONES, ALCOHOL, AND VIOLENCE

There is clearly a complex impact of alcohol on levels of violence in U.S. and U.K. populations—the associations are generated by alcohol influencing a variety of processes and perceptions (Brain, 1986). Acute or chronic alcohol ingestion has major impacts on neurophysiology and endocrinology (Berry and Brain, 1986). With particular relevance to the present study, ethanol changes the levels of endogenous opioids and other peptides, lowers blood calcium level, alters carbohydrate metabolism, increases pituitary adrenocortical function, and profoundly reduces the secretion of LHRF, LH, and testosterone. The last finding suggests that alcohol does not increase aggression by augmenting androgen release, but combinations of alcohol and androgens may be especially lethal. It is likely that certain alcohol-related increases in aggression and violence are more concerned with inappropriate processing of information or signaling rather than disinhibiting aggression. There is evidence that some endocrine disorders predispose individuals to ingest alcohol.

CONCLUSION

These are exciting times for studies attempting to relate hormones to aggressive and hostile behaviors. The approaches are moving away from the highly simplistic view that particular hormones switch aggressiveness on and off, to acceptance that these chemical factors play complex roles at a variety of stages of development in particular species and in particular contexts by altering the predisposition to produce activities that are likely to receive the labels aggressive or hostile. Although such associations are much more plastic than the old truisms, they are no less important.

Genuine progress in this area (which has lagged in sophistication behind developments in the study of sexual behavior) is likely if the quality of the behavioral analysis (which should be as direct and as detailed as possible) is matched to the sophistication of the endocrine manipulations and measures. The techniques for hormone assays are well developed for most current purposes, but it is obvious that we should look at a much wider range of hormones. Although still a long way from realization, nuclear magnetic resonance and magnetometry techniques might eventually hold the key to noninvasively estimating what is going on *in vivo* in clinical studies. In terms of research directions in the area of hormones and violence, it is suggested that studies on nonprimates, infrahuman primates, and humans all have utility to our understanding of the phenomena encountered in this research area. In nonprimates, it is suggested that one can examine the range of behaviors receiving the label aggression and can use molecular biologic and neurophysiologic techniques (rarely possible in other animals) to consider chemical transformations and hormone-brain interactions in greater detail. Such animals also provide valuable pilots for assessing the impact of currently little-investigated hormonal systems on aggression, as well as looking at associations among genes/hormones/behavior, drugs/hormones/behavior, and environment/hormones/behavior. Given current technologies, it should be possible to noninvasively examine heart rate as an indicator of arousal in many of these models. It should not be necessary to remind the panel that such studies should not be interpreted as providing support for a mechanism of violence. Studies with infrahuman primates are important not only because such animals have an evolutionary affinity to our own species, but because they enable us to examine the involvement of hormones over the life span in organisms that operate in complex social organizations. It is strongly advocated that early influences of hormones on subsequent behavioral potential be reexamined by using such species in well-controlled investigations that employ reasonable numbers of animals and look at alternative explanations (e.g., do primate mothers treat their androgenized female offspring differently?). The impact of hormonal changes around puberty should also be reinvestigated along with a more systematic evaluation of the impact of a range of hormonal and antihormonal treatments. So far as humans are concerned, there is an urgent need to perform multiple hormonal measurements (preferably using saliva samples to reduce stress effects) in well-evaluated, homogeneous groups of individuals. Multiple behavioral measurements concerning ongoing behavior are distinctly advantageous. Measuring hormones over the life span might be a valid aim, but it is difficult to see how this could be achieved without great personal interference and ethical problems (do you jail a person for excess hormone production?).

Finally, it should not be assumed that one can consider the impact of hormones on such behavior without being cognizant of the contributions of environmental factors, social experience and other biological factors (e.g., genes, neural circuits, and drugs) to the generation of those activities that (rightly) cause such current concern.

The major public policy implication of the review is that there is not a simple relationship between any hormone and behaviors that will receive the labels aggression or violence. "Raging hormones" do not cause violence. This having been said, there are diverse and subtle influences of hormones on the developing and developed individual, which can alter the predisposition for showing particular responses; authorities should be prepared to consider the involvement of these powerful messenger molecules in particular human processes (e.g., development of gender differences in behavior and onset of puberty) and recognize the possibility of mitigation in individuals with profound pathologies of the endocrine system. We should, however, move away from the highly simplistic view that hormones simply switch aggression on and off.

Another policy implication (because of the powerful and lasting effects of some hormones on morphology, physiology, and mood) is that we should be concerned about the impact of some clinical treatments (e.g., therapies of pregnant women) and borderline "misuse" (e.g., employment of anabolic steroids and human growth hormone by sports people and others). In the latter case, we should be especially concerned about hormone and alcohol combinations.

Because of their chronic effects, hormones and hormone derivatives offer the possibility of the development of therapies to assist individuals in coping with overwhelming behavioral problems. There is an urgent need to establish (a) which therapies are useful and (b) which subgroups of conditions are appropriate for treatment (we have already noted that rape, homicide, and violence are not homogeneous categories). Hormone treatments should never be considered as alternatives to other therapies but as an option. It is felt that appropriate hormone-based treatments should be combined with psychotherapy, counseling, and empathy training to achieve lasting results. The simple view of the relationship between hormones and violence leads to such therapies being regarded as a form of punishment or as a means of curing a disease. Finally, one should comment about the ethical mine field evident historically in all areas of biology. Legal professionals have to give serious consideration to the appropriateness of such treatments—because it is extremely difficult, especially when dealing with rapists, murderers, and assaulters, to establish informed consent. It would be a great pity simply to ban treatments that could be used to the benefit of individuals and society, but there have to be safeguards.

REFERENCES

Adams, D.B.
   1980   Motivational systems of agonistic behavior in muroid rodents: A comparative review and neural model. *Aggressive Behavior* 6:295-346.
Albin, J., J. Vitteck, G.G. Gordon, K. Altman, J. Olivo, and A.L. Southren
   1973   On the mechanism of the anti-androgenic effect of medroxyprogesterone acetate. *Endocrinology* 93:417.
Andrew, R.J.
   1975   Effects of testosterone on the behaviour of the domestic chick 1. Effects present in males but not in females. *Animal Behaviour* 23:139-155.
Archer, J.
   1976   The organization of aggression and fear in vertebrates. In P.P.G. Bateson and P. Klopfer, eds., *Perspectives in Ethology*. New York: Plenum Press.
   1977   Testosterone and persistence in mice. *Animal Behaviour* 25:479-488.
   1988   *The Behavioural Biology of Aggression*. Cambridge, England, Cambridge University Press.
   1990   The influence of androgens on human aggression. *British Journal of Psychology* 82:1-28.
Aron, C., D. Chateau, C. Schaeffer, and J. Roos
   1990   Heterotypic sexual behaviour in male mammals: The rat as an experimental model. Pp 98-126 in M. Haug, P.F. Brain, and C. Aron, eds., *Heterotypical Behaviour in Man and Animals*. London: Chapman and Hall.
Averill, J.R.
   1982   *Anger and Aggression: An Essay in Emotion*. New York: Springer-Verlag.
Bain, J., R. Langevin, R. Dickey, and M. Ben-Aron
   1987   Sex hormones in murderers and assaulters. *Behavioral Science and the Law* 5:95-101.
Bain, J., R. Langevin, R. Dickey, S. Hucker, and P. Wright
   1988   Hormones in sexually aggressive men I. Baseline values for eight hormones II. The ACTH test. *Annals of Sex Research* 1:63-78.

Balthazart, J., and J.C. Hendrick
1977 Hormonal control of behaviour and of testes growth in the quail *Coturnix c. japonica. Comptes Rendus des Seances de la biologie Societie* 171:656-663.

Balthazart, J., and M. Stevens
1975 Effects of testosterone propionate on the social behaviour of groups of male domestic ducklings *Anas platyrhynchos L. Animal Behaviour* 23:926-931.

Bardwick, J.M.
1976 Psychological correlates of the menstrual cycle and oral contraceptive medication. Pp 95-103 in E.J. Sachard, ed., *Hormones, Behavior and Psychopathology.* New York: Raven Press.

Barkley, M.S., and B.D. Goldman
1977 A quantitative study of serum testosterone, sex accessary organ weight growth and the development of intermale aggression in the mouse. *Hormones and Behavior* 8:208-218.

Barnett, S.A.
1967 Attack and defence in animal societies. Pp. 35-36 in C.D. Clemente, and D.B. Lindsley, eds., *Aggression and Defence: Neural Mechanisms and Social Patterns.* Berkley: University of California Press.
1975 *The Rat: A Study in Behaviour,* rev. ed. Chicago: University of Chicago Press.

Barr, G.A., J.L. Gibbons, and K.E. Moyer
1976 Male-female differences and the influence of neonatal and adult testosterone on intraspecies aggression in rats. *Journal of Comparative and Physiological Psychology* 90:1169-1183.

Benson, R.M., and C.J. Migeon
1975 Physiological and pathological puberty and human behavior. Pp. 155-184 in B.E. Eleftheriou and R.L. Sprott, eds., *Hormonal Correlates of Behavior.* New York: Plenum Press.

Bernard, B.K., E.R. Finkelstein, and G.M. Everett
1975 Alterations in mouse aggressive behavior and brain monoamine dynamics as a function of age. *Physiology and Behavior* 15:731-736.

Berner, W., G. Brownstone, and W. Sluga
1983 The cyproteroneacetate treatment of sexual offenders. *Neuroscience and Biobehavioral Reviews* 7:441-443.

Berry, M.S., and P.F. Brain
1986 Neurophysiological and endocrinological consequences of alcohol. Pp. 19-54 in P.F. Brain, ed., *Alcohol and Aggression.* London: Croom Helm.

Blanchard, R.J., P.F. Brain, D.C. Blanchard, and S. Parmigiani, eds.
1989 *Ethoexperimental Approaches to the Study of Behavior.* Dordrecht, Holland: Kluwer Academic Publishers.

Bowman, S.P., A. Leake, and I.D. Morris
1982 Hypothalamic, pituitary and uterine cytoplasmic and nuclear

oestrogen receptors and their relationships to the serum concentration of tamoxifen and its metabolite, 4-hydroxy tamoxifen, in the ovariectomized rat. *Journal of Endocrinology* 94:167-175.

Brahams, D.
1988   Voluntary chemical castration of a mental patient. *The Lancet* June 4:1291-1292.

Brain, P.F.
1977   *Hormones and Aggression*, Vol. 1. Montreal, Canada: Eden Press.
1978   *Hormones and Aggression*, Vol. 2. Montreal, Canada: Eden Press.
1979a   *Hormones, Drugs and Aggression*, Vol. 3. Montreal, Canada: Eden Press.
1979b   Effects of the hormones of the pituitary-gonadal axis on behaviour. Pp. 255-328 in K. Brown and S.J. Cooper, eds., *Chemical Influences on Behaviour*. London: Academic Press.
1979c   Effects of hormones of the pituitary-adrenocortical axis on behaviour. Pp. 329-371 in K. Brown and S.J. Cooper, eds., *Chemical Influences on Behaviour*. London: Academic Press.
1981   Differentiating types of attack and defense in rodents. Pp. 53-78 in P.F. Brain and D. Benton, eds., *Multidisciplinary Approaches to Aggression Research*. Amsterdam, Holland: Elsevier/North-Holland.
1983   Pituitary-gonadal influences and intermale aggressive behavior. Pp. 3-15 in B.B. Svare, ed., *Hormones and Aggressive Behavior*. New York: Plenum Press.
1984a   Comments on laboratory-based "aggression" tests. *Animal Behaviour* 32:1256-1257.
1984b   Biological explanations of human aggression and the resulting therapies offered by such approaches: A critical evaluation. Pp. 63-102 in R.J. Blanchard and D.C. Blanchard, eds., *Advances in the Study of Aggression*, Vol. 1. New York: Academic Press.
1986   Multidisciplinary examinations of the "causes" of crime: The case of the link between alcohol and violence. *Alcohol and Alcoholism* 21:237-240.
1989a   *The Nature and Control of Aggression*. Oxford, England: Oxford Project for Peace Studies No. 19.
1989b   The adaptiveness of house mouse aggression. Pp. 1-21 in P.F. Brain, D. Mainardi, and S. Parmigiani, eds., *House Mouse Aggression*. Chur, Switzerland: Harwood Academic Publishers gmbh.
1989c   An ethoexperimental approach to behavioural endocrinology. Pp. 539-557 in R.J. Blanchard, P.F. Brain, D.C. Blanchard, and S. Parmigiani, eds., *Ethoexperimental Analysis of Behavior*. Dordrecht, Holland: Kluwer Academic Publishers.
1990a   Stress in agonistic contexts in rodents. Pp. 73-85 in R. Dantzer

and R. Zayan, eds., *Social Stress in Domestic Animals*. Dordrecht, Holland: Kluwer Academic Publishers.

1990b *Mindless Violence? The Nature and Biology of Aggression*. Swansea, U.K.: University College of Swansea.

Brain, P.F., and D. Benton
1983 Conditions of housing, hormones and aggressive behavior. Pp. 349-372 in B.B. Svare, ed., *Hormones and Aggressive Behavior*. New York: Plenum Press.

Brain, P.F., and N.J. Bowden
1979 Sex steroid control of intermale fighting in mice. Pp. 403-465 in W.B. Essman and L. Valzelli, eds., *Current Developments in Psychopharmacology*, Vol. 5. New York: Spectrum Publications.

Brain, P.F., and M. Haug
1990 Are behaviours specific to animals of particular sex? Pp. 1-15 in M. Haug, P.F. Brain, and C. Aron, eds., *Heterotypical Behaviour in Man and Animals*. London: Chapman and Hall.

Brain, P.F., and N.W Nowell
1969 Some endocrine and behavioral changes in the development of the albino laboratory mouse. *Communications in Behavioral Biology* 4:203-220.

Brain P.F., M. Haug, and Alias bin Kamis
1983 Hormones and different tests for "aggression" with particular reference to the effects of testosterone metabolites. Pp. 290-304 in J. Balthazart, E. Prove, and R. Gilles, eds., *Hormones and Behaviour in Higher Vertebrates*. Berlin: Springer-Verlag.

Brain, P.F., V. Simon, S. Hasan, M. Martinez, and D. Castano
1988 The potential of antiestrogens as centrally-acting anti-hostility agents: Recent animal data. *International of Journal of Neuroscience* 41:169-177.

Browne, K.D., and J. Archer
1989 *Human Aggression: Naturalistic Approaches*. London: Routledge.

Buss, A.H.
1971 Aggression pays. Pp. 7-18 in J.L. Singer, ed., *The Control of Aggression and Violence: Cognitive and Physiological Factors*. New York: Academic Press.

Buss, A.H., and A. Durkee
1957 An inventory for assessing different types of hostility. *Journal of Consulting Psychology* 21:343-349.

Cairns, R.B.
1972 Fighting and punishment from a developmental perspective. Pp. 59-124 in J.K. Cole and D.D. Jensen, eds., *Nebraska Symposium on Motivation*. Lincoln: Nebraska University Press.

Carlson, N.R.
1977 *Physiology of Behavior*, 3rd ed. Boston: Allyn and Bacon Inc.

Chatz, T.L.
1972   Recognizing and treating dangerous sex offenders. *International Journal of Offenders and Therapy* 2:109-115.
Chevins, P.F.D.
1989   Early environmental influences on fear and defence in rodents. Pp. 269-288 in P.F. Brain, S. Parmigiani, D. Mainardi, and R.J. Blanchard, eds., *Fear and Defence*. Chur, Switzerland: Harwood Academic Press gmbh.
Clemens, L.G., B.A. Gladue, and L.F. Coniglio
1978   Prenatal endogenous androgenic influences on masculine sexual behavior and genital morphology in male and female rats. *Hormones and Behavior* 10:40-53.
Conner, R.L., A.P. Constantino, and G.C. Scheuch
1983   Hormonal influences on shock-induced fighting. Pp. 119-144 in B.B. Svare, ed., *Hormones and Aggressive Behavior*. New York: Plenum Press.
Dalton, K.
1964   *The Premenstrual Syndrome*. London: William Heinemann.
Daruna, J.H.
1978   Patterns of brain monoamine activity and aggressive behavior. *Neuroscience and Biobehavioral Reviews* 2:101-113.
Darwin, C.
1872   *The Expression of the Emotions in Man and Animals*. London: D. Appleton and Co.
Dixson, A.F.
1980   Androgens and aggressive behavior in primates: A review. *Aggressive Behavior* 6:37-67.
Dixson, A.F., and J. Herbert
1977   Testosterone, aggressive behavior and dominance rank in captive adult male talapoin monkeys *Miopithecus talapoin*. *Physiology and Behavior* 18:539-543.
Doering, C.H., P.R. McGinnis, H.C Kraemer, and D.A. Hamburg
1980   Hormonal and behavioral response of male chimpanzees to a long-acting analogue of gonadotropin-releasing hormone. *Archives of Sexual Behavior* 9:441-450.
Dunn, C.W.
1941   Stilbestrol-induced testicular degeneration in hypersexual males. *Journal of Clinical Endocrinology and Metabolism* 1:643-648.
Eaton, G.G., R.W. Goy, and C.H. Phoenix
1973   Effects of testosterone treatment in adulthood on sexual behaviour of female pseudohermaphrodite rhesus monkeys. *Nature* (London) 242:119-120.
Eccles, J.S., C. Miller, M.L. Tucker, J. Becker, W. Schramm, R. Midgley, W. Holmes, L. Pasch, and M. Miller
1988   Hormones and Affect at Early Adolescence. Paper presented at the biannual meeting of the Society for Research on Adolescence, Alexandria, Va.

Edwards, D.A.
1968 Mice: Fighting by neonatally androgenized females. *Science* 161:1027-1028.
1969 Early androgen stimulation and aggressive behavior in male and female mice. *Physiology and Behavior* 4:333-338.
1971 Neonatal administration of androstenedione, testosterone or testosterone propionate: Effects on ovulation, sexual receptivity, and aggressive behavior in female mice. *Physiology and Behavior* 6:223-228.
Ehrhardt, A.A.
1969 Zur wirkung fötaler Hormone auf Intelligenz und geschlechtsspezifisches Verhalten. Doctoral thesis, Universitat Dusseldorf.
Ehrhardt, A.A., and S.W. Baker
1974 Fetal androgen, human CNS differentiation and behavior sex differences. Pp. 53-76 in R.C. Friedman, R.M. Richart, R.L. Van de Wiele, eds., *Sex Differences in Behavior*. New York: John Wiley and Sons.
Ehrhardt, A.A., and J. Money
1967 Progestin-induced hermaphroditism: IQ and psychosexual identity in a study of ten girls. *Journal of Sex Research* 3:83-100.
Elias, M.
1981 Serum cortisol, testosterone, and testosterone-binding globulin responses to competitive fighting in human males. *Aggressive Behavior* 7:215-224.
Epple, G.
1978 Lack of effects of castration on scent marking displays and aggression in a South American primate *Saguinus fuscicollis*. *Hormones and Behavior* 11:139-150.
Farrington, D.P.
1989 Early predictors of adolescent aggression and adult violence. *Violence and Victims* 4:79-100.
Field, L.H., and M. Williams
1970 The hormonal treatment of sexual offenders. *Medicine, Science, and the Law* 10:27-34.
Floody, O.R., and D.W. Pfaff
1977 The hormonal basis for fluctuations in female aggressiveness correlated with estrous state. *Journal of Comparative and Physiological Psychology* 91:443-464.
Frodi, A.
1977 Sexual arousal, situational restrictiveness, and aggressive behavior. *Journal of Research in Personality* 11:48-58.
Gadelman, R.
1981 Androgens and fighting behavior. Pp. 215-230 in P.F. Brain and D. Benton, eds., *The Biology of Aggression*. Alphen aan den Rijn, Holland: Sijthoff and Noordhoof b.v.

Gladue, B.A., M. Boechler, and K.D. McCaul
1989 Hormonal response to competition in human males. *Aggressive Behavior* 15:409-422.

Goldman, L., and H.H. Swanson
1975 Developmental changes in pre-adult behavior in confined colonies of Golden hamsters. *Developmental Psychology* 8:137-150.

Goldstein, J.H.
1986 *Aggression and Crimes of Violence.* New York: Oxford University Press.

Gotz, F., W. Rohde, and C. Dorner
1990 Neuroendocrine differentiation of sex-specific gonadotrophin secretion, sexual orientation, and gender role behavior. Pp. 167-194 in M. Haug, P.F. Brain, and C. Aron, eds., *Heterotypical Behavior in Man and Animals.* London: Chapman and Hall.

Goy, R.W.
1968 Organizing effects of androgen on the behavior of rhesus monkeys. Pp. 12-31 in R.P. Michael, ed., *Endocrinology and Human Behavior.* London: Oxford University Press.

Goy, R.W., and M. Roy
1990 Heterotypic sexual behavior in female mammals. Pp. 71-97 in M. Haug, P.F. Brain, and C. Aron, eds., *Heterotypical Behaviour in Man and Animals.* London: Chapman and Hall.

Green, R., R.E. Whalen, B. Butley, and C. Battie
1972 Dominance hierarchy in squirrel monkeys: Role of gonads and androgen on genital display and feeding order. *Folia Primatologia* 18:185-195.

Groth, A.N.
1979 *Men Who Rape: The Psychology of the Offender.* New York: Plenum Press.

Hamburg, D.A.
1971 Recent research on hormonal factors relevant to human aggressiveness. *International Social Science Journal* 23:36-47.

Hammer, R.P., K.M. Hori, P. Cholvanich, D.C. Blanchard, and R.J. Blanchard
1990 Opiate, serotonin, and benzodiazepine receptor systems in rat brain defense circuits. Pp. 201-217 in P.F. Brain, S. Parmigiani, R. Blanchard, and D. Mainardi, eds., *Fear and Defence.* Chur, Switzerland: Harwood Academic Press, gmbh.

Harding, C.F.
1981 Social modulation of circulating hormone levels in the male. *American Zoologist* 21:223-231.
1989 Interactions of androgens and estrogens in the modulation of social behavior in male songbirds. Pp. 558-579 in R.J. Blanchard, P.F. Brain, D.C. Blanchard, and S. Parmigiani, eds., *Ethoexperimental Approaches to the Study of Behavior.* Dordrecht, Holland: Kluwer Academic Press.

Hasan, S.A., P.F. Brain, and D. Castano
1988 Studies of the effects of Tamoxifen ICI 46474 on agonistic en-

counters between pairs of intact mice. *Hormones and Behavior* 22:178-185.

Haug, M., and P.F. Brain
1989 Psychobiological influences of attack on lactating females: A varient on "typical" house mouse aggression. Pp. 205-222 in P.F. Brain, D. Mainardi, and S. Parmigiani, eds., *House Mouse Aggression*. Chur, Switzerland: Harwood Academic Publishers gmbh.

Hays, S.E.
1978 Strategies for psychoendocrine studies of puberty. *Psychoeuroendocrinology* 3:1-15.

Hellhammer, D.H., W. Hubert, and T. Schurmeyer
1985 Changes in saliva testosterone after psychological stimulation in men. *Psychoeuroendocrinology* 10:77-81.

Hucklebridge, F.H., L. Gamal el Din, and P.F. Brain
1981 Social status and the adrenal medulla in the house mouse (*Mus musculus* L.). *Behavioral and Neural Biology* 33:345-363.

Huntingford, F., and A. Turner
1987 *Animal Conflict*. London: Chapman and Hall.

Jeffcoate, W.J., N.B. Lincoln, C. Selby, and M. Herbert
1986 Correlation between anxiety and serum prolactin in humans. *Journal of Psychosomatic Research* 30:217-222.

Jones, S.E., and P.F. Brain
1987 Performances of inbred and outbred laboratory mice in putative tests on aggression. *Behavior Genetics* 17:87-96.

Joslyn, W.D.
1973 Androgen-induced social dominance in infant female rhesus monkeys. *Journal of Child Psychology and Psychiatry* 14:137-145.

Julian, T., and P.C. McKenry
1989 Relationship of testosterone to men's family functioning at midlife: A research note. *Aggressive Behavior* 15:281-289.

Kaplan, H., and S.O. Hyland
1972 Behavioral development in the Mongolian gerbil *Meriones unguiculatus*. *Animal Behaviour* 20:147-154.

Karli, P.
1981 Conceptual and methological problems associated with the study of brain mechanisms underlying aggressive behavior. Pp. 323-362 in P.F. Brain and D. Benton, eds., *The Biology of Aggression*. Alphen aan den Rijn, Holland: Sijthoff and Noordhoof b.v.

Kelley, D.B.
1981 Social signals—an overview. *American Zoologist* 21:111-116.

Kesler, P., R. Green, S.J. Finch, and K. Williams
1980 Prenatal "female hormone" administration and psychosexual development in human males. *Psychoeuroendocrinology* 5:269-285.

Klama, J.
1988 Aggression: *Conflict in Animals and Humans Reconsidered.* Burnt Mill, Harlow, England: Longman Group.

Kutash, I.L., S.B. Kutash, L.B. Schlesinger and Associates, eds.
1978 *Violence: Perspectives on Murder and Aggression.* San Francisco: Jossey-Bass Publishers.

Langevin, R., J. Bain, M. Ben-Aron, R. Coulthard, D. Day, L. Handy, G. Heasman, S. Hucker, J. Purins, V. Roper, A. Russan, C. Webster, and G. Wortzman
1985 Sexual aggression: Constructing a predictive equation. Pp. 50-93 in R. Langevin, ed., *Erotic Preference Gender Identity and Aggression in Men.* Hillsdale, N.J.: Lawrence Erlbaum Associates.

Larsson, K., P. Sodersten, and C. Beyer
1973 Induction of male sexual behavior by oestradiol benzoate in conjunction with dihydrotestosterone. *Journal of Endocrinology* 57:563-564.

Laskowski, W.
1954 Einige verhaltensstudien an *Platypoecilus variatus. Biologisches Zentralblatt* 73:429-438.

Lehrman, D.S.
1965 Interaction between internal and external environments in the regulation of the reproductive cycle of the ring dove. Pp. 355-380 in F.A. Beach, ed., *Sex and Behavior.* New York: John Wiley & Sons.

Leshner, A.I.
1980 Interaction of experience and neuroendocrine factors in determining behavioral adaptations to aggression. Pp. 427-438 in P.S. McConnell et al., eds., *Progress in Brain Research,* Vol. 53. Amsterdam, Holland: Elsevier/North-Holland.

1981 The role of hormones in the control of submissiveness. Pp. 309-322 in P.F. Brain and D. Benton, eds., *Multidisciplinary Approaches to Aggression Research.* Amsterdam, Holland: Elsevier/North-Holland.

Leshner, A.I., and K.E. Roche
1977 Comparison of the effects of ACTH and lysine vasopressin on avoidance-of-attack in mice. *Physiology and Behavior* 18:879-833.

Levander, S., A. Mattsson, D. Schalling, and A. Dalteg
1987 Psychoendocrine patterns within a group of male juvenile delinquents as related to early psychosocial stress, diagnostic classification, and follow-up data. Pp. 235-252 in D. Magnusson and A. Ohman, eds., *Psychopathology: An International Perspective.* Orlando, Florida: Academic Press.

Lincoln, G.A.
1987 Long-term stimulatory effects of a continuous infusion of LHRH

agonist on testicular function in male red deer (*Cervus elaphus*). *Journal of Reproduction and Fertility* 80:257-261.

Lloyd, C.W., and J. Weisz
1975   Hormones and aggression. Pp. 92-113 in W.S. Fields and W.H. Sweet, eds., *Neural Bases of Violence and Aggression*. St. Louis, Missouri: Warren H. Green.

Magnusson, D.
1987   Adult delinquency in the light of conduct and physiology at an early age: A longitudinal study. Pp. 221-234 in D. Magnusson and A. Ohman, eds., *Psychopathology: An International Perspective*. Orlando, Florida: Academic Press.
1988   *Individual Development from an Interactional Perspective: A Longitudinal Study*. Hillsdale, N.J.: Lawrence Erlbaum Associates.

Mainwaring, W.I.P.
1975   A review of the formation and binding of 5α dihydrotestosterone in the mechanism of action of androgens in the prostate of the rat and other speicies. *Journal of Reproduction and Fertility* 44:377-393.

Massa, R., L. Bottoni, and V. Lucini
1983   Brain testosterone metabolism and sexual behavior in birds. Pp. 230-236 in J. Balthazart, E. Prove, and R. Gilles, eds., *Hormones and Behavior in Higher Vertebrates*. Berlin: Springer-Verlag.

Mattsson, A., D. Schalling, D. Olweus, H. Low, and J. Svensson
1980   Plasma testosterone, aggressive behavior and personality dimensions in young male delinquents. *Journal of the American Academy of Child Psychiatry* 19:476-490.

Mazur, A., and T.A. Lamb
1980   Testosterone, status and mood in human males. *Hormones and Behavior* 14:236-246.

McEwen, B.S., and D.W. Pfaff
1985   Hormone effects on hypothalamic neurons: Analysing gene expression and neuromodulator actions. *Trends in Neuroscience* 8:105-110.

McEwen, B.S., D.W. Pfaff, and R.E. Zigmond
1970   Factors influencing sex hormone uptake by rate brain regions: Effects of competing steroids on testosterone uptake. *Brain Research* 21:29-38.

McKinney, T.D., and C. Desjardins
1973   Postnatal development of the testis, fighting behavior and fertility in house mice. *Biology of Reproduction* 9:279-294.

Meyer-Bahlburg, H.F.L.
1980   Androgens and human aggression. Pp. 263-290 in P.F. Brain and D. Benton, eds., *The Biology of Aggression*. Alphen aan den Rijn, Holland: Sijthoff and Noordhoof b.v.

Meyer-Bahlburg, H.F.L., and A.A. Ehrhardt
1982 Prenatal sex hormone and human aggression: A review and new data on progestogen effects. *Aggressive Behavior* 8:39-62.

Meyerson, B.J.
1967 Relationship between the anesthetic and gestagenic action and oestrous behavior-inducing activity of different progestins. *Endocrinology* 81:369-374.
1983 Endorphin-monoamine interaction and steroid-dependent behavior. Pp. 111-117 in J. Balthazart, E. Prove, and R. Gilles, eds., *Hormones and Behavior in Higher Vertebrates*. Berlin: Springer-Verlag.

Miczek, K.A., and M. Krsiak
1981 Pharmacological analysis of attack and flight. Pp. 341-354 in P.F. Brain and D. Benton, eds., *Multidisciplinary Approaches to Aggression Research*. Amsterdam, Holland: Elsevier/North Holland.

Mirsky, A.F.
1955 The influence of sex hormones on social behavior in monkeys. *The Journal of Comparative and Physiological Psychology* 48:327-335.

Money, J.
1990 The development of sexuality and eroticism in humankind. Pp. 127-166 in M. Haug, P.F. Brain, and C. Aron, eds., *Heterotypical Behavior in Man and Animals*. London: Chapman and Hall.

Money, J., and A.A. Ehrhardt
1972 *Man and Woman, Boy and Girl*. Baltimore, Md.: Johns Hopkins University Press.

Money, J., and C. Ogunro
1974 Behavioral sexology: Ten cases of genetic male intersexuality with impaired prenatal and pubertal androgenization. *Archives of Sexual Behavior* 3:181-205.

Money, J., and M. Schwartz
1976 Fetal androgens in the early treated adrenogenital syndrome of 46 hermaphroditism: Influence on assertive and aggressive types of behavior. *Aggressive Behavior*: 2:19-30.

Monti, P.M., W.A. Brown, and D.P. Corriveau
1977 Testosterone and components of aggressive and sexual behavior in man. *American Journal of Psychiatry* 134:692-694.

Mugford, R.A.
1974 Androgenic stimulation of aggression eliciting cues in adult opponent mice castrated at birth, weaning or maturity. *Hormones and Behavior* 5:93-102.

Naess, O., and A. Attramadal
1974 Uptake and binding of androgens in the anterior pituitary gland, hypothalmus, preoptic and brain cortex of rats. *Acta endocrinologica Kbn.* 76:417-430.

Naftolin, F., and R.J. Ryan
1975   The metabolism of androgen in central neuroendocrine tissues. *Journal of Steroid Biochemistry* 6:993-997.
O'Connor, M., and H.W.G. Baker
1983   Depo-medroxy progesterone acetate as an adjunctive treatment in three aggressive schizophrenic patients. *Acta Psychiatrica Scandinavica* 67:399-402.
Olweus, D.
1984   Development of stable aggressive reaction patterns in males. Pp. 103-137 in R.J. Blanchard and D.C. Blanchard, eds., *Advances in the Study of Aggression*, Vol. 1. Orlando, Florida.: Academic Press.
1986   Aggression and hormones: Behavioral relationship with testosterone and adrenaline. Pp. 51-72 in D. Olweus, J. Block, and M. Radke-Yarrow, eds., *Development of Antisocial and Prosocial Behavior: Research, Theories, and Issues*. New York: Academic Press.
Olweus, D., A. Mattsson, D. Schalling, and H. Low
1980   Testosterone, aggression, physical and personality dimensions in normal adolescent males. *Psychosomatic Medicine* 42:253-269.
1988   Circulating testosterone levels and aggression in adolescent males. *Psychosomatic Medicine* 50:261-272.
Owens, N.W.
1975   A comparison of aggressive play and aggression in free-living baboons *Papio anubis*. *Animal Behaviour* 23:757-765.
P'an, S.Y., and G.D. Laubach
1964   Steroid central depressants. Pp. 415-475 in R.I. Dorfman, ed., *Methods in Hormone Research*, Vol. 3. New York: Academic Press.
Parrott, R.F.
1975   Aromatizable and 5α-reduced androgens: Differentiation between central and peripheral effects on male rat sexual behavior. *Hormones and Behavior* 6:99-108.
1976   Homotypical sexual behavior in gonadectomized female and male rats treated with 5α-19-hydroxytestosterone: Comparison with related androgens. *Hormones and Behavior* 7:207-215.
Payne, A.P.
1976   A comparison of the effects of neonatally administered testosterone, testosterone propionate and dihydrotestosterone on aggressive and sexual behavior in female Golden hamster. *Journal of Endocrinology* 69:23-31.
Payne, A.P., and H.H. Swanson
1972   The effect of sex hormones on the agonistic behavior of the male Golden hamster. *Physiology and Behavior* 8:687-691.
Pfaff, D.W.
1971   Steroid sex hormones in the rate brain: Specificity of uptake

and physiological effects. Pp. 103-112 in C.H. Sawyer and R.A. Gorski, eds., *Steroid Hormones and Brain Function.* Berkeley.: University of California Press.

Poole, T.B.
    1985   *Social Behavior in Mammals.* New York: Chapman and Hall.

Popova, N.K., N.N. Voitenko, S.I. Pavlova, E.V. Naumenko, and D.K. Belyaev
    1980   Genetics and phenogenetics of hormonal characteristics in animals. VII. Relationship between brain serotonin and hypothalamic-pituitary-adrenal axis under emotional stress in domesticated and non-domesticated silver foxes. *Genetics* 16:1865-1870 (in Russian).

Poshivalov, V.P.
    1982   Ethological analysis of neuropeptides and psychotropic drugs: Effects on intraspecies aggression and sociability of isolated mice. *Aggressive Behavior* 8:355-369.

Reinisch, J.M.
    1981   Prenatal exposure to synthetic progestins increases potential for aggression in humans. *Science* 211:1171-1173.

Resko, J.A.
    1974   The relationship between fetal hormones and the differentiation of the central nervous system in primates. Pp. 211-222 in W. Montagna and W.A. Sadler, eds., *Reproductive Behavior.* New York: Plenum Press.

Roberts, S.S.
    1990   Murder, mayhem, and other joys of youth. *The Journal of NIH Research* 2:67-72.

Rose, R.M., T.P. Gordon, and I.S. Bernstein
    1978   Diurnal variation in plasma testosterone and cortisol in rhesus monkeys living in social groups. *Journal of Endocrinology* 76:67-74.

Salvador, A., V. Simon, F. Suay, and L. Llorens
    1987   Testosterone and cortisol responses to competitive fighting in human males: A pilot study. *Aggressive Behavior* 13:9-13.

Schallert, T.
    1977   Reactivity to food odours during hypothalamic stimulation in rats not experienced with stimulation-induced eating. *Physiology and Behavior* 18:1061-1066.

Scott, J.P.
    1981   The evolution of function in agonistic behavior. Pp. 129-157 in P.F. Brain and D. Benton, eds., *Multidisciplinary Approaches to Aggression Research.* Amsterdam, Holland: Elsevier/North-Holland Biomedical Press.

Schultz, L., and R. Lore
    1987   Jolly fat rats? The effects of diet-induced obesity on fighting. *Aggressive Behavior* 13:359-366.

Selmanoff, M.K., J.E. Jumonville, S.C. Maxson, and B.E. Ginsburg
1975 Evidence for a Y-chromosome contribution to an aggressive phenotype in inbred mice. *Nature* 253-529-530.
Sharp, P.J., J. Culbert, and J.W. Wells
1977 Variations in stored and plasma concentrations of androgens and luteinizing hormone during sexual development in the cockerel. *Journal of Endocrinology* 74:467-476.
Sheard, M.H.
1987 Psychopharmacology of aggression in humans. Pp. 257-266 in B. Olivier, J. Mos, and P.F. Brain, eds., *Ethopharmacology of Agonistic Behavior in Animals and Humans.* Dordecht, Holland: Martinus Nijhoff Publishers.
Sherwin, B.B., M.M. Gelfand, and B. Brender
1985 Androgen enhances sexual motivation in females: A prospective crossover study of sex steroid administration in the surgical menopause. *Psychosomatic Medicine* 47:339-351.
Siegel, J.M.
1986 The multidimensional anger inventory. *Journal of Personality and Social Psychology* 51:191-200.
Sigg, E.B., C. Day, and C. Colombo
1966 Endocrine factors in isolation-induced aggressiveness in rodents. *Endocrinology* 78:679-684.
Silver, R.
1983 Biparental care in birds: Mechanisms controlling incubation bout duration. Pp. 451-462 in J. Balthazart, E. Prove, and R. Gilles, eds., *Hormones and Behavior in Higher Vertebrates.* Berlin: Springer-Verlag.
Simon, N.G., and R.E. Whalen
1986 Hormonal regulation of aggression: Evidence for a relationship among genotype, receptor binding and behavior sensitivity to androgen and estrogen. *Aggressive Behavior* 12:255-266.
Simon, V., M. Martinez, D. Castano, P.F. Brain, and S. Hasan
1987 Studies on the effects of the anti-androgen cyproterone acetate on social encounters between pairs of male mice. *International Journal of Neuroscience* 41:231-240.
Steiner, M., and B.J. Carroll
1977 The psychobiology of premenstrual dysphoria: Review of theories and treatments. *Psychoneuroendocrinology* 2:321-325.
Straus, M.
1979 Measuring intrafamily conflict violence: The conflict tactics CT scales. *Journal of Marriage and the Family* 41:75-88.
Suchowsky, G.K., L. Pegrassi, and A. Bonsignori
1969 The effect of steroids on aggressive behavior in isolated male mice. Pp. 161-171 in S. Garattini and E.B. Sigg, eds., *Aggressive Behaviour.* Amsterdam, Holland: Excerpta Medica Foundation.
Suffrin, G., and D.S. Coffey
1973 A new model for studying the effect of drugs on prostatic growth

I. Antiandrogens and DNA synthesis. *Investigative Urology* II:45-54.

Susman, E.J., G. Inoff-Germain, E.D. Nottelmann, D.L. Loriaux, G.B. Cutler, and G.P. Chrousos
  1987 Hormones, emotional dispositions, and aggressive attributes in young adolescents. *Child Development* 58:1114-1134.

Svare, B.
  1977 Maternal aggression in mice: Influence of the young. *Biobehavioral Reviews* I:151-164.
  1989 Recent advances in the study of female aggressive behavior in mice. Pp. 135-159 in P.F. Brain, D. Mainardi, and S. Parmigiani, eds., *House Mouse Aggression*. Chur, Switzerland: Harwood Academic Publishers gmbh.

Svare, B.B., and M.A. Mann
  1983 Hormonal influences on maternal aggression. Pp. 91-104 in B.B. Svare, ed., *Hormones and Aggressive Behavior*. New York: Plenum Press.

Symons, J.
  1973 Aggressive Play in a Free-Ranging Group of Rhesus Monkeys *Macaca mulatta*. Ph.D. thesis, University of California, Berkeley.

Thomas, D.A., R.J. Talala, and R.J. Barfield
  1981 Effect of devocalization of the male on mating behavior in rats. *Journal of Comparative and Physiological Psychology* 95:630-637.

Tiwary, C.M.
  1974 Testosterone, LHRH, and behavior. *Lancet* May 18:993.

Tonks, C.M.
  1977 Psychiatric aspects of endocrine disorders. *Practioner* 218:526-531.

Valenti, G., and M. Mainardi
  1989 Aggressiveness in mice and thyroid hormones. Pp. 293-309 in P.F. Brain, D. Mainardi, and S. Parmigiani, eds., *House Mouse Aggression*. Chur, Switzerland: Harwood Academic Publishers gmbh.

van de Poll, N.E., and H. van Dis
  1977 Hormone induced lordosis and its relation to masculine sexual activity in male rats. *Hormones and Behavior* 8:17-7.

van de Poll, N.E., J.P.C. de Bruin, H. van Dis, and H.G. Van Oyen
  1978 Gonadal hormones and the differentiation of sexual and aggressive behavior and learning in the rat. Pp. 309-327 in *Progress in Brain Research*, Vol. 18. Amsterdam: Elsevier/North-Holland.

van de Poll, N.E., F. de Jonge, H.G. Van Oyen, J. Van Pelt, and J.P.C. de Bruin
  1981 Failure to find sex differences in testosterone-activated aggression in two strains of rats. *Hormones and Behavior* 15:94-105.

vom Saal, F.S.
   1983   Models of early hormonal effects on intersex aggression in mice. Pp. 197-222 in B.B. Svare, ed., *Hormones and Aggressive Behavior*. New York: Plenum Press.

Neurochemistry and Pharmacotherapeutic Management of Agggression and Violence

Klaus A. Miczek, Margaret Haney, Jennifer Tidey, Jeffrey Vivian, and Elise Weerts

NEUROSCIENCE PERSPECTIVE

Violence and aggression like all other behaviors are ultimately a function of brain activity. The evolution of brain mechanisms that mediate aggressive and violent behaviors may be traced from humans to other animal species, and most of the neurochemical and neuropharmacologic evidence stems from studies with nonhuman species. The relevant neurochemical systems start with genetic instructions, undergo critical maturation periods, and—as evidence during the past two decades demonstrates—environmental, social, nutritional, and experiential factors modulate these systems continuously.

Insight into the neurochemical mechanisms of violence in humans has been obtained only indirectly by correlating biochemical markers in peripheral fluids or in the spinal cord with past behavioral events. In the meantime, an explosion of neuroscience research continuously informs on highly discrete neuroanatomical processes, pools of synthetic and metabolic enzymes, exquisitely regulated neural receptor populations, and transducer systems. None of these newly developed research methods have been applied to the issues of violence as of yet.

Klaus Miczek, Margaret Haney, Jennifer Tidey, Jeffrey Vivian, and Elise Weerts are at the Department of Psychology, Tufts University.

Up to the 1960s, the canonical transmitter substances such as norepinephrine (NE), dopamine (DA), serotonin (5-hydroxytryptamine, 5-HT), and acetylcholine (ACh) were the major focus of neuroscience research. Accelerating since the 1970s has been the research on receptor subtypes for endogenous neurotransmitters and neuromodulators and for psychoactive drugs. The discovery of peptides and steroids in the brain, as well as their neural receptors, prompts the consideration of possible new mechanisms that may be relevant to aggressive and violent behavior. In the early zeal, neuroscience research attempted to discover the "chemical code" of specific behavioral functions; noradrenergic feeding and cholinergic drinking were initial examples of normal homeostatic functions, the dopamine hypothesis of schizophrenia was advanced, and serotonin was sometimes referred to as a "civilizing neurohumor" keeping sex and aggression under control. However, by now, nearly every neurotransmitter has been implicated in the neural mechanisms for these complex physiologic and behavioral phenomena, and this applies also to aggressive and violent behavior. It is highly unlikely that the problem of violence can be reduced to a dysfunction in a single enzyme, receptor, or molecular component of a nerve cell. The present framework for studies on neurochemical mechanisms of violence distinguishes a neurochemical profile of individuals with an aggressive "trait" from those events that mediate the initiation, execution, and termination of aggressive and violent acts on a moment-to-moment "state" basis. The latter are significant in the development of rational therapeutic interventions. In general, clinical studies focus on biochemical markers of aggression, or violence as a trait, whereas experimental studies in animals provide mostly data on the proximal antecedents and consequences of aggressive behavior (state). Genetic studies of aggressive traits in animals have only rarely included concurrent assessments of their biochemical basis (see Carey, in this volume).

It has become a truism to point out that each type of violent and aggressive behavior is associated with distinctive neurochemical changes, and more selective logical interventions modulate these different behavior patterns in an increasingly specific manner. In order to appreciate the range of aggressive and violent behaviors at the animal and human level that have been studied for their neurochemical basis, it will be useful to briefly summarize the major animal models as well as clinical types of aggression and violence.

TYPES OF AGGRESSIVE AND VIOLENT BEHAVIOR

In the psychiatric clinic, violent and aggressive behaviors are not very well defined, although these behavior patterns may be symptoms of many disorders (e.g., Eichelman, 1986). According to the terminology and criteria of the revised *Diagnostic and Statistical Manual of Mental Disorders* (DSM-IIIR) (American Psychiatric Association, 1987), these may include conduct disorder in adolescents, isolated or intermittent explosive disorder in adults, parent-child problem in certain cases of child abuse, dementia, schizophrenia, alcohol and substance abuse, depression, mania, antisocial personality disorder, mental retardation, and attention-deficit disorder.

Several neurological diseases feature in their symptomatology violent or pathological aggressive behavior; most noteworthy are aggressive and violent outbursts in some patients with Gilles de la Tourette's syndrome, Down's syndrome, Lesch-Nyhan syndrome, epilepsy, and limbic as well as hypothalamic tumors (see Mirsky and Siegel, in this volume).

Ethological, experimental-psychological, and neurophysiologic concepts and methods have contributed to the development of preclinical models of aggressive behavior that have been investigated for their neurochemical and neuropharmacologic bases (e.g., Miczek, 1987). Several schemes have been proposed to categorize the different types of animal aggression in terms of

(1) the experimental manipulations, either pervasive (e.g., isolated housing) or discrete (e.g., exposure to pain stimuli, omission of scheduled reinforcement, brain stimulation, brain lesion);

(2) the type of behavioral phenomena (e.g., affective defense, killing); or

(3) the potential function (e.g., territorial defense, maternal aggression, dominance-related aggression).

Table 1* summarizes the major experimental models of animal aggression in laboratory research by differentiating those that are based on (A) aversive *environmental* manipulations, (B) *brain* manipulations, and (C) *ethological situations*. *Killing* (D) high-

---

*The tables appear at the end of this paper, beginning on page 349.

lights the difficulties of these categorical schemes; since variants of this behavior have been referred to as a form of "predatory aggression" (ethological) or "irritable aggression" (aversive environmental manipulations), it may be produced by brain stimulation or brain lesions (brain manipulations) and it may be self-reinforcing as in the case of "excess" killing. The attempt to assign biologic functions to animal models of aggression demonstrates the ambiguities associated with most of these models (last column in Table 1), and the difficulties in relating many types of animal aggression to the phenomena of human violence, as defined legally or clinically, are important for the present discussion.

<h2 style="text-align:center">CONCLUDING STATEMENT</h2>

Clinical and preclinical definitions of violent and aggressive behavior range across a variety of behavioral phenomena that differ in terms of distal and proximal antecedents, intensity and frequency of behavioral acts, and functions. During the past 15 years, animal aggression research, influenced by an ethological framework, has begun to focus on adaptive patterns of behavior in biologically meaningful contexts, while clinical research is concerned with aggressive and violent acts as "behavioral pathologies," viewing aggression alternatively as a trait or a state. In order to trace the evolutionary origins of aggressive behavior at the behavioral, physiologic, and neurobiologic levels, detailed functional and structural analyses at each level are needed; this need is particularly acute at the behavioral and diagnostic levels.

NEUROCHEMISTRY AND NEUROPHARMACOLOGY OF AGGRESSION AND VIOLENCE

Until the development during the last decade of microdissection and imaging techniques for neural tissue, as well as techniques for in vivo microdialysis and improved sensitivity of biochemical assay, the evidence for the involvement of ACh, gamma-aminobutyric acid (GABA), NE, DA, and 5-HT in neural mechanisms of animal aggression was based entirely on single measures that summarized an experimental subject's entire brain activity at one time point. In humans, access to the central nervous system (CNS) is even more limited, so clinical researchers have relied on more readily collected indirect measures such as blood and urine; a somewhat more invasive technique is spinal punctures to obtain cerebrospinal fluid (CSF). Again, these indirect indices are single values, totally reflecting the activity of many anatomically differentiated, functionally opposing, and interacting systems that follow a daily rhythm and are greatly influenced by environmental and nutritional factors.

For present purposes, the most frequently and thoroughly investigated of the more than 50 identified neurotransmitter and neuromodulator substances are surveyed. The evidence that is examined links (1) direct neurochemical measures, as well as (2) neuropharmacologic manipulations of norepinephrine, dopamine, serotonin, acetylcholine, and GABA to aggressive and violent behavior both in animals and in humans. (3) Major pharmacotherapeutic interventions are reviewed and evaluated for their effectiveness and selectivity in modulating aggressive and violent behavior. Key features of the cited empirical studies are summarized in tabular form.

CATECHOLAMINES

Noradrenergic Correlates of Animal Aggression

Massive adrenergic activity in the sympathetic nervous system and in the adrenal gland accompanies intense emotional behavior, including aggressive and violent behavior (e.g., Lamprecht et al., 1972; Stoddard et al., 1986; Barrett et al., 1990). However, the focus here is less on the autonomic correlates and consequences, then on levels of *brain* norepinephrine, the noradrenergic neuronal pathways, the *alpha-* and *beta*-adrenergic receptor subtypes, and their respective role in violent and aggressive behavior (Table 2, section A).

Divergent changes are reported for *whole brain* levels of NE, as well as indices of NE turnover and synthesis in animals, just before or after they have engaged in a range of aggressive behaviors. In lobsters, rainbow trout, and pheasants, octopamine (the invertebrate counterpart to NE) and NE are decreased in the more aggressive dominant member in comparison to the subordinate member (Kravitz et al., 1981; McIntyre et al., 1979; McIntyre and Chew, 1983). In mice, whole brain NE is elevated after isolated housing that renders many animals aggressive (Welch and Welch, 1965) or after they have just fought (Modigh, 1973). NE turnover is either increased or decreased in isolated, presumably aggressive mice (Valzelli, 1973; Rolinski, 1975) or immediately after a fight (Modigh, 1973). Either aggressive strains of mice do not differ from less aggressive ones in terms of their NE turnover (Karczmar et al., 1973; Goldberg et al., 1973) or turnover is increased in the more aggressive strains (Bernard, 1975). When rats are reacting defensively to electric shock, their diencephalic and mesencephalic NE turnover is increased (Stolk et al., 1974). Cats as well as rats that rage after acute brain stem transection or after septal lesions show elevated hindbrain NE turnover (Reis and Fuxe, 1964, 1968; Salama and Goldberg, 1973b), but "rage" due to amygdaloid stimulation lowers NE levels in their brain stem (Reis and Gunne, 1965). Hypothalamic and amygdaloid levels of the NE metabolite MHPG (3-methoxy-4-hydroxyphenylglycol) were also reduced in rats that engaged in stress-induced biting (Tsuda et al., 1988). When rats have just killed a mouse, their forebrain NE turnover is increased (Goldberg and Salama, 1969; Salama and Goldberg, 1973b; Tani et al., 1987).

Anatomically more discrete measurements of noradrenergic activity in aggressive animals often reveal opposite changes in different brain regions. Increased synaptosomal uptake of cortical NE was measured in mice after intense fighting (Hendley et al., 1973; Hadfield and Weber, 1975). Isolated mice of particularly aggressive strains show increased turnover of NE in three brain areas (frontal cortex, caudate, hypothalamus; Tizabi et al., 1979). After exhibiting fighting behavior they have less NE in olfactory tubercle and substantia nigra, but increased NE in the septal forebrain (Tizabi et al., 1980). Increased levels of NE were also found in the hypothalamus of rats that kill mice (Tani et al., 1987). However, many investigations fail to detect any changes in NE levels, turnover, or synthesis in brain regions of animals exhibiting aggressive behavior (e.g., Payne et al., 1984, 1985).

Brain norepinephrine undergoes large changes before, during, and after different kinds of aggressive and defensive behavior in animals; these changes are, however, localized in specific brain regions that even within the limbic system appear to exert opposing behavioral effects. At present, it is not yet possible on the basis of experimental evidence from animal models to map a "noradrenergic neurochemical profile" of different brain regions that are critically important just preceding or consequent to an aggressive act.

Dopaminergic Correlates of Animal Aggression

As detailed in Table 2, section B, levels of DA and measures of DA synthesis and turnover in the *whole brain* have been found to increase in aggressive strains of mice and in mice that have just engaged in aggressive behavior (e.g., Bernard et al., 1975; Modigh, 1973). With regard to *specific brain regions*, isolation-induced aggressive behavior in mice has been reported to increase DA levels in the striatum (Tizabi et al., 1979); DA uptake in the prefrontal cortex, but not striatum (Hadfield, 1981, 1983); and DA turnover in striatum (Hutchins et al., 1975), frontal cortex, and hypothalamus (Tizabi et al., 1979); hypothalamic DA levels were also elevated in attacking rats (Barr et al., 1979). In mice attacking for the first time, DA turnover in the nucleus accumbens is increased, but not after multiple aggressive experiences (Haney et al., 1990).

When mice or rats defend against attacks, several limbic forebrain structures show elevated metabolite levels of DA (Mos and van Valkenburg, 1979; Louilot et al., 1986; Puglisi-Allegra and Cabib, 1990). Defensive reactions to electric shock are also correlated with increased DA uptake in striatum (Hadfield and Rigby, 1976), and increased DA turnover in cortical and limbic areas (Dantzer et al., 1984).

Rats that kill mice do not significantly differ from so-called nonkillers in limbic DA but may differ slightly in hippocampal DA (Broderick et al., 1985; Barr et al., 1979); muricidal rats may also show increased DA metabolite levels (Tani et al., 1987).

The activity of brain dopamine undergoes large changes subsequent to either aggressive or defensive behavior. At present, different experimental preparations have implicated all three major forebrain dopamine systems (i.e., nigrostriatal, mesolimbic, and mesocortical). Brain dopamine systems appear to be particularly significant in (1) the reinforcing or rewarding aspects of violence and aggression, possibly via the mesolimbic and mesocortical DA systems, and/or (2) the neural mechanisms for initiation, execution, and termination of violent or aggressive behavior patterns, possibly via the nigrostriatal and mesolimbic DA systems. In order to assess these possibilities, it will be important to apply methodology with greater temporal, anatomically, and behaviorally differentiating resolution.

Catecholaminergic Correlates of Human Aggression and Violence

The evidence from studies with humans on the role of NE in neural mechanisms responsible for violent and aggressive behavior is limited to measurements of noradrenergic activity in the

CSF, blood, or urine (see Table 2, section C). In military personnel rated as highly aggressive in terms of nine categories of lifestyle, the MHPG level in CSF was positively correlated with average "aggression score" (Brown et al., 1979). However, NE turnover rates in the CSF of men convicted of violent crimes did not differ among those that were judged to be premeditated versus those considered to be impulsive (Linnoila et al., 1983). Similarly, DA levels and turnover in CSF of five XYY patients arrested for assaults did not differ from controls (Bioulac et al., 1980).

Several studies attempted to identify indices of catecholamine activity in blood or urine that may characterize aggressive or violent individuals. For example in one series of studies, higher urinary NE values, particularly in response to an upcoming experimental stressful event, appear to be more prevalent in violent incarcerated male patients in a maximum security hospital setting (Woodman et al., 1977; Woodman and Hinton, 1978a,b; Woodman, 1979) than in nonviolent controls. Violent male offenders also differ in their levels of free and conjugated plasma phenylacetic acid, although one study finds increases and another, decreases (Sandler et al., 1978; Boulton et al., 1983).

These correlative studies of indices of catecholamine activity in CSF, blood, or urine provide little support for brain NE as a specific "marker" for aggressive or violent behavior. A promising diagnostic strategy is to examine an individual's catecholamine response to an environmental or pharmacologic challenge rather than to rely on basal levels undergoing circadian rhythmic oscillations. NE, DA, and their metabolites are highly compartmentalized in the brain, and their concentrations are relatively low compared to those in other organs of the body. Conclusions about brain catecholamines and the propensity to aggressive and violent behavior on the basis of peripheral measures are to be considered very tenuous.

Neuropharmacologic Manipulations of Catecholamines

The pharmacologic evidence from animal and human studies suggests a permissive role for catecholamines in aggressive and violent behavior. One type of experimental strategy is to compromise catecholamine *synthesis, storage,* or *release;* these manipulations reliably reduce aggressive and defensive behavior in animals ranging from mice to monkeys (e.g., Eichelman, 1981; Torda, 1976). Of course, brain catecholamine (CA) systems are of critical significance in a large variety of basic physiologic and behavioral processes such as sleep/wakefulness rhythmicity, homeostatic and motor functions, and a range of active and reactive behavior patterns. The critical issue in these data is the relative lack of specificity with which these pharmacologic interventions reduce aggressive behavior. Pharmacologic inhibition of catecholamine synthesis, presynaptic storage, or release profoundly alters all active behavior, including aggressive acts. Consistent evidence during the past three decades repeatedly demonstrates that inhibition of the synthetic enzymes tyrosine hydroxylase or dopamine ß-oxidase, as well as depletion of storage sites, decrease many behavioral initiatives, including attacks and threats in mice, rats, cats, and monkeys (see Table 2, section D; e.g., Redmond et al., 1971a,b; Torda, 1976; Katz and Thomas, 1976; Diringer et al., 1982). This evidence emphasizes the necessity of intact catecholamine synthesis, storage, and release for aggressive behavior to occur, but does not establish a specific role for catecholamines in these types of behavior patterns.

A further approach in assessing the role of brain catecholamines in animal aggression is to produce degenerations of catecholamine-containing neurons or, more specifically, those neurons that contain either dopamine or norepinephrine with selective *cytotoxic* agents and subsequently to measure alterations in aggressive behavior patterns. Rage-like reactions and heightened irritability may be produced by CA-depleting doses of the cytotoxic agent 6-hydroxydopamine (6-OHDA) in laboratory rats, and the indiscriminate biting and defensive reactions can further be amplified by exposure to pain stimuli (e.g., Eichelman et al., 1972; Eichelman and Thoa, 1973; Nakamura and Thoenen, 1972; Geyer and Segal, 1974; Pucilowski and Valzelli, 1986; Beleslin et al., 1986; see Table 2, section D). In contrast to these observations are the suppressive effects of 6-OHDA on aggressive behavior in monkeys when confronting conspecifics (Redmond et al., 1973) or in cats preying on rats (Dubinsky et al., 1973). Of course, destruction of brain catecholamine-containing neurons renders an organism severely impaired in a wide range of important bodily functions, which in turn may be indirectly leading to a hyperreactive defensive mode of behavior.

Another strategy consists of modulating aggressive behavior by the administration of catecholamine *precursors*. During the 1960s and 1970s the "ℓ-dopa-rage" phenomenon attracted attention, and it continues to serve as evidence for an important role of brain dopamine in aggressive behavior (e.g., Eichelman, 1981, 1987). In laboratory rats and mice, administration of very large doses of the CA precursor, ℓ-dopa (ℓ-dehydroxyphenylalanine) facilitates or induces indiscriminate biting and other defensive reactions. These reactions are further intensified if the animals are exposed to chronic cannabis, are withdrawn from opiates, have sustained CA neurotoxicity or depletion, or have inhibition of CA synthesis or of monoamine oxidase (see Table 2, section D; Everett, 1961; Vander Wende and Spoerlein, 1962; Randrup and Munkvad, 1966, 1969a,b; Ernst, 1967; Lammers and van Rossum, 1968; Zetler and Otten, 1969; Yen et al., 1970; Lal and Puri, 1971; Benkert et al., 1973; Rolinski, 1973). The relevance of the experimental ℓ-dopa-rage phenomenon to aggressive behavior in animals or human violence is, however, tenuous because it occurs only after massive pharmacologic interventions and consists of behavioral fragments of uncertain significance (e.g., Krsiak, 1974b). L-Dopa actually suppresses fighting behavior in mice but increases defensive responses to painful stimuli (e.g., Karczmar and Scudder, 1969; Thoa et al., 1972a). The amino acid precursors ℓ-tyrosine and ℓ-phenylalanine, if added to the diet, may transiently increase aggressive behavior in mice (Thurmond et al., 1979, 1980). DA, when given directly into the cerebral ventricles, may also increase pain-induced defensive responses in rats (Geyer and Segal, 1974).

Most of the evidence on brain NE and DA derives from studies with increasingly selectively acting *receptor agonists* and *antagonists*. Initial evidence indicated that the nonselective DA receptor agonist, apomorphine, results in hyperdefensive responses similar to those seen after ℓ-dopa in mice and rats, particularly under conditions in which brain dopamine receptors are unusually sensitive (see Table 2, section D; e.g., Senault, 1968; McKenzie, 1971; Thoa et al., 1972a,b; Lal and Puri, 1971; Torda, 1976; Baggio and Ferrari, 1980; Pucilowski et al., 1986, 1987). By contrast, in situations requiring coordinated pursuit, threat, and attack, apomorphine exerts suppressive effects on aggressive behavior in mice (e.g., Hodge and Butcher, 1975; Lassen, 1978; Baggio and Ferrari, 1980). These studies suggested a clear pharmacologic differentiation between offensive aggression and exaggerated defense. Recently developed selective agonists for the D1 and D2 receptor subtypes mimic the effects of apomorphine in terms of hyperdefensive and indiscriminate biting reactions in laboratory rodents (e.g., Puglisi-Allegra and Cabib, 1988, 1990; Cabib and Puglisi-Allegra, 1989). A large number of studies have consistently documented the inhibitory effects of catecholaminergic and particularly dopaminergic receptor agonists on killing behavior by omnivorous rats and carnivores (see Table 2, section D; e.g., Schmidt, 1979, 1983; Bandler,

1970, 1971a; Rolinski, 1975; Berzsenyi et al., 1983; Molina et al., 1987; Isel and Mandel, 1989). Although the literature on human violence uses the term "predatory" in analogy to stalking and killing in carnivorous animal species, the relationship between the predatory behavior of certain animal species and human aggressive or violent behavior remains to be explored.

Dopamine receptor antagonists have been studied extensively for their antiaggressive effects; however, their selectivity as antiaggressive drugs remains unsatisfactory (e.g., Eichelman, 1986; Miczek, 1987; Miczek et al., 1994). Most of these substances have been developed as potential antipsychotic or neuroleptic drugs, and this literature is reviewed below (Table 7). Recently, selective antagonists for D1 and D2 receptors have been developed (e.g., McMillen et al., 1989; Redolat et al., 1991). Initial evidence indicates that blockade of either dopamine receptor subtype potently decreases aggressive behavior in mice and monkeys, albeit with limited behavioral specificity (Ellenbroek and Cools, 1990; Tidey and Miczek, 1992; Miczek et al., 1994). Future studies will have to identify the dopamine receptor populations that are most relevant in the initiation and execution of aggressive and defensive behavior patterns in animals in order to develop a rational basis for clinical trials in humans.

The successful use of beta-adrenergic receptor blockers in the management of violent patients identifies these substances as potential therapeutic options (e.g., Ratey et al., 1986, 1987). The clinical evidence on beta-blockers is reviewed below. When the prototypical beta-blocker, propranolol, was found to be beneficial in calming violent individuals who are unresponsive to other medications (e.g., Elliott, 1977), its therapeutic value was thought to derive from its blockade of noradrenergic beta-receptors. In the meantime, propranolol, pindolol, nadolol, and similar substances, which have been found to show high affinity for 5-HT_{1A}, act as antagonists (Olivier et al., 1990), and it is this serotonergic mechanism of action that may be the basis for the antiaggressive effects of beta-blockers.

SEROTONIN

No other neurotransmitter has been more intimately implicated in the neurobiologic mechanisms of aggressive and violent behavior than 5-HT (e.g., Brown et al., 1979; Valzelli, 1981; van Praag et al., 1987; Roy and Linnoila, 1988; Coccaro, 1989; Miczek and Donat, 1989). A major theme in the biological psychiatry literature during the last decade is the proposed role of brain serotonin in impulse control as manifested in an individual's tendency toward alcoholism, obsessive-compulsive disorders, suicide, irritability, hostile feelings, and violent outbursts (e.g., Asberg et al., 1987). Yet, the evidence on brain 5-HT systems and different kinds of aggressive and violent activities in animals and humans argues against a single direct link, and requires an evaluation that differentiates neural 5-HT pathways and their receptor subtypes in a range of aggressive and violent activities.

5-HT Correlates of Animal Aggression

Brain 5-HT or its major metabolite 5-hydroxyindoleacetic acid (5-HIAA) have been assayed in animals that are subjected to conditions in which aggression is likely to occur or that have just engaged in aggressive behavior, but with varying outcomes (Table 3, section A). In whole brain, 5-HT has been found to increase after fighting or not to change in mice that were isolated for aggression (Modigh, 1973, 1974 versus Goldberg et al., 1973; Garattini et al., 1967). The metabolite 5-HIAA decreased, increased, or did not change in whole brains of mice isolated for aggression (Garattini et al., 1967; Goldberg et al., 1973; Lasley and Thurmond, 1985). Of course, isolated housing does not invariably lead to aggressive behavior; only a varying percentage of isolated mice will engage in aggressive behavior, whereas others remain nonaggressive or even develop heightened escape and defensive reactions ("isolation-induced timidity"; Krsiak, 1975b).

Large increases in indices of *amygdaloid* 5-HT turnover were found in mice that attacked for the first time (Haney et al., 1990), in rats that were muricidal (Broderick et al., 1985; but see Tani et al., 1987), and in group-housed mice after olfactory bulbectomy (Garris et al., 1984). In a carnivorous species such as mink, the elevated 5-HIAA level in hypothalamus and amygdala was associated with a sated state during which the animal was slow to initiate a predatory kill (Nikulina and Popova, 1988). No changes in 5-HT or 5-HIAA were detected in several hypothalamic, limbic, and mesencephalic regions of aggressive hamsters or in isolated mice after prolonged aggressive and defensive behavior (Payne et al., 1984, 1985; Hadfield and Milio, 1988). When rats react defensively to electric shock pulses, their 5-HT levels in raphe and striatum as well as their 5-HIAA in hippocampus decrease (Lee et al., 1987).

It appears possible that the activity of 5-HT in the amygdala versus mesencephalic and striatal regions is differentially mediating aggressive versus defensive behavior patterns. Such specificity with regard to brain region and behavior pattern should prompt the development of more selective diagnostic behavioral assessment and pharmacotherapeutic intervention; it also casts doubt on single indices of 5-HT that summarily attempt to represent 5-HT activity in the entire brain.

Neuropharmacologic Manipulations of 5-HT in Animals

Impairments of brain 5-HT systems by removing the dietary precursor ℓ-tryptophan, blocking the synthetic enzyme tryptophan hydroxylase, depleting the 5-HT vesicular storage, or cytotoxically or electrolytically destroying serotonin neurons lead mostly to suppression of *attack* and *threat* behavior by isolated mice (see Table 3, section B; e.g., Poschlova et al., 1975; Rolinski, 1975; Eichelman, 1981; Payne et al., 1984; Svare and Mann, 1983; Ieni and Thurmond, 1985). However, pharmacologic manipulations with opposite biochemical effects such as tryptophan loads, administration of the precursor 5-HTP (5-hydroxytryptophan) or releasing agents, blockade of enzymatic inactivation of 5-HT with MAO (monoamine oxidase) inhibitors, or uptake inhibition, either acutely or chronically also decreased attack and threat behavior (Table 3, section B; Thurmond et al., 1979; Eichelman, 1981; Chamberlain et al., 1987). Most of these manipulations lack behavioral specificity in that sedation and motor incapacitation accompany the antiaggressive effects.

Defensive-aggressive responses in rats reacting to painful electrical shock pulses may be facilitated through impairing 5-HT by omitting ℓ-tryptophan from the diet, inhibiting 5-HT synthesis, or cytotoxically or electrolytically destroying 5-HT-containing neurons, at least under some experimental conditions (see Table 3, section B; e.g., Ellison and Bresler, 1974; Benkert et al., 1973; Eichelman, 1981; Rolinski and Herbut, 1981; Pucilowski and Valzelli, 1986). A reliable facilitation of defensive reactions, but not attack behavior, is seen after chronic inhibition of 5-HT reuptake or MAO with antidepressants (see below; e.g., Delini-Stula and Vassout, 1981; Mogilnicka and Przewlocka, 1981; Prasad and Sheard, 1983a,b). In cats, but not in monkeys, inhibition of 5-HT synthesis amplified affective defense (e.g., MacDonnell et al., 1971; Redmond et al., 1971a).

By far, the strongest evidence for an inhibitory role of 5-HT in animal aggression has been accrued by neuropharmacologic studies of 5-HT and *killing* behavior by laboratory rats, usually directed toward a mouse—"muricide" (see review by Miczek and Donat, 1989; also Table 3, section B). Because of its similarity to killing in the sequence of predatory stalking and hunting by carnivores, muricide by omnivorous rats has been termed "predatory aggression." From an ethological view, it is conceptually inconsistent to combine predation and aggression, since the causative and functional dimensions of these behaviors differ fundamentally. More than 70 studies during the past 25 years demonstrate that laboratory rats that have not killed a mouse previously are more likely to do so after electrolytic or neurotoxic insults to the serotonin-containing raphe nuclei, omission of ℓ-tryptophan from the diet, or inhibition of 5-HT synthesis (see Table 3, section B; DiChiara et al., 1971; Eichelman and Thoa, 1973; Isel and Mandel, 1989; Banerjee, 1974; Breese and Cooper, 1975; Vergnes et al., 1973, 1988; Gibbons et al., 1978). Conversely, tryptophan loads in the diet, precursor administration, and inhibition of enzymatic inactivation with MAO inhibitors or reuptake blockers effectively suppress the killing response (Table 3, section B; e.g., Kulkarni, 1970; Bocknik and Kulkarni, 1974; Gibbons et al., 1978, 1981).

However, the evidence for a close link between 5-HT and killing needs to be qualified: (1) Many severely 5-HT-depleted rats fail to show the killing response, and others without any detectable abnormality in or even increased levels of 5-HT or 5-HIAA engage regularly in this behavior (e.g., Salama and Goldberg, 1973b; Miczek et al., 1975; Broderick et al., 1985). (2) It has not been possible to specify a threshold value of 5-HT impairment or, alternatively, to relate 5-HT depletion to the probability of killing in a systematically graded dose-effect manner. (3) Rats that have been previously exposed to the potential prey, will not develop the killing response after insults to 5-HT activity (e.g., Marks et al., 1977; Vergnes et al., 1977; Vergnes and Kempf, 1981). (4) Once the killing behavior has become part of the animal's repertoire, it persists in the absence of any changes in 5-HT levels, synthesis, or metabolism (e.g., Vergnes and Kempf, 1981). (5) Some carnivores such as cats, ferrets, or grasshopper mice are actually impaired in their killing behavior when 5-HT synthesis is blocked, and the killing response cannot be blocked by antidepressants that are 5-HT reuptake inhibitors (McCarty et al., 1976; Leaf et al., 1978; Schmidt and Meierl, 1980; Schmidt, 1980).

The neurobiological mechanisms of killing behavior include an important role of 5-HT, particularly in rats that do not exhibit this behavior normally. At the same time, additional mechanisms override or modulate 5-HT activity in the mediation of killing behavior, especially when this behavior is already part of the organism's repertoire or when the organism is habituated to the provocative stimulus. The predatory killing by carnivores appears not to be critically dependent on 5-HT, which suggests a different interpretation of the mouse-killing response by the laboratory rat, possibly as a model of pathology (e.g., Karli, 1981; Valzelli, 1985).

Investigations of the role of postsynaptic 5-HT *receptors* in aggressive behavior, particularly killing, in animals have been relatively secondary to the presynaptic events until the important discoveries in the last decade of differentiated receptor subtypes for 5-HT (Peroutka, 1988). The first-generation agonists and antagonists with nonselective affinity for all 5-HT receptors suppressed every type of aggressive behavior in a more or less specific manner (e.g., Table 3, section B; e.g, Malick and Barnett, 1976; Weinstock and Weiss, 1980; Sheard, 1981; Miczek and DeBold, 1983; Svare and Mann, 1983; Ieni and Thurmond, 1985; Winslow and Miczek, 1983; Lindgen and Kantak, 1987).

A new phase of investigating the role of 5-HT receptor subtypes in different aggressive behavior patterns began with the increasing availability of agonists and antagonists that are more selective for each receptor subtype. Agonists at $5-HT_{1A}$ receptors, such as 8-OH-DPAT (8-hydroxy-2-(dl-n-propylamino)tetralin), buspirone, and ipsapirone, reduce attack and threat behavior by a male rat confronting an intruder or a female rat defending her litter, but also produce some sedation (Olivier et al., 1989). Newer agents such as eltoprazine, a mixed $5-HT_{1A/B}$ agonist, or TFMPP (trifluorometaphenylpiperazine), a more specific $5-HT_{1B}$ agonist, appear to have a more behaviorally specific antiaggressive activity in resident male mice and rats, and in lactating female and brain-stimulated male rats (Kruk et al., 1987; Olivier et al., 1985, 1989, 1990; Miczek et al., 1989). At present, no data exist for the effects of $5-HT_{1C}$ and $5-HT_{1D}$ receptor agonists on aggressive behavior. The $5-HT_2$ receptor antagonist ketanserin reduces attack and threat behavior by isolated resident mice and rats effectively, although with limited behavioral specificity (Haney and Miczek, 1989; Olivier et al., 1989). Initial data on the effects of experimental $5-HT_3$ receptor antagonists show no specific effects on aggressive behavior in isolated mice or lactating female rats (Mos et al., 1990).

An evolutionary approach to the question of serotonin and aggression seeks to determine a functional constancy or divergence across different phyla and order. The activity of serotonin-containing neurons has been studied in invertebrates, fish, birds, and a range of mammals, including nonhuman primates before, during, or after performance of aggressive behavior (e.g., Miczek and Donat, 1989). For example, injection of 5-HT into lobsters triggers an "aggressive" looking stance, although cytotoxic destruction of 5-HT neurons is without behavioral effect (Livingstone et al., 1980). Similarly, administration of 5-HTP increased the proportion of ants (Formica rufa) fighting with each other (Kostowski and Tarchalska, 1972); however, intraventricular injections of 5-HT or 5-HT reuptake blockers into South American electric fish (Gymnotidae) decrease their aggressive signaling (Maler and Ellis, 1987). Intense defensive reactions by the prosimian tree-shrews (Tupaia belangeri) are accompanied by strong elevation of the firing of 5-HT-containing raphe cells (Walletschek and Raab, 1982). These observations, together with the earlier data from mice, hamsters, and rats, do not provide evidence for a uniform functional role of 5-HT activity in the postulated inhibitory mechanisms of aggressive behavior that may be generalized across animal species.

5-HT and nonhuman primate aggression have been studied in vervet monkeys (Cercopithecus aethiops sabaeus), talapoin monkeys (Miopithecus talapoin), rhesus macaques (Macaca mulatta), and squirrel monkeys (Saimiri sciureus) (Table 3, section B). A series of studies in vervet monkeys found consistently elevated levels of 5-HT in whole blood or in blood platelets of dominant group members as defined by success in aggressive interactions (Raleigh et al., 1980, 1983). However, at present, it is unclear how measurements of whole blood 5-HT relate to the complexities of the various 5-HT cell bodies, neuronal pathways, and receptors in brain. Preliminary data from vervet monkeys also suggested *higher* 5-HIAA in CSF of dominant group members than in subordinates (Raleigh et al., 1983). Inconsistent and unreliable correlations between CSF 5-HIAA and aggressive behavior were found in several extensive series of studies in talapoin, rhesus macaque, and squirrel monkeys (Yodyingyuad et al., 1985; Kraemer, 1985; Green et al., personal communication). A particularly instructive example involves studies of talapoin monkeys in which the day-to-day variation in number of attacks or number of threats did not correlate with CSF 5-HIAA and resulted in highly variable scattergrams (Yodyingyuad et al., 1985). When squirrel monkeys of higher or lower social rank are examined, only MHPG, but not 5-HIAA, was elevated in subordinate males, and active conflict led to further elevations of MHPG (Green et al., personal communication, cited in Miczek and Donat, 1989). A series of recent studies in squirrel monkeys and rhesus monkeys with chronic administration of the hallucinogenic "designer" amphetamine derivatives "ecstasy" and "eve," which have profound cytotoxic actions on brain 5-HT and markedly lower CSF 5-HIAA, has revealed no evidence for increased violent activity (e.g., Ricaurte et al., 1985, 1989; Molliver 1987).

To interpret the significance of CSF 5-HIAA data, the relative contribution of anatomically distinctive pools of 5-HT neurons in different brain regions to CSF metabolite levels needs to be determined. Seasonal and circadian rhythmicity, activity levels, and nutritional status, in addition to the propensity to engage in aggressive behavior, are among the prominent determinants of synthesis and metabolism of 5-HT and 5-HIAA. These variables are not appropriately reflected in single measurements at a singe time.

5-HT Correlates of Human Aggression and Violence

During the late 1970s, Brown, Goodwin, and their associates reported that two samples of institutionalized Navy men showed an inverse relationship ($r = -.77, -.78$) between ratings of a life history of events that were thought to reflect "aggression" and CSF 5-HIAA (Table 7; Brown et al., 1979, 1982). These aggression ratings were also related to the number of suicide attempts, which—in an earlier influential Scandinavian study—were found to be inversely related to the concentration of 5-HIAA in CSF of patients with unipolar depression (Asberg et al., 1976). Low CSF values for 5-HIAA were subsequently measured in certain samples of alcoholic violent offenders with "impulsive" personality, homicidal men (Linnoila et al., 1983; Lidberg et al., 1985), and impulsive arsonists (Virkkunen et al., 1989a). These latter studies led to the proposal that low CSF 5-HIAA concentration reflects a disorder of poor "impulse control rather than aggressiveness or violence as such" (Virkkunen et al., 1989a; Linnoila et al., 1989).

So far, low 5-HIAA in CSF has been inversely correlated not only with the original subpopulation of violent suicide attempters (Asberg et al., 1976), but also with clinician-rated or self-reported ratings of a life history of aggression (Brown et al., 1979, 1982; Linnoila et al., 1983; Lidberg et al., 1985); Rorschach ratings of hostility and anxiety (Rydin et al., 1982); outwardly directed hostility (Van Praag, 1982; Roy and Linnoila, 1988); criminal behavior (Linnoila et al., 1983; Lidberg et al., 1985; Van Praag, 1982; Virkkunen et al., 1989a,b); self-reported behavior problems during childhood (Kruesi et al., 1990); and preoccupation with violent thoughts (Leckman et al., 1990). The correlation coefficients usually range between $r = -.46$ and $r = -.78$ (Table 7). There is considerable controversy in interpreting the significance of 5-HIAA measurements from CSF (e.g., Eriksson and Humble, 1990). It is difficult to account for the rostrocaudal gradient in 5-HIAA concentration: the highest concentration is in the lateral cerebral ventricles, but most measurements stem from lumbar regions of the spinal cord. How exactly CSF 5-HIAA concentrations relate to the anatomically differentiated 5-HT neuronal pathways and how this measure reflects 5-HT turnover or monoamine oxidase activity remain to be specified.

CSF 5-HIAA and MAO in blood platelets show a weak correlation—in some samples a positive correlation and in others a negative one (e.g., Asberg et al., 1987). It is unclear whether and how platelet MAO-B on the one hand, and A- and B-type MAO in brain are related. Ellis (1991) has summarized the correlative studies suggesting that low platelet MAO levels may serve as markers for suicidal behavior, increased sensation seeking, impulsiveness, childhood hyperactivity, alcoholism, and criminality. However, many failures to replicate these reports and the lack of relationship between blood and brain measures need to be resolved before low platelet MAO and low CSF 5-HIAA can be accepted as biological markers for violence (Asberg et al., 1987).

Alterations in 5-HT activity have also been related to aggressive behaviors by resorting to measurements of blood chemistry (Table 3, section C). These peripheral indices include whole blood levels, plasma tryptophan/neutral amino acid ratio, and binding of tritiated imipramine to blood platelets. One study reported an inverse relationship between plasma 5-hydroxyindole and hyperactivity/aggression in 24 mentally retarded patients (Greenberg and Coleman, 1976). A further study reports that 11 out of 15 male outpatients seeking treatment for frequent bouts of verbal and physical aggression had slightly lower 5-HT uptake into blood platelets than matched controls (Kent et al., 1988; Brown et al., 1989). Although one study found a positive correlation between plasma 5-HT and conduct ratings in adolescent males (Pliszka et al., 1988), another study reported no difference in whole blood 5-HT between violent offenders and normal controls (Virkkunen and Narvanen, 1987). A modest inverse correlation between imipramine binding to 5-HT uptake sites on blood platelets and parent-rated aggressiveness in children with conduct disorder was also reported (Stoff et al., 1987). The interpretation of these peripheral blood-borne indices of 5-HT function is even more problematic than the CSF 5-HIAA measures. At present, it is unclear whether or not these peripheral indices are direct reflections of 5-HT activity in brain. It would be highly premature to propose the use of these measures from CSF or blood as diagnostic or prognostic tools for the propensity to engage in violent or aggressive behavior.

An alternative test of 5-HT activity, particularly in the hypothalamus, is to challenge individuals with a 5-HT agonist and measure the subsequent increased release in prolactin from the anterior pituitary. "Assault" and "irritability" ratings in patients with DSM-III axis 11 personality disorders inversely correlated with the peak prolactin response to the 5-HT-releasing agent fenfluramine (Coccaro et al., 1989). Again, future studies will have to determine the mechanisms for the reduced prolactin response and delineate more precisely the relevant patient population.

In a series of follow-up studies, Virkkunen et al. (1989a) and Linnoila et al. (1989) studied Finnish alcohol-abusing males, previously imprisoned for arson or manslaughter, and classified them as either impulsive or nonimpulsive on the basis of premeditation, monetary gain, and familiarity with the victim. The nadir of blood glucose during a glucose tolerance test predicted recidivism in 3 out of 13 cases and nonrecividism in 43 out of 44 cases. Adding the variable "CSF 5-HIAA" concentration as a second step to the discriminant analysis improved the predictive classification by two more cases. By itself, CSF 5-HIAA concentration failed to predict any recidivism, but CSF 5-HIAA and MHPG concentration classified more than 70 percent correctly for a history of suicide attempts. These studies point to a complex interplay between responsivity to glucose challenge, aspects of monoamine metabolism as reflected in CSF concentrations, and familial and current alcoholism as they relate to suicide history, impulsive homicide, and criminal behavior.

CONCLUDING STATEMENT

Anatomically discrete brain serotonin systems contribute importantly to brain mechanisms subserving aggressive, defensive, and muricidal behavior in animals and to a range of impulsive and violent behaviors in humans. The critical brain regions extend from 5-HT cell groups in the raphe nuclei in the mesencephalic region to the hypothalamic area and further to the hippocampal, amygdaloid, striatal, and cortical areas. It is highly problematic to generalize from single peripheral measures such as blood or CSF 5-HIAA to the complexity of 5-HT activity in a violent or aggressive individual. At the cellular level, pharmacologic manipulations that impair presynaptic events such as precursor uptake, synthesis, storage, release or metabolism, increase the probability of the muricidal reaction and, less reliably, of defensive responses in rats. By contrast, postsynaptic 5-HT receptor manipulations, particularly at the 5-HT$_{1A}$ subtype, are effective antiaggressive drugs in animal models (see below). An important qualification of the results from animal models of aggression is the impressive evolutionary variation in 5-HT functions, rendering extrapolation from a specific animal species to another one, including human, problematic.

GABA

At the cellular level, GABA causes hyperpolarization of neurons in the mammalian central nervous system, primarily in short inhibitory interneurons. Whether or not this inhibitory role at the cellular level, occurring in about one-third of the synapses in the brain, may be extrapolated to the behavioral level remains unknown, although it has been postulated with regard to aggression (e.g., Mandel et al., 1981). During the past decade, the study of GABA receptors intensified when it was discovered that benzodiazepine-type anxiolytic drugs achieve their physiologic and behavioral effects by action on a supramolecular benzodiazepine receptor-GABA$_A$-chloride ionophore complex. Before discussing the pharmacotherapeutic applications and associated risks of these types of drugs in the management of violent patients, it will be useful to summarize the evidence on GABA, aggression, and violence in animals and in humans (see Table 4).

GABA AND AGGRESSION

GABA Correlates of Animal Aggression

Several neurochemical studies found an inverse relationship between brain GABA levels, particularly in the olfactory bulb and striatum, and aggressive behavior in mice (Table 4, section A; Earley and Leonard, 1977; Simler et al., 1982). Also, GABA content in olfactory bulbs of muricidal rats was lower than in nonkiller rats (Mack et al., 1975; Mandel et al., 1979). Lower GABA content was also measured in the striatum and hypothalamus of Spanish fighting bulls versus nonaggressive Friesian bulls (Muñoz-Blanco et al., 1986).

By contrast, aggressive female mice confronting a lactating intruder had higher whole-brain GABA levels (Haug et al., 1980), and more specifically, GABA was elevated in the hypothalamus, olfactory bulbs, and amygdala of very aggressive mouse strains (Haug et al., 1984). With regard to GABA receptors, small increases in GABA binding in limbic brain areas were noted in aggressive female hamsters (Potegal et al., 1982), and amygdaloid muscimol binding was increased in muricidal rats (DaVanzo et al., 1986).

The pattern of GABA concentrations and GABA receptor binding in animals with a high propensity to engage in several aggressive behaviors highlights the large changes in opposite direction in discrete brain regions. The present neurochemical evidence does not warrant a generalization that attributes to GABA an inhibitory influence on aggressive behavior in animals and violent behavior in humans.

Neuropharmacologic Modulation of GABA

As detailed in Table 4, section B, activation of GABA-$_A$ receptors with systemic muscimol decreases the percentage of rats that kill (Delini-Stula and Vassout, 1978; Mandel et al., 1979; Molina et al., 1986), but facilitates muricide and irritability when given into the septum (Potegal et al., 1983). Activation of GABA$_A$ and $_B$ receptors suppressed pain-induced defensive responses in rats (Rodgers and Depaulis, 1982), aggressive responses in mice (Puglisi-Allegra and Mandel, 1980), and muricidal behavior in rats (Delini-Stula and Vassout, 1978). Under conditions in which normally no aggressive behavior is likely, the GABA agonist THIP (4,5,6,7-tetrahydroisoxazolo[5,4-c]pyridin-3-ol HC1) may induce aggressive behavior in rats, whereas the antagonist bicuculline suppresses aggressive and facilitates defensive responses in rats (Depaulis and Vernes, 1985). The receptor agonist, THIP, may enhance or suppress the killing response. Intracerebral blockade of GABA$_A$ receptors with bicuculline suppressed muricide (Depaulis and Vernes, 1983, 1984; Molina et al., 1986), and when this receptor antagonist was injected into the hypothalamus and amygdala, it reduced maternal aggression in rats (Hansen and Ferreira, 1986). In nonkiller rats, the GABA antagonists bicuculline and picrotoxin may induce the muricidal response (Mandel et al., 1979).

Inhibition of GABA transaminase, mostly by valproic acid, decreased attack behavior by resident male mice toward intruders (DaVanzo and Sydow, 1979; Puglisi-Allegra and Mandel, 1980; Haug et al., 1980; Poshivalov, 1981; Sulcova et al., 1981; Puglisi-Allegra et al., 1979, 1981; Simler et al., 1983), concurrent with an increase in GABA levels in several brain areas, particularly in olfactory bulbs and striatum. Also, GABA or valproate, given either systemically or directly to the olfactory bulbs, blocked mouse-killing behavior (Mack et al., 1975; Mandel et al., 1979; Molina et al., 1986), but a GABA agonist can induce killing behavior (Depaulis and Vergnes, 1983).

Defensive behaviors by rats are enhanced by GABA receptor antagonists such as picrotoxin administered either systemically, intracerebroventricularly, or directly into the periaqueductal gray area; inhibition of GABA degradation does not increase defensive behavior (Rodgers and Depaulis, 1982; Depaulis and Vergnes, 1985, 1986).

Given the ubiquity of GABA in many brain areas and its role in various physiological and behavioral functions, particularly in motor and convulsive disorders, it is not altogether surprising that the interpretation of the evidence on GABA's role in different kinds of aggressive behavior ranges from inhibition to facilitation. The classic view attributes an inhibitory role to GABA, particularly with regard to mouse killing in rats, attack behavior by isolated mice, and defensive reactions in rats (e.g., Mandel et al., 1979, 1981). It is now evident that agonists at benzodiazepine receptors lead to increased GABA transmission due to activation of $GABA_A$ receptors, which in turn increases chloride flux causing hyperpolarization. The data on benzodiazepine-type anxiolytics (see below) increasing and decreasing aggressive behavior in the contexts of maternal defense, dominance, and territorial fighting in animals and in several clinical cases of violent outbursts point to a modulatory role of $GABA_A$ receptors.

ACETYLCHOLINE

Early evidence implicated brain acetylcholine in the induction of killing behavior in laboratory rats and in "rage" and defensive reactions in cats and rodents (see reviews by Romaniuk, 1974; Allikmets, 1974). However, evidence on brain acetylcholine and human aggression and violence is chiefly limited to the effects of nicotine (e.g., Bell et al., 1985, Eichelman, 1986, 1987).

Cholinergic agonists at muscarinic receptors such as arecoline, carbachol, muscarine, or acetylcholine when given directly into certain forebrain structures evoke rage-like and defensive responses in male and female cats (e.g., Hernandez-Peon et al., 1963; Grossman, 1963; Baxter, 1968a; Beleslin and Samardzic, 1979; Brudzynski, 1981a,b); these responses are blocked by muscarinic antagonists such as atropine or scopolamine (see Table 5). A second set of findings links activation of muscarinic receptors to killing behavior in laboratory rats and domestic cats (Table 5). Systemic administration or intrahypothalamic injection of muscarinic agonists or acetylcholinesterase (AChE) inhibitors induces animals to kill (e.g., Bandler, 1969, 1970; Smith et al., 1970; Berntson and Leibowitz, 1973). Killing behavior, whether induced by cholinergic agonists or part of the animals' repertoire, is blocked by antimuscarinic drugs (see Table 5). A third set of studies reports on increased pain-induced defensive responses in rats that have been given carbachol or the AChE inhibitor physostigmine in the lateral hypothalamus or basolateral amygdala (Rodgers et al., 1976; Bell and Brown, 1980).

It is not too surprising that comparable data on heightened aggressive or violent behavior in humans after muscarinic receptor activation have not been forthcoming. These substances are too toxic and have potent effects on many autonomic functions that are severely compromising.

Nicotine has been found to exert relatively specific antiaggressive effects in several animal species. A subcutaneous (s.c.) "smoking dose" of nicotine (25/µg/kg) specifically reduced aggressive acts and postures in rats, and pain-induced biting and postures in squirrel monkeys and rats (Driscoll and Baettig, 1981; Emley and Hutchinson, 1983; Waldbillig, 1980). Nicotine also decreased killing behavior by rats and cats (Bernston et al., 1976; Waldbillig, 1980), although predatory killing by ferrets remained unaffected (Meierl and Schmidt, 1982).

Experienced aggressive-type smokers titrate their nicotine intake in part to reduce their anger, and withdrawal from smoking is associated with increased hostility and irritability (e.g., Bell et al., 1985). Nicotine cigarettes decrease experimental measures of aggression in a competitive task for human subjects (Cherek, 1981, 1984). The therapeutic potential of nicotine in the control and management of aggressive and violent behavior has not been dissociated from the enormous health risks that are associated with tobacco smoking.

PHARMACOTHERAPEUTIC INTERVENTIONS

"Chemical restraint" and "symptom management lacking curative actions" are longstanding and frequently voiced criticisms of pharmacotherapeutic interventions in violent individuals. Although these sweeping criticisms are mostly of historical relevance, they do remind us of the trial-and-error approach of the early clinical psychopharmacologists. In recent years, more rational pharmacotherapeutic developments have been based on a more adequate understanding of the drugs' sites and mechanisms of action and the more selective neurochemical systems that are targeted. The essence of these increasingly selective pharmacologic tools for violence research remains the assessment of their behavioral specificity: How specifically and permanently are the excessive aggressive and violent acts controlled without compromising physiologic functions or the remaining behavioral repertoire? The following examination of the evidence for the management of violent individuals with major classes of therapeutic drugs demonstrates that (1) there is no universally effective antiviolence drug, but rather specific agents exert efficacious antiaggressive effects in specific types of violent individuals; (2) acute and long-term management are achieved with different types of agents; and (3) the multiply interacting neurochemical systems mediating aggressive and violent acts offer several targets for pharmacologic interventions.

ANTIPSYCHOTICS

The management of violent psychiatric patients, whether requiring institutionalization or not, relies primarily on antipsychotic drugs (e.g., Itil and Wadud, 1975; Tupin, 1985; Eichelman, 1986, 1987; Yudofsky et al., 1987). The clinical effectiveness of most typical antipsychotic or neuroleptic drugs is believed to be due primarily to their action on the D2 subtype of brain dopamine receptors (see above; e.g., Carlsson, 1987; Seeman et al., 1976). Under appropriate treatment conditions, these substances effectively control the florid, positive symptoms of schizophrenia (Kellam et al., 1967; Ebert et al., 1977). Social dysfunctions as part of the negative symptomatology may be based on separate neurochemical mechanisms and have not been treated as successfully. Neuroleptic drugs are effective in reducing aggressive and violent behavior not only in patients with an affective pathology, but also in individuals without diagnosed disorder.

During the past 30 years, experimental evidence from studies on animal aggression has repeatedly demonstrated that phenothiazines, butyrophenones, and thioxanthines as well as the more recent atypical neuroleptics effectively decrease aggressive behavior by isolated mice, by mice or rats exposed to various noxious and provocative stimuli, by rats rendered irritably due to brain injury or neurotoxicity, by brain-stimulated cats, by mice or rats confronting an intruder, and by fish, mice, rats, cats, pigs, and monkeys establishing and maintaining their dominance (see Table 7A; e.g., Yen et al., 1959; Tedeschi et al., 1959a; Horovitz et al., 1963; Lister et al., 1971; Lal et al., 1975; Kido et al., 1967; Olivier and van Dalen, 1982). This substantial literature defines several features of the antiaggressive effects of neuroleptics. One of the most important characteristics of neuroleptics is their potent sedative and tranquilizing effect. The antiaggressive effects of these drugs appear to be part of the tranquil state. Depending on the demand that is placed on the organism, assessments of the motor capacities show that neuroleptics suppress a range of *active* forms of behavior, but preserve *reactive* response capacities. A characteristic effect of moderate doses of these drugs is a lengthening of the time to initiate behavioral action, including aggressive acts. Detailed ethological analyses of behavioral sequences reveal a shift from rapid short-duration acts to prolonged postures (e.g., Silverman, 1965a,b, 1966; Schmidt and Apfelbach, 1977). A further increase in neuroleptic dose renders animals incapable of engaging in coordinated motor behavior.

Defensive reactions to brain stimulation or painful environmental events have proven relatively immune to the suppressive effects of neuroleptic drugs (see Table 7A; e.g., Dubinsky and Goldberg, 1971; Andy and Velamati, 1978; Tedeschi et al., 1969). Studies in mice and rats point to potent modulatory effects on escape and defensive responses by treatment with D2 receptor antagonists (Silverman, 1965a,b) and agonists (Puglisi-Allegra and Cabib, 1988; Cabib and Puglisi-Allegra, 1989).

The clinical experiences with neuroleptics confirm their *immediate* effectiveness in controlling aggressive and violent behavior, which makes these drugs suitable for emergency treatment (e.g., Ananth et al., 1972; Poldinger, 1981; Tardiff, 1982, 1984; Sheard, 1983; Conn and Lion, 1984; Tupin, 1985; Clinton et al., 1987; Itil and Reisberg, 1978). As summarized in Table 7B, neuroleptics potently decrease a *wide spectrum* of aggressive and violent acts ranging from inpatients and outpatients with amphetamine or endogenous psychosis (e.g., Itil and Wadud, 1975; Yesavage, 1982; Sheard, 1983; Klar and Siever, 1984), to hostile depressive patients

(Overall et al., 1964), mentally retarded patients (e.g., Hacke, 1980; Read and Batchelor, 1986), and opiate addicts or alcoholics during withdrawal (e.g., Itil and Seaman, 1978). The generality of the antiaggressive effects of neuroleptics does not extend to patients with intermittent explosive and epileptic disorders. Neuroleptics are not recommended for use in controlling aggressive behavior in epileptics (Itil, 1981; Sheard, 1983). Overall, antipsychotic drugs, along with benzodiazepines (see below), may be considered relatively safe emergency medications in managing violent outbursts. However, the neuroleptics' impressive efficacy and wide generality, with relatively few nonresponders or contraindications, should not be confused with a specific action on neurobiologic mechanisms mediating aggressive and violent behavior. As a matter of fact, neuroleptics engender serious compromising autonomic and neurologic conditions in the long-term treatment of aggressive, violent, hostile, and explosive individuals that argue against their use as maintenance therapies of choice (e.g., Lion, 1975; Leventhal and Brodie, 1981; Tupin, 1985; Yudofsky et al., 1987; Itil and Reisberg, 1978).

Aggressive and destructive behavior in children, usually diagnosed with conduct disorder, minimal brain damage, or mental retardation, has been successfully decreased with thioridazine and similar piperidylalkylphenothiazines (see Table 7B; e.g., Alderton and Hoddinott, 1964; Alexandris and Lundell, 1968). In addition to the reduction in aggressive behavior, these neuroleptics also had pronounced effects on the level of motor activity, not only decreasing hyperactivity, but often lengthening reaction times and sometimes leading to apathy and drowsiness (e.g., Shaw et al., 1963; Alexandris and Lundell, 1968; Le Vann, 1971; Campbell et al., 1972; Campbell, 1987). These observations parallel those in animal preparation and again question the specificity of these drugs for the treatment of aggressive behavior. Winsberg et al. (1976) recommend psychostimulants and tricyclic antidepressants in preference to antipsychotics as pharmacotherapy for hyperactive and aggressive children.

Very violent psychiatric inmates, 44 of whom had committed violent crimes, were considered to have benefited in terms of their functioning and social adjustment when maintained on fluphenazine deconate, thioridazine, or other long-lasting antipsychotics (Scarnati, 1986). It appears that this most violent psychiatric inmate population requires higher than normal doses of antipsychotic medication, despite the risk of tardive dyskinesias.

Several decades of clinical experience and research history have documented the strong sedative effects of neuroleptics and revealed their most significant long-term side effects. Tardive dyskinesias and dystonias may be managed by dosage adjustments and usage of depo formulations (e.g., Leventhal and Brodie, 1981; Itil, 1981). In a small subpopulation of individuals, the risk of hypotension or agranulocytosis limits the use of certain phenothiazines or the atypical neuroleptic, clozapine.

Concluding Statement

The neurobiologic mechanisms for the antiaggressive and violence-controlling effects of neuroleptic or antipsychotic drugs are linked to their action on dopaminergic and noradrenergic receptors (discussed earlier). These drugs fundamentally modulate the way in which antecedent and consequent events impact on behavior, and the way behavioral activities are initiated and patterned. The potent antiaggressive effects of antipsychotics appear to be part of the profound action of these drugs; they are not specific to aggressive and violent types of behavior. Given the lack of behavioral specificity and the risk of neurologic and autonomic side effects, neuroleptics should be used only in emergency situations. Novel drugs with selective action on dopamine receptor subtypes may feature more specific behavioral effects. These agents need to be investigated for their therapeutic effects in aggressive and violent individuals.

ANTIDEPRESSANTS

The potential common origin for aggressive, hostile, and violent acts and feelings and for depressive pathologies, particularly in those with suicidal tendencies, has been a long-standing postulate (e.g., Freud, 1917). During the past 15 years, low 5-HT concentrations and metabolite levels in brain and spinal cord as well as in blood platelets in subgroups of depressive patients with poor impulse control and in violent alcoholics have been interpreted as consistent with early psychoanalytic postulates (e.g., Asberg et al., 1987; Van Praag, 1982; see above). Recent efforts in antidepressant research have focused on the inhibition of uptake sites and receptor subtypes for serotonin, blurring the distinction between pharmacotherapies for certain types of anxiety and depression. The classic antidepressant drugs include such chemically diverse substances as imipramine-type tricyclics, monoamine oxidase inhibitors, and lithium.

Noradrenergic and Serotonergic Reuptake Blockers and MAO Inhibitors

The most consistent effect of antidepressant drugs in animal preparations is the blockade of killing behavior in laboratory rats (see Table 8A). The efficacious and potent suppression of mouse killing (antimuricidal effect) by drugs that are clinically useful as antidepressants is exploited as a screening device in drug development research, but has provided little insights into the neurobiology of killing behavior. The antimuricidal effect is seen with relatively low doses of antidepressants that act either by noradrenergic or serotonergic reuptake inhibition or as MAO inhibitors (e.g., Horovitz et al., 1966; Valzelli and Bernasconi, 1976; Shibata et al., 1984). Microinjection studies suggest the amygdala and the posterior and lateral hypothalamus as the most effective sites for tricyclics to achieve blockade of killing behavior in rats and cats (Leaf et al., 1969; Dubinsky et al., 1973; Watanabe et al., 1979; Hara et al., 1983). The neurobiologic mechanisms for killing behavior by rats and those for predatory killing by carnivores appear to differ. Blockade of noradrenergic or serotonergic uptake sites by tricyclic antidepressants or by fluoxetine does not modulate predatory killing by cats or ferrets (Karli et al., 1968; Leaf et al., 1978; Schmidt, 1979, 1980; Schmidt and Meierl, 1980). These findings question to what extent antidepressant blockade of killing by rats can be extrapolated to killing behavior in other animal species or possibly humans.

As summarized in Table 8A, much of the work with experimental preparations of animal aggression and antidepressants has employed acute administration of these drugs. The clinical relevance of this considerable literature is to caution against the use of tricyclics or MAO inhibitors in acute emergencies for controlling violent individuals. The studies in animals prove these agents not to be particularly selective or reliable in decreasing aggressive responses. Acute administration of tricyclic as well as MAO-inhibiting antidepressants most often decreased aggressive behavior of isolated mice, rats in competition tests, or rats rendered irritable by neurotoxicity or neural lesions, often only at sedative and motor-impairing doses (e.g., DaVanzo et al., 1966; Sofia, 1969b; Isel et al., 1988; Isel and Mandel, 1989). Antidepressants with 5-HT reuptake-blocking properties decrease aggressive elements of behavior, but increase defensive responses in several animal preparations (Table 8A; e.g., Poshivalov, 1981; Olivier and van Dalen, 1982; Carlini and Lindsey, 1982). Antidepressants, especially when given chronically to rats, may actually *increase* aggressive-defensive responses in reaction to noxious stimuli such as pain or sleep deprivation (see Table 8A; e.g., Eichelman and Barchas, 1975; Carlini et al., 1976; Prasad and Sheard, 1982, 1983a,b; Willner et al., 1981; Mogilnicka et al., 1983). The increased sensitivity to aggression-provoking aversive events during chronic treatment with antidepressants in mice and rats may be based on regulatory changes at noradrenergic or serotonergic receptors (e.g., Eichelman, 1979; Sheard, 1981). It is conceivable that noradrenergic receptor hypersensitivity may be responsible for the occasional instances of so-called paradoxical rage in human patients (e.g., Pallmeyer and Petti, 1979; Rampling, 1978; Tupin, 1985).

Similar to the overall pattern of results with animal preparations, in clinical studies of inpatients and outpatients, imipramine and similar antidepressants are of poor and inconsistent efficacy in reducing aggressive behavior (e.g., Itil and Seaman, 1978; see Table 8B). In a small number of obsessive-compulsive patients, clomipramine reduces aggressive thoughts (Rapoport, 1989). The most clear effects on aggressive behavior by imipramine and amitriptyline are observed in hyperactive aggressive children (see Table 8A; e.g., Winsberg et al., 1972; Yepes et al., 1977; Puig-Antich, 1982). In this group of patients, tricyclic antidepressants may be a suitable alternative to the more commonly used methylphenidate-type medications.

Lithium

More than 40 years ago, Cade (1949) recommended lithium as a beneficial treatment to "control . . . restless impulses and ungovernable tempers" in preference over prefrontal leukotomy! There is remarkably consistent evidence for the selective antiaggressive effects of lithium from experimental studies in several animal preparations and from psychiatric inpatient studies with various diagnoses (see Tables 9A and 9B). The most relevant evidence comes from experimental studies in violent prisoners that show long-lasting decrements in assaultive episodes while being maintained on lithium (e.g., Sheard, 1983, 1988).

The experimental evidence from animal studies delineates several important characteristics of lithium's effects on aggressive behavior (see Table 9A). (1) The suppression of aggressive behavior is most reliable when lithium is given *chronically* such as in drinking water or food. (2) It is essential to monitor lithium in the blood to define the therapeutic window for a given individual.

Very low doses of lithium either are ineffective or can actually increase certain aggressive responses, and high doses induce renal and thyroid toxicity (e.g., Ozawa et al., 1975; Broderick and Lynch, 1982; Sovner and Hurley, 1981; Glenn et al., 1989; Craft et al., 1987). (3) Lithium is particularly effective in reducing attack behavior by isolated mice; threats and attacks by resident mice, hamsters, or fish toward an intruder; defensive responses to noxious stimuli; and irritability due to brain injury (see Table 9A; e.g., Weischer, 1969; Sheard, 1973; Eichelman et al., 1977; Oehler et al., 1985a). However, other types of aggressive behavior, such as maternal aggression and predatory killing, remain unaltered by lithium, even when given chronically (e.g., Brain and Al-Maliki, 1979).

The animal studies also highlight important limitations of the use of lithium in the management of violent and aggressive individuals. The potential for renal and thyroid toxicity at too high lithium doses requires close monitoring of adequate dosing and blood levels. Lithium may induce nausea that is readily conditioned to features of the environment surrounding the drug administration. For example, conditioned taste aversion, initially demonstrated in laboratory rats, inhibits coyotes from attacking sheep after they have consumed as little as a single meal of lithium-laced sheep meat (e.g., Gustavson et al., 1974; Krames et al., 1973; O'Boyle et al., 1973). Lithium-induced nausea may be a further reason for the poor compliance with this type of medication in outpatients.

There are numerous demonstrations of antiaggressive effects in lithium responders among institutionalized individuals with diagnoses ranging from mental retardation to epilepsy, psychosis, and antisocial personality (see Table 9B; e.g., Dostal and Zvolsky, 1970; Tupin, 1972; Goetzl et al., 1977; Dale, 1980; Craft et al., 1987; Glenn et al., 1989; Luchins and Dojka, 1989). The most convincing evidence for lithium's effectiveness as an antiaggressive medication was gathered in violent prisoners (see Table 9B; Sheard, 1971, 1977b; Sheard and Marini, 1978; Marini and Sheard, 1976, 1977; Tupin et al., 1973). Sheard's initial placebo-controlled study on 12 male volunteer prisoners in whom lithium given three times a day decreased assaultive behavior and verbal hostility as assessed by the prison staff was confirmed in a larger double-blind study with 66 highly aggressive prisoners (Sheard et al., 1976). In both a Connecticut and a California sample of violent convicts, lithium achieved a near elimination of aggressive feelings and violent behavior over 3-18 months of treatment (Tupin et al., 1973;

Sheard and Marini, 1978). These therapeutic effects were obtained with lithium doses that did not produce cognitive impairment or physiological toxicity.

Concluding Statement

Antidepressants that produce their therapeutic effect as a result of noradrenergic or serotonergic uptake blockade are relatively weak and inconsistent antiaggressive agents. The only possible exception may be beneficial effects in patients with compulsive-obsessive disorder in whom serotonergic uptake blockers reduce aggressive behavior.

The preclinical and clinical evidence establishes lithium as an effective long-term antiaggressive substance. Due to its poor compliance the antiaggressive effects are more readily seen in institutionalized individuals in whom medication delivery is supervised and blood levels are monitored. The major limitations of lithium therapy are the potential for toxic reactions in the thyroid and renal systems, nausea, and complete lack of therapeutic response. Only a more satisfactory understanding of the mechanisms by which lithium produces its antiaggressive effects would aid in diagnosing lithium responders more readily.

ANXIOLYTICS

The management of aggressive and violent behavior with antianxiety drugs emerged as a therapeutic option within the past three decades when benzodiazepines and serotonin receptor agonists became the most effective and widely used substances for generalized anxiety disorders. Earlier generations of sedative drugs with antianxiety effects such as alcohol and barbiturates are highly problematic in their actions on violent and aggressive behavior because of their aggression-heightening properties in a considerable proportion of individuals (e.g., Miczek, 1987; Miczek et al., in this volume).

Ever since benzodiazepines were introduced in the early 1960s and prior to an adequate understanding of their mechanisms of action, they have been successfully used in clinical practice not only for primary indications in anxiety and sleep disorders, muscle relaxation and sedation, but also for the acute management of aggressive and violent individuals. A new era of benzodiazepine research began in 1977 with the biochemical identification of the benzodiazepine receptor in several organs, including brain (Braestrup and Squires, 1977; Möhler and Okada, 1977). This receptor has been localized on certain subunits of the GABA$_A$-benzodiazepine receptor-chloride channel complex (e.g., Schwartz, 1988; Haefely, 1990). One may anticipate the identification of endogenous chemical factors that activate or inhibit benzodiazepine receptors and play a significant role in the neural mechanisms of intense affect, extending also to aggressive and violent behavior. At present, a wide range of substances has been synthesized that alter the activity of benzodiazepine receptors and have considerable therapeutic promise in the acute as well as the long-term treatment of violent and aggressive individuals.

GABA$_A$-BENZODIAZEPINE RECEPTOR-CHLORIDE CHANNEL COMPLEX, AGGRESSION, AND VIOLENCE

One of the first identified major properties of benzodiazepines was their taming and antiaggressive effect on wild and domesticated animals ranging from large primates to various zoo and laboratory animals (Table 10A; e.g., Heise and Boff, 1961; Heuschele, 1961; Langfeldt and Ursin, 1971). In the past 30 years, benzodiazepines' effects on many types of aggressive behavior in animals, as well as on hostile, aggressive and violent behavior in clinical populations, have been assessed (for reviews, see DiMascio, 1973; Tupin, 1985; Miczek, 1987; Brizer, 1988). Several important pharmacologic, clinical, and behavioral insights about benzodiazepines and aggressive or violent behavior have emerged.

Benzodiazepines effectively reduce aggressive behavior in animals and humans that is primarily of a *defensive* nature. Consistent evidence documents that chlordiazepoxide, diazepam, oxazepam, and other benzodiazepines effectively reduce retaliatory, defensive, and flight reactions in various animal preparations. As detailed in Table 10A, when feral animals are provoked by an approaching experimenter, or when laboratory rats or cats are exposed to painful electric shock pulses or electrical stimulation of limbic and hypothalamic structures, they react with retaliative and defensive acts (e.g., Christmas and Maxwell, 1970; Malick, 1970; Langfeldt and Ursin, 1971; Fukuda and Tsumagari, 1983; Blanchard et al., 1989; Sulcova and Krsiak, 1989; Kalin and Shelton, 1989). Benzodiazepines decrease these defensive reactions on acute administration in a dose range that is lower than that sufficient to cause sedation or muscle relaxation.

Offensive and charging aggressive behavior in animals as well as assaultive and combative behavior in humans may also be reduced by benzodiazepines, as demonstrated by the early studies with chlordiazepoxide and diazepam, and more recently with lorazepam, oxazepam, and midazolam (e.g., Lion, 1979; Bond et al., 1989; see Tables 10A and 10B, 11A and 11B). For example in animal studies, chlordiazepoxide, diazepam, alprazolam, or other benzodiazepines cause isolated mice to pursue, threaten, and bite opponents less frequently (e.g., Da Vanzo et al., 1966; Poshivalov et al., 1987; Krsiak and Sulcova, 1990), and reduce the aggression of dominant rhesus monkeys toward group members (e.g., Delgado et al., 1976). In clinical studies, different kinds of benzodiazepines reduce various types of aggressive and violent behavior ranging from from assaults, homicidal violence, and destructive or abusive behavior, to temper tantrums, postictal aggressive outbursts, agitation, and feelings of hostility and irritability (see Tables 11A and 11B; e.g., Tobin and Lewis, 1960; Monroe and Dale, 1967; Rickels and Downing, 1974; Lion, 1979; Keats and Mukherjee, 1988).

Shortly after the introduction of chlordiazepoxide (Librium®) and diazepam (Valium®), these substances were evaluated in the medication of patient inmates in penal institutions and juvenile detention centers (Kalina, 1964; Gleser et al., 1965). Hostility, abusiveness, belligerence, destructiveness, and assaultiveness were found to be markedly reduced in these open, nonplacebo-controlled trials with both benzodiazepines. Even on acute administration, midazolam stopped temper tantrums, assaults, and self-injurious behavior in certain mentally retarded patients (Bond et al., 1989).

A significant limitation of the antiaggressive effects of acutely administered benzodiazepines is their strongly sedating and muscle-relaxant side effects. Several extensive dose-effect studies in animal preparations indicate that the dose range for the antiaggressive effects overlaps considerably that for sedation (see Table 10A). Although these side effects are acceptable in an emergency situation with a violent individual, they indicate that the antiaggressive effects of acutely administered benzodiazepines cannot be considered behaviorally specific. With repeated administration of benzodiazepines, substantial tolerance to the sedative and muscle-relaxant effects is seen (File, 1985; Sepinwall et al., 1978).

The most problematic feature of benzodiazepines and aggressive behavior is their potential to *increase* this behavior under several conditions in a considerable proportion of animals and in humans (see Tables 10 and 11). Starting in the 1960s a series of case reports, as well as experimentally well-controlled double-blind studies, alerted to the paradoxical aggression-heightening effects of benzodiazepines in certain individuals (Lion et al., 1975a;

Salzman et al., 1969, 1974; DiMascio et al., 1969; Kochansky et al., 1975, 1977; Griffiths et al., 1983; Lipman et al., 1986). For example, clonazepam treatment had to be discontinued in epileptic patients due to the emergence of aggressive outbursts and temper tantrums (Guldenpfennig, 1973). Also, several case reports indicate that alprazolam may induce hostile, agitated, irritable, and physically assaultive behavior in panic disorder patients (e.g., Rosenbaum et al., 1984; Strahan et al., 1985; Gardner and Cowdry, 1985; Pecknold and Fleury, 1986).

Every review during the past two decades points to the increased hostility and violent episodes (paradoxical rage) in a certain proportion of individuals treated with benzodiazepines, although opinions differ as to the frequency or rarity of these "paradoxical" responses (see, Tables 11A and 11B; e.g., Rickels and Downing, 1974; Zisook et al., 1978; Lipman et al., 1986; Dietch and Jennings, 1988). Violent outbursts are more likely in certain individuals as a result of toxic reactions to diazepam, chlordiazepoxide, clonazepam, and alprazolam, but are apparently absent with oxazepam (Bond and Lader, 1979; Sheard, 1983; Eichelman, 1987; Dietch and Jennings, 1988). The benzodiazepine dose is a critical determinant of whether or not irritability, hostility, or violent acts may occur (e.g., Azcarte, 1975). Systematic dose-effect determinations in animal aggression preparations demonstrate the aggression-increasing effects of lower diazepam doses and the opposite, aggression-decreasing effects at higher doses (Table 10A). There is also evidence that during the course of chronic benzodiazepine treatment, and on withdrawal from prolonged treatment, increased irritability and aggressive behavior may emerge in a certain proportion of individuals (Table 10B and 11B; e.g., File, 1986a,b; Yoshimura et al., 1987; Yudofsky et al., 1987).

Because benzodiazepines are the most widely prescribed psychoactive drugs worldwide, an unambiguous assessment of their violence-controlling efficacy as well as their so-called paradoxical aggression-enhancing effects is urgently needed. Of particular urgency is the improved psychiatric and neurologic diagnosis of those patients who are prone to exhibit increased hostility and violent outbursts. It is feasible that the regulation of the $GABA_A$-benzodiazepine receptor is critically determined by genetic predispositions and modulated by life experiences with affective aggressive behaviors. Pharmacologic probes of the $GABA_A$-benzodiazepine receptor complex should be investigated for its utility as a diagnostic tool in individuals with various anxiety disorders and also in those with a propensity for violent outbursts.

In recent years, the range of substances activating and blocking the $GABA_A$-benzodiazepine receptor complex, fully or in part, has increased greatly in number and in selectivity (Haefely, 1990). Promising evidence indicates that benzodiazepine receptor antagonists such as flumazenil effectively reduce the heightened aggressive behavior after alcohol in rats and monkeys (Weerts et al., 1993). The clinical potential of benzodiazepine receptor partial agonists and antagonists in the treatment of violent individuals needs to be explored.

5-HT Anxiolytics

During the last decade, buspirone emerged as prototypic substance for a new class of anxiolytic drugs with the 5-HT_{1A} receptor subtype as primary site of action (Taylor, 1988). These drugs have begun to be explored for their use in the pharmacotherapeutic management of aggressive and violent individuals. In a few available animal studies, buspirone, ipsapirone, and gepirone reduced aggressive and defensive responses with modest specificity for aggressive elements of behavior (Table 10A; e.g., Olivier et al., 1984, 1989). Two recent studies in 24 mentally retarded patients showed that buspirone eventually decreases aggressive and self-injurious behavior without sedation or tolerance (Ratey et al., 1989; Ratey and Driscoll, 1989). It is important to point to the requirement of repeated drug administrations over days and weeks before the therapeutic effects of buspirone and similar 5-HT anxiolytics fully emerge (e.g., Barrett and Witkin, 1991). Whether or not these new 5-HT_{1A} anxiolytics prove to be more effective in controlling intense assaultive and combative behavior and less problematic in terms of side effects than existing substances remains to be demonstrated. Systematic evidence needs to be gathered as to which type of aggressive or violent behavior is most effectively reduced by these anxiolytics, and which treatment conditions are required for therapeutic success.

Beta-*Blockers*

Ever since Elliott (1977) resorted to the cardiovascular drug propranolol to reduce aggressive outbursts in seven belligerent patients, a series of open clinical trials confirmed the clinical success in several patient populations. Specifically, propranolol and other *beta*-blockers reduce irritability, self-injurious behavior, violent outbursts, and assaultive and destructive behavior in patients diagnosed as mentally retarded, brain damaged, schizophrenics, psychotics, autistic, Korsakoff, or organic brain disease (see Table

11; e.g., Yudofsky et al., 1981, 1984; Williams et al., 1982; Ratey et al., 1983, 1986, 1987; Greendyke et al., 1984; Luchins and Dojka, 1989). It is most remarkable that these clinical improvements were achieved in a patient subpopulation that had been treated unsuccessfully with anxiolytic, antidepressant, and neuroleptic agents. Several animal studies also found relatively selective antiaggressive effects of propranolol (see Table 10A; e.g., Delini-Stula and Vassout, 1979; Miczek and DeBold, 1983; Yoshimura and Ogawa, 1985). In recent placebo-controlled, double-blind studies, Ratey and coworkers (1990; Lindem et al., 1990) demonstrated significant decreases in destructive behavior and verbal outbursts of mentally retarded inpatients with the newer agents nadolol and pindolol. At present, it is unclear whether or not these substances derive their clinical benefit in aggressive individuals from their antagonistic action on *beta*-adrenergic receptors or their action on 5-HT_{1A} receptors.

The most significant limitation of treatment with *beta*-blockers is their effect on the cardiovascular system. For example, Sheard (1984) points to low blood pressure, headaches, dizziness, fatigue, insomnia, and depression as the most serious side effects of *beta*-blockers. Most clinical studies also find a reemergence of aggressive behavior after *beta*-blocking agents are discontinued (Schreier, 1979; Horn, 1987).

Concluding Statement

The use of anxiolytic agents in the pharmacotherapeutic management of violent individuals, although widespread, is not without problems. In emergency situations, injections of benzodiazepines effectively calm violent individuals. In these types of situations, sedation, loss of motor coordination, and other debilitating effects are rarely a concern. Prolonged treatment of violent individuals within the clinical population typically produces tolerance of the sedative effects of benzodiazepines without reducing their therapeutic effects on violent or aggressive behavior. The selective 5-HT_{1A} anxiolytics show considerable promise as nonsedative antiaggressive agents.

An additional concern for clinician is the "paradoxical rage" response observed in a portion of the patients treated with benzodiazepines. The present diagnostic tools do not reliably identify individuals that are prone to these aggressive outbursts. An important objective of future research is to delineate unique characteristics of the $GABA_A$-benzodiazepine receptor complex or 5-HT receptors that predict propensity to engage in violent behavior.

REFERENCES

Albert, D.J., and S.E. Richmond
1977   Reactivity and aggression in the rat: Induced by α-adrenergic blocking agents injected ventral to anterior septum but not into lateral septum. *Journal of Comparative and Physiological Psychology* 91:886-896.

Alderton, H.R., and B.A. Hoddinott
1964   A controlled study of the use of thioridazine in the treatment of hyperactive and aggressive children in a children's psychiatric hospital. *Canadian Psychiatric Journal* 9:239-247.

Alexandris, A., and F.W. Lundell
1968   Effect of thioridazine, amphetamine and placebo on the hyperkinetic syndrome and cognitive area in mentally deficient children. *Canadian Medical Association Journal* 98:92-96.

Al-Khatib, I.M.H., M. Fujiwara, K. Iwasaki, Y. Kataoka, and S. Ueki
1987   The role of brain catecholamines in the exhibition of muricide induced by nucleus accumbens lesions and the effect of antidepressants in rats. *Pharmacology Biochemistry and Behavior* 26:351-355.

Allikmets, L.K.
1974   Cholinergic mechanisms in aggressive behavior. *Medical Biology* 52:19-30.

Allikmets, L., and J.M.R. Delgado
1968   Injection of antidepressants in the amygdala of awake monkeys. *Archives Internationales de Pharmacodynamie et de Therapie* 175:170-178.

Allikmets, L.K., and I.P. Lapin
1967   Influence of lesions of the amygdaloid complex on behaviour and on effects of antidepressants in rats. *International Journal of Neuropharmacology* 6:99-108.

Allikmets, L.K., J.M.R. Delgado, and S.A. Richards
1968   Intra-mesencephalic injection if imipramine, promazine and chlorprothixene in awake monkeys. *International Journal of Neuropharmacology* 7:185-193.

Allikments, L.K., V.A. Vahing, and I.P. Lapin
1969   Dissimilar influences of imipramine, benactyzine and promazine on effects of micro-injections of noradrenaline, acetylcholine and serotonin into the amygdala in the cat. *Psychopharmacologia* 15:392-403.

Allikments, L.K., M. Stanley, and S. Gershon
1979   The effect of lithium on chronic haloperidol enhanced apomorphine aggression in rats. *Life Sciences* 25:165.

Altshuler, K.Z.
1977   Lithium and aggressive behavior in patients with early total deafness. *Diseases of the Nervous System* 38:521-524.

Aman, M.G., and N.N. Singh
1980   The usefulness of thioridazine for treating childhood disorders—Fact or folklore? *American Journal of Mental Deficiency* 84:331-338.

Aman, M.G., and J.S. Werry
1975   The effects of methylphenidate and haloperidol on the heart rate and blood pressure of hyperactive children with special reference to time of action. *Psychopharmacologia* 43:163-168.

American Psychiatric Association
1987   *Diagnostic and Statistical Manual of Mental Disorders*, 3rd ed., revised. Washington, D.C.: American Psychiatric Association.

Anand, M., G.P. Gupta, and K.P. Bhargava
1977   Modification of electroshock fighting by drugs known to interact with dopaminergic and noradrenergic neurons in normal and brain lesioned rats. *Journal of Pharmacy and Pharmacology* 29:437-439.

Anand, M., S. Mehrotra, K. Gopal, S.N. Sur, and S.V. Chandra
1985   Role of neurotransmitters in endosulfan-induced aggressive behaviour in normal and lesioned rats. *Toxicology Letters* 24:79-84.

Anand, M., A. Gulati, K. Goapl, G.S.D. Gupta, and S.V. Chandra
1989   Role of neurotransmitters in fenitrothion-induced aggressive behaviour in normal and lesioned rats. *Toxicology Letters* 45:215-220.

Ananth, J.V., M. Salib, T.A. Ban, and H.E. Lehmann
1972   Propericiazine in psychiatric emergencies. *Canadian Psychiatric Association Journal* 17:143-145.

Andy, O.J., and S. Velamati
1978   Limbic system seizures and aggressive behavior (superkindling effects). *Pavlovian Journal of Biological Science* 12:152-164.

Apfelbach, R.
1978   Instinctive predatory behavior of the ferret (*Putorius putorius furo* L.) modified by chlordiazepoxide hydrochloride (Librium). *Psychopharmacology* 59:179-182.

Appelbaum, P.S., A.H. Jackson, R.I. Shader
1983   Psychiatrists' responses to violence: Pharmacologic management of psychiatric inpatients. *American Journal of Psychiatry* 140-301-304.

Applegate, C.D.
1980   5,7-Dihydroxytryptamine-induced mouse killing and behavioral reversal with ventricular administration of serotonin in rats. *Behavioral and Neural Biology* 30:178-190.

Arnone, M., and R. Dantzer
1980   Effects of diazepam on extinction induced aggression in pigs. *Pharmacology Biochemistry and Behavior* 13:27-30.

Arnt, J., and J. Scheel-Krüger
1980 Intracranial GABA antagonists produce dopamine-independent biting in rats. *European Journal of Pharmacology* 62:51-61.

Asberg, M., P. Thoren, and L. Traskman
1976 "Serotonin depression"—A biochemical subgroup within the affective disorders? *Science* 191:478-480.

Asberg, M., D. Schalling, L. Traskman-Bendz, and A. Wagner
1987 Psychobiology of suicide, impulsivity, and related phenomena. Pp. 655-668 in H.Y. Meltzer, ed., *Psychopharmacology: The Third Generation of Progress*. New York: Plenum.

Atkinson, J.H.
1982 Managing the violent patient in the general hospital. *Postgraduate Medicine* 71:193-201.

Avis, H.H., and H.V.S. Peeke
1979 The effects of pargyline, scopolamine, and imipramine on territorial aggression in the convict cichlid (*Cichlasoma nigrofasciatum*). *Psychopharmacology* 66:1-2.

Ayitey-Smith, E., and I. Addae-Mensah
1983 Effects of wisanine and dihydrowisanine on aggressive behaviour in chicks. *European Journal of Pharmacology* 95:139-141.

Azcarate, C.L.
1975 Minor tranquilizers in the treatment of aggression. *Journal of Nervous and Mental Disease* 160:100-107.

Baggio, G., and F. Ferrari
1980 Role of brain dopaminergic mechanisms in rodent aggressive behavior: Influence of ± N-n-propyl-norapomorphine on three experimental models. *Psychopharmacology* 70:63-68.

Bainbridge, J.G., and D.T. Greenwood
1971 Tranquillizing effects of propranolol demonstrated in rats. *Neuropharmacology* 10:453-458.

Bandler, R.J.
1969 Facilitation of aggressive behavior in rats by direct cholinergic stimulation of the hypothalamus. *Nature* 224:1035-1036.
1970 Cholinergic synapses in the lateral hypothalamus for the control of predatory aggression in the rat. *Brain Research* 20:409-424.
1971a Direct chemical stimulation of the thalamus: Effects on aggressive behavior in the rat. *Brain Research* 26:81-93.
1971b Chemical stimulation of the rat midbrain and aggressive behavior. *Nature New Biology* 229:221-223.

Banerjee, U.
1974 Modification of the isolation-induced abnormal behavior in male Wistar rats by destructive manipulation of the central monoaminergic systems. *Behavioral Biology* 11:573-579.

Barkov, N.K.
1973 Effect of neuroleptics on aggressive behavior (translated from the Russian in *Zhurnal Neuropatol. Psikhiatri.* 1972, 72:108-111). *Neuroscience and Behavioral Physiology* 6:119-121.

Barnett, A., R.I. Taber, and F.E. Roth
1969 Activity of antihistamines in laboratory antidepressant tests. *International Journal of Neuropharmacology* 8:73-79.
Barnett, A., R.I. Taber, and S.S. Steiner
1974 The behavioral pharmacology of Sch12679, a new psychoactive agent. *Psychopharmacologia* 36:381-290.
Barnett, J.A., M.B. Shaikh, H. Edinger, and A. Siegel
1987 The effects of intrahypothalamic injections of norepinephrine upon affective defense behavior in the cat. *Brain Research* 426:381-384.
Barr, G.A., K.E. Moyer, and J.L. Gibbons
1976 Effects of imipramine, *d*-amphetamine, and tripelennamine on mouse and frog killing by the rat. *Physiology and Behavior* 16:267-169.
Barr, G.A., J.L. Gibbons, and W.H. Bridger
1979 A comparison of the effects of acute and subacute administration of ß-phenylethylamine and *d*-amphetamine on mouse killing behavior of rats. *Pharmacology Biochemistry and Behavior* 11:419-422.
Barrett, J.A., H. Edinger, and A. Siegel
1990 Intrahypothalamic injections of norepinephrine facilitate feline affective aggression via α_2-adrenoceptors. *Brain Research* 525:285-293.
Barrett, J.E., and J.M. Witkin
1991 Buspirone in animal models of anxiety. Pp. 37-79 in G. Tunnicliff et al., eds., *Buspirone: Mechanisms and Clinical Aspects*. New York: Academic Press.
Barrett, J.E., J.A. Stanley, L.S. Brady, R.S. Mansbach, and J.M. Witkin
1986 Behavioral studies with anxiolytic drugs. II. Interactions of zopiclone with ethyl-ß-carboline-3-carboxylate and Ro 15-1788 in squirrel monkeys. *Journal of Pharmacology and Experimental Therapeutics* 236:313-319.
Bartunkova, Z., L. Cerny, J. Drtilova, and J. Sturma
1972 Propericiazine, diazepam, chlorpromazine and placebo in a doubleblind trial in pedopsychiatric therapy. *Acta Nervosa Superior* 14:83-84.
Barzaghi, F., R. Fournex, and P. Mantegazza
1973 Pharmacological and toxicological properties of clobazam (1-phenyl-5 methyl-8-chloro-1,2,4,5-tetrahydro-2,4-diketo-3H-1,5-benzodiazepine), a new psychotherapeutic agent. *Arzneimittel-Forschung/Drug Research* 23:683-686.
Bauen, A., and G.J. Possanza
1970 The mink as a psychopharmacological model. *Archives Internationales de Pharmacodynamie et de Therapie* 186:133-136.
Baxter, B.L.
1964 The effect of chlordiazepoxide on the hissing response elicited via hypothalamic stimulation. *Life Sciences* 3:531-537.

1968a Elicitation of emotional behavior by electrical or chemical stimulation applied at the same loci in cat mesencephalon. *Experimental Neurology* 21:1-10.

1968b The effect of selected drugs on the "emotional" behavior elicited via hypothalamic stimulation. *International Journal of Neuropharmacology* 7:45-54.

Bean, N.J., K. Loman, and R. Conner
1978 Effects of benzazepine (Sch-12679) on shock-induced fighting and locomotor behavior in rats. *Psychopharmacology* 59:189-192.

Beattie, C.W., H.I. Chernov, P.S. Bernard, and F.H. Glenny
1969 Pharmacological alteration of hyper-reactivity in rats with septal and hypothalamic lesions. *International Journal of Neuropharmacology* 8:365-371.

Beck, C.H.M., and S.J. Cooper
1986 Beta-carboline FG 7142-reduced aggression in male rats: Reversed by the benzodiazepine receptor antagonist, Ro 15-1788. *Pharmacology Biochemistry and Behavior* 24:1645-1649.

Beleslin, D.B., and R. Samardzic
1979 The pharmacology of aggressive behavioural phenomena elicited by muscarine injected into the cerebral ventricles of conscious cats. *Psychopharmacology* 60:155-158.

Beleslin, D.B., R. Samardzic, and S.K. Krstic
1986 6-Hydroxydopamine-induced aggression in cats: Effects of various drugs. *Pharmacology Biochemistry and Behavior* 24:1821-1823.

Bell, R., and K. Brown
1979 The effects of two "anti-aggressive" compounds, an indenopyridine and a benzothiazepin, on shock-induced defensive fighting in rats. *Progress in Neuro-Psychopharmacology* 3:399-402.

1980 Shock-induced defensive fighting in the rat: Evidence for cholinergic mediation in the lateral hypothalamus. *Pharmacology Biochemistry and Behavior* 12:487-491.

Bell, R., D.M. Warburton, and K. Brown
1985 Drugs as research tools in psychology: Cholinergic drugs and aggression. *Neuropsychobiology* 14:181-192.

Benkert, O., H. Gluba, and N. Matussek
1973 Dopamine, noradrenaline, and 5-hydroxy-tryptamine in relation to motor activity, fighting and mounting behaviour. I. L-DOPA and DL-threo-dihydroxphenylserine in combination with Ro 4-4602, pargyline and reserpine. *Neuropharmacology* 12:177-186.

Benton, D.
1984 The long-term effects of naloxone, dibutyryl cyclic CMP, and chlorpromazine on aggression in mice monitored by an automated device. *Aggressive Behavior* 10:79-89.

1985 Mu and kappa opiate receptor involvement in agonistic behaviour in mice. *Pharmacology Biochemistry and Behavior* 23:871-876.

Bernard, B.K.
1975 Aggression and the brain monoamines: What are the answers, but of more importance, what are the questions? Pp. 71-84 in B.J. Bernhard, ed., *Aminergic Hypotheses of Behavior: Reality or Choice?* NIDA Research Monograph, Vol. 3. Rockville, Md.: National Institute on Drug Abuse.

Bernard, B.K., E.R. Finkelstein, and G.M. Everett
1975 Alterations in mouse aggressive behavior and brain monoamine dynamics as a function of age. *Physiology and Behavior* 15:731-736.

Berntson, G.G., and S.F. Leibowitz
1973 Biting attack in cats: Evidence for central muscarinic mediation. *Brain Research* 51:366-370.

Berntson, G.G., M.S. Beattie, and J.M. Walker
1976 Effects of nicotine and muscarinic compounds on biting attack in the cat. *Pharmacology Biochemistry and Behavior* 5:235-239.

Berzsenyi, P., E. Galateo, and L. Valzelli
1983 Fluoxetine activity on muricidal aggression induced in rats by *p*-chlorophenylalanine. *Aggressive Behavior* 9:333-338.

Bioulac, B., M. Benezech, B. Renaud, B. Noel, and D. Roche
1980 Serotoninergic dysfunction in the 47,XYY syndrome. *Biological Psychiatry* 15(6):917-923.

Bisbee, D.S., and D.D. Cahoon
1973 The effects of induced nausea upon shock-elicited aggression. *Bulletin of the Psychonomic Society* 1:19-21.

Blanchard, D.C., K. Hori, R.J. Rodgers, C.A. Hendrie, and R.J. Blanchard
1989 Attenuation of defensive threat and attack in wild rats (*Rattus rattus*) by benzodiazepines. *Psychopharmacology* 97:392-401.

Blyther, S., and A.S. Marriott
1969 The effects of drugs on the hyper-reactivity of rats with bilateral anterior hypothalamic lesions. *British Journal of Pharmacology* 37:507-508.

Bocknik, S.E., and A.S. Kulkarni
1974 Effect of a decarboxylase inhibitor (Ro 4-4602) on 5-HTP induced muricide blockade in rats. *Neuropharmacology* 13:279-281.

Boissier, J.R., S. Grasset, and P. Simon
1968 Effect of some psychotropic drugs on mice from a spontaneously aggressive strain. *Journal of Pharmacy and Pharmacology* 20:972-973.

Bond, A., and M. Lader
1979 Benzodiazepines and aggression. Pp. 173-182 in M. Sandler, ed., *Psychopharmacology of Aggression.* New York: Plenum.

Bond, W.S., L.A. Mandos, and M.B. Kurtz
1989 Midazolam for aggressivity and violence in three mentally retarded patients. *American Journal of Psychiatry* 146:925-926.

Boulton, A.A., B.A. Davis, P.H. Yu, J.S. Wormith, and D. Addington
1983   Trace acid levels in the plasma and MAO activity in the platelets of violent offenders. *Psychiatry Research* 8:19-23.
Boyle, D., and J.M. Tobin
1961   Pharmaceutical management of behavior disorders. *Journal of the Medical Society of New Jersey* 58:427-429.
Braestrup, C., and R.F. Squires
1977   Brain specific benzodiazepine receptors in rats characterized by high affinity 3H-diazepam binding. *Proceedings of the National Academy of Sciences* 74:3805.
Brain, P.F.
1972   Oral lithium chloride, endocrine function and isolation-induced agonistic behaviour in male albino mice. *Journal of Endocrinology* 55:1-2.
Brain, P.F., and S. Al-Maliki
1979   Effects of lithium chloride injections on rank-related fighting, maternal aggression and locust-killing responses in naive and experienced "TO" strain mice. *Pharmacology Biochemistry and Behavior* 10:663-669.
Brain, P.F., S. Al-Maliki, and D. Benton
1981   Attempts to determine the status of electroshock-induced attack in male laboratory mice. *Behavioral Processes* 6:171-189.
Brain, P.F., S.E. Jones, S. Brain, and D. Benton
1984   Sequence analysis of social behaviour illustrating the action of two antagonists of endogenous opioids. Pp. 43-58 in K.A. Miczek, M.R. Kruk, and B. Oliver, eds., *Ethopharmacological Aggression Research*. New York: Alan R. Liss.
Brain, P.F., R. Smoothy, and D. Benton
1985   An ethological analysis of the effects of tifluadom on social encounters in male albino mice. *Pharmacology Biochemistry and Behavior* 23:979-985.
Breese, G.R., and B.R. Cooper
1975   Behavior and biochemical interactions of 5,7-dihydroxytryptamine with various drugs when administered intracisternally to adult and developing rats. *Brain Research* 98:517-527.
Breese, G.R., K.L. Hulebak, T.C. Napier, A. Baumeister, G. Frye, and R.A. Mueller
1987   Enhanced muscimol-induced behavioral responses after 6-OHDA lesions: Relevance to susceptibility for self-mutilation behavior in neonatally lesioned rats. *Psychopharmacology* 91:356-362.
Brenner, H.D., L. Alberti, F. Keller, and L. Schaffner
1984   Pharmacotherapy of agitational states of psychiatric gerontology: Double-blind study: Febarbamate-pipamperone. *Neuropsychobiology* 11:187-190.
Brizer, D.A.
1988   Psychopharmacology and the management of violent patients. *Psychiatric Clinics of North America* 11:551-568.

Broderick, P., and V. Lynch
1982 Behavioral and biochemical changes induced by lithium and L-Tryptophan in muricidal rats. *Psychopharmacology* 21:671-679.

Broderick, P.A., G.A. Barr, N.S. Sharpless, and W.H. Bridger
1985 Biogenic amine alterations in limbic brain regions of muricidal rats. *Research Communications in Chemical Pathology and Pharmacology* 48:3-15.

Brown, C.S., T.A. Kent, S.G. Bryant, R.M. Gevedon, J.L. Campbell, A.R. Felthous, E.S. Barratt, and R.M. Rose
1989 Blood platelet uptake of serotonin in episodic aggression. *Psychiatry Research* 27:5-12.

Brown, G.L., and F.K. Goodwin
1986 Cerebrospinal fluid correlates of suicide attempts and aggression. Pp. 175-220 in J.J. Mann and M. Stanley, eds., *Psychobiology of Suicidal Behavior.* (Annals of the New York Academy of Sciences, Vol. 487.) New York: New York Academy of Sciences.

Brown, G.L., F.K. Goodwin, J.C. Ballenger, P.F. Goyer, and L.F. Major
1979 Aggression in humans correlates with cerebrospinal fluid amine metabolites. *Psychiatry Research* 1:131-139.

Brown, G.L., M.H. Ebert, P.F. Goyer, D.C. Jimerson, W.J. Klein, W.E. Bunney, and F.K. Goodwin
1982 Aggression, suicide, and serotonin-relationships to CSF amine metabolites. *American Journal of Psychiatry* 139:741-746.

Brudzynski, S.M.
1981a Carbachol-induced agonistic behavior in cats: Aggressive or defensive response? *Acta Neurobiologiae Experimentalis* 41:15-32.

1981b Growing component of vocalization as a quantitative index of carbachol-induced emotional-defensive response in cats. *Acta Neurobiologiae Experimentalis* 41:33-51.

Brunaud, M., and G. Siou
1959 Action de substances psychotropes, chez le rat, sur un etat d'agressivite provoquee. Pp. 282-286 in P.B. Bradley, P. Deniker, and C. Radouco-Thomas, eds., *Neuro-Psychopharmacology.* Amsterdam: Elsevier.

Bryson, G.
1971 Biogenic amines in normal and abnormal behavioral states. *Clinical Chemistry* 17:5-26.

Bryson, G., and F. Bischoff
1971 A scattered-jump syndrome with gnawing and fighting induced by L-DOPA in mice. *Research Communications in Pathology and Pharmacology* 2:469-476.

Buck, O.D., and P. Havey
1986 Combined carbamazepine and lithium therapy for violent behavior. *American Journal of Psychiatry* 143:1487.

Burov, Y.V.
1975 The influence of psychotropic drugs upon emotions. *CNS and Behavioural Pharmacology* 3:197-205.

Burrowes, K.L., R.E. Hales, and E. Arrington
1988 Research on the biologic aspects of violence. *Psychiatric Clinics of North America* 11:499-509.
Cabib, S., and S. Puglisi-Allegra
1989 Genotype-dependent modulation of LY 171555-induced defensive behavior in the mouse. *Psychopharmacology* 97:166-168.
Cade, J.F.J.
1949 Lithium salts in the treatment of psychotic excitement. *Medical Journal of Australia* 2:349-352.
Cairns, R.B., and S.D. Scholz
1973 Fighting in mice: Dyadic escalation and what is learned. *Journal of Comparative and Physiological Psychology* 85:540-550.
Campbell, M.
1987 The effect of neuroleptics on cognition and diagnosis, and their influence on stereotypes. *Journal of Mental Deficiency Research* 31:220-222.
Campbell, M., B. Fish, J. Korein, T. Shapiro, P. Collins, and C. Koh
1972 Lithium and chlorpromazine: A controlled crossover study of hyperactive severely disturbed young children. *Journal of Autism and Childhood Schizophrenia* 2:234-263.
Campbell, M., I.L. Cohen, and A.M. Small
1982 Drugs in aggressive behavior. *Journal of the American Academy of Child Psychiatry* 21:107-117.
Carlini, E.A., and C.J. Lindsey
1982 Effect of serotonergic drugs on the aggressiveness induced by Δ^9-tetrahydrocannabinol in REM-sleep-deprived rats. *Brazilian Journal of Medical and Biological Research* 15:281-283.
Carlini, E.A., C.J. Lindsey, and S. Tufik
1976 Environmental and drug interference with effects of marihuana. Pp. 229-242 in E.S. Vesell and M.C. Braude, eds., *Interactions of Drugs of Abuse* (Annals of the New York Academy of Sciences, Vol. 281). New York: New York Academy of Sciences.
Carlsson, A.
1987 Perspectives on the discovery of central monoaminergic neurotransmission. *Annual Review of Neuroscience* 10:19-40.
Chamberlain, B., F.R. Ervin, R.O. Pihl, and S.N. Young
1987 The effect of raising or lowering tryptophan levels on aggression in vervet monkeys. *Pharmacology Biochemistry and Behavior* 28:503-510.
Chen, G., B. Bohner, and A.C. Bratton
1963 The influence of certain central depressants on fighting behavior of mice. *Archives Internationales de Pharmacodynamie et de Therapie* 142:30-34.
Cherek, D.R.
1981 Effects of smoking different doses of nicotine on human aggressive behavior. *Psychopharmacology* 75:339-345.
1984 Effects of cigarette smoking on human aggressive behavior. Pp.

333-344 in K.J. Flannelly, R.J. Blanchard, and D.C. Blanchard, eds., *Biological Perspectives on Aggression*. New York: Alan R. Liss.

Cherek, D.R., and J.L. Steinberg
1987 Effects of drugs on human aggressive behavior. Pp. 239-290 in G.D. Burrows and J.S. Werry, eds., *Advances in Human Psychopharmacology*. Greenwich, Conn.: JAI Press.

Cherek, D.R., J.L. Steinberg, and T.H. Kelly
1987 Effects of diazepam on human laboratory aggression: Correlations with alcohol effects and hostility measures. Pp. 95-101 in L.S. Harris, ed., *Problems of Drug Dependence 1986*. NIDA Research Monograph 76. Rockville, Md.: National Institute on Drug Abuse.

Cherek, D.R., J.L. Steinberg, T.H. Kelly, D.E. Robinson, and R. Spiga
1990 Effects of acute administration of diazepam and *d*-amphetamine on aggressive and escape responding of normal male subjects. *Psychopharmacology* 100:173-181.

Cherek, D.R., R. Spiga, J.D. Roache, R.A. Meisch, and K.A. Cowan
1991 Effects of triazolam on human multi-operant responding. Unpublished manuscript.

Christ, H.
1985 Effects of Met-ENK, substance P and SRIF on the behavior of *Hemichromis bimaculatus*. *Peptides* 6:139-148.

Christmas, A.J., and D.R. Maxwell
1970 A comparison of the effects of some benzodiazepines and other drugs on aggressive and exploratory behaviour in mice and rats. *Neuropharmacology* 9:17-29.

Clinton, J.E., S. Sterner, Z. Stelmachers, and E. Ruiz
1987 Haloperidol for sedation of disruptive emergency patients. *Annals of Emergency Medicine* 16:319-322.

Clody, D.E., and J.R. Vogel
1973 Drug-induced conditioned aversion to mouse-killing in rats. *Pharmacology Biochemistry and Behavior* 1:477-481.

Coccaro, E.F.
1989 Central serotonin and impulsive aggression. *British Journal of Psychiatry* 155:52-62.

Coccaro, E.F., L.J. Siever, H.M. Klar, G. Maurer, K. Cochrane, T.B. Cooper, R.C. Mohs, and K.L. Davis
1989 Serotonergic studies in patients with affective and personality disorders. *Archives of General Psychiatry* 46:587-599.

Cohen, M., G. Oaks, N. Freedman, D.M. Engelhardt, and R.A. Margolis
1968 Family interaction patterns, drug treatment, and change in social aggression. *Archives of General Psychiatry* 19:50-56.

Cole, H.F., and H.H. Wolf
1966 The effects of some psychotropic drugs on conditioned avoidance and aggressive behaviors. *Psychopharmacologia* 8:389-396.
1970 Laboratory evaluation of aggressive behavior of the grasshopper

mouse (*Onychomys*). *Journal of Pharmaceutical Sciences* 59:969-971.

Conn, L., and J. Lion
1984   Pharmacologic approaches to violence. *Psychiatric Clinics of North America* 7:879-886.

Conner, R.L., J.M. Stolk, J.D. Barchas, W.C. Dement, and S. Levine
1970   The effect of parachlorophenylalanine (PCPA) on shock-induced fighting behavior in rats. *Physiology and Behavior* 5:1221-1224.

Cook, L., and E. Wiedley
1960   Effects of a series of psychopharmacological agents on isolation induced attack behavior in mice. *Federation Proceedings* 19:22.

Copenhaver, J.
1989   Relationship between preexposure to prey and PCPA-induced filicidal activity in Sprague-Dawley rats. *Physiology and Behavior* 45:343-346.

Coscina, D.V., J. Goodman, D.D. Dodse, and H.C. Stancer
1975   Taming effects of handling on 6-hydroxydopamine induced rage. *Pharmacology Biochemistry and Behavior* 3:525-528.

Craft, M., I.A. Ismail, D. Krishnamurti, J. Matthews, A. Regan, R.V. Seth, and P.M. North
1987   Lithium in the treatment of aggression in mentally handicapped patients. *British Journal of Psychiatry* 150:685-689.

Crescimanno, G., P. Piazza, A. Benigno, and G. Amato
1986   Effects of substantia nigra stimulation on hypothalamic rage reaction in cats. *Physiology and Behavior* 37:129-133.

Crowley, T.J.
1972   Dose-dependent facilitation or suppression of rat fighting by methamphetamine, phenobarbital, or imipramine. *Psychopharmacologia* 27:213-222.

Crowley, T.J., and C.O. Rutledge
1974   Chronic methamphetamine, imipramine and phenobarbital effects on shock-induced aggression in rats. Pp. 65-80 in J. Singh and H. Lal, eds., *Drug Addiction*, Vol. 3. Miami, Florida: Symposia Specialists.

Cunningham, M.A., V. Pillai, and W.J. Blachford Rogers
1968   Haloperidol in the treatment of children with severe behaviour disorders. *British Journal of Psychiatry* 114:845-854.

Cutler, M.G., and A.K. Dixon
1988   Effects of ipsapirone on the behavior of mice during social encounters. *Neuropharmacology* 27:1039-1044.

Cutler, N., and J.F. Heiser
1978   Retrospective diagnosis of hypomania following successful treatment of episodic violence with lithium: A case report. *American Journal of Psychiatry* 135:753-754.

Dale, P.G.
1980   Lithium therapy in aggressive mentally subnormal patients. *British Journal of Psychiatry* 137:469-474.

Dantzer, R., D. Guilloneau, P. Mormede, J.P. Herman, and M. LeMoal
1984 Influence of shock-induced fighting and social factors on dopamine turnover in cortical and limbic areas in the rat. *Pharmacology Biochemistry and Behavior* 20:331-334.

Daruna, J.H.
1978 Patterns of brain monamine activity and aggressive behavior. *Neuroscience and Biobehavioral Reviews* 2:101-113.

Das, N.N., S.R. Dasgupta, and G. Werner
1954 Changes of behaviour and electroencephalogram in rhesus monkeys caused by chlorpromazine. *Archives Internationales de Pharmacologie et de Therapie* 99:451-457.

DaVanzo, J.P.
1969 Observation related to drug-induced alterations in aggressive behaviour. Pp. 263-272 in S. Garattini and E.B. Sigg, eds., *Aggressive Behaviour*. Amsterdam: Excerpta Medica Foundation.

DaVanzo, J.P., and M. Sydow
1979 Inhibition of isolation-induced aggressive behavior with GABA transaminase inhibitors. *Psychopharmacology* 62:23-27.

DaVanzo, J.P., M. Daugherty, R. Ruckart, and L. Kang
1966 Pharmacological and biochemical studies in isolation-induced fighting mice. *Psychopharmacologia* 9:210-219.

DaVanzo, J.P., J.K. Chamberlain, D.R. Garris, and M.S. Swanson
1986 Regional [^3H]testosterone uptake in the brain of isolated non-aggressive mice. *Brain Research* 369:224-230.

Deberdt, R.
1976 Pipamperone (Dipiperon) in the treatment of behaviour disorders. *Acta Psychiatrica Belgica* 76:157-166.

DeCuyper, H., H.M. Van Praag, and D. Verstraeten
1985 The effect of milenperone on the aggressive behavior of oligophrenic patients. *Neuropsychobiology* 13:101-105.

DeFeo, G., R. Lisciani, L. Pavan, M. Samarelli, and P. Valeri
1983 Possible dopaminergic involvement in biting compulsion induced by large doses of clonidine. *Pharmacological Research Communications* 15:613-619.

Delgado, J.M.R.
1973 Antiaggressive effects of chlordiazepoxide. Pp. 419-432 in S. Garattini, E. Mussini, and L.O. Randall, eds., *The Benzodiazepines*. New York: Raven Press.

Delgado, J.M.R., and F.V. DeFeudis
1969 Effects of lithium injections into the amygdala and hippocampus of awake monkeys. *Experimental Neurology* 25:255-267.

Delgado, J.M.R., C. Grau, J.M. Delgado-Garcia, and J.M. Rodero
1976 Effects of diazepam related to social hierarchy in rhesus monkeys. *Neuropharmacology* 15:409-414.

Delini-Stula, A., and A. Vassout
1978 Influence of baclofen and GABA-mimetic agents of spontaneous and olfactory-bulb-ablation-induced muricidal behaviour in the rat. *Arzneimittel-Forschung/Drug Research* 28:1508-1509.

1979 Differential effects of psychoactive drugs on aggressive responses in mice and rats. Pp. 41-60 in M. Sandler, ed., *Psychopharmacology of Aggression.* New York: Raven Press.

1981 The effects of antidepressants on aggressiveness induced by social deprivation in mice. *Pharmacology Biochemistry and Behavior* 14(S1):33-41.

Depaulis, A., and M. Vergnes
1983 Induction of mouse-killing in the rat by intraventricular injection of a GABA-agonist. *Physiology and Behavior* 30:383-388.

1984 Gabaergic modulation of mouse-killing in the rat. *Psychopharmacology* 83:367-372.

1985 Elicitation of conspecific attack or defense in the male rat by intraventricular injection of a GABA agonist of antagonist. *Physiology and Behavior* 35:447-453.

1986 Elicitation of intraspecific defensive behaviors in the rat by microinjection of picrotoxin, a gamma-aminobutyric acid antagonist, into the midbrain periaqueductal gray matter. *Brain Research* 367:87-95.

Depue, R.A., and M.R. Spoont
1986 Conceptualizing a serotonin trait: A behavioral dimension of constraint. Pp. 47-62 in J.J. Mann and M. Stanley, eds., *Psychobiology of Suicidal Behavior* (Annals of the New York Academy of Sciences, Vol. 487). New York: New York Academy of Sciences.

Diaz, J.-L., and M. Asai
1990 Dominant mice show much lower concentrations of methionine-enkephalin in brain tissue than subordinates: Cause or effect? *Behavioural Brain Research* 39:275-280.

DiChiara, G., R. Camba, and P.F. Spano
1971 Evidence for inhibition by brain serotonin of mouse killing behaviour in rats. *Nature* 233:272-273.

Didiergeorges, F., M. Vergnes, and P.Karli
1968 Sur le mode d'action d'une influence inhibitrice d'origine olfactive s'exercant sur l'agressivite interspecifique du rat. *Comptes rendus des seances de la Societe de Biologie* 162:276-270.

Dietch, J.T., and R.K. Jennings
1988 Aggressive dyscontrol in patients treated with benzodiazepines. *Journal of Clinical Psychiatry* 49:184-188.

DiMascio, A.
1973 The effects of benzodiazepines on aggression: Reduced or increased? *Psychopharmacologia* 30:95-102.

DiMascio, A., R.I. Shader, and J. Harmatz
1969 Psychotropic drugs and induced hostility. *Psychosomatics* 10:46-47.

Diringer, M.N., N.R. Kramarcy, J.W. Brown, and J.B. Thurmond
1982 Effect of fusaric acid on aggression, motor activity, and brain monoamines in mice. *Pharmacology Biochemistry and Behavior* 16:73-79.

Dixon, A.K.
1975   Effects of psychoactive drugs and hormones on the social behaviour of mice: An ethopharmacological analysis. *Experimental Brain Research* 23:59.
1982   A possible olfactory component in the effects of diazepam on social behavior of mice. *Psychopharmacology* 77:246-252.
Dominguez, M., and V.G. Longo
1969   Taming effects of *para*-chlorophenylalanine on septal rats. *Physiology and Behavior* 4:1031-1033.
1970   Effects of *p*-chlorophenylalanine, α-methylparatyrosine and of other indol- and catechol-amine depletors on the hyperirritability syndrome of septal rats. *Physiology and Behavior* 5:607-610.
Donat, P., and M. Krsiak
1985   Effects of a combination of diazepam and scopolamine in animal model of anxiety and aggression. *Activitas Nervosa Superior* 27:307-308.
Dostal, T., and P. Zvolsky
1970   Antiaggressive effect of lithium salts in severely mentally retarded adolescents. *International Pharmacopsychiatry* 5:203-207.
Dotson, L.E., L.S. Robertson, and B. Tuchfeld
1975   Plasma alcohol, smoking, hormone concentrations and self-reported aggression. *Journal of Studies on Alcohol* 36:578-586.
Driscoll, P., and K. Baettig
1981   Selective inhibition by nicotine of shock-induced fighting in the rat. *Pharmacology Biochemistry and Behavior* 14:175-179.
Dubinsky, B., and M.E. Goldberg
1970   The selective blockade by imipramine of hypothalamically-induced attack in cats. *Pharmacologist* 12:207.
1971   The effect of imipramine and selected drugs on attack elicited by hypothalamic stimulation in the cat. *Neuropharmacology* 10:537-454.
Dubinsky, B., J.K. Karpowicz, and M.E. Goldberg
1973   Effects of tricyclic antidepressants on attack elicited by hypothalamic stimulation: To brain biogenic amines. *Journal of Pharmacology and Experimental Therapeutics* 187:550-557.
Earley, C.J., and B.E. Leonard
1977   The effect of testosterone and cyproterone acetate on the concentration of aminobutyric acid in brain areas of aggressive and non-aggressive mice. *Pharmacology Biochemistry and Behavior* 6:409-413.
Ebert, J.N., J.H. Ewing, M.H. Rogers, and D.J. Reynolds
1977   Changes in primary process expression in hospitalized schizophrenics treated with phenothiazines: Two projective tasks compared. *Journal of Genetic Psychology* 130:83-94.
Eichelman, B.
1977   Pharmacological treatment of aggressive disturbances. Pp. 260-

269 in J. Barchas, R. Berger, R. Ciaranello, and G. Elliott, eds., *Psychopharmacology: From Theory to Practice.* New York: Oxford University Press.

1979 Role of biogenic amines in aggressive behavior. Pp. 61-93 in M. Sandler, ed., *Psychopharmacology of Aggression.* New York: Raven Press.

1981 Neurochemical correlates of aggressive behavior. *Psychopharmacology Bulletin* 17(1):58-62.

1986 The biology and somatic experimental treatment of aggressive disorders. Pp. 651-678 in P.A. Berger and H.K.H. Brodie, eds., *The American Handbook of Psychiatry.* New York: Basic Books.

1987 Neurochemical and psychopharmacologic aspects of aggressive behavior. Pp. 697-704 in H.Y. Meltzer, ed., *Psychopharmacology: The Third Generation of Progress.* New York: Raven Press.

Eichelman, B., and J. Barchas
1975 Facilitated shock-induced aggression following anti-depressive medication in the rat. *Pharmacology Biochemistry and Behavior* 3:601-604.

Eichelman, B.S., Jr., and N.B. Thoa
1973 The aggressive monamines. *Biological Psychiatry* 6:143-164.

Eichelman, B.S., Jr., N.B. Thoa, and K.Y. Ng
1972 Facilitated aggression in the rat following 6-hydroxydopamine administration. *Physiology and Behavior* 8:1-3.

Eichelman, B., E. Seagraves, and J. Barchas
1977 Alkali metal cations: Effects on isolation-induced aggression in the mouse. *Pharmacology Biochemistry and Behavior* 7:407-409.

Eichelman, B., G.R. Elliott, and J.D. Barchas
1981 Biochemical, pharmacological, and genetic aspects of aggression. Pp. 51-84 in D.A. Hamburg and M.B. Trudeau, eds., *Biobehavioral Aspects of Aggression.* New York: Alan R. Liss.

Eisenstein, N., L.C. Iorio, and D.E. Clody
1982 Role of serotonin in the blockade of muricidal behavior by tricyclic antidepressants. *Pharmacology Biochemistry and Behavior* 17:847-849.

Ekkers, C.L.
1975 Catecholamine excretion, conscience function and aggressive behaviour. *Biological Psychology* 3:15-30.

Elie, R., Y. Langlois, S.F. Cooper, G. Gravel, and J. Albert
1980 Comparison of Sch-12679 and thioridazine in aggressive mental retardates. *Canadian Journal of Psychiatry* 25:484-491.

Ellenbroek, B.A., and A.R. Cools
1990 Animal models with construct validity for schizophrenia. *Behavioral Pharmacology* 1:469-490.

Elliott, F.A.
1977 Propanolol for the control of belligerent behavior following acute brain damage. *Annual Neurology* 1:489-491.

Elliott, R.L.
  1986 Lithium treatment and cognitive changes in two mentally re-
        tarded patients. *Journal of Nervous and Mental Disease* 174(11):689-
        692.
Ellis, L.
  1991 Monamine oxidase and criminality: Identifying an apparent bio-
        logical marker for antisocial behavior. *Journal of Research in
        Crime and Delinquency* 28:227-251.
Ellison, G.
  1976 Monamine neurotoxins: Selective and delayed effects on be-
        havior in colonies of laboratory rats. *Brain Research* 103:81-92.
Ellison, G.D., and D.E. Bresler
  1974 Tests of emotional behavior in rats following depletion of nore-
        pinephrine, of serotonin, or of both. *Psychopharmacologia* 34:275-
        288.
Elsworth, J.D., D.E. Redmond, Jr., C.R.J. Ruthven, and M. Sandler
  1985 Phenylacetic acid production in dominant and non-dominant
        vervet monkeys. *Life Sciences* 37:1727-1730.
Emley, G.S., and R.R. Hutchinson
  1971 Similar and selective actions of chlorpromazine, chlordiazepoxide,
        and nicotine on shock-produced aggressive and anticipatory motor
        responses in the squirrel monkeys. *Proceedings of the 79th
        Annual Convention of the American Psychological Association*
        759-760.
  1972 Basis of behavioral influence of chlorpromazine. *Life Sciences*
        11:43-47.
  1983 Unique influences of ten drugs upon post-shock biting attack
        and pre-shock manual responding. *Pharmacology Biochemistry
        and Behavior* 19:5-12.
Eriksson, E., and M. Humble
  1990 Serotonin in psychiatric pathophysiology. Pp. 66-119 in R. Pohl
        and S. Gershon, eds., *Biological Basis of Psychiatric Treatment*.
        Basel: Medical and Scientific Publishers.
Ernst, A.M.
  1967 Mode of action of apomorphine and dexamphetamine in gnaw-
        ing compulsion in rats. *Psychopharmacologia* 10:316-323.
Everett, G.M.
  1961 Some electrophysiological and biochemical correlates of motor
        activity and aggressive behavior. Pp. 479-487 in E. Rothlin, ed.,
        Neuro-Psychopharmacology. Amsterdam: Elsevier.
Extein, I.
  1980 Psychopharmacology in psychiatric emergencies. *International
        Journal of Psychiatry in Medicine* 10:189-204.
Fanselow, M.S., and R.A. Sigmundi
  1982 The enhancement and reduction of defensive fighting by naloxone
        pretreatment. *Physiological Psychology* 10(3):313-316.

Fanselow, M.S., R.A. Sigmundi, and R.C. Bolles
1980 Naloxone pretreatment enhances shock-elicited aggression. *Physiological Psychology* 8:369-371.
Faretra, G., L. Dooher, and J. Dowling
1970 Comparison of haloperidol and fluphenazine in disturbed children. *American Journal of Psychiatry* 126:1670-1673.
Feldman, P.E.
1962 An analysis of efficacy of diazepam. *Journal of Neuropsychiatry* 3:S62-S67.
Fernandez-Tome, M.P., J.A. Fuentes, R. Madronero, and J. del Rio
1975 Pharmacological properties of 6,7-tetramethylene-5-phenyl-1,2-dihydro-^3H-thieno[2,3-*e*](1,4)-diazepin-2-one (QM-6008, thiadipone), a new psychotropic drug. *Arzneimittel-Forschung/Drug Research* 25:926-934.
Ferrini, R., G. Miragoli, and B. Taccardi
1974 Neuro-pharmacological studies on SB 5833, a new psychotherapeutic agent of the benzodiazepine class. *Arzneimittel-Forschung/Drug Research* 24:2029-2032.
Fielding, S., and I. Hoffman
1979 Pharmacology of anti-anxiety drugs with special reference to clobazam. *Journal of Clinical Pharmacology* 7:7-15.
File, S.E.
1982 Colong aggression: Effects of benzodiazepines on intruder behavior. *Physiological Psychology* 10:413-416.
1985 Tolerance to the behavioral actions of benzodiazepines. *Neuroscience and Biobehaviorial Reviews* 9:113-121.
1986a Effects of neonatal administration of diazepam and lorazepam on performance of adolescent rats in tests of anxiety, aggression, learning and convulsions. *Neurobehavioral Toxicology and Teratology* 8:301-306.
1986b The effects of neonatal administration of clonazepam on passive avoidance and on social, aggressive and exploratory behavior of adolescent male rats. *Neurobehavioral Toxicology and Teratology* 8:447-452.
File, S.E., and A.J. Johnson
1989 Lack of effects of 5HT$_3$ receptor antagonists in the social interaction and elevated plus-maze tests of anxiety in the rat. *Psychopharmacology* 99:248-251.
File, S.E., and P.S. Mabbutt
1990 Chronic ethanol and chlordiazepoxide—Contrasting effects on reward learning. *Alcohol* 7:307-310.
File, S.E., and J.C. Tucker
1983 Lorazepam treatment in the neonatal rat alters submissive behavior in adulthood. *Neurobehavioral Toxicology and Teratology* 5:289-294.
Fishbein, D.H., D. Lozovsky, and J.H. Jaffe
1989 Impulsivity, aggression, and neuroendocrine responses to sero-

tonergic stimulation in substance abusers. *Biological Psychiatry* 25:1049-1066.

Fog, R.
1969   Rage reactions produced in rats by a combination of thymoleptics and monoamine oxidase inhibitors. *Pharmacological Research Communications* 1:79-83.

Fox, K.A., and R.L. Snyder
1969   Effect of sustained low doses of diazepam on aggression and mortality in grouped male mice. *Journal of Comparative and Physiological Psychology* 69:663-666.

Fox, K.A., J.C. Webster, and F.J. Guerriero
1972   Increased aggression among grouped male mice fed nitrazepam and flurazepam. *Pharmacological Research Communications* 4:157-162.

Freinhar, J.P., and W.A. Alvarez
1985   Lithium treatment of four "affect-related" disorders. *Psychosomatics* 26:332-337.

Freud, S.
1917   Trauer and melancholie. Pp. 428-446 in A. Freud, E. Bibring, W. Hoffer, E. Kris, and O. Isakower, eds., *Sigmund Freud: Gesammelte Werke*, 2nd ed., Vol. X. Frankfurt/Main: Fischer (1973).

Fugham, E., A. Schillinger, J.B. Andersen, B.E. Belstad, D. Jensen, F. Miller, K.J. Muller, B. Schulstad, and K. Elgen
1989   Zuclopenthixol and haloperidol/levopromazine in the treatment of elderly patients with symptoms of aggressiveness and agitation: A double-blind, multi-center study. *Pharmatherapeutics* 5:285-291.

Fujiwara, M., and S. Ueki
1974   Correlation of Δ^9-tetrahydrocannabinol-induced muricide to biogenic amines in the rat brain. *Japanese Journal of Pharmacology* 24:54.

Fujiwara, M., N. Ibii, Y. Kataoka, and S. Ueki
1980   Effects of psychotropic drugs on delta Δ^9-tetrahydrocannabinol induced long-lasting muricide. *Psychopharmacology* 68:7-13.

Fukuda, T., and T. Tsumagari
1983   Effects of psychotropic drugs on the rage responses induced by electrical stimulation of the medial hypothalamus in cats. *Japanese Journal of Pharmacology* 33:885-890.

Funderburk, W.H., M.H. Foxwell, and M.W. Hakala
1970   Effects of psychotherapeutic drugs on hypothalamic-induced hissing in cats. *Neuropharmacology* 9:1-7.

Garattini, S., E. Giacalone, and L. Valzelli
1967   Isolation, aggressiveness and brain 5-hydroxytryptamine turnover. *Journal of Pharmacy and Pharmacology* 19:338-339.

Gardner, C.R., and A.P. Guy
1984   A social interaction model of anxiety sensitive to acutely ad-

ministered benzodiazepines. *Drug Development Research* 4:207-216.

Gardner, D.L., and R.W. Crowdry
1985 Alprazolam-induced dyscontrol in borderline personality disorder. *American Journal of Psychiatry* 142:98-100.

Gardos, G., A. DiMascio, C. Salzman, and R.I. Shader
1968 Differential actions of chlordiazepoxide and oxazepam on hostility. *Archives of General Psychiatry* 18:757-760.

Garris, D.R., J.K. Chamberlain, and J.P. DaVanzo
1984 Histofluorescent identification of indoleamine-concentrating brain loci associated with intraspecies, reflexive biting and locomotor behavior in olfactory-bulbectomized mice. *Brain Research* 294:385-389.

Gay, P.E., and L.D. Clark
1976 Effects of some physiological and pharmacological manipulations on shock-facilitated mouse killing by *Onychomys leucogaster* (northern grasshopper mouse). *Aggressive Behavior* 2:107-121.

Gelders, Y.G., A.J. Reyntjens, and T.J. Aerts
1984 Milenperone: A selective drug for the control of behavioral disorders in epileptic and alcoholic patients. *Acta Psychiatrica Belgica* 84:151-159.

George, D.T., P. Anderson, D.J. Nutt, and M. Linnoila
1989 Aggressive thoughts and behavior: Another symptom of panic disorder? *Acta Psychiatrica Scandanavica* 79:500-502.

Geyer, M.A., and D.S. Segal
1974 Shock-induced aggression: Opposite effects of intraventricularly infused dopamine and norepinephrine. *Behavioral Biology* 10:99-104.

Gianutsos, G., and H. Lal
1976 Blockage of apomorphine-induced aggression by morphine or neuroleptics: Differential alteration by antimuscarinics and naloxone. *Pharmacology Biochemistry and Behavior* 4:639-642.

Gianutsos, G., R.B. Drawbaugh, M.D. Hynes, and H. Lal
1974 Behavioral evidence for dopaminergic supersensitivity and chronic haloperidol. *Life Science* 14:887-898.

Gibbons, J.L., G.A. Barr, W.H. Bridger, and S.F. Leibowitz
1978 Effects of *para*—chlorophenylalanine and 5-hydroxytryptophan on mouse killing behavior in killer rats. *Pharmacology Biochemistry and Behavior* 9:91-98.

1979 Manipulations of dietary tryptophan: Effects on mouse killing and brain serotonin in the rat. *Brain Research* 169:139-153.

1981 L-Tryptophan's effects on mouse killing, feeding, drinking, locomotion, and brain serotonin. *Pharmacology Biochemistry and Behavior* 15:201-206.

Gittelman-Klein, R., D.F. Klein, S. Katz, K. Saraf, and E. Pollack
1976 Comparative effects of methylphenidate and thioridazine in

hyperkinetic children. *Archives of General Psychiatry* 33:1217-1231.

Glenn, M.B., B. Wroblewski, J. Parziale, L. Levine, J. Whyte, and M. Rosenthal
  1989 Lithium carbonate for aggressive behavior or affective instability in ten brain-injured patients. *American Journal of Physical Medicine and Rehabilitation* 68(5):221-226.

Gleser, G.C., L.A. Gottschalk, R. Fox, and W. Lippert
  1965 Immediate changes in affect with chlordiazepoxide. *Archives of General Psychiatry* 13:291-295.

Goddard, P., and V.G. Lokare
  1970 Diazepam in the management of epilepsy. *British Journal of Psychiatry* 117:213-214.

Goetzl, U., F. Grunberg, and B. Berkowitz
  1977 Lithium carbonate in the management of hyperactive aggressive behavior of the mentally retarded. *Comprehensive Psychiatry* 18(6):599-606.

Goldberg, M.E.
  1970 Pharmacologic activity of a new class of agents which selectively inhibit aggressive behavior in rats. *Archives Internationales de Pharmacodynamie et de Therapie* 186:287-297.

Goldberg, M.E., and A.I. Salama
  1969 Norepinephrine turnover and brain monoamine levels in aggressive mouse-killing rats. *Biochemical Pharmacology* 18:532-534.

Goldberg, M.E., J.R. Insalaco, M.A. Hefner, and A.I. Salama
  1973 Effect of prolonged isolation on learning, biogenic amine turnover and aggressive behaviour in three strains of mice. *Neuropharmacology* 12:1049-1058.

Goldstein, J.M., and J. Siegel
  1980 Suppression of attack behavior in cats by stimulation of ventral tegmental area and nucleus accumbens. *Brain Research* 183:181-192.

Gonyou, H.W., K.A. Parget, D.B. Anderson, and R.D. Olson
  1988 Effects of amperozide and azaperone on aggression and productivity of growing-finishing pigs. *Journal of Animal Science* 66:2856-2864.

Goodwin, F.K., D.L. Murphy, H.K.H. Brodie, and W.E. Bunney, Jr.
  1970 L-DOPA, catecholamines, and behavior: A clinical and biochemical study in depressed patients. *Biological Psychiatry* 2:341-366.

Gorelick, D.A., M.L. Elliott, and R.J. Sbordone
  1981 Naloxone increases shock-elicited aggression in rats. *Research Communications in Substances of Abuse* 2:419-422.

Gottschalk, L.A., G.C. Gleser, H.W. Wylie, Jr., and S.M. Kaplan
  1965 Effects of impramine on anxiety and hostility levels. *Psychopharmacologia* 7:303-310.

Gray, W.D., A.C. Osterberg, and C.E. Rauh
  1961 Neuropharmacological actions of mephenoxalone. *Archives*

Internationales de Pharmacodynamie et de Therapie 134:198-215.

Greenberg, A.S., and M. Coleman
1976 Depressed 5-hydroxyindole levels associated with hyperactive and aggressive behavior. *Archives of General Psychiatry* 33:331-336.

Greenblatt, D.J., R.I. Shader, and J. Koch-Weser
1975 Flurazepam hydrochloride. *Clinical Pharmacology and Therapeutics* 17:1-14.

Greendyke, R.M., D.B. Schuster, and J.A. Wooton
1984 Propranolol in the treatment of assaultive patients with organic brain disease. *Journal of Clinical Psychopharmacology* 4:282-285.

Griffiths, R.R., G.E. Bigelow, and I. Liebson
1983 Differential effects of diazepam and pentobarbital on mood and behavior. *Archives of General Psychiatry* 409:865-873.

Grimm, V.E., and A. Zelikovich
1982 Aspects of social interaction in "antiaggression-drugs" treated mice and in their nontreated opponents. *Aggressive Behavior* 8:168-171.

Grossman, S.P.
1963 Chemically induced epileptiform seizures in the cat. *Science* 142:409-411.

Guaitani, A., F. Marucci, and S. Garattini
1971 Increased aggression and toxicity in grouped male mice treated with tranquilizing benzodiazepines. *Psychopharmacologia* 19:241-245.

Guldenpfennig, W.M.
1973 Clinical experience with a new benzodiazepine in the treatment of epilepsy. *South African Medical Journal* 47:998-1000.

Gunn, J.
1979 Drugs in the violence clinic. Pp. 183-195 in M. Sandler, ed., *Psychopharmacology of Aggression*. New York: Raven Press.

Gustavson, C.R., J. Garcia, W.G. Hankins, and K.W. Rusiniak
1974 Coyote predation control by aversive conditioning. *Science* 184:581-583.

Hacke, W.
1980 Die pharmakologische Beeinflussung aggressiven und autoaggressiven Verhaltens bei Geistigbehinderten mit Melperone. *Pharmakopsychiatrie und Nueropsychopharmakologie* 13:20-24.

Hadfield, M.G.
1981 Mesocortical versus nigrostriatal dopamine uptake in isolated fighting mice. *Brain Research* 222:172-176.
1983 Dopamine: Mesocortical vs. nigrostriatal uptake in isolated fighting mice and control. *Behavioural Brain Research* 7:269-281.

Hadfield, M.G., and C. Milio
1988 Isolation-induced fighting in mice and regional brain monoamine utilization. *Behavioural Brain Research* 31:93-96.

Hadfield, M.G., and W.F.C. Rigby
1976 Dopamine-adaptive uptake changes in striatal synaptosomes after 30 seconds of shock-induced fighting. *Biochemical Pharmacology* 25:2752-2754.

Hadfield, M.G., and N.E. Weber
1975 Effect of fighting and diphenylhydantoin on the uptake of ^3H-l-norepinephrine in vitro in synaptosomes isolated from retired male breeding mice. *Biochemical Pharmacology* 24:1538-1540.

Haefely, W.
1990 GABA and benzodiazepine receptor subtypes: Molecular biology, pharmacology, and clinical aspects. Pp. 231-234 in G. Biggio and E. Costa, eds., *GABA and Benzodiazepine Receptor Subtypes*. New York: Raven Press.

Hahn, R.A., M.D. Hynes, and R.W. Fuller
1982 Apomorphine-induced aggression in rats chronically treated with oral clonidine: Modulation by central serotonergic mechanisms. *Journal of Pharmacology and Experimental Therapeutics* 220:389-393.

Haney, M., and K.A. Miczek
1989 Morphine effects on maternal aggression, pup care and analgesia in mice. *Psychopharmacology* 98:68-74.

Haney, M., K. Noda, R. Kream, and K.A. Miczek
1990 Regional 5-HT and dopamine activity: Sensitivity to amphetamine and aggressive behavior in mice. *Aggressive Behavior* 16:259-270.

Hansen, S., and A. Ferreira
1986 Effects of bicuculline infusions in the ventromedial hypothalamus and amygdaloid complex on food intake and affective behavior in mother rats. *Behavioral Neuroscience* 100:410-415.

Hara, C., S. Wantanabe, and S. Ueki
1983 Effects of psychotropic drugs microinjected into the hypothalmus on muricide, catalepsy and cortical EEG in OB rats. *Pharmacology Biochemistry and Behavior* 18:423-431.

1984 Anti-muricide mechanisms of chlorpromazine and imipramine in OB rats: Andrenoceptors and hypothalamic functions. *Pharmacology Biochemistry and Behavior* 21:267-272.

Hasselager, E., Z. Rolinski, and A. Randrup
1972 Specific antagonism by dopamine inhibitors of items of amphetamine induced aggressive behaviour. *Psychopharmacologia* 24:485-495.

Haug, M., S. Simler, L. Kim, and P. Mandel
1980 Studies on the involvement of GABA in the aggression directed by groups of intact or gonadectomized male and female mice towards lactating intruders. *Pharmacology Biochemistry and Behavior* 12:189-193.

Haug, M., S. Simler, L. Ciesielski, P. Mandel, and R. Moutier
1984 Influence of castration and brain GABA levels in three strains

of mice on aggression towards lactating intruders. *Physiology and Behavior* 32:767-770.

Haug, M. P.F. Brain, and A.B. Kamis
1986   A brief review comparing the effects of sex steroids on two forms of aggression in laboratory mice. *Neuroscience and Biobehavioral Reviews* 10:463-468.

Haug, M., M.L. Ouss-Schlegel, J.F. Spertz, D. Benton, P.F. Brain, P. Mandel, L. Ciesielski, and S. Simler
1987   An attempt to correlate attack on lactating females and brain GABA levels in C57 and C3H strains and their reciprocal hybrids. *Biogenic Amines* 4:83-94.

Haug, M., L. Wallian, and P.F. Brain
1990   Effects of 8-OH-DPAT and fluoxetine on activity and attack by female mice towards lactating intruders. *General Pharmacology* 21:845-849.

Hegstrand, L.R., and B. Eichelman
1983   Increased shock-induced fighting with supersensitive ß-adrenergic receptors. *Pharmacology Biochemistry and Behavior* 19:313-320.

Heilman, R.D., E.W. Bauer, and J.P. DaVanzo
1974   Pharmacologic studies with triflubazam (ORF-8063):   A new psychotherapeutic agent. *Current Therapeutic Research* 16:1022-1032.

Heimstra, N.W.
1961   Effects of chlorpromazine on dominance and fighting behavior in rats. *Behavior* 18:313-321.

Heise, G.A., and E. Boff
1961   Taming action of chlordiazepoxide. *Federation Proceedings* 20:393.

Hendley, E.D., B. Moisser, and B.L. Welch
1973   Catecholamine uptake in cerebral cortex: Adaptive change induced by fighting. *Science* 180:1050-1052.

Herbut, M., and Z. Rolinski
1985   The cholinergic influences on aggression in isolated mice. *Polish Journal of Pharmacology and Pharmacy* 37:1-10.

Herink, J., V. Golda, V. Hrdina, and S. Nemecek
1973   The effect of three anticholinergics on the evoked aggressivity in normal and septal rats. *Activitas Nervosa Superior (Praha)* 15:131.

Hernandez-Peon, R., G. Chavez-Ibarra, P.J. Morgane, and C. Timo-Iaria
1963   Limbic cholinergic pathways involved in sleep and emotional behavior. *Experimental Neurology* 8:93-111.

Herrera, J.N., J.J. Sramek, J.F. Costa, S. Roy, C.W. Heh, and R.N. Nguyen
1988   High potency neuroleptics and violence in schizophrenics. *Journal of Nervous and Mental Disease* 176:558-561.

Heuschele, W.P.
1961   Chlordiazepoxide for calming zoo animals. *Journal of the American Veterinary Medical Association* 139:996-998.

Hirose, K., A. Matsushita, M. Eigyo, H. Jyoyama, A. Fujita, Y. Tsukinoki, T. Shiomi, and K. Matasubara
1981   Pharmacology of 2-0-chlorobenzoyl-4-chloro-N-methyl-Nα-glycylglycinanilide hydrate (45-0088-S), a compound with benzodiazepine-like properties. *Arzneimittel-Forschung/Drug Research* 31:63-69.

Hitchens, J.T., R. Orzechowski, S. Goldstein, and I. Shemano
1972   Pharmacological evaluation of aletamine (α-allylphenethylamine hydrochloride) as an antidepressant. *Toxicology and Applied Pharmacology* 21:302-314.

Hodge, G.K., and L.L. Butcher
1975   Catecholamine correlates of isolation-induced aggression in mice. *European Journal of Pharmacology* 31:81-93.

Hoffmeister, F., and W. Wuttke
1969   On the actions of psychotropic drugs on the attack- and aggressive-defensive behaviour of mice and cats. Pp. 273-280 in S. Garattini and E.B. Sigg, eds., *Aggressive Behaviour*. Amsterdam: Excerpta Medica Foundation.

Hong, K.W., W.S. Lee, and B.Y. Rhim
1987   Role of central α_2-adrenoceptors on the development of muricidal behavior in olfactory bulbectomized rats: Effects of α_2-adrenoceptor antagonists. *Physiology and Behavior* 39:535-539.

Horn, L.J.
1987   "Atypical" medications for the treatment of disruptive, aggressive behavior in the brain-injured patient. *Journal of Head Trauma Rehabilitation* 2:18-28.

Horovitz, Z.P., A.R. Furgiuele, L.J. Brannick, J.C. Burke, and B.N. Craver
1963   A new chemical structure with specific depressant effects on the amygdala and on the hyper-irritability of the septal rat. *Nature* 200:369-370.

Horovitz, Z.P., R.W. Ragozzino, and R.C. Leaf
1965   Selective block of rat mouse-killing by antidepressants. *Life Sciences* 4:1909-1912.

Horovitz, Z.P., J.J. Piala, J.P. High, J.C. Burke, and R.C. Leaf
1966   Effects of drugs on the mouse-killing (muricide) test and its relationship to amygdaloid function. *International Journal of Neuropharmacology* 5:405-411.

Humber, L.G., F.T. Bruderlein, A.H. Philipp, M. Gotz, and K. Voith
1979   Mapping the dopamine receptor. 1. Features derived from modifications in ring E of the neuroleptic butaclamol. *Journal of Medicinal Chemistry* 22:761-767.

Hutchins, D.A., J.D.M. Pearson, and D.F. Sharman
1974   An altered metabolism of dopamine in the striatal tissue of mice made aggressive by isolation. *British Journal of Pharmacology* 51:115p-116p.
1975   Striatal metabolism of dopamine in mice made aggressive by isolation. *Journal of Neurochemistry* 24:1151-1154.

Ieni, J.R., and J.B. Thurmond
1985   Maternal aggression in mice: Effects of treatments with PCPA, 5-HTP and 5-HT receptor antagonists. *European Journal of Pharmacology* 111:211-220.
Ingram, I.M., and G.C. Timbury
1960   Side-effects of librium. *Lancet* 2:766.
Irwin, S., R. Kinohi, M. Van Sloten, and M.P. Workman
1971   Drug effects on distress-evoked behavior in mice: Methodology and drug class comparisons. *Psychopharmacologia* 20:172-185.
Isel, F., and P. Mandel
1989   Alterations of serotonin neurotransmission and inhibition of mouse-killing behavior: III. Effects of minaprine, CM 30366 and SR 95191. *Pharmacology Biochemistry and Behavior* 33:655-662.
Isel, F., L. Ciesielski, S. Gobaille, V. Molina, and P. Mandel
1988   Alterations of serotonin neurotransmission and inhibition of mouse killing behavior: II. Effects of selective and reversible monoamine oxidase inhibitors of type A. *Pharmacology Biochemistry and Behavior* 29:97-104.
Itil, T.M.
1981   Drug therapy in the management of aggression. Pp. 489-501 in P.F. Brain and D. Benton, eds., *Multidisciplinary Approaches to Aggression Research*. New York: Elseview/North Holland Biomedical.
Itil, T.M., and S. Mukhopadhyay
1978   Pharmacological management of human violence. Pp. 139-158 in L. Vlazelli, T.A. Ban, F.A. Freyhan, and P. Pichot, eds., *Psychopharmacology of Aggression (Modern Problems of Pharmacopsychiatry*, Vol. 13.) Basel, Switzerland: Karger.
Itil, T.M., and B. Reisberg
1978   Pharmacologic treatment of aggressive syndromes. *Current Psychiatric Therapies* 18:137-142.
Itil, T.M., and P. Seaman
1978   Drug treatment of human aggression. *Progress in Neuro-Psychopharmacology* 2:659.
Itil, T.M., and A. Wadud
1975   Treatment of human aggression with major tranquilizers, antidepressants, and newer psychotic drugs. *Journal of Nervous and Mental Disease* 160:83-99.
Itil, T.M., N. Polvan, and W. Hsu
1972   Clinical and EEF effects of GB-94, a "tetracyclic" antidepressant (EEG model in discovery of a new psychotropic drug). *Current Therapeutic Research* 14:395-413.
Iwasaki, K., M. Fujiwara, S. Shibata, and S. Ueki
1986   Changes in brain catecholamine levels following olfactory bulbectomy and the effect of acute and chronic administration

of desipramine in rats. *Pharmacology Biochemistry and Behavior* 24:1715-1719.

Jacobs, B.L., and A. Cohen
1976 Differential behavioral effects of lesions of the median or dorsal raphe nuclei in rats: Open field and pain-elicited aggression. *Journal of Comparative and Physiological Psychology* 90:102-108.

Jain, K., and F.S.K. Barar
1986 Brain acetylcholine content in experimentally induced aggression in mice and its modification by testosterone, diethylstilbestrol and norgestrel. *Indian Journal of Medical Research* 84:635-639.

Janssen, P.A.J., A.H. Jageneau, and J.E. Niemegeers
1960 Effects of various drugs on isolation-induced fighting behavior of male mice. *Journal of Pharmacology and Experimental Therapeutics* 129:471-475.

Janssen, P.A.J., C.J.E. Niemegeers, and F.J. Verbruggen
1962 A propos d'une methode d'investigation de substances susceptibles de modifier le comportement agressif inne du rat blanc vis-a-vis de la souris blanche. *Psychopharmacologia* 3:114-123.

Jarvis, M.F., M. Krieger, G. Cohen, and G.C. Wagner
1985 The effects of phencyclidine and chlordiazepoxide on target biting of confined male mice. *Aggressive Behavior* 11:201-205.

Jefferson, J.W.
1982 The use of lithium in childhood and adolescence: An overview. *Journal of Clinical Psychiatry* 43:174-177.

Jimerson, D., and D.J. Reis
1973 Effects of intrahypothalamic injection of 6-hydroxydopamine on predatory aggression in rat. *Brain Research* 61:141-152.

Jirgl, M., J. Drtil, and J. Cepelak
1970 The influence of propericiazine on the behavior of difficult delinquents. *Activitas Nervosa Superior* 12:134-135.

Johansson, G.
1974 Relation of biogenic amines to aggressive behaviour. *Medical Biology* 52:189-192.

Johansson, G., T. Pääkkönen, S. Ruusunen, M. Sandström, and M. Huttunen
1974 Effect of 6-hydroxydopamine on agonistic behaviour elicited by stimulation of the hypothalamus in the cat. *Medical Biology* 52:406-414.

Jumber, L.G., F.T. Bruderlein, A.H. Philipp, and M. Gotz
1979 Mapping the dopamine receptor. I. Features derived from modifications in ring E of the neuroleptic butaclamol. *Journal of Medicinal Chemistry* 22:761-767.

Kalin, N.H., and S.E. Shelton
1989 Defensive behavior in infant rhesus monkeys: Environmental cues and neurochemical regulation. *Science* 243:1718-1721.

Kalina, R.K.
1964   Diazepam: Its role in a prison setting. *Diseases of the Nervous System* 25:101-107.

Kamioka, T., I. Nakayama, S. Akiyama, and H. Takagi
1977   Effects of oxazolam, cloxazolam, and CS-386, new anti-anxiety drugs, on socially induced suppression and aggression in pairs of monkeys. *Psychopharmacology* 52:17-23.

Kampov-Polevoi, A.B.
1978   Effect of drugs on domination-subordination relationships in pairs of rats. *Byulleten' Eksperimental'noi Biologii i Meditsiny* 86:306-308.

Karczmar, A.G., and C.L. Scudder
1969   Aggression and neurochemical changes in different strains and genera of mice. Pp. 209-227 in S. Garattini and E.B. Sigg, eds., *Aggressive Behaviour*. Amsterdam: Excerpta Medica Foundation.

Karczmar, A.G., C.L. Scudder, and D.L. Richardson
1973   Interdisciplinary approach to the study of behavior in related mice types. *Neuroscience Research* 5:159-245.

Karli, P.
1958   Action de l'amphetamine et de la chlorpromazine sur l'agressivite interspecifique rat-souris. *Comptes Rendus de Societe de Biologie* 152:1796-1798.

1959a   Action de substances dites "tranquillisantes" sur l'agressivite interspecifique rat-souris. *Comptes Rendus de Societe de Biologie* 153:467-469.

1959b   Recherches pharmacologiques sur de comportment d'agression rat-souris. *Comptes Rendus de Societe de Biologie* 153:497-498.

1981   Conceptual and methodological problems associated with the study of brain mechanisms underlying aggressive behavior. Pp. 322-361 in P.F. Brain and D. Benton, eds., *The Biology of Aggression*, Vol. 1. Amsterdam: Sijthoff & Noordhoff International Publishers.

Karli, P., M. Vergnes, and F. Didiergeorges
1969   Rat-mouse interspecific aggressive behaviour and its manipulation by brain ablation and by brain stimulation. Pp. 47-55 in S. Garattini and E.B. Sigg, eds., *Aggressive Behaviour*. Amsterdam: Excerpta Medica Foundation.

Katz, R.J., and E. Thomas
1975   Effects of scopolamine and α-methylparatyrosine upon predatory attack in cats. *Psychopharmacologia* 42:153-157.

1976   Effects of *para*-chlorophenylalanine upon brain stimulated affective attack in the cat. *Pharmacology Biochemistry and Behavior* 5:391-394.

Kazdin, A.E.
1987 Treatment of antisocial behavior in children: Current status and future directions. *Psychological Bulletin* 102:187-203.

Keats, M.M., and S. Mukherjee
1988 Antiaggressive effect of adjunctive clonazepam in schizophrenia associated with seizure disorder. *Journal of Clinical Psychiatry* 49:117-118.

Keckich, W.A.
1978 Neuroleptics: Violence as a manifestation of akathesia. *Journal of the American Medical Association* 240:2185.

Kellam, S.G., S.C. Goldberg, N.R. Schooler, A. Berman, and J.L. Shmelzer
1967 Ward atmosphere and outcome of treatment of acute schizophrenia. *Journal of Psychiatric Research* 5:145-163.

Kelly, J.T., M. Koch, and D. Buegel
1976 Lithium carbonate in juvenile manic-depressive illness. *Diseases of the Nervous System* 37:90-92.

Kennett, G.A., P. Whitton, K. Shah, and G. Curzon
1989 Anxiogenic-like effects of mCPP and TFMPP in animal models are opposed by 5-HT$_{1C}$ receptor antagonists. *European Journal of Pharmacology* 164:445-454.

Kent, T.A., C.S. Brown, S.G. Bryant, E.S. Barratt, A.R. Felthous, and R.M. Rose
1988 Blood platelet uptake of serotonin in episodic aggression: Correlation with red blood cell protein Tl and impulsivity. *Psychopharmacology Bulletin* 24:454-457.

Kermani, E.J.
1969 "Aggression": Biophysiological aspects. *Diseases of the Nervous System* 30:407-414.

Kerr, W.C.
1976 Lithium salts in the management of a child batterer. *Medical Journal of Australia* 2:414-415.

Kido, R., K. Hirose, D.-I. Yamamoto, and A. Matsushita
1967 Effects of some drugs on aggressive behaviour and the electrical activity of the limbic system. Pp. 365-387 in W.R. Adley and P. Tokizane, eds., *Progress in Brain Research*. Amsterdam: Elsevier.

Klar, H., and L.J. Siever
1984 The psychopharmacologic treatment of personality disorders. *Psychiatric Clinics of North America* 7:791-801.

Kletztin, M.
1969 An experimental analysis of aggressive-defensive behavior in mice. Pp. 253-262 in S. Garattini and E.B. Sigg, eds., *Aggressive Behaviour*. Amsterdam: Excerpta Medica Foundation.

Klunder, C.S., and M. O'Boyle
1979 Suppression of predatory behaviors in laboratory mice following lithium chloride injections or electric shock. *Animal Learning and Behavior* 7:13-16.

Kochansky, G.E., C. Salzman, R.I. Shader, J.S. Harmatz, and A.M. Ogeltree
1975   The differential effects of chlordiazepoxide and oxazepam on hostility in a small group setting. *American Journal of Psychiatry* 132:861-863.
1977   Effects of chlordiazepoxide and oxazepam administration on verbal hostility. *Archives of General Psychiatry* 34:1457-1459.
Kocur, J., Z. Rydzynski, S. Duszyk, and W. Trendak
1984   Bromazepam in behavioural disturbances in children. *Activitas Nervosa Superior* 26:258-259.
Kono, R., N. Tashiro, and H. Nakao
1986   Inhibitory effects of acetylcholine on aggressive-defense reaction induced by electrical stimulation of the hypothalamus in cats. *Brain Research Bulletin* 16:491-495.
Kostowski, W.
1966   A note on the effects of some psychotropic drugs on the aggressive behavior in the ant, *Formica rufa*. *Journal of Pharmacy and Pharmacology* 18:747-749.
Kostowski, W., and A. Czlonkowski
1973   The activity of some neuroleptic drugs and amphetamine in normal and isolated rats. *Pharmacology* 10:82-87.
Kostowski, W., and B. Tarchalska
1972   The effects of some drugs affecting brain 5-HT on the aggressive behaviour and spontaneous electrical activity of the central nervous system of the ant, *Formica* rufa. *Brain Research* 38:143-149.
Kostowski, W., L. Valzelli, and W. Kozak
1983   Chlordiazepoxide antagonizes locus coeruleus-mediated suppression of muricidal aggression. *European Journal of Pharmacology* 91:329-330.
Kostowski, W., W. Danysz, A. Plaznik, and E. Nowakowska
1984   Studies on the locus coeruleus system in an animal model for antidepressive activity. *Journal of Pharmacology and Pharmacy* 36:523-530.
Kostowski, W., L. Valzelli, and G. Baiguerra
1986   Effect of chronic administration of alprazolam and adinazolam on clonidine- or apomorphine-induced aggression in laboratory rodents. *Neuropharmacology* 25:757-761.
Kozak, W., L. Valzelli, and S. Garattini
1984   Anxiolytic activity on locus coeruleus-mediated suppression of muricidal aggression. *European Journal of Pharmacology* 105:323-326.
Kraemer, G.E.
1985   The primate social environment, brain neurochemical changes and psychopathology. *Trends in Neurosciences* 8:339-340.
Krames, L., N.W. Milgram, and D.P. Christie
1973   Predatory aggression: Differential suppression of killing and feeding. *Behavioral Biology* 9:641-647.

Kravitz, E.A., S. Glusman, M.S. Livingstone, and R.M. Harris-Warrick
  1981   Serotonin and octopamine in the lobster nervous system: Mechanism of action at neuromuscular junctions and preliminary behavioral studies. Pp. 189-210 in B.L. Jacobs and A. Gelperin, eds., *Serotonin Neurotransmission and Behavior.* Cambridge, Mass.: MIT Press.
Krsiak, M.
  1974a Isolation-induced timidity in mice as a measure of anxiolytic activity of drugs. *Activitas Nervosa Superior* 16:241-242.
  1974b Behavioral changes and aggressivity evoked by drugs in mice. *Research Communications in Chemical Pathology and Pharmacology* 7:237-257.
  1975a Tail rattling in aggressive mice as a measure of tranquillizing activity of drugs. *Activitas Nervosa Superior* 17:225-226.
  1975b Timid singly-housed mice: Their value in prediction of psychotropic activity of drugs. *British Journal of Pharmacology* 55:141-150.
  1979   Effects of drugs on behaviour of aggressive mice. *British Journal of Pharmacology* 65:525-533.
Krsiak, M., and H. Steinberg
  1969   Psychopharmacological aspects of aggression: A review of the literature and some new experiments. *Journal of Psychosomatic Research* 13:243-252.
Krsiak, M., and A. Sulcova
  1990   Differential effects of six structurally related benzodiazepines on some ethological measures of timidity, aggression and locomotion in mice. *Psychopharmacology* 101:396-402.
Krsiak, M., A. Sulcova, Z. Tomasikova, N. Dlohozkova, E. Kosar, and K. Masek
  1981   Drug effects on attack, defense, and escape in mice. *Pharmacology Biochemistry and Behavior* 14(suppl. 1):47-52.
Kruesi, M.J.P., J.L. Rapoport, S. Hamburger, E. Hibbs, W.Z. Potter, M. Lenane, and G.R. Brown
  1990   Cerebrospinal fluid monoamine metabolites, aggression, and impulsivity in disruptive behavior disorders of children and adolescents. *Archives of General Psychiatry* 47:419-426.
Kruk, M.R., A.M. van der Poel, J.H.C.M. Lammers, T. Hagg, A.M.D.M. de Hey, and S. Oostwegel
  1987   Ethopharmacology of hypothalamic aggression in the rat. Pp. 35-45 in B. Olivier, J. Mos, and P.F. Brain, eds., *Ethopharmacology of Agonistic Behaviour in Animals and Humans.* Dordrecht, The Netherlands: Martinus Nijhoff.
Kulkarni, A.S.
  1968   Muricidal block produced by 5-hydroxytryptophan and various drugs. *Life Science* 7:125-128.
  1970   Decarboxylase inhibitor on 5-HTP induced blockade of mouse killing (abstract). *Pharmacologist* 12:207.

Lal, H., and S.K. Puri
1971   Morphine-withdrawal aggression: Role of dopaminergic stimu-
       lation. Pp. 301-310 in J.M. Singh, L. Miller, and H. Lal, eds.,
       Drug Addiction: Experimental Pharmacology. Mount Kisco,
       N.Y.: Futura Publishing Company.
Lal, H., G. Gianutsos, and S.K. Puri
1975   A comparison of narcotic analgesics with neuroleptics on be-
       havioral measures of dopaminergic activity. *Life Sciences* 17:29-
       32.
Lammers, A.J.J.C., and J.M. van Rossum
1968   Bizarre social behavior in rats induced by a combination of a
       peripheral decarboxylase inhibitor and DOPA. *European Jour-
       nal of Pharmacology* 5:103-106.
Lamprecht, F., B. Eichelman, N.B. Thoa, R.B. Williams, and I.J. Kopin
1972   Rat fighting behavior: Serum dopamine-ß-hydroxylase and hy-
       pothalamic tyrosine hydroxylase. *Science* 177:1214-1215.
Langfeldt, T.
1974   Diazepam-induced play behavior in cats during prey killing.
       Psychopharmacologia 36:181-184.
Langfeldt, T., and H. Ursin
1971   Differential action of diazepam on flight and defense behavior
       in the cat. *Psychopharmacologia* 19:61-66.
Langley, W.
1981   Failure of food-aversion conditioning to suppress predatory at-
       tack of the grasshopper mouse, *Onychomys leucogaster.* *Be-
       havioral and Neural Biology* 33:317-333.
Lapin, I.P.
1962   Qualitative and quantitative relationships between the effects
       of imipramine and chlorpromazine on amphetamine group tox-
       icity. *Psychopharmacologia* 3:413-422.
1967   Simple pharmacological procedures to differentiate antidepres-
       sants and cholinolyties in mice and rats. *Psychopharmacologia*
       11:79-87.
Lasley, S.M., and J.B. Thurmond
1985   Interaction of dietary tryptophan and social isolation on territo-
       rial aggression, motor activity, and neurochemistry in mice.
       Psychopharmacology 87:313-321.
Lassen, J.B.
1978   Piperoxan reduces the effects of clonidine on aggression in mice
       and noradrenaline dependent hypermobility in rats. *European
       Journal of Pharmacology* 47:45-49.
Laverty, R., and K.M. Taylor
1969   Behavioural and biochemical effects of 2-(2,6-dichlorophenylamino)-
       2-imidazoline hydrochloride (St 155) on the central nervous sys-
       tem. *British Journal of Pharmacology* 35:253-264.

Leaf, R.C., and D.J. Wnek
1978 Pilocarpine, food deprivation, and induction of mouse killing by cats. *Pharmacology Biochemistry and Behavior* 9:439-444.
Leaf, R.C., L. Lerner, and Z.P. Horovitz
1969 The role of the amygdala in the pharmacological and endocrinological manipulation of aggression. Pp. 120-131 in S. Garattini and E.B. Sigg, eds., *Aggressive Behaviour*. Amsterdam: Excerpta Medica Foundation.
Leaf, R.C., D.J. Wnek, P.E. Gay, R.M. Corcia, and S. Lamon
1975 Chlordiazepoxide and diazepam induced mouse killing by rats. *Psychopharmacologia* 44:23-28.
Leaf, R.C., D.J. Wnek, and S. Lamon
1978 Despite various drugs, cats continue to kill mice. *Pharmacology Biochemistry and Behavior* 9:445-452.
1984 Oxazepam induced mouse killing by rats. *Pharmacology Biochemistry and Behavior* 20:311-313
Leckman, J.F., W.K. Goodman, M.A. Riddle, M.T. Hardin, and G.M. Anderson
1990 Low CSF 5HIAA and obsession of violence: Report of two cases. *Psychiatry Research* 33:95-99.
Le Douarec, J.C., and L. Broussy
1969 Dissociation of the aggressive behaviour in mice produced by certain drugs. Pp. 281-295 in S. Garattini and E.B. Sigg, eds., *Aggressive Behaviour*. Amsterdam: Excerpta Medica Foundation.
Lee, E.H.Y., H.H. Lin, and H.M. Yin
1987 Differential influences of different stressors upon midbrain raphe neurons in rats. *Neuroscience Letters* 80:115-119.
Lena, B.
1979 Lithium in child and adolescent psychiatry. *Archives of General Psychiatry* 36:854-855.
Le Vann, L.J.
1971 Clinical comparison of haloperidol with chlorpromazine in mentally retarded children. *American Journal of Mental Deficiency* 6:719-723.
Leventhal, B.L., and H.K.H. Brodie
1981 The pharmacology of violence. Pp. 85-106 in D.A. Hamburg and M.B. Trudeau, eds., *Biobehavioral Aspects of Aggression*. New York: Alan R. Liss.
Lidberg, L., J.R. Tuck, M. Asberg, G.P. Scalia-Tomba, and L. Bertilsson
1985 Homicide suicide and CSF 5-HIAA. *Acta Psychiatrica Scandinavica* 71:230-236.
Lindem, K., J. Fletcher, M. Blumenkrantz, and J. Ratey
1990 Pindolol to treat aggression and self-injury in mentally retarded adults. Unpublished manuscript.
Lindgren, T., and K.M. Kantak
1987 Effects of serotonin receptor agonists and antagonists on offensive aggression in mice. *Aggressive Behavior* 13:87-96.

Ling, G.S.F., R. Simantov, J.A. Clark, and G.W. Pasternak
1986   Naloxonazine actions in vivo. *European Journal of Pharmacology* 129:33-38.

Linnoila, M., M. Virkkunen, M. Scheinin, A. Nuutila, R. Rimon, and F.K. Goodwin
1983   Low cerebrospinal fluid 5-hydroxyindoleacetic acid concentration differentiates impulsive from nonimpulsive violent behavior. *Life Science* 33:2609-2614.

Linnoila, M., J. De Jong, and M. Virkkunen
1989   Family history of alcoholism in violent offenders and impulsive fire setters. *Archives of General Psychiatry* 46:613-616.

Lion, J.R.
1975   Conceptual issues in the use of drugs for the treatment of aggression in man. *Journal of Nervous and Mental Disease* 160-76-82.
1979   Benzodiazepines in the treatment of aggressive patient. *Journal of Clinical Psychiatry* 40:70-71.
1981   Medical treatment of violent individuals. Pp. 343-351 in J. Ray Hays, T.K. Roberts, and K.S. Solway, eds., *Violence and the Violent Individual*. New York: SP Medical and Scientific Books.

Lion, J.R., C. Azcarate, and H. Hoepke
1975a "Paradoxical rage reactions" during psychotropic medication. *Diseases of the Nervous System* 36:557-558.

Lion, J.R., J. Hill, and D.J. Madden
1975b Lithium carbonate and aggression: A case report. *Diseases of the Nervous System* 36:97-98.

Liou, S.Y., S. Shibata, and S. Ueki
1985   The role of central noradrenergic neurons in electroconvulsive shock-induced muricide inhibition to olfactory bulbectomized rats. *Pharmacology Biochemistry and Behavior* 23:65-70.

Lipman, R.S., L. Covi, K. Rickels, D.M. McNair, R. Downing, R.J. Kahn, V.K. Lasseter, and V. Faden
1986   Imipramine and chlordiazepoxide in depressive and anxiety disorders. *Archives of General Psychiatry* 43:68-77.

Lister, R.E., I.A. Beattie, and P.A. Berry
1971   Effects of drugs on the social behaviour of baboons. Pp. 299-303 in O. Vinar, Z. Votava, and P.B. Bradley, eds., *Advances in Neuropsychopharmacology* Amsterdam/London: North-Holland Publishing Co.

Livingstone, M.S., R.M. Harris-Warrick, and E.A. Kravitz
1980   Serotonin and octopamine produce opposite postures in lobsters. *Science* 208:76-79.

Llorente, A.F.
1969   The management of behavior disorders with thioridazine in the mentally retarded. *Journal of the Maine Medical Association* 70:229-231.

Loiselle, R.H., and H.V. Capparell
  1967   Effects of chlorpromazine HC1 and chlordiazepoxide HC1 on "instinctual" aggressive behavior in rats. *Psychiatric Communications* 9:29-33.
Loizzo, A., and M. Massotti
  1973   Taming effect of nonnarcotic analgesics on the septal syndrome in rats. *Pharmacology Biochemistry and Behavior* 1:367-370.
Lonowski, D.J., R.A. Levitt, and S.D. Larson
  1973   Effects of cholinergic brain injections on mouse killing or carrying by rats. *Physiological Psychology* 1:341-345.
Lonowski, D.J., R.A. Levitt, and W.A. Dickinson
  1975   Carbachol-elicited mouse killing by rats: Circadian rhythm and dose response. *Bulletin of the Psychonomic Society* 6:601-604.
Louilot, A., M. LeMoal, and H. Simon
  1986   Differential reactivity of dopaminergic neurons in the nucleus accumbens in response to different behavioral situations. An in vivo voltammetric study in free moving rats. *Brain Research* 397:395-400.
Luchins, D.J., and D. Dojka
  1989   Lithium and propranolol in aggression and self-injurious behavior in the mentally retarded. *Psychopharmacology Bulletin* 25:372-375.
MacDonnell, M.F., L. Fessock, and S.H. Brown
  1971   Aggression and associated neural events in cats. Effects of *p*-chlorophenylalanine compared with alcohol. *Quarterly Journal of Studies on Alcohol* 32:748-763.
Mack, G., S. Simler, and P. Mandel
  1975   Systeme inhibiteur gabanergique dans l'agressivite interspecifique rat-souris. *Journal de Physiologie* 71:162a.
Maeda, H., and S. Maki
  1986   Dopaminergic facilitation of recovery from amygdaloid lesions which affect hypothalamic defensive attack in cats. *Brain Research* 363:135-140.
Maengwyn-Davies, G.D., D.G. Johnson, N.B. Thoa, V.K. Weise, and I.J. Kopin
  1973   Influence of isolation and of fighting on adrenal tyrosine hydroxylase and phenylethanolamine-*N*-methyltransferase activities in three strains of mice. *Psychopharmacologia* 28:399-350.
Maj, J.
  1980   Studies on the action of antidepressant drugs of second generation. *Polish Journal of Pharmacology and Pharmacy* 32:437-449.
Maj, J., E. Mogilnicka, and A. Kordecka
  1979   Chronic treatment with antidepressant drugs: Potentiation of apomorphine-induced aggressive behaviour in rats. *Neuroscience Letters* 13:337-341.

Maj, J., E. Mogilnicka, and A. Kordecka-Magiera
1980 Effects of chronic administration of antidepressant drugs on aggressive behavior induced by clonidine in mice. *Pharmacology Biochemistry and Behavior* 13:153-154.
Maj, J., E. Mogilnicka, V. Klimek, and A. Kordecka-Magiera
1981 Chronic treatment with antidepressants: Potentiation of clonidine-induced aggression in mice via noradrenergic mechanism. *Neural Transmission* 52:189-197.
Maj, J., Z. Rogoz, G. Skuza, and H. Sowinska
1982 Effects of chronic treatment with antidepressants on aggressiveness induced by clonidine. *Journal of Neural Transmission* 55:19-25.
Maj, J., Z. Rogoz, and G. Skuza
1983 (+)Oxaprotiline but not (-)oxaprotiline given chronically potentiates the aggressive behaviour induced by clonidine. *Journal of Pharmacy and Pharmacology* 35:180-181.
Maj, J., V. Klimek, A. Lewandowska, and M. Zazula
1987 Central ß and α-adrenolytic activities of adimolol. *Polish Journal of Pharmacology and Pharmacy* 39:81-90.
Maler, L., and W.G. Ellis
1987 Inter-male aggressive signals in weakly electric fish are modulated by monoamines. *Behavioural Brain Research* 25:75-81.
Maletzky, B.M.
1973 The episodic dyscontrol syndrome. *Diseases of the Nervous System* 34:178-185.
Malick, J.B.
1970 Effects of selected drugs on stimulus-bound emotional behavior elicited by hypothalamic stimulation in the cat. *Archives Internationales de Pharmacodynamie et de Therapie* 186:137-141.
1976 Pharmacological antagonism of mouse-killing behavior in the olfactory bulb lesion-induced killer rat. *Aggressive Behavior* 2:123-130.
1978a Selective antagonism of isolation-induced aggression in mice by diazepam following chronic administration. *Pharmacology Biochemistry and Behavior* 8:497-499.
1978b Inhibition of fighting in isolated mice following repeated administration of lithium chloride. *Pharmacology Biochemistry and Behavior* 8:579-581.
1979 The pharmacology of isolation-induced aggressive behavior in mice. Pp. 1-27 in W.B. Essman and L. Valzelli, eds., *Current Developments in Psychopharmacology*. New York: SP Medical and Scientific Books.
Malick, J.B., and A. Barnett
1976 The role of serotonergic pathways in isolation-induced aggression in mice. *Pharmacology Biochemistry and Behavior* 5:55-61.

Malick, J.B., R.D. Sofia, and M.E. Goldberg
1969   A comparative study of the effects of selected psychoactive agents upon three lesion-induced models of aggression in the rat. *Archives Internationales de Pharmacodynamie et de Therapie* 181:459-465.

Mandel, P., G. Mack, and E. Kempf
1979   Molecular basis of some models of aggressive behavior. Pp. 95-110 in M. Sandler, ed., *Psychopharmacology of Aggression*. New York: Raven Press.

Mandel, P., L. Ciesielski, M. Maitre, S. Simler, E. Kempf, and G. Mack
1981   Inhibitory amino acids, aggressiveness, and convulsions. Pp. 1-9 in F.V. De Feudis and P. Mandel, eds., *Amino Acid Neurotransmitters*. New York: Raven Press.

Manning, F.J., and T.F. Elsmore
1972   Shock-elicited fight and Δ^9-tetrahydrocannabinol. *Psychopharmacologia* 25:218-228.

Marini, J.L., and M.H. Sheard
1976   Sustained-release lithium carbonate in a double-blind study: Serum lithium levels, side effects, and placebo response. *Journal of Clinical Pharmacology* 16:276-283.
1977   Antiaggressive effect of lithium ion in man. *Acta Psychiatrica Scandinavica* 55:269-286.

Marini, J.L., M.H. Sheard, and T. Kosten
1979   Study of the role of serotonin in lithium action using shock-elicited fighting. *Communications in Psychopharmacology* 3:225-233.

Marks, P.C., M. O'Brien, and G. Paxinos
1977   5,7-DHT-induced muricide: Inhibition as a result of preoperative exposure of rats to mice. *Brain Research* 135:383-388.
1978   Chlorimipramine inhibition of muricide: The role of the ascending 5-HT projection. *Brain Research* 149:270-273.

Martensz, N.D., S.V. Vellucci, E.B. Keverne, and J. Herbert
1986   Beta-endorphin levels in the cerebrospinal fluid of male talapoin monkeys in social groups related to dominance status and the luteinizing hormone response to naloxone. *Neuroscience* 18:651-658.

Martorano, J.T.
1972   Target symptoms in lithium carbonate therapy. *Comprehensive Psychiatry* 13:533-537.

Matte, A.C.
1982   The effect of 5,7-dihydroxytryptamine on motor activity, aggression, and "emotionality" in isolated wild male mice. *Aggressive Behavior* 8:198-200.

Mattes, J.A.
1986   Psychopharmacology of temper outbursts. *Journal of Nervous and Mental Disease* 174:464-470.

McCarthy, D.
1966   Mouse-killing induced in rats treated with pilocarpine. *Federation Proceedings* 25:385.

McCarty, R.C., G.H. Whitesides, and T.K. Tomosky
1976   Effects of *para*-chlorophenylalanine on the predatory behavior of *Onychomyts torridus*. *Pharmacology Biochemistry and Behavior* 4:217-220.

McDonald, R.L.
1967   The effects of personality type on drug response. *Archives of General Psychiatry* 17:680-686.

McGivern, R.F., N.J. Lobaugh, and A.C. Collier
1981   Effect of naloxone and housing conditions on shock-elicited reflexive fighting: Influence of immediate prior stress. *Physiological Psychology* 9:251-256.

McGlone, J.J., S. Ritter, and K.W. Kelley
1980   The antiaggressive effect of lithium is abolished by area postrema lesion. *Physiology and Behavior* 24:1095-1100.

McIntyre, D.C., and G.L. Chew
1983   Relation between social rank, submissive behavior, and brain catecholamine levels in ring-necked pheasants (*Phasianus colchicus*). *Behavioral Neuroscience* 97:595-601.

McIntyre, D.C., L.M. Healy, and M. Saari
1979   Intraspecies aggression and monoamine levels in rainbow trout (*Salmo gairdneri*) fingerlings. *Behavioral and Neural Biology* 25:90-98.

McKenzie, G.M.
1971   Apomorphine-induced aggression in the rat. *Brain Research* 34:323-330.

McLain, W.C., B.T. Cole, R. Schrieber, and D.A. Powell
1974   Central catechol and indolamine systems and aggression. *Pharmacology Biochemistry and Behavior* 2:123-126.

McMillen, B.A., S.M. Scott, H.L. Williams, and M.K. Sanghera
1987   Effects of gepirone, an aryl-piperazine anxiolytic drug, on aggressive behavior and brain monoaminergic neurotransmission. *Naunyn-Schmiedeberg's Archives of Pharmacology* 335:454-464.

McMillen, B.A., E.A. DaVanzo, S.M. Scott, and A.H. Song
1988   *N*-alkyl-substituted aryl-piperazine drugs: Relationship between affinity for serotonin receptors and inhibition of aggression. *Drug Development Research* 12:53-62.

McMillen, B.A., E.A. DaVanzo, A.H. Song, S.M. Scott, and M.E. Rodriguez
1989   Effects of classical and atypical antipsychotic drugs on isolation-induced aggression in male mice. *European Journal of Pharmacology* 160:149-153.

McNaughton, N., and S.T. Mason
1980   The neuropsychology and neuropharmacology of the dorsal ascending noradrenergic bundle. *Progress in Neurobiology* 14:157-219.

Meierl, G., and W.J. Schmidt
1982 No evidence for cholinergic mechanisms in the control of spontaneous predatory behavior of the ferret. *Pharmacology Biochemistry and Behavior* 16:677-681.
Meller, R.E., E.B. Keverne, and J. Herbert
1980 Behavioural and endocrine effects of naltrexone in male talapoin monkeys. *Pharmacology Biochemistry and Behavior* 13:663-672.
Micev, V., and D.M. Lynch
1974 Effect of lithium on disturbed severely mentally retarded patients. *British Journal of Psychiatry* 125:110.
Miczek, K.A.
1974 Intraspecies aggression in rats: Effects of *d*-amphetamine and chlordiazepoxide. *Psychopharmacologia* 39:275-301.
1976 Mouse-killing and motor activity: Effects of chronic Δ^9-tetrahydrocannabinol and pilocarpine. *Psychopharmacology* 47:59-64.
1977 Effects of L-DOPA, *d*-amphetamine and cocaine on intruder evoked aggression in rats and mice. *Progress in Neuro-Psychopharmacology* 1:272-277.
1981 Pharmacological evidence for catecholamine involvement in animal aggression. *Psychopharmacology Bulletin* 17:60-62.
1985 Alcohol and aggressive behavior in rats: Interaction with benzodiazepines. *Society for Neuroscience Abstracts* 11:1290.
1987 The psychopharmacology of aggression. Pp. 183-238 in L.L. Iversen, S.D. Iversen, and S.H. Snyder, eds., *Handbook of Psychopharmacology*, Vol. 19: *New Directions in Behavioral Pharmacology*. New York: Plenum.
Miczek, K.A., and H. Barry, III
1976 Pharmacology of sex and aggression. Pp. 176-257 in S.D. Glick and J. Goldfarb, eds., *Behavioral Pharmacology*. St. Louis, Mo.: C.V. Mosby.
Miczek, K.A., and J.F. DeBold
1983 Hormone-drug interactions and their influence on aggressive behavior. Pp. 313-347 in B.B. Svare, ed., *Hormones and Aggressive Behavior*. New York: Plenum Press.
Miczek, K.A., and P. Donat
1989 Brain 5-HT systems and inhibition of aggressive behavior. Pp. 117-144 in P. Bevan, A.R. Cools, and T. Archer, eds., *Behavioral Pharmacology of 5-HT*. Hillsdale, N.J.: Lawrence Erlbaum Associates.
Miczek, K.A., and J.M. O'Donnell
1978 Intruder-evoked aggression in isolated and nonisolated mice: Effects of psychomotor stimulants and L-DOPA. *Psychopharmacology* 57:47-55.
Miczek, K.A., and J.T. Winslow
1987 Psychopharmacological research on aggressive behavior. Pp. 27-113 in A.J. Greenshaw and C.T. Dourish, eds., *Experimental Psychopharmacology*. Clifton, N.J.: Humana Press.

Miczek, K.A., and H. Yoshimura
1982 Disruption of primate social behavior by *d*-amphetamine and cocaine: Differential antagonism by antipsychotics. *Psychopharmacology* 76:163-171.

Miczek, K.A., J.L. Altman, J.B. Appel, and W.O. Boggan
1975 *Para*-chlorophenylalanine, serotonin and killing behavior. *Pharmacology Biochemistry and Behavior* 3:355-361.

Miczek, K.A., J. Mos, and B. Olivier
1989 Brain 5-HT and inhibition of aggressive behavior in animals: 5-HIAA and receptor subtypes. *Psychopharmacology Bulletin* 25:399-403.

Miczek, K.A., E.M. Weerts, M. Haney, and J.W. Tidey
1994 Neurobiological mechanisms controlling aggression: Preclinical developments for pharmacotherapeutic interventions. *Neuroscience and Biobehavioral Reviews* 18:97-110.

Minde, K., G. Weiss, and N. Mendelson
1972 A 5-year follow-up study of 91 hyperactive school children. *Journal of the American Academy of Child Psychiatry* 11:595-610.

Modigh, K.
1973 Effects of isolation and fighting in mice on the rate of synthesis of noradrenaline, dopamine and 5-hydroxytryptamine in the brain. *Psychopharmacologia* 33:1-17.
1974 Effects of social stress on the turnover of brain catecholamines and 5-hydroxytryptamine in mice. *Acta Pharmacologica et Toxicologica* 34:97-105.

Mogilnicka, E., and B. Przewlocka
1981 Facilitated shock-induced aggression after chronic treatment with antidepressant drugs in the rat. *Pharmacology Biochemistry and Behavior* 14:129-132.

Mogilnicka, E., C.G. Boissard, P.C. Waldmeier, and A. Delini-Stula
1983 The effects of single and repeated doses of maprotiline, oxaprotiline and its enantiomers on foot-shock induced fighting in rats. *Pharmacology Biochemistry and Behavior* 19:719-723.

Möhler, H., and T. Okada
1977 Benzodiazepine receptor: Demonstration in the central nervous system. *Science* 198:849-851.

Molina, V., L. Ciesielsski, S. Gobaille, and P. Mandel
1986 Effects of the potentiation of the GABAergic neurotransmission in the olfactory bulbs on mouse-killing behavior. *Pharmacology Biochemistry and Behavior* 24:657-664.

Molina, V., L. Ciesielski, S. Gobaille, F. Isel, and P. Mandel
1987 Inhibition of mouse killing behavior by serotonin-mimetic drugs: Effects of partial alterations of serotonin neurotransmission. *Pharmacology Biochemistry and Behavior* 27:123-131.

Molliver, M.E.
1987 Serotonergic neuronal systems: What their anatomic organiza-

tion tells us about function. *Journal of Clinical Psychopharmacology* 7(S6):3S-23S.

Monroe, R.R., and R. Dale
1967 Chlordiazepoxide in the treatment of patients with "activated EEG's." *Diseases of the Nervous System* 28:390-396.

Moore, M.S., R.L. Tychson, and D.M. Thompson
1976 Extinction-induced mirror responding as a baseline for studying drug effects on aggression. *Pharmacology Biochemistry and Behavior* 4:99-102.

Mos, J., and B. Olivier
1988 Differential effects of selected psychoactive drugs on dominant and subordinate male rats housed in a colony. *Neuroscience Research Communications* 2:29-36.
1989 Quantitative and comparative analyses of pro-aggressive actions of benzodiazepines in maternal aggression of rats. *Psychopharmacology* 97:152-153.

Mos, J., and C.F.M. Van Valkenburg
1979 Specific effect on social stress and aggression on regional dopamine metabolism in rat brain. *Neuroscience Letters* 15:325-327.

Mos, J., B. Olivier, and A.M. Van der Poel
1987 Modulatory actions of benzodiazepine receptor ligands on agonistic behaviour. *Physiology and Behavior* 41:265-278.

Mos, J., B. Olivier, and R. van Oorschot
1990 Behavioural and neuropharmacological aspects of maternal aggression in rodents. *Aggressive Behavior* 16:145-163.

Mueller, K., and W.L. Nyhan
1982 Pharmacologic control of pemoline induced self-injurious behavior in rats. *Pharmacology Biochemistry and Behavior* 16:957-963.

Mueller, K., S. Saboda, R. Palmour, and W.L. Nyhan
1982 Self-injurious behavior produced in rats by daily caffeine and continuous amphetamine. *Pharmacology Biochemistry and Behavior* 17:613-617.

Mühlbauer, H.D.
1985 Human aggression and the role of central serotonin. *Pharmacopsychiatry* 18:218-221.

Mukherjee, B.P., and S.N. Pradhan
1976a Effects of lithium on foot shock-induced aggressive behavior in rats. *Archives Internationales de Pharmacodynamie et de Therapie* 222:125-131.
1976b Effects of lithium on septal hyperexcitability and muricidal behavior in rats. *Research Communications in Psychology, Psychiatry and Behavior* 1:241-247.

Müller-Oerlinghausen, B.
1985 Lithium long-term treatment—Does it act via serotonin? *Pharmacopsychiatry* 18:214-217.

Muñoz-Blanco, J., B. Yusta, and F. Cordoba
1986 Differential distribution of neurotransmitter amino acids from the limbic system of aggressive and non-aggressive bull strains. *Pharmacology Biochemistry and Behavior* 25:71-75.

Munro, A.D.
1986 The effects of apomorphine, *d*-amphetamine and chlorpromazine on the aggressiveness of isolated *Aequidens pulcher* (Teleostei, Cichlidae). *Psychopharmacology* 88:124-128.

Murasaki, M., T. Hara, M. Oguchi, and Y. Ikeda
1976 Action of enpiprazole on emotional behavior induced by hypothalamic stimulation in rats and cats. *Psychopharmacology* 49:271-274.

Murray, N.
1962 Covert effects of chlordiazepoxide therapy. *Journal of Neuropsychiatry* 3:168-170.

Nagy, J., and L. Decsi
1974 Simultaneous chemical stimulation of the hypothalamus and dorsal hippocampus in the waking cat. *Pharmacology Biochemistry and Behavior* 2:285-292.

Nagy, J., K. Zambo, and L. Desci
1979 Anti-anxiety action of diazepam after intraamygdaloid application in the rat. *Neuropharmacology* 18:573-576.

Nakamura, K., and H. Thoenen
1972 Increased irritability: A permanent behavior change induced in the rat by intraventricular administration of 6-hydroxydropamine. *Psychopharmacologia* 24:359-372.

Nakao, K., T. Higashio, and T. Inukai
1985 Antagonism of picrotoxin against the taming effect of carbamazepine on footshock induced fighting behavior in mice. *Japanese Journal of Pharmacology* 39:281-283.

Niemegeers, C.J.E., J.M. Van Nueten, and P.A.J. Janssen
1974 Azaperone, a sedative neuroleptic of the butyrophenone series with pronounced anti-aggressive and anti-shock activity in animals. *Arzeimittel-Forschung/Drug Research* 1798-1806.

Nikulina, E.M., and N.K. Popova
1988 Predatory aggression in the mind (*Mustela vison*): Roles of serotonin and food satiation. *Aggressive Behavior* 14:77-84.

O'Boyle, M., T.A. Looney, and P.S. Cohen
1973 Suppression and recovery of mouse killing in rats following immediate lithium-chloride injections. *Bulletin of Psychonomic Society* 1:250-252.

Oehler, J., M. Jähkel, and J. Schmidt
1985a Einfluss von Lithium, Carbamazepin, Ca-Valproat und Diazepam auf isolationsbedingte Verhaltensaenderungen der Maus. *Biomedica et Biochimica Acta* 44:1523-1530.
1985b The influences of chronic treatment with psychotropic drugs on

behavioral changes by social isolation. *Polish Journal of Pharmacology and Pharmacy* 37:841-849.

Ogren, S.O., A.C. Holm, A.L. Renyi, and S.B. Ross
1980 Anti-aggressive effect of zimelidine in isolated mice. *Acta Pharmacologica et Toxicologica* 47:71-74.

Ojeda, P.A.
1970 Treatment with thioridazine of emotionally disturbed children in a day hospital. *Michigan Medicine* 69:215-217.

Olivier, B., and J. Mos
1986 Serenics and aggression. *Stress Medicine* 2:197-209.

Olivier, B., and D. van Dalen
1982 Social behaviour in rats and mice: An ethologically based model for differentiating psychoactive drugs. *Aggressive Behavior* 8:163-168.

Olivier, B., H. van Aken, I. Jaarsma, R. van Oorshot, T. Zethof, and D. Bradford
1984 Behavioural effects of psychoactive drugs on agonistic behaviour of male territorial rats (resident-intruder model). Pp. 137-156 in K.A. Miczek, M.R. Kruk, and B. Olivier, eds., *Ethopharmacological Aggression Research*. New York: Alan R. Liss.

Olivier, B., J. Mos, and R. van Oorschot
1985 Maternal aggression in rats: Effects of chlordiazepoxide and fluprazine. *Psychopharmacology* 86:68-76.

Olivier, B., J. Mos, J. Van der Heyden, and J. Hartog
1989 Serotonergic modulation of social interactions in isolated male mice. *Psychopharmacology* 97:154-156.

Olivier, B., J. Mos, and D. Rasmussen
1990 Behavioural pharmacology of the serenic, eltoprazine. *Reviews on Drug Metabolism and Drug Interactions* 8:31-83.

Olivier, B., M. Mos, and K.A. Miczek
1991 Ethopharmacological studies of anxiolytics and aggression. *European Neuropsychopharmacology* 1:97-100.

Otsuke, M., T. Tsuchiya, and S. Kitagawa
1973 Electroencephalographic and behavioral studies on the central action of nimetazepam (S-1530) in cats. *Arzneimittel-Forschung/Drug Research* 23:645-652.

Overall, J.E., L.E. Hollister, F. Meyer, I. Kimbell, and J. Shelton
1964 Imipramine and thioridazine in depressed and schizophrenic patients. *Journal of the American Medical Association* 189:605-608.

Ozawa, H., T. Miyauchi, and K. Sugawara
1975 Potentiating effect of lithium chloride on aggressive behaviour induced in mice by nialamide plus L-DOPA and by clonidine. *European Journal of Pharmacology* 34:169-179.

Palermo Neto, J., J.F. Numes, and F.V. Carvalho
1975 The effects of chronic cannabis treatment upon brain 5-hydroxytryptamine, plasma corticosterone and aggressive behavior in

female rats with different hormonal status. *Psychopharmacologia* 42:195-200.

Pallmeyer, T.P., and T.A. Petti
1979 Effects of imipramine on aggression and dejection in depressed children. *American Journal of Psychiatry* 136:1472-1473.

Panksepp, J.
1971 Drugs and stimulus-bound attack. *Physiology and Behavior* 6:317-320.

Panter, B.M.
1977 Lithium in the treatment of a child abuser. *American Journal of Psychiatry* 134:1436-1437.

Payne, A.P., M.J. Andrews, and C.A. Wilson
1984 Housing, fighting and biogenic amines in the midbrain and hypothalamus of the golden hamster. Pp. 227-247 in K.A. Miczek, M. Kruk, and B. Olivier, eds., *Ethopharmacological Aggression Research.* New York: Alan R. Liss
1985 The effects of isolation, grouping and aggressive interaction on indole and catecholamine levels and apparent turnover in the hypothalamus and midbrain of the male golden hamster. *Physiology and Behavior* 34:911-916.

Pecknold, J.C., and D. Fleury
1986 Alprazolam-induced manic episode in two patients with panic disorder. *American Journal of Psychiatry* 143:652-653.

Pellis, S.M., D.P. O'Brien, V.C. Pellis, P. Teitelbaum, D.L. Wolgin, and S. Kennedy
1988 Escalation of feline predation along a gradient from avoidance through "play" to killing. *Behavioral Neuroscience* 102:760-777.

Penaloza-Rojas, J.H., G. Bach-y-Rita, H.F. Rubio-Chevannier, and R. Hernandez-Peon
1961 Effects of imipramine on hypothalamic and amygdaloid excitability. *Experimental Neurology* 4:205-213.

Penot, C., M. Vergnes, G. Mack, and E. Kempf
1978 Comportement d'agression interspecifique et reactivite chez le rat: Etude comparative des effets de lesions electrolytiques du raphe et d'injections intraventriculaires de 5,7-DHT. *Biology of Behavior* 3:71-85.

Perini, C., F.B. Müller, U. Rauchfleisch, R. Battegay, and F.R. Bühler
1986 Hyperadrenergic borderline hypertension is characterized by suppressed aggression. *Journal of Cardiovascular Pharmacology* 8:S53-S56.

Peroutka, S.J.
1988 5-Hydroxytryptamine receptor subtypes: Molecular, biochemical and physiological characterization. *Tins* 11:496-500.

Phillip, A.H., L.G. Humber, and K. Voith
1979 Mapping the dopamine receptor. 2. Features derived from modi-

fications in the rings A/B region of the neuroleptic butaclamol. *Journal of Medicinal Chemistry* 22:768-773.

Pinder, R.M., R.N. Brogden, T.M. Speight, and G.S. Avery
1977 Maprotiline: A review of its pharmacological properties and therapeutic efficacy in mental depressive states. *Drugs* 13:321-352.

Platt, J.E., M. Campbell, W.H. Green, R. Perry, and I.L. Cohen
1981 Effects of lithium carbonate and haloperidol on cognition in aggressive hospitalized school-age children. *Journal of Clinical Psychopharmacology* 1:8-13.

Platt, J.E., M. Campbell, W.H. Green, and D.M. Grega
1984a Cognitive effects of lithium carbonate and haloperidol in treatment-resistant aggressive children. *Archives of General Psychiatry* 41:657-662.

Platt, J.E., M. Campbell, D.M. Grega, and W.H. Green
1984b Cognitive effects of haloperidol and lithium in aggressive conduct-disorder children. *Psychopharmacology Bulletin* 20:93-97.

Pliszka, S.R., G.A. Rogeness, P. Renner, and J. Sherman
1988 Plasma neurochemistry in juvenile offenders. *Journal of the American Academy of Child and Adolescent Psychiatry* 27(5):588-594.

Plotnikoff, N.P., A.J. Kastin, D.H. Coy, C.W. Christensen, A.V. Schally, and M.A. Spirtes
1976 Neuropharmacological actions of enkephalin after systemic administration. *Life Sciences* 19:1283-1288.

Plummer, H.K., III, and I.V. Holt
1987 Effects of alprazolam and triazolam on isolation-induced aggression in rats. *Ohio Journal of Science* 4:107-111.

Podobnikar, I.G.
1971 Implementation of psychotherapy by librium in a pioneering rural-industrial psychiatric practice. *Psychosomatics* 12:205-209.

Polakoff, S.A., P.J. Sorgi, and J.J. Ratey
1986 The treatment of impulsive and aggressive behavior with nadolol. *Journal of Clinical Psychopharmacology* 6:125-126.

Polc, P., J.P. Laurent, R. Scherschlicht, and W. Haefely
1981 Electrophysiological studies on the specific benzodiazepine antagonist Ro 15-1788. *Naunyn-Schmiedeberg's Archives of Pharmacology* 316:317-325.

Poldinger, W.
1981 Pharmakotherapie der aggressivitat. *Schweizerische Archiv der Neurologie, Neurochirurgie und Psychiatrie* 129:147-155.

Poole, T.B.
1973 Some studies on the incidence of chlordiazepoxide on the social interaction of golden hamsters (*Mesocricetus auratus*). *British Journal of Pharmacology* 48:538-545.

Poschlova, N., K. Masek, and M. Krsiak
1975   The effect of 6-hydroxydopamine and 5,6-dihydroxytryptamine on social behavior in mice. *Activitas Nervosa Superior* 17:254-255.

Poshivalov, V.P.
1973   Pharmacological analysis of aggressive behaviour of mice induced by isolation (Russian). *Journal of Higher Nervous Activity* 24:1079-1081.
1974   Pharmacological analysis of aggressive behaviour of mice induced by isolation. *Journal of the Higher Nervous Activity* 24:1079-1081.
1978   Ethological analysis of the action exerted by medazepam and diazepam on the zoosocial behavior of isolated mice (Russian). *Farmacologiya Toksikologiya* 41:263-266.
1980   The integrity of the social hierarchy in mice following administration of psychotropic drugs. *British Journal of Pharmacology* 70:367-373.
1981   Pharmaco-ethological analysis of social behaviour of isolated mice. *Pharmacology Biochemistry and Behavior* 14(S1):53-59.
1982   Ethological analysis of neuropeptides and psychotropic drugs: Effects on intraspecies aggression and sociability of isolated mice. *Aggressive Behavior* 8:355-369.

Poshivalov, V.P., S.A. Nieminen, and M.M. Airaksinen
1987   Ethopharmacological studies of the effects of ß-carbolines and benzodiazepines on murine aggression. *Aggressive Behavior* 13:141-147.

Potegal, M., A.S. Perumal, A.I. Barkai, G.E. Cannova, and A.D. Blau
1982   GABA binding in the brains of aggressive and non-aggressive female hamsters. *Brain Research* 247:315-324.

Potegal, M., B. Yoburn, and M. Glusman
1983   Disinhibition of muricide and irritability by intraseptal muscimol. *Pharmacology Biochemistry and Behavior* 19:663-669.

Pott, C.B., S.Z. Kramer, and A. Siegel
1987   Central gray modulation of affective defense is differentially sensitive to naloxone. *Physiology and Behavior* 40:207-213.

Powell, D.A., W.L. Milligan, and K. Walters
1973   The effects of muscarinic cholinergic blockade upon shock-elicited aggression. *Pharmacology Biochemistry and Behavior* 1:394-398.

Prasad, V., and M.H. Sheard
1982   Effect of lithium upon desipramine enhanced shock-elicited fighting in rats. *Pharmacology Biochemistry and Behavior* 17:337-378.
1983a Time course of chronic desipramine on shock-elicited fighting in rats. *Agressologie* 24:15-17.
1983b Synergistic effect of propranolol and quipazine on desipramine enhanced shock-elicited fighting in rats. *Pharmacology Biochemistry and Behavior* 19:419-421.

Pucilowski, O.
1987 Monoaminergic control of affective aggression. *Acta Neurobiologiae Experimentalis* 47:213-238.

Pucilowski, O., and W. Kostowski
1983 Aggressive behaviour and the central serotonergic systems. *Behavioural Brain Research* 9:33-48.
1988 Diltiazem suppresses apomorphine-induced fighting and pro-aggressive effect of withdrawal from chronic ethanol or haloperidol in rats. *Neuroscience Letters* 93:96-100.

Pucilowski, O., and L. Valzelli
1986 Chemical lesions of the nucleus accumbens septi in rats: Effects on muricide and apomorphine-induced aggression. *Behavioural Brain Research* 19:171-178.

Pucilowski, O., A. Plaznik, and W. Kostowski
1985 Aggressive behavior inhibition by serotonin and quipazine injected into the amygdala in the rat. *Behavioral and Neural Biology* 43:58-68.

Pucilowski, O., W. Kozak, and L. Valzelli
1986 Effect of 6-OHDA injected into the locus coeruleus on apomorphine-induced aggression. *Pharmacology Biochemistry and Behavior* 24:773-775.

Pucilowski, O., E. Trzaskowska, W. Kostowski, and L. Valzelli
1987 Norepinephrine-mediated suppression of apomorphine-induced aggression and locomotor activity in the rat amygdala. *Pharmacology Biochemistry and Behavior* 26:217-222.

Puech, A.J., P. Simon, R. Chermat, and J.R. Boisseir
1974 Profil neuropsychopharmacologique de l'apomorphine. *Journal of Pharmacology* 5:241-254.

Puglisi-Allegra, S., and S. Cabib
1988 Pharmacological evidence for a role of D_2 dopamine receptors in the defensive behavior of the mouse. *Behavioral and Neural Biology* 50:98-111.
1990 Effects of defeat experiences on dopamine metabolism in different brain areas of the mouse. *Aggressive Behavior* 16:271-284.

Puglisi-Allegra, S., and P. Mandel
1980 Effects of sodium *n*-dipropylacetate, muscimol hydrobromide and (*R,S*)nipecotic acid amide on isolation-induced aggressive behavior in mice. *Psychopharmacology* 70:287-290.

Puglisi-Allegra, S., G. Mack, A. Oliverio, and P. Mandel
1979 Effects of apomorphine and sodium di-*n*-propylacetate on the aggressive behaviour of three strains of mice. *Progress in Neuro-Psychopharmacology* 3:491-502.

Puglisi-Allegra, S., S. Simler, E. Kempf, and P. Mandel
1981 Involvement of the GABAergic system on shock-induced aggressive behavior in two strains of mice. *Pharmacology Biochemistry and Behavior* 14(S1):13-18.

Puglisi-Allegra, S., P. Carletti, and S. Cabib
1990   LY 171555-induced catalepsy and defensive behavior in four strains of mice suggest the involvement of different D_2 dopamine receptor systems. *Pharmacology Biochemistry and Behavior* 36:327-331.

Puig-Antich, J.
1982   Major depression and conduct disorder in prepuberty. *Journal of the American Academy of Child Psychiatry* 21:118-128.

Quenzer, L.F., and R.S. Feldman
1975   The mechanism of anti-muricidal effects of chlordiazepoxide. *Pharmacology Biochemistry and Behavior* 3:567-571.

Quenzer, L.F., R.S. Feldman, and J.W. Moore
1974   Toward a mechanism of the anti-aggression effects of chlordiazepoxide in rats. *Psychopharmacologia* 34:81-94.

Rada, R.T., and P.T. Donlon
1972   Piperacetazine vs. thioridazine for the control of schizophrenia in outpatients. *Psychosomatics* 13:373-376.

Raisanen, M.J., M. Virkkunen, M.O. Huttunen, B. Furman, and J. Karkkainen
1984   Increased urinary excretion of bufotenin by violent offenders with paranoid symptoms and family violence. *Lancet* 2:700-701.

Raleigh, M.J., G.L. Brammer, A. Yuwiler, J.W. Flannery, M.T. McGuire, and E. Geller
1980   Serotonergic influences on the social behavior of vervet monkeys (*Cercopithecus aethiops* sabaeus). *Experimental Neurology* 68:322-334.

Raleigh, M.J., G.L. Brammer, and M.T. McGuire
1983   Male dominance, serotonergic systems, and the behavioral and physiological effects of drugs in vervet monkeys (*Cercopithecus aethiops* sabaeus). Pp. 184-198 in K.A. Miczek, ed., *Ethopharmacology: Primate Models of Neuropsychiatric Disorders*. New York: Alan R. Liss.

Raleigh, M.J., G.L. Brammer, E.R. Ritvo, E. Geller, M.T. McGuire, and A. Yuwiler
1986   Effects of chronic fenfluramine on blood serotonin, cerebrospinal fluid metabolites, and behavior in monkeys. *Psychopharmacology* 90:503-508.

Rampling, D.
1978   Aggression: A paradoxical response to tricyclic antidepressants. *American Journal of Psychiatry* 135:117-118.

Randall, L.O., W. Schallek, G.A. Heise, E.F. Keith, and R.E. Bagdon
1960   The psychosedative properties of methaminodiazepoxide. *Journal of Pharmacology and Clinical Therapeutics* 129:163-171.

Randall, L.L., C.L. Scheckel, and R.F. Banziger
1965   Pharmacology of the metabolites of chlordiazepoxide and diazepam. *Current Therapeutic Research* 7:590-606.

Randrup, A., and I. Munkvad
1966 DOPA and other naturally occurring substances as causes of stereotypy and rage in rats. *Acta Psychiatrica Scandinavica* 42:191-193.
1969a Pharmacological studies on the brain mechanisms underlying two forms of behavioral excitation stereotyped hyperactivity and rage. *Annals of the New York Academy of Sciences* 159:928-938.
1969b Relation of brain catecholamines to aggressiveness and other forms of behavioral excitation. Pp. 228-235 in S. Garattini and E.B. Sigg, eds., *Aggressive Behaviour*. Amsterdam: Excerpta Medica Foundation.
Randt, C.T., D.A. Blizard, and E. Friedman
1975 Early life undernutrition and aggression in two mouse strains. *Developmental Psychobiology* 8:275-279.
Rapoport, J.L.
1989 The biology of obsessions and compulsions. *Scientific American* 260:83-89.
Rapoport, J.L., P.O. Quinn, G. Bradbard, K.D. Riddle, and E. Brooks
1974 Imipramine and methlyphenidate treatments of hyperactive boys. *Archives of General Psychiatry* 30:789-793.
Ratey, J.J., and G.A. O'Driscoll
1989 Buspirone as a habilitative drug for patients with a dual diagnosis. *Family Practice Recertification* 11:38-45.
Ratey, J.J., R. Morrill, and G. Oxenkrug
1983 Use of propranolol for provoked and unprovoked episodes or rage. *American Journal of Psychiatry* 140:1356-1357.
Ratey, J.J., E.J. Mikkelsen, G.B. Smith, A. Upadhyaya, H.S. Zuckerman, D. Martell, P. Sorgi, S. Polakoff, and J. Bemporad
1986 Beta-blockers in the severely and profoundly mentally retarded. *Journal of Clinical Psychopharmacology* 6:103-107.
Ratey, J.J., E. Mikkelsen, P. Sorgi, H.S. Zuckerman, S. Polakoff, J. Bemporad, P. Bick, and W. Kadish
1987 Autism: The treatment of aggressive behaviors. *Journal of Clinical Psychopharmacology* 7:35-41.
Ratey, J.J., R. Sovner, E. Mikkelsen, and H.E. Chmielinski
1989 Buspirone therapy for maladaptive behavior and anxiety in developmentally disabled persons. *Journal of Clinical Psychiatry* 50:382-384.
Ratey, J.J., P. Sorgi, K.J. Lindem, G.A. O'Driscoll, M.L. Daehler, J.R. Fletcher, W. Kadish, G. Spruiell, S. Polakoff, S. Sands, J.R. Bemporad, and L. Richardson
1990 Nadolol to treat aggression and psychiatric symptomatology in chronic inpatients: A double-blind placebo-controlled study. Unpublished manuscript.
Ray, A., K.K. Sharma, M. Alkondon, and P. Sen
1983 Possible interrelationship between the biogenic amines involved

in the modulation of footshock aggression in rats. *Archives Internationales de Pharmacodynamie et de Therapie* 265:36-41.

Read, S.G., and D.H. Batchelor
1986 Violent and self-injurious behaviour in mentally handicapped patients—Psychopharmacological control. *International Clinical Psychopharmacology* 1:63-74.

Redmond, D.E., J.W. Maas, A. Kling, C.W. Graham, and H. Dekirmenjian
1971a Social behavior of monkeys selectively depleted of monoamines. *Science* 174:428-431.

Redmond, D.E., Jr., J.W. Maas, A. Kling, and H. Dekirmenjian
1971b Changes in primate social behavior after treatment with α-methyl-*para*-tyrosine. *Psychosomatic Medicine* 33:97-113.

Redmond, D.E., Jr., R.L. Hinrichs, J.W. Maas, and A. Kling
1973 Behavior of free-ranging macaques after intraventricular 6-hydroxydopamine. *Science* 181:1256-1258.

Redolat, R., P.F. Brain, and V.M. Simon
1991 Sulpiride has an antiaggressive effect in mice without markedly depressing motor activity. *Neuropharmacology* 30:41-46.

Reis, D.J.
1974 Consideration of some problems encountered in relating specific neurotransmitters to specific behaviors or disease. *Journal of Psychiatric Research* 11:145-148.

Reis, D.J., and K. Fuxe
1964 Brain norepinephrine: Evidence that neuronal release is essential for sham rage behavior following brainstem transection in cat. *Proceedings of the National Academy of Sciences* 64:108-112.
1968 Depletion of noradrenaline in brainstem neurons during sham rage behaviour produced by acute brainstem transection in cat. *Brain Research* 7:448-451.

Reis, D.J., and L.M. Gunne
1965 Brain catecholamines: Relation to the defense reaction evoked by amygdaloid stimulation in the cat. *Science* 149:450-451.

Reis, D.J., D.T. Moorhead, and N. Merlino
1970 DOPA-induced excitement in the cat. *Archives of Neurology* 22:31-39.

Renzi, P.
1982 Increased shock-induced attack after repeated chlordiazepoxide administration in mice. *Aggressive Behaviour* 8:172-174.

Ricaurte, G., G. Bryan, L. Strauss, L. Seiden, and C. Schuster
1985 Hallucinogenic amphetamine selectively destroys brain serotonin nerve terminals. *Science* 229:986-988.

Ricaurte, G.A., L.E. DeLanney, I. Irwin, J.M. Witkin, J.L. Katz, and J.W. Langston
1989 Evaluation of the neurotoxic potential of N,N-dimethylamphetamine: An illicit analog of methamphetamine. *Brain Research* 490:301-306.

Rickels, K., and R.W. Downing
1974   Chlordiazepoxide and hostility in anxious outpatients. *American Journal of Psychiatry* 131:442-444.
Rifkin, A., F. Quitkin, C. Carrillo, A.G. Blumberg, and D.F. Klein
1972   Lithium carbonate in emotionally unstable character disorder. *Archives of General Psychiatry* 27:519-523.
Robichaud, R.C., and M.E. Goldberg
1974   Pharmacological properties of two chlordiazepoxide metabolites following microsomal enzyme inhibition. *Archives of International Pharmacodynamics* 211:165-173.
Robichaud, R.C., J.A. Gylys, K.A. Sledge, and I.W. Hillyard
1970   The pharmacology of prazepam, a new benzodiazepine derivative. *Archives Internationales de Pharmacodynamie et de Therapie* 185:213-227.
Rodgers, R.J.
1977   The medial amygdala: Serotonergic inhibition of shock-induced aggression and pain sensitivity in rats. *Aggressive Behavior* 3:277-288.
Rodgers, R.J., and A. Depaulis
1982   Gabaergic influences on defensive fighting in rats. *Pharmacology Biochemistry and Behavior* 17:451-456.
Rodgers, R.J., J.M. Semple, S.J. Cooper, and K. Brown
1976   Shock-induced aggression and pain sensitivity in the rat: Catecholamine involvement in the corticomedial amygdala. *Aggressive Behavior* 2:193-204.
Rolinski, Z.
1973   Analysis of aggressiveness-stereotypy complex induced in mice by amphetamine or nialamide and L-DOPA. *Polish Journal of Pharmacology and Pharmacy* 25:552-558.
1974   Analysis of the aggressiveness-stereotypy complex induced in mice by amphetamine or D,L-DOPA. *Polish Journal of Pharmacology and Pharmacy* 26:369-378.
1975   Interspecies aggressiveness of rats towards mice after the application of p-chlorophenylalanine. *Polish Journal of Pharmacology and Pharmacy* 27:223-229.
Rolinski, Z., and M. Herbut
1979   Determination of the role of serotonergic and cholinergic systems in apomorphine-induced aggressiveness in rats. *Polish Journal of Pharmacology and Pharmacy* 31:97-106.
1981   The role of the serotonergic system in foot shock-induced behavior in mice. *Psychopharmacology* 73:246-251.
1985   The significance of central nicotine receptors in the aggression of isolated mice. *Polish Journal of Pharmacology and Pharmacy* 37:479-486.
Rolinski, Z., and W. Kozak
1979   The role of the catecholaminergic system of footshock-induced fighting in mice. *Psychopharmacology* 65:285-290.

Romaniuk, A.
  1974 Neurochemical bases of defensive behavior in animals. *Acta Neurobiologiae Experimentalis* 34:205-214.
Romaniuk, A., S. Brudzynski, and J. Gronska
  1973 The effect of chemical blockade of hypothalamus cholinergic system on defensive reactions in cats. *Acta Physica Polonica* 24:809-816.
  1974 The effects of intrahypothalamic injections of cholinergic and adrenergic agents on defensive behavior in cats. *Acta Physica Polonica* 25:297-305.
Romaniuk, A., M. Filipczak, and J. Fryczak
  1987 The influence of injection of 5,6-dihydroxytryptamine to the dorsal raphe nucleus on carbachol-induced defensive behavior and regional brain amine content in the cat. *Polish Journal of Pharmacology and Pharmacy* 39:17-25.
Rosenbaum, J.F., S.W. Woods, J.E. Groves, and G.L. Klerman
  1984 Emergence of hostility during alprazolam treatment. *American Journal of Psychiatry* 141:792-793.
Rothballer, A.B.
  1967 Aggression, defense, and neurohumors. Pp. 135-170 in C.D. Clemente and D.B. Lindsley, eds., *Aggression and Defense*. Los Angeles: University of California Press.
Roy, A., and M. Linnoila
  1988 Suicidal behavior, impulsiveness and serotonin. *Acta Psychiatrica Scandinavica* 78:529-535.
Roy, A., M. Virkkunen, and M. Linnoila
  1987 Reduced central serotonin turnover in a subgroup of alcoholics. *Progress in Psychopharmacology and Biological Psychiatry* 11:173-177.
Rush, J., and J. Mendels
  1975 Effects of lithium chloride on muricidal behavior in rats. *Pharmacology Biochemistry and Behavior* 3:795-797.
Rydin, E., D. Schalling, and M. Asberg
  1982 Rohrshach ratings in depressed and suicidal patients with low levels of 5-hydroxyindolacetic acid in cerebrospinal fluid. *Psychiatric Research* 7:229-243.
Sahakian, B.J., G.S. Sarna, D.B. Kantamaneni, A. Jackson, P.H. Hutson, and G. Curzon
  1986 CSF tryptophan and transmitter amine turnover may predict social behaviour in the normal rat. *Brain Research* 399:162-166.
Salama, A.I., and M.E. Goldberg
  1970 Neurochemical effects of imipramine and amphetamine in aggressive mouse-killing (muricidal) rats. *Biochemical Pharmacology* 19:2023-2032.
  1973a Enhanced locomotor activity following amphetamine in mouse-

killing rats. *Archives of International Pharmacodynamics* 204:162-169.

1973b Norepinephrine turnover and brain monoamine. *Life Sciences* 12:521-526.

Saletu, B., M. Saletu, J. Simeon, G. Viamontes, and T.M. Itil

1975 Comparative symptomatological and evoked potential studies with *d*-amphetamine, thioridazine, and placebo in hyperkinetic children. *Biological Psychiatry* 10:253-275.

Salzman, C.

1988 Use of benzodiazepines to control disruptive behavior in inpatients. *Journal of Clinical Psychiatry* 49:13-15.

Salzman, C., A. DiMascio, R.I. Shader, and J.S. Harmatz

1969 Chlordiazepoxide, expectation and hostility. *Psychopharmacologia* 14:38-45.

Salzman, C., G.E. Kochansky, R.I. Shader, L.J. Porrino, J.S. Harmatz, and C.P. Swett, Jr.

1974 Chlordiazepoxide-induced hostility in a small group setting. *Archives of General Psychiatry* 31:401-405.

Salzman, C., G.E. Kochansky, R.I. Shader, J.S. Harmatz, and A.M. Ogletree

1975 Is oxazepam associated with hostility? *Diseases of the Nervous System* 36:30-32.

Sandler, M., C.R.J. Ruthven, B.L. Goodwin, H. Field, and M. Rhys

1978 Phenylethylamine overproduction in aggressive psychopaths. *Lancet* 12/16:1269-1270.

Scarnati, R.A.

1986 Most-violent psychiatric inmates and neuroleptics. *Journal of Psychiatry and Law* 447-468.

Scheckel, C.L., and E. Boff

1966 Effects of drugs in aggressive behavior in monkeys. *Excerpta Medica Internationales* 129:789-795.

1968 The effects of drugs on conditioned avoidance and aggressive behavior. Pp. 301-312 in H. Vagtborg, ed., *Use of Nonhuman Primates in Drug Evaluation.* Austin: University of Texas Press.

Scheel-Kruger, J., and A. Randrup

1968 Aggressive behavior provoked by pargyline in rats pretreated with diethyldithiocarbamate. *Journal of Pharmacy and Pharmacology* 29:948-949.

Schmidt, W., and R. Apfelbach

1977 Psychopharmakologische Beeinflussung des Beutefangverhaltens beim Frettchen (*Putorius furo* L.). *Psychopharmacology* 51:147-152.

Schmidt, W.J.

1979 Effects of *d*-amphetamine, maprotiline, *L*-DOPA, and haloperidol on the components of the predatory behavior of the ferret, *Putorius furo* L. *Psychopharmacology* 64:355-359.

1980 Unlike rats, ferrets do kill under antidepressants. *Naturwissenschaften* 67:262-263.

1983 Involvement of dopaminergic neurotransmission in the control of goal-directed movements. *Psychopharmacology* 80:360-364.

Schmidt, W.J., and G. Meierl

1980 Antidepressants and the control of predatory behavior. *Physiology and Behavior* 25:17-19.

Schou, M.

1979 Lithium in the treatment of other psychiatric and nonpsychiatric disorders. *Archives of General Psychiatry* 36:856-859.

Schreier, H.A.

1979 Use of propranolol in the treatment of postencephalitic psychosis. *American Journal of Psychiatry* 136:840-841.

Schrold, J.

1970 Aggressive behaviour in chicks induced by tricyclic antidepressants. *Psychopharmacologia* 17:225-233.

Schulte, J.L.

1985 Homicide and suicide associated with akathisia and haloperidol. *American Journal of Forensic Psychiatry* 3-8.

Schwartz, R.D.

1988 The GABAa receptor-gated ion channel: Biochemical and pharmacological studies of structure and function. *Biochemical Pharmacology* 37:3369-3375.

Scriabine, A., and M. Blake

1962 Evaluation of centrally acting drugs in mice with fighting behavior induced by isolation. *Psychopharmacologia* 2:224-226.

Seeman, P., T. Lee, M. Chau-Wong, and K. Wong

1976 Antipsychotic drug doses and neuroleptic/dopamine receptors. *Nature* 261-717-719.

Senault, B.

1968 Syndrome agressif induit par l'apomorphine chez le rat. *Journal de Physiologie* 60:543-544.

1970 Comportement d'aggresivite intraspecifique induit par l'apomorphine chez le rat. *Psychopharmacologia* 18:271-287.

1971 Influence de l'isolement sur de comportement d'agressivite intraspecifique induit par l'apomorphine chez le rat. *Psychopharmacologia* 20:389-394.

1972 Influence de la surrenalectomie, de l'hypophysectomie, de la thyroidectomie, de la castration ainsi que de la testosterone sur la comportement d'agressivite intraspecifique induit par l'apomorphine chez le rat. *Psychopharmacologia* 24:476-484.

1973 Effects de lesions du septum, de l'amygdala, du striatum, de la substantia nigra et de l'ablation des bulbes olfactifs sur le comportement d'agressivite intraspecifique induit par l'apomorphine chez le rat. *Psychopharmacology* 28:13-25.

1974 Amines cerebrales et comportement d'agressivite intraspecifique induit par l'apomorphine chez le rat. *Psychopharmacologia* 34:143-154.

Sepinwall, J., F.S. Grodsky, and L. Cook
1978 Conflict behavior in the squirrel monkey: Effects of chlordiazepoxide, diazepam and N-desmethyldiazepam. *Journal of Pharmacology and Experimental Therapeutics* 204:88-102.

Shader, R.I., A.H. Jackson, and L.M. Dodes
1974 The antiaggressive effects of lithium in man. *Psychopharmacologia* 40:17-24.

Shaikh, M.B., A.B. Shaikh, and A. Siegel
1988 Opioid peptides within the midbrain periaqueductal gray suppress affective defense behavior in the cat. *Peptides* 9:999-1004.

Shaikh, M.B., M. Dalsass, and A. Siegel
1990 Opiodergic mechanisms mediating aggressive behavior in the cat. *Aggressive Behavior* 16:191-206.

Sharon, L.
1984 Benzodiazepines: Guidelines for use in correctional facilities. *Psychosomatics* 25:784-788.

Sharma, V.N., R.L. Mital, S.P. Banerjee, and H.L. Sharma
1969 Pharmacological studies with some newly synthesized phenothiazines exhibiting lesser extrapyramidal reactions. *Japanese Journal of Pharmacology* 19:211-223.

Shaw, C.R., H.J. Lockett, A.R. Lucas, C.H. Lamontagne, and F. Grimm
1963 Tranquilizer drugs in the treatment of emotionally disturbed children: I. Inpatients in a residential treatment center. *Journal of American Academy of Child Psychiatry* 2:725-742.

Sheard, M.H.
1970a Behavioral effects of p-chlorophenylalanine in rats: Inhibition by lithium. *Communications in Behavioral Biology* 5:1-3.

1970b Effect of lithium on foot shock aggression in rats. *Nature* 228:284-285.

1971 Effect of lithium on human aggression. *Nature* 230:113-114.

1973 Aggressive behavior: Modification by amphetamine, p-chlorophenylalanine and lithium in rats. *Agressologie* 14:323-326.

1975 Lithium in the treatment of aggression. *Journal of Nervous and Mental Disease* 160:108-118.

1976 The effect of parachloroamphetamine (PCA) on behavior. *Psychopharmacology Bulletin* 12:59-61.

1977a Animal models of aggressive behavior. Pp. 247-257 in I. Hanin and E. Usdin, eds., *Animal Models in Psychiatry and Neurology.* Oxford: Pergamon.

1977b The role of drugs in precipitating or inhibiting human aggression. *Psychopharmacology Bulletin* 13:23-25.

1978 The effect of lithium and other ions on aggressive behavior. Pp. 53-68 in L. Valzelli, ed., *Psychopharmacology of Aggression. Modern Problems of Pharmopsychiatry.* Basel, Switzerland: Krager.

1981　Shock-induced fighting (SIF): Psychopharmacological studies. *Aggressive Behavior* 7:41-49.

1983　Psychopharmacology of aggression. Pp. 188-201 in H. Hippius and G. Winokur, eds., *Clinical Psychopharmacology*. Amsterdam: Excerpta Medica.

1984　Clinical pharmacology of aggressive behavior. *Clinical Neuropharmacology* 7:173-183.

1988　Clinical pharmacology of aggressive behavior. *Clinical Neuropharmacology* 11:483-492.

Sheard, M.H., and J.L. Marini

1978　Treatment of human aggressive behavior: Four case studies of the effect of lithium. *Comprehensive Psychiatry* 19:37-45.

Sheard, M.H., J.L. Marini, C.I. Bridges, and E. Wagner

1976　The effect of lithium on impulsive aggressive behavior in man. *American Journal of Psychiatry* 133:1409-1413.

Sheard, M.H., D.I. Astrachan, and M. Davis

1977　The effect of *d*-lysergic acid diethylamide (LSD) upon shock-elicited fighting in rats. *Life Sciences* 20:427-430.

Shibata, S., D. Suwandi, T. Yamamoto, and S. Ueki

1982　Effects of medial amygdaloid lesions on the initiation and the maintenance of muricide in olfactory bulbectomized rats. *Physiology and Behavior* 29:939-941.

Shibata, S., S. Watanabe, S.Y. Liou, and S. Ueki

1983　Effects of adrenergic blockers on the inhibition of muricide by desipramine and noradrenaline injected into the amygdala in olfactory bulbectomized rats. *Pharmacology Biochemistry and Behavior* 18:203-207.

Shibata, S., H. Nakanishi, S. Watanabe, and S. Ueki

1984　Effects of chronic administration of antidepressants on mouse-killing behavior (muricide) in olfactory bulbectomized rats. *Pharmacology Biochemistry and Behavior* 21:225-230.

Shintomi, K.

1975　Effects of psychotropic drugs on methamphetamine-induced behavioral excitation in grouped mice. *European Journal of Pharmacology* 32(2):195-206.

Sieber, B., H.-R. Frischknecht, and P. Waser

1982　Behavioural effects of hashish in mice in comparison with psychoactive drugs. *General Pharmacology* 13:315-320.

Siegel, H.I., A.L. Giordano, C.M. Mallafre, and J.S. Rosenblatt

1983　Maternal aggression in hamsters: Effects of stage of lactation, presence of pups, and repeated testing. *Hormones and Behavior* 17:86-93.

Silverman, A.P.

1965a Ethological and statistical analysis of drug effects on the social behaviour of laboratory rats. *British Journal of Pharmacology* 24:579-590.

1965b Social behaviour of rats and the action of chlorpromazine. *Neuropsychopharmacology* 4:346-351.

1966 The social behaviour of laboratory rats and the action of chlorpromazine and other drugs. *Behaviour* 27:1-38.

1971 Behaviour of rats given a "smoking dose" of nicotine. *Animal Behaviour* 19:67-74.

1972 Effects of various drugs on the behaviour of laboratory rats. In J. van Noordwijk, ed., *Animal Behaviors Under the Influence of Psychoactive Drugs*. Bilthoven, The Netherlands: National Institute of Public Health.

Simler, S., S. Puglisi-Allegra, and P. Mandel

1982 Gamma-aminobutyric acid in brain areas of isolated aggressive or non-aggressive inbred strains of mice. *Pharmacology Biochemistry and Behavior* 16:57-61.

1983 Effects of n-di-propylacetate on aggressive behavior and brain GABA level in isolated mice. *Pharmacology Biochemistry and Behavior* 18:717-720.

Simonds, J.F., and J. Kashani

1979 Drug abuse and criminal behavior in delinquent boys committed to a training school. *American Journal of Psychiatry* 136:1444-1448.

Singhal, R.L., and J.I. Telner

1983 A perspective: Psychopharmacological aspects of aggression in animals and man. *Psychiatric Journal of the University of Ottawa* 8:145-153.

Skolnick, P., G.F. Reed, and S.M. Paul

1985 Benzodiazepine-receptor mediated inhibition of isolation-induced aggression in mice. *Pharmacology Biochemistry and Behavior* 23:17-20.

Smith, D.E., M.B. King, and B.G. Hoebel

1970 Lateral hypothalamic control of killing: Evidence for a cholinoceptive mechanism. *Science* 167:900-901.

Sofia, R.D.

1969a Structural relationship and potency of agents which selectively block mouse killing (muricide) behavior in rats. *Life Sciences* 8:1201-1210.

1969b Effects of centrally active drugs on four models of experimentally-induced aggression in rodents. *Life Sciences* 8:705-716.

Sorenson, C.A., and G.D. Ellison

1973 Nonlinear changes in activity and emotional reactivity scores following central noradrenergic lesions in rats. *Psychopharmacologia* 32:313-325.

Sorgi, P.J., J.J. Ratey, and S. Polakoff

1986 Beta-adrenergic blockers for the control of aggressive behaviors in patients with chronic schizophrenia. *American Journal of Psychiatry* 143:775-776.

Sovner, R., and A. Hurley
1981   The management of chronic behavior disorders in mentally re-tarded adults with lithium carbonate. *Journal of Nervous and Mental Disease* 169:191-195.

Sprague, R.L., K.R. Barnes, and J.S. Werry
1970   Methylphenidate and thioridazine: Learning, reaction time, ac-tivity, and classroom behavior in disturbed children. *American Journal of Orthopsychiatry* 40:615-628.

Stark, P., and J.K. Henderson
1972   Central cholinergic suppression of hyper-reactivity and aggres-sion in septal-lesioned rats. *Neuropharmacology* 11:839-847.

Stoddard, S.L., V.K. Bergdall, D.W. Townsend, and B.E. Levin
1986   Plasma catecholamines associated with hypothalamically-elic-ited defense behavior. *Physiology and Behavior* 36:867-873.

Stoff, D.M., L. Pollock, B. Vitiello, and D. Behar
1987   Reduction of (3H)imipramine binding sites on platelets of con-duct-disorder children. *Neuropsychopharmacology* 1(1):55-62.

Stolk, J.M., R.L. Conner, and J.D. Barchas
1974   Social environment and brain biogenic amine metabolism in rats. *Journal of Comparative and Physiological Psychology* 87:203-207.

Strahan, A., J. Rosenthal, M. Kaswan, and A. Winston
1985   Three case reports of acute paroxysmal excitement associated with alprazolam treatment. *American Journal of Psychiatry* 142:859-861.

Strickland, J.A., and J.P. DaVanzo
1986   Must antidepressants be anticholinergic to inhibit muricide? *Pharmacology Biochemistry and Behavior* 24:135-137.

Subrahmanyam, S.
1975   Role of biogenic amines in certain pathological conditions. *Brain Research* 87:355-362.

Sulcova, A.
1985   Tranquilizing effects of alprazolam in animal models of agonis-tic behavior. *Activitas Nervosa Superior* 27:310-311.

Sulcova, A., and M. Krsiak
1980   Effect of piracetam on agonistic behaviour in mice. *Activitas Nervosa Superior* 22:200-201.
1981   Effects of castration on aggressive and defensive-escape compo-nents of agonistic behavior in male mice. *Activitas Nervosa Superior* 23:317-318.
1984   The benzodiazepine-receptor antagonist Ro 15-1788 antagonizes effects of diazepam on aggressive and timid behaviour in mice. *Activitas Nervosa Superior* 26:255-256.
1986   Beta-carbolines (ß-CCE,FG 7142) and diazepam: Synergistic ef-fects on aggression and antagonistic effects of timidity in mice. *Activitas Nervosa Superior* 28:312-316.

1989 Differences among nine 1,4-benzodiazepines: An ethopharmacological evaluation in mice. *Psychopharmacology* 97:157-159.

Sulcova, A., M. Krsiak, and K. Masek
1976 Effect of repeated administration of chlorpromazine and diazepam on isolation-induced timidity in mice. *Activitas Nervosa Superior* 18:233-234.

Sulcova, A., M. Krsiak, K. Masek, and R.U. Ostrovskaya
1979 Bicuculline antagonized effects of diazepam on aggressive and timid behaviour in mice. *Activitas Nervosa Superior* 21:179-180.

Sulcova, A., M. Krsiak, and K. Masek
1981 Effects of calcium valproate and aminooxyacetic acid on agonistic behaviour in mice. *Activitas Nervosa Superior* 23:287-289.

Svare, B.B., and M.A. Mann
1983 Hormonal influences on maternal aggression. Pp. 91-104 in B.B. Svare, ed., *Hormones and Aggressive Behavior*. New York: Plenum Press.

Sweidan, S., H. Edinger, and A. Siegel
1990 The role of D_1 and D_2 receptors in dopamine agonist-induced modulation of affective defense behavior in the cat. *Pharmacology Biochemistry and Behavior* 36:491-499.

Tani, Y., Y. Kataoka, Y. Sakurai, K. Yamashita, M. Ushio, and S. Ueki
1987 Changes of brain monoamine contents in three models of experimentally induced muricide in rats. *Pharmacology Biochemistry and Behavior* 26:725-729.

Tardiff, K.
1982 The use of medication for assaultive patients. *Hospital and Community Psychiatry* 33:307-308.
1983 A survey of drugs used in the management of assaultive inpatients. *Bulletin of the American Academy of Psychiatry and the Law* 11:215-222.
1984 Medication for violent psychiatric patients. *American Journal of Social Psychiatry* 4:45-48.

Taylor, D.P.
1988 Buspirone, a new approach to the treatment of anxiety. *Federation of American Societies for Experimental Biology Journal* 2:2445-2452.

Tazi, A., R. Dantzer, P. Mormede, and M. Le Moal
1983 Effects of post-trial administration of naloxone and ß-endorphin on shock-induced fighting in rats. *Behavioral and Neural Biology* 39:192-202.

Tedeschi, D.H., R.E. Tedeschi, and E.J. Fellows
1959a The effects of tryptamine on the central nervous system including a pharmacological procedure for the evaluation of iproniazid-like drugs. *Journal of Pharmacology and Experimental Therapeutics* 126:223-232.

Tedeschi, R.E., D.H. Tedeschi, A. Mucha, L. Cook, P.A. Mattis, and E.J. Fellows
1959b Effects of various centrally acting drugs on fighting behavior of mice. *Journal of Pharmacology and Experimental Therapeutics* 125:28-34.

Tedeschi, D.H., P.J. Fowler, E.B. Miller, and E. Macko
1969 Pharmacological analysis of footshock-induced fighting behaviour. Pp. 245-252 in S. Garattini and E.B. Sigg, eds., *Aggressive Behaviour*. Amsterdam: Excerpta Medica Foundation.

Thoa, N.B., B. Eichelman, and L.K.Y. Ng
1972a Shock-induced aggression: Effects of 6-hydroxydopamine and other pharmacological agents. *Brain Research* 43:467-475.

Thoa, N.B., B. Eichelman, J.S. Richardson, and D. Jacobowitz
1972b 6-Hydroxydopa depletion of brain norepinephrine and the facilitation of aggressive behavior. *Science* 178:75-77.

Thoa, N.B., Y. Tizabi, and D.M. Jacobowitz
1977 The effect of isolation on catecholamine concentration and turnover in discrete areas of the rat brain. *Brain Research* 131:259-269.

Thompson, T.
1961 Effect of chlorpromazine on "aggressive" responding in the rat. *Journal of Comparative and Physiological Psychology* 54:398-400.
1962 The effect of two phenothiazines and a barbiturate on extinction-induced rate increase of a free operant. *Journal of Comparative and Physiological Psychology* 55:714-718.

Thor, D.H., and W.B. Ghiselli
1975 Suppression of mouse killing and apomorphine-induced social aggression in rats by local anesthesia of the mystacial vibrissae. *Journal of Comparative and Physiological Psychology* 88:40-46.

Thurmond, J.B.
1975 Technique for producing and measuring territorial aggression using laboratory mice. *Physiology and Behavior* 14:879-881.

Thurmond, J.B., S.M. Lasley, A.L. Conkin, and J.W. Brown
1977 Effects of dietary tyrosine, phenylalanine, and trytophan on aggression in mice. *Pharmacology Biochemistry and Behavior* 6:475-478.

Thurmond, J.B., S.M. Lasley, N.R. Kramarcy, and J.W. Brown
1979 Differential tolerance to dietary amino acid-induced changes in aggressive behavior and locomotor activity in mice. *Psychopharmacology* 66:301-308.

Thurmond, J.B., N.R. Kramarcy, S.M. Lasley, and J.W. Brown
1980 Dietary amino acid precursors: Effects of central monoamines, aggression, and locomotor activity in the mouse. *Pharmacology Biochemistry and Behavior* 12:525-532.

Tidey, J.W., and K.A. Miczek
1991 Effects of SKF38393 and quinpirole on patterns of aggressive,

motor and schedule-controlled behaviors in mice. *Behavioural Pharmacology* 3:553-656.

Tizabi, Y., N.B. Thoa, G.D. Maengwyn-Davies, I.J. Kopin, and D.M. Jacobowitz
1979 Behavioral correlation of catecholamine concentration and turnover in discrete brain areas of three strains of mice. *Brain Research* 166:199-205.

Tizabi, Y., V.J. Massari, and D.M. Jacobowitz
1980 Isolation induced aggression and catecholamine variations in discrete brain areas of the mouse. *Brain Research Bulletin* 5:81-86.

Tobe, A., and T. Kobayashi
1976 Pharmacological studies on triazine derivatives. 5. Sedative and neuroleptic actions of 2-amino-2-[4-(2-hydroxyethyl)-piperazin-1yl]-6-triflouromethyl-*s*-triazine (TR-10). *Japanese Journal of Pharmacology* 26:559-570.

Tobe, A., Y. Yoshida, H. Ikoma, S. Tonomura, and R. Kikumoto
1981 Pharmacological evaluation of 2-(4-methylaminobutoxy)diphenylmethane hydrochloride (MCI-2016), a new psychotropic drug with antidepressant activity. *Arzneimittel-Forschung/Drug Research* 31:1278-1285.

Tobin, J.M., and N.D.C. Lewis
1960 New psychotherapeutic agent, chlordiazepoxide. *Journal of the American Medical Association* 174:1242-1249.

Tompkins, E.C., A.J. Clemento, D.P. Taylor, and J.L. Perhach, Jr.
1980 Inhibition of aggressive behavior in rhesus monkeys by buspirone. *Research Communications in Psychology, Psychiatry and Behavior* 5:337-352.

Torda, C.
1976 Effects of catecholamines on behavior. *Journal of Neuroscience Research* 2:193-202.

Traversa, U., L. De Angelis, R. Della Loggia, M. Bertolissi, G. Nardini, and R. Vertua
1985 Effects of caffeine and chlor-desmethyldiazepam on fighting behavior of mice with different reactivity baselines. *Pharmacology Biochemistry and Behavior* 23:237-241.

Troncone, L.R.P., T.M.S. Ferreira, S. Braz, N.G.S. Filho, and S. Tufik
1988 Reversal of the increase in amomophine-induced stereotypy and aggression in REM sleep-deprived rats by dopamine agonist pretreatments. *Psychopharmacology* 94:79-83.

Tsuda, A., M. Tanaka, Y. Ida, I. Shirao, Y. Gondoh, M. Oguchi, and M. Yoshida
1988 Expression of aggression attenuates stress-induced increases in rat brain noradrenaline turnover. *Brain Research* 474:174-180.

Tsumagari, T., A. Nakajima, T. Fukuda, S. Shuto, T. Kenjo, Y. Morimoto, and Y. Takigawa
1978 Pharmacological properties of 6-(o-chlorophenyl)-8-methyl-1-methyl-4*H*-*s*-triazolo[3,4-*c*]thieno-[2,3-*e*]-1,4-diazepine (Y-7131), a new

anti-anxiety drug. *Arzneimittel-Forschung/Drug Research* 28:1158-1164.

Tupin, J.P.
1972 Lithium use in nonmanic depressive conditions. *Comprehensive Psychiatry* 13:209-214.
1985 Psychopharmacology and aggression. Pp. 83-99 in L.H. Roth, ed., *Clinical Treatment of the Violent Person.* Washington, D.C.: U.S. Department of Health and Human Services.

Tupin, J.P., D.B. Smith, T.L. Clanon, L.I. Kim, A. Nugent, and A. Groupe
1973 The long-term use of lithium in aggressive prisoners. *Comprehensive Psychiatry* 14:311-317.

Tyrer, S.P., A. Walsh, D.E. Edwards, T.P. Berney, and D.A. Stephens
1984 Factors associated with a good response to lithium in aggressive mentally handicapped subjects. *Progress in Neuro-psychopharmacology and Biological Psychiatry* 8:751-755.

Ucer, E., and K.C. Kreger
1969 A double-blind study comparing haloperidol with thioridazine in emotionally disturbed, mentally retarded children. *Current Therapeutic Research* 11:278-283.

Ueki, S., S. Murimoto, and N. Ogawa
1972 Effects of psychotropic drugs on emotional behavior in rats with limbic lesions, with special reference to olfactory bulb ablations. *Folia Psychiatrica Neurologica Japonica* 26:246-255.

Valdman, A.V., and V.P. Poshivalov
1986 Pharmaco-ethological analysis of antidepressant drug effects. *Pharmacology Biochemistry and Behavior* 25:515-519.

Valenca, M.M., and R.L. Falcao-Valenca
1988 Role of endogenous opiates in aggressive behavior in *Betta splendens.* *Medical Science Research* 16:625.

Valzelli, L.
1967 Drugs and aggressiveness. *Advances in Pharmacology* 5:79-108.
1971 Further aspects of the exploratory behaviour in aggressive mice. *Psychopharmacologia* 19:91-94.
1972 Psychoactive drugs and brain neurochemical transmitters. *Archives Internationales de Pharmacodynamie et de Therapie* 196:221-228.
1973 Activity of benzodiazepines on aggressive behavior in rats and mice. Pp. 405-417 in S. Garattini, E. Mussini, and L.O. Randall, eds., *The Benzodiazepines.* New York: Raven Press.
1979 Effect of sedatives and anxiolytics on aggressivity. Pp. 143-156 in J.R. Boissier, ed., *Differential Psychopharmacology of Anxiolytics and Sedatives.* Basel, Switzerland. S. Krager.
1981 Psychopharmacology of aggression: An overview. *International Pharmacopsychiatry* 16:39-48.
1985 Animal models of behavioral pathology and violent aggression.

Methods and Findings in Experimental and Clinical Pharmacology 7:189-193.

Valzelli, L., and S. Bernasconi
1971   Differential activity of some psychotropic drugs as a function of emotion levels in animals. *Psychopharmacologia* 20:91-96.
1976   Psychoactive drug effect on behavioural changes induced by prolonged socio-environmental deprivation in rats. *Psychological Medicine* 6:271-276.

Valzelli, L., and E. Galateo
1984   Serotonergic control of experimental aggression. *Polish Journal of Pharmacology and Pharmacy* 36:495-530.

Valzelli, L., and S. Garattini
1968   Behavioral changes and 5-hydroxytryptamine turnover in animals. *Advances in Pharmacology* 6B:249-260.

Valzelli, L., E. Giacalone, and S. Garattini
1967   Pharmacological control of aggressive behavior in mice. *European Journal of Pharmacology* 2:144-146.

Van der Poel, A.M., and M. Remmelts
1971   The effect of anticholinergics on the behaviour of the rat in a solitary and in a social situation. *Archives Internationales de Pharmacodynamie et de Therapie* 189:394-396.

Vander Wende, C., and M.T. Spoerlein
1962   Psychotic symptoms induced in mice by the intravenous administration of solutions of 3,4-dihydroxyphenylalanine (DOPA). *Archives Internationales de Pharmacodynamie et de Therapie* 137:145-155.

Van Praag, H.M.
1982   Depression, suicide and the metabolism of serotonin in the brain. *Journal of Affective Disorders* 4:275-290.

Van Praag, H.M., R. Plutchik, and H. Conte
1986   The serotonin hypothesis of (auto)aggression. Critical appraisal of the evidence. Pp. 150-167 in J.J. Mann and M. Stanley, eds., *Psychobiology of Suicidal Behavior*. (Annals of the New York Academy of Sciences, Vol. 487). New York: New York Academy of Sciences.

Van Praag, H.M., C. Lemus, and R. Kahn
1987   Hormonal probes of central serotonergic activity: Do they really exist? *Biological Psychiatry* 22:86-98.

Van Putten, T., and D.G. Sanders
1975   Lithium in treatment failures. *Journal of Nervous and Mental Disease* 161:255-264.

Van Riezen, H., W.J. Van der Burg, H. Berendsen, and M.-L. Jaspar
1973   OI 77, a new tricyclic antidepressant. *Arzneimittel-Forschung/Drug Research* 23:1295-1302.

Vassout, A., and A. Delini-Stula
1977   Effects of ß-bloqueurs (propranolol et oxprenolol) et du diazepam

sur differents modeles d'agressivite chez le rat. *Journal de Pharmacologie* 8:5-14.

Verebey, K., J. Volavka, and D. Clouet
1978 Endorphins in psychiatry. *Archives of General Psychiatry* 35:877-888.

Vergnes, M., and P. Karli
1963 Declenchement on comportement d'agression interspecifique rat-souris par ablation bilaterale des bulbes olfactifs. Action de l'hydroxyzine sur cette agressivite provoquee. *Comptes Rendus des Seances de la Societe de Biologie* 157:1061-1063.

Vergnes, M., and E. Kempf
1981 Tryptophan deprivation: Effects on mouse-killing and reactivity in the rat. *Pharmacology Biochemistry and Behavior* 14(suppl. 1):19-23.

Vergnes, M., G. Mack, and E. Kempf
1973 Lesions du raphe et reaction d'agression interspecifique rat souris effets comportementaux et biochimiques. *Brain Research* 57:67-74.

Vergnes, M., C. Penot, E. Kempf, and G. Mack
1977 Lesion selective des neurones serotonergiques du raphe par 5,7-dihydroxytryptamine: effets sur de comportement d'agression interspecifique du rat. *Brain Research* 133:167-171.

Vergnes, M., A. Depaulis, A. Boehrer, and E. Kempf
1988 Selective increase of offensive behavior in the rat following intrahypothalamic 5,7-DHT-induced serotonin depletion. *Behavioural Brain Research* 29:85-91.

Vessey, S.
1967 Effects of chlorpromazine on aggression in laboratory populations of wild house mice. *Ecology* 48:367-376.

Vetro, A., I. Szentistvanyi, L. Pallag, M. Vargha, and J. Szilard
1985 Therapeutic experience with lithium in childhood aggression. *Pharmacopsychiatry* 14:121-127.

Vialatte, J.
1966 Troubles du comportement chez l'enfant interet du traitement symptomatique par un psycholeptique. *Annals de Pediatrie* 45:733-735.

Virkkunen, M., and S. Narvanen
1987 Plasma insulin, tryptophan and serotonin levels during the glucose tolerance test among habitually violent and impulsive offenders. *Neuropsychobiology* 17:19-23.

Virkkunen, M, J. De Jong, J. Barko, F.K. Goodwin, and M. Linnoila
1989a Relationship of psychobiological variables to recidivism in violent offenders and impulsive fire setters. *Archives of General Psychiatry* 46:600-603.

Virkkunen, M., J. De Jong, J. Barko, and M. Linnoila
1989b Psychobiological concomitants of history of suicide attempts

among violent offenders and impulsive fire setters. *Archives of General Psychiatry* 46:604-606.

Vogel, G., P. Harley, D. Neill, M. Hagler, and D. Kors
1988 Animal depression model by neonatal clomipramine: Reduction of shock induced aggression. *Pharmacology Biochemistry and Behavior* 31:103-106.

Volavka, J., M. Crowner, D. Brizer, A. Convit, H. Van Praag, and R.F. Suckow
1990 Tryptophan treatment of aggressive psychiatric inpatients. *Biological Psychiatry* 28:728-732.

Waizer, J., S.P. Hoffman, P. Polizos, and D.M. Engelhardt
1974 Outpatient treatment of hyperactive school children with imipramine. *American Journal of Psychiatry* 131:587-591.

Walaszek, E.J., and L.G. Abood
1956 Effect of tranquilizing drugs on fighting response of Siamese fighting fish. *Science* 124:440-441.

Waldbillig, R.J.
1980 Suppressive effects of intraperitoneal and intraventricular injections of nicotine on muricide and shock-induced attack on conspecifics. *Pharmacology Biochemistry and Behavior* 12:619-623.

Walletschek, H., and A. Raab
1982 Spontaneous activity of dorsal raphe neurons during defensive and offensive encounters in the tree-shrew. *Physiology and Behavior* 28:697-705.

Ward, M.E., S.R. Saklad, and L. Ereshefsky
1986 Lorazepam for the treatment of psychotic agitation. *American Journal of Psychiatry* 143:1195-1196.

Watanabe, S., M. Inoue, and S. Ueki
1979 Effects of psychotropic drugs injected into the limbic structures on mouse-killing behavior in the rat with olfactory bulb ablations. *Japanese Journal of Pharmacology* 29:493-496.

Weerts, E.M., W. Tornatzky, and K.A. Miczek
1988 Alcohol-benzodiazepine receptor interactions: Aggressive behavior and motor activity in rats and squirrel monkeys. *Society for Neuroscience Abstracts* 14(2):1261.
1993 Prevention of the proaggressive effects of alcohol by benzodiazepine receptor antagonists in rats and in squirrel monkeys. *Psychopharmacology* 111:144-152.

Weinstock, M., and C. Weiss
1980 Antagonism by propranolol of isolation-induced aggression in mice: Correlation with 5-hydroxytryptamine receptor blockade. *Neuropharmacology* 19:653-656.

Weischer, M.-L.
1969 Über die antiaggressive Wirkung von Lithium. *Psychopharmacologia* 15:245-254.

Weischer, M.-L., and K. Opitz
1972 Einfluss von Fenfluramin, Chlorphentermin und verwandten

Verbindungen auf das Verhalten von aggressiven Mausen. *Archives Internationales de Pharmacodynamie et de Therapie* 195:252-259.

Welch, A.S., and B.L. Welch
1971 Isolation, reactivity and aggression: Evidence for an involvement of brain catecholamines and serotonin. Pp. 91-142 in B.E. Eleftheriou and J.P. Scott, eds., *Physiology of Aggression and Defeat*. New York: Plenum.

Welch, B.L., and A.M. Goldberg
1973 Adrenal choline acetyltransferase activity: Sustained effects of chronic intermittent psychological and psychosocial stimulation. *International Journal of Neuroscience* 5:95-99.

Welch, B.L., and A.S. Welch
1965 Effect of grouping on the level of brain norepinephrine in white Swiss mice. *Life Sciences* 4:1011-1018.

1968a Rapid modification of isolation-induced aggressive behavior and elevation of brain catecholamines and serotonin by the quick-acting monoamine-oxidase inhibitor pargyline. *Communications in Behavioral Biology* 1:347-351.

1968b Greater lowering of brain and adrenal catecholamines in group-housed than in individually-housed mice administered DL-α-methyltyrosine. *Journal of Pharmacy and Pharmacology* 20:244-246.

1969a Aggression and the biogenic amine neurohumors. Pp. 188-202 in S. Garattini and E.B. Sigg, eds., *Aggressive Behaviour*. Amsterdam: Excerpta Medica Foundation.

1969b Fighting: Preferential lowering of norepinephrine and dopamine in the brainstem, concomitant with a depletion of epinephrine from the adrenal medulla. *Communications in Behavioral Biology* 3:125-130.

1970 Control of brain catecholamines and serotonin during acute stress and after *d*-amphetamine by natural inhibition of monoamine oxidase: An hypothesis. Pp. 415-445 in E. Costa and S. Garattini, eds., *International Symposium on Amphetamines and Related Compounds*. New York: Raven Press.

1973 Chronic social stimulation and tolerance to amphetamine: Interacting effects of amphetamine and natural nervous stimulation upon brain amines and behavior. Pp. 107-115 in E.H. Ellinwood and S. Cohen, eds., *Current Concepts in Amphetamine Abuse*. Washington, D.C.: U.S. Government Printing Office.

Wenzl, H., E. Graf, and A. Sieck
1978 Central nervous effects of a new tricyclic antidepressant (amitriptylinoxide). *Arzneimittel-Forschung/Drug Research* 28:1874-1879.

Werry, J.S.
1981 Drugs and learning. *Journal of Child Psychology* 22:283-290.

Werry, J.S., and M.G. Aman
 1975 Methylphenidate and haloperidol in children. *Archives of General Psychiatry* 32:790-795.
Whitman, J.R., G.J. Maier, and B. Eichelman
 1987 Beta-adrenergic blockers for aggressive behavior in schizophrenia. *American Journal of Psychiatry* 144:538.
Wickham, E.A., and J.V. Reed
 1987 Lithium for the control of aggressive and self-mutilating behaviour. *International Clinical Psychopharmacology* 2:181-190.
Wiedeking, C., C.R. Lake, M. Ziegler, E. Muske, and G. Jorgensen
 1977 Plasma noradrenalin and dopamine-ß-hydroxylase during behavioral testing of sexually deviant XYY and XXY males. *Human Genetics* 37:243-247.
Wilkinson, C.J.
 1985 Effects of diazepam (Valium) and trait anxiety on human physical aggression and emotional state. *Journal of Behavioral Medicine* 8:101-114.
Williams, D.T., R. Mehl, S. Yudofsky, D. Adams, and B. Roseman
 1982 The effect of propranolol on uncontrolled rage outbursts in children and adolescents with organic brain dysfunction. *Journal of the American Academy of Child Psychiatry* 21:129-135.
Willner, P., A. Theodorou, and A. Montgomery
 1981 Subchronic treatment with the tricyclic antidepressant DMI increases isolation-induced fighting in rats. *Pharmacology Biochemistry and Behavior* 14:475-479.
Winsberg, B.G., I. Bialer, S. Kupietz, and J. Tobias
 1972 Effects of imipramine and dextroamphetamine on behavior of neuropsychiatrically impaired children. *American Journal of Psychiatry* 128:1425-1431.
Winsberg, B.G., L.E. Yepes, and I. Bialer
 1976 Pharmacologic management of children with hyperactive/aggressive/inattentive behavior disorders. *Clinical Pediatrics* 15:471-477.
Winslow, J.T., and K.A. Miczek
 1983 Habituation of aggression in mice: Pharmacological evidence of catecholaminergic and serotonergic mediation. *Psychopharmacology* 81:286-291.
Wood, D.R., F.W. Reimherr, P.H. Wender, and G.E. Johnson
 1976 Diagnosis and treatment of minimal brain dysfunction in adults. *Archives of General Psychiatry* 33:1453-1460.
Woodman, D.
 1979 Evidence of a permanent imbalance in catecholamine secretion in violent social deviants. *Journal of Psychosomatic Research* 23:155-157.
Woodman, D., and J. Hinton
 1978a Catecholamine balance during stress anticipation: An abnormality in maximum security hospital patients. *Journal of Psychosomatic Research* 22:477-483.

1978b Anomalies of cyclic AMP excretion in some abnormal offenders. *Biological Psychology* 7:103-108.

Woodman, D., J. Hinton, and M. O'Neill
1977 Relationship between violence and catecholamines. *Perceptual and Motor Skills* 45:702.

Worrall, E.P., J.P. Moody, and G.J. Naylor
1975 Lithium in non-manic-depressives: Antiaggressive effect and red blood cell lithium values. *British Journal of Psychiatry* 126:464-468.

Yamamoto, T., and S. Ueki
1978 Effects of drugs on hyperactivity and aggression induced by raphe lesions in rats. *Pharmacology Biochemistry and Behavior* 9:821-826.

Yamamoto, T., H. Araki, Y. Abe, and S. Ueki
1985 Effects of chronic LiCl and RbCl on muricide induced by midbrain raphe lesions in rats. *Pharmacology Biochemistry and Behavior* 22:559-563.

Yamamoto, T., M. Ohno, K. Takao, and S. Ueki
1988 Anti-serotonin action in combination with noradrenaline-stimulating action is important for inhibiting muricide in midbrain raphe-lesioned rats. *Neuropharmacology* 27:123-127.

Yaryura-Tobias, J.A., and F. Naziroglu
1978 Compulsions, aggression, and self-mutilation: A hypothalamic disorder? *Journal of Orthomolecular Psychiatry* 7:114-117.

Yen, H.C.Y., R.L. Stanger, and N. Millman
1958 Isolation-induced aggressive behavior in ataractic tests. *Journal of Pharmacology and Experimental Therapeutics* 122:85A.
1959 Ataractic suppression of isolation-induced aggressive behavior. *Archives Internationales de Pharmacodynamie et de Therapie* 123:179-185.

Yen, H.C.Y., M.H. Katz, and S. Krop
1970 Effects of various drugs on 3,4-dihydroxyphenylalanine (DL-DOPA)-induced excitation (aggressive behavior) in mice. *Toxicology and Applied Pharmacology* 17:597-604.

Yepes, L.E., E.B. Balka, B.G. Winsberg, and I. Bialer
1977 Amitriptyline and methylphenidate treatment of behaviorally disordered children. *Journal of Child Psychology and Psychiatry* 18:39-52.

Yesavage, J.A.
1982 Inpatient violence and the schizophrenic patient: An inverse correlation between danger-related events and neuroleptic levels. *Biological Psychiatry* 17:1331-1337.

Yodyingyuad, U., C. de la Riva, D.H. Abbott, J. Herbert, and E.B. Keverne
1985 Relationship between dominance hierarchy, cerebrospinal fluid levels of amine transmitter metabolites (5-hydroxyindole acetic acid and homovanillic acid) and plasma cortisol in monkeys. *Neuroscience* 16:851-858.

Yoshimura, H.
1987　Studies contrasting drug effects on reproduction induced agonistic behavior in male and female mice. Pp. 94-109 in B. Olivier, J. Mos, and P.F. Brain, eds., *Ethopharmacology of Agonistic Behavior in Animals and Humans.* Dordrecht, The Netherlands: Martinus Nijhoff.
Yoshimura, H., and N. Ogawa
1984　Pharmaco-ethological analysis of agonistic behavior between resident and intruder mice: Effects of psychotropic drugs. *Folia Pharmacologica Japonica* 84:221-228.
1985　Pharmaco-ethological analysis of agonistic behavior between resident and intruder mice: Effects of adrenergic ß-blockers. *Japanese Journal of Psychopharmacology* 5:223-229.
1989　Acute and chronic effects of psychotropic drugs on maternal aggression in mice. *Psychopharmacology* 97:339-342.
Yoshimura, H., V. Kihara, and N. Ogawa
1987　Psychotropic effects of adrenergic ß-blockers on agonistic behavior between resident and intruder mice. *Psychopharmacology* 91:445-450.
Yudofsky, S., D. Williams, and J. Gorman
1981　Propranolol in the treatment of rage and violent behavior in patients with chronic brain syndromes. *American Journal of Psychiatry* 138:218-220.
Yudofsky, S.C., L. Stevens, J. Silver, J. Barsa, and D. Williams
1984　Propranolol in the treatment of rage and violent behavior associated with Korsakoffs's psychosis. *American Journal of Psychiatry* 141:114-115.
Yudofsky, S.C., J.M. Silver, and S.E. Schneider
1987　Pharmacologic treatment of aggression. *Psychiatric Annals* 17:397.
Zetler, G., and B. Hauer
1975　Pharmacological dissociation between vocalization and biting produced in rats by the combination of imipramine and isocarboxazid. *Psychopharmacologia* 45:73-77.
Zelter, G., and U. Otten
1969　Aggressivitat der Ratte nach kombinierter Behandlung mit Monoaminoxydase-Inhibitoren und anderen psychotropen Pharmaka, insbesondere Thymoleptica. *Naunyn-Schmiedeberg's Archiv der Experimentellen Pathologie und Phamakologie* 264:32-54.
Zisook, S., and A. DeVaul
1977　Adverse behavioral effects of benzodiazepines. *Journal of Family Practice* 5:963-966.
Zisook, S., P.J. Rogers, T.R. Faschingbauer, and R.A. DeVaul
1978　Absence of hostility in outpatients after administration of halazepam: A new benzodiazepine. *Journal of Clinical Psychiatry* 39:683-686.
Zwirner, P.P., R.D. Porsolt, and D.M. Loew
1975　Inter-group aggression in mice. A new method for testing the effects of centrally active drugs. *Psychopharmacologia* 43:133-138.

TABLE 1: Major Experimental Models of Aggression in Laboratory Animals

Model and Species	Procedure	Behavioral Topography	Biological Function
A. Aversive environmental manipulations			
Isolation-induced aggression, mostly in mice	Isolated housing before confrontation with another isolate or group-housed animal	Complete agonistic behavior pattern: isolates attack, threaten, pursue opponent	Territorial defense or compulsive, abnormal, pathological behavior
Pain-elicited or shock-induced aggression, mostly in rats, also in monkeys	Pairs of animals are exposed to pulses of electric shock delivered through grid floor or to the tail	Defensive reactions, including upright postures, bites toward face of opponent, audible vocalizations; bites toward inanimate targets	Some similarity to reaction toward predator or toward large opponent
Aggression due to omission of reward, mostly in pigeons, also in monkeys	Conditioning history; schedule-controlled operant behavior; omitted or infrequent reinforcement	Attack bites or pecks, threat displays towards suitable object or conspecific	Competition for resources such as food, sex, protected niches (?)
B. Brain manipulations			
Brain lesion-induced aggression, mostly in rats, also in cats	Destruction of neural tissue, and subsequent social or environmental challenges	Defensive reactions, biting	Neurological disease
Brain stimulation-induced aggression, mostly in cats, also in rats	Electrical excitation of tissue in diencephalon and mesencephalon, also in other limbic or cerebellar areas	(1) Defensive reactions accompanied by autonomic arousal (2) Predatory attack and killing	Defense against attacker Predation
C. Ethological situations			
Aggression by resident toward intruder, in most species and in both sexes	Confrontation with an unfamiliar adult member of the species	Full repertoire of agonistic behavior (attack and threat vs. defense, submission, and flight)	Territorial or group defense (?); rivalry among males and among females
Female aggression, mostly in maternal rodents	Lactating female, in the presence of litter, confronting an intruder male	Species-specific repertoire of attack and threat behavior toward intruder	Defense of young, competition for resources and territory

Dominance-related aggression, mostly in monkeys, mice, and rats	Formation or maintenance of a social group	Species-specific repertoire of signals (displays, sounds, odors) between group members of different social rank; low level and intensity of agonism	Social cohesion and dispersion

D. Killing

Muricide, mostly in rats, cats	Presence of prey, food deprivation	Stalking, seizing, killing, sometimes consuming prey	Food source; "killer instinct"

Source: Adapted from Miczek and DeBold (1983).

TABLE 2: Catecholamines

References	Methods and Procedures	Results and Conclusions
A. Noradrenergic Correlates of Animal Aggression		
Whole brain measurements		
Isolation-induced Aggression		
Welch and Welch 1965	Comparison of isolated and group-housed male mice. No aggression measured	Isolated mice had higher whole brain levels of NE (mg/g) than group-housed mice.
Welch and Welch 1968a, 1970	Male mice grouped in a neutral environment for varying lengths of time (10 mins-2.5 hrs). Animals were sacrificed immediately after the aggressive interaction. Whole brain NE (ng/g) and DA were assayed.	The earlier study reported no difference in NE levels as a function of aggressive experience, while the later study reported decreases (8%) in NE following a 30 min fight.
Modigh 1974	Male mice placed together in a neutral environment for 30 min. Animals were sacrificed immediately after the aggressive interaction. AMPT or NSD 1015 (aromatic amino acid decarboxylase inhibitor) was used to estimate turnover in whole brain.	Fight experience resulted in increased AMPT or NSD 1015-induced depletion of NE (μg/g).
Hutchins et al. 1974; 1975	Male mice placed together in a neutral environment. Animals were sacrificed immediately after the behavioral interaction. HVA and DOPAC were assayed in the striatum.	DOPAC levels (μg/g) in the striatum were higher in isolated mice transferred to a new environment for 15 min, with or without an agonistic interaction, in comparison to isolated mice left undisturbed. DOPAC levels in group-housed mice were not altered when they were placed in a new cage. Striatal concentrations of DA was not affected by housing or transfer.
Bernard et al. 1975	Male mice tested in a neutral environment against a similarly treated conspecific. Strain (BALB, ICR and C57B1/6J), housing, (group or isolated) and age (isolated 6 or 30 weeks) were varied. Catecholamine dynamics in whole brain were measured 24 hrs after agonistic interaction. Turnover was estimated with AMPT.	C57B1/6J did not fight and had higher NE (ng/g) levels than the other strains. Age or strain related differences in NE rate and utility constants did not vary as a function of aggressivity. Differential housing did not alter any of the biochemical measures.

351

| Karczmar et al. 1973; Goldberg et al. 1973 | Comparison of neurochemistry in different strains of male mice; turnover was estimated using AMPT. | Neural levels or turnover rates of NE (µg/g) are not correlated with aggression levels across strains. |

Dominance-related Aggression

| McIntyre et al. 1979 | Comparison of whole brain NE in dominant and submissive rainbow trout. | Submissive fish that are rarely attacked have comparable NE to dominant fish. Submissive fish that are attacked have increased NE compared to dominant fish |

| McIntyre and Chew 1983 | Comparison of whole brain NE in dominant and submissive pheasants. | Dominance status was not correlated with catecholamine concentrations (ng/g). |

| Hadfield and Weber 1975 | Pairs of group-housed mice were tested in a neutral arena. NE uptake was measured in whole brain synaptosomes immediately following the agonistic interaction | Fighting increases K_m and V_{max} for NE uptake compared to non-fighting controls. |

Regional brain measurements
Isolation-induced Aggression

| Welch and Welch 1969b | Male mice grouped in a neutral environment for varying lengths of time (5, 45, 60, 150 mins). Animals were sacrificed immediately after the aggressive interaction. NE (ng/g) was assayed in the metencephalon, mesencephalon and telencephalon. | Fighting experience was associated with decreased NE in the brain stem at all time points. NE was elevated in the telencephalon after a 150 min interaction. |

| Modigh 1973 | Male mice placed together in a neutral environment for 30 min. Animals were sacrificed immediately after the aggressive interaction. NE and its precursors and metabolites were assayed in the striatum, cerebral hemispheres and the rest of the brain. NSD 1015 (aromatic amino acid decarboxylase inhibitor) was used to estimate turnover | In animals administered NSD 1015, fight experience resulted in decreased NE, and increased tyrosine, and DOPA accumulation in each brain region, compared to isolates with no fight experience. Whole brain concentrations of HVA were also lower following attack experience. |

352

Reference	Methodology	Results
Hutchins et al. 1974, 1975	Male mice placed together in a neutral environment. Animals were sacrificed immediately after the behavioral interaction. HVA and DOPAC were assayed in the striatum.	DOPAC levels (µg/g) in the striatum were higher in isolated mice transferred to a new environment for 15 min, with or without an agonistic interaction, in comparison to isolated mice left undisturbed. DOPAC levels in group-housed mice were not altered when they were placed in a new cage. Striatal concentrations of HVA or NE were not affected by housing or transfer.
Tizabi et al. 1979	Comparison of neurochemistry and aggression in 3 strains of male mice. NE was assayed in 14 brain regions. Turnover was estimated using AMPT.	The most aggressive strain had higher steady state levels and turnover rate of NE in the frontal cortex, caudate nucleus and hypothalamus compared to the least aggressive strain.
Tizabi et al. 1980	Male mice placed together in a neutral arena were sacrificed 48 hrs after the last behavioral interaction. NE was assayed in 17 nuclei. AMPT was used to estimate turnover.	Mice that attacked had higher levels of NE (pg/vg) in the septum and lower levels in the olfactory tubercle and substantia nigra than mice that did not attack. Aggressive mice had increased NE turnover in the A10 region.
Hadfield and Milio 1988	Pairs of male mice tested in a neutral arena. Animals were sacrificed immediately after behavioral interactions. NE and its metabolites were assayed in 10 brain regions.	MHPG/NE levels were not significantly altered as a function of fighting experience.
Thoa et al. 1977	Comparison of neurochemistry in isolated and group-housed male rats. Rats were isolated for 13 weeks. NE was estimated in 23 nuclei using AMPT.	Following isolation, steady state NE (ng/mg protein) was decreased in the hippocampus and n. amygdala centralis and increased in the entorhinal cortex compared to group-housed controls. NE turnover was decreased in the cingulate cortex, caudate nucleus, stria terminalis and paraventricular nucleus.

Pain-induced Aggression and Defense

Reference	Methodology	Results
Stolk et al. 1974	Electric foot shock aggression in male rats. Animals were sacrificed immediately after the aggressive interaction. NE was assayed in the brain stem, diencephalon and telencephalon	Rats that were shocked but not given the opportunity to fight with another animal had decreased NE (µg/g) and increased normetanephrine in the brain stem; in rats that fought, brainstem concentrations of NE and metabolites were comparable to controls.

353

Tsuda et al. 1988	Restraint stress-induced target biting in male rats. Animals were sacrificed immediately or 50 min after restraint. Plasma corticosterone was measured. NE and MHPG were assayed in 8 brain regions	50 min after restraint, rats given the opportunity to bite had lower plasma corticosterone (mg/dl) and lower MHPG in the hypothalamus, amygdala, thalamus, and basal ganglia than rats not given the opportunity to bite.

Brain lesion-induced Aggression

Reis and Fuxe 1964; 1968	"Sham rage" in cats following acute brainstem transection. NE histofluorescence was measured in the hindbrain. H 44/68 was used to estimate NE turnover.	Decerebration above the superior colliculus was associated with both "sham rage" and increased NE turnover. Midcollicular transected cats, which did not manifest "sham rage" had comparable NE turnover to controls. Haloperidol blocked this effect.
Salama and Goldberg 1973a	Comparison of neurochemistry in septal-lesioned and sham-lesioned male rats. NE was measured in the forebrain and hindbrain. Turnover was estimated using AMPT	Septal-lesioned rats had increases in steady-state NE (μg/g) and NE turnover rate (μg/g/hr) in the hindbrain compared to controls

Defensive Aggression induced by brain stimulation

Reis and Gunne 1965	"Sham rage" following electrical stimulation of the amygdala in male and female cats. Animals were sacrificed immediately after the behavioral interaction. NE was measured in the brainstem, telencephalon and adrenal gland.	Animals that became defensive had decreased NE (ng/g) in the brain and decreased NE in the adrenal compared to non-stimulated controls or animals that did not become defensive upon amygdala stimulation.

Aggression by Resident toward an Intruder

Payne et al. 1984, 1985	Isolated and group-housed male hamsters sacrificed after the behavioral interaction. NE was assayed in the midbrain and hypothalamus.	More aggressive isolates do not differ from less aggressive group-housed hamsters in NE levels (ng/100 mg) in the hypothalamus and midbrain. Attack experience did not alter the percentage change in NE following pargyline or AMPT, respectively. Residents defeated by intruders had amine changes similar to attackers.

Killing

Goldberg and Salama 1969	Muricide by male rats. Forebrain and hindbrain NE was measured 24 hrs following muricide. AMPT was used to estimate turnover	Muricidal rats had increased NE (μg/g) in the forebrain compared to non-muricidal rats. The effects of AMPT were similar in muricidal and non-muricidal groups.
Barr et al. 1979	Muricide by isolated male rats. Animals were sacrificed 1 week after attack experience. NE was assayed in the hypothalamus, amygdala and olfactory bulbs	Concentrations of NE (ng/mg protein) were similar in muricidal and non-muricidal rats in all 3 brain regions
Salama and Goldberg 1973b	Muricide by male rats. Forebrain NE and turnover were measured at 2, 24, 48 hrs or 1 week following muricide; Radioactively labelled NE was used to estimate turnover	Forebrain NE levels (μg/g) and turnover rate were elevated 2 or 24 hrs after muricide, but not 48 hrs or 1 week later.
Tani et al. 1987	Muricide by male rats with olfactory bulbectomy or n. raphe or n. accumbens lesioning. Animals were sacrificed immediately after muricide. NE and its metabolites were assayed in the frontal cortex and hypothalamic and amygdaloid nuclei.	Muricidal rats had increased NE (ng/mg protein) in the hypothalamus and increased HVA and DOPAC in the frontal cortex and LH in comparison to intact non-muricidal rats. Type of lesion did not affect neurochemical measures.

B. Dopaminergic Correlates of Animal Aggression

Whole brain measurements

Isolation-induced Aggression

Welch and Welch 1968b, 1970	Male mice grouped in a neutral environment for varying lengths of time (10 mins-2.5 hrs). Animals were sacrificed immediately after the aggressive interaction.	DA levels (ng/g) were not altered as a function of aggressive experience.
Modigh 1974	Male mice placed together in a neutral environment for 30 min. Animals were sacrificed immediately after the aggressive interaction. AMPT or NSD 1015 (aromatic amino acid decarboxylase inhibitor) was used to estimate turnover in whole brain.	Fight experience resulted in increased AMPT but not NSD 1015-induced depletion of DA (μg/g).

355

Reference	Methods	Results
Bernard et al. 1975	Male mice tested in a neutral environment against a similarly treated conspecific. Strain (BALB, ICR and C57B1/6J), housing, (group or isolated) and age (isolated 6 or 30 weeks) were varied. Catecholamine dynamics in whole brain were measured 24 hrs after agonistic interaction. Turnover was estimated with AMPT.	BALB mice fought at 6 and 30 weeks of isolation and had higher DA levels than the other strains at either time point. Age or strain related differences in DA rate and utility constants did not vary as a function of aggressivity. Differential housing did not alter any of the biochemical measures.
Karczmar et al. 1973; Goldberg et al. 1973	Comparison of neurochemistry in different strains of male mice; turnover was estimated using AMPT.	Neural levels or turnover rates of DA (μg/g) are not correlated with aggression levels across strains.
Lasley and Thurmond 1985	Compared NE, DA, 5HT and major metabolites in isolated and group-housed mice confronting a male conspecific in a neutral arena; not clear when amines were measured in relation to the agonistic experience.	Mice that were isolated for 5 days were more aggressive and had higher DOPAC and HVA levels (μg/g) than group-housed mice. After 14 days of isolation, frequency of attack was comparable in isolated and group-housed mice, but the isolated group had elevated DA turnover (HVA:DA).
Pain-induced Aggression and Defense		
Anand et al. 1985	Electric foot shock in female rats with substantia nigra, septal, or amygdala lesions. Timing of neurochemical measures in relation to behavior not specified	Aggression frequency was increased by substantia nigra or septum lesions and decreased by amygdala lesions. DA (μg/g) was only decreased in the septal-lesion group.
Dominance-related Aggression		
McIntyre et al. 1979	Comparison of whole brain DA in dominant and submissive rainbow trout.	Submissive fish that are rarely attacked have DA levels comparable to dominant fish. Submissive fish that are attacked have decreased DA compared to dominant fish
McIntyre and Chew 1983	Comparison of whole brain DA in dominant and submissive pheasants.	Dominance status was not correlated with altered catecholamine concentrations (ng/g). Lower rank in the social hierarchy was associated with decreased DA in the neostriatum

Regional brain measurements

Isolation-induced Aggression

Reference	Methods	Results
Welch and Welch 1969b	Male mice grouped in a neutral environment for varying lengths of time (5, 45, 60, 150 mins). Animals were sacrificed immediately after the aggressive interaction. DA was assayed in the metencephalon, mesencephalon and telencephalon.	Fighting experience was associated with decreased DA in the brain stem. DA was elevated in the telencephalon after a 150 min interaction.
Modigh 1973	Male mice placed together in a neutral environment for 30 min. Animals were sacrificed immediately after the aggressive interaction. DA and its precursors and metabolites were assayed in the striatum, cerebral hemispheres and the rest of the brain. NSD 1015 (aromatic amino acid decarboxylase inhibitor) was used to estimate turnover.	In animals administered NSD 1015, fight experience resulted in increased tyrosine and DOPA accumulation in each brain region compared to isolates with no fight experience. Whole brain concentrations of HVA were also lower following attack experience.
Hutchins et al. 1974, 1975	Male mice placed together in a neutral environment. Animals were sacrificed immediately after the behavioral interaction. HVA and DOPAC were assayed in the striatum.	DOPAC levels ($\mu g/g$) in the striatum were higher in isolated mice transferred to a new environment for 15 min, with or without an agonistic interaction, in comparison to isolated mice left undisturbed. DOPAC levels in group-housed mice were not altered when they were placed in a new cage. Striatal concentrations of DA were not affected by housing or transfer.
Tizabi et al. 1979	Comparison of neurochemistry and aggression in 3 strains of male mice. DA was assayed in 14 brain regions. Turnover was estimated using AMPT	The most aggressive strain had higher steady state levels and turnover rate of DA in the frontal cortex, caudate nucleus and hypothalamus compared to the least aggressive strain.
Tizabi et al. 1980	Male mice placed together in a neutral arena were sacrificed 48 hrs after the last behavioral interaction. DA was assayed in 17 nuclei. AMPT was used to estimate turnover.	DA levels in aggressive mice were lower in the olfactory tubercle and higher in the caudate putamen. Aggressive mice had decreased DA turnover in the olfactory tubercle and caudate-putamen.

Hadfield 1981, 1983	Groups of male mice tested in a neutral arena. Animals were sacrificed immediately after the behavioral interaction. DA uptake was measured in the prefrontal cortex and caudate-putamen.	Fighting increased DA uptake (K_m and V_{max}) in the prefrontal cortex but not in the caudate-putamen.
Hadfield and Milio 1988	Pairs of male mice tested in a neutral arena. Animals were sacrificed immediately after behavioral interactions. DA and its metabolites were assayed in 10 brain regions.	There was an overall increase in DOPAC:DA ratios following fighting; changes in DA utilization within individual tissues were not significant.
Thoa et al. 1977	Comparison of neurochemistry in isolated and group-housed male rats. Rats were isolated for 13 weeks. DA turnover was estimated in 23 nuclei using AMPT	Steady state DA was decreased in the n. amygdala centralis and increased in the olfactory tubercle of isolated rats. DA turnover was also decreased in the n. amygdala centralis.

Aggression by Resident toward an Intruder

Barr et al. 1979	Isolated male rats were sacrificed 1 week after agonistic experience. DA was assayed in the hypothalamus, amygdala and olfactory bulbs.	Rats that attacked an intruder had higher hypothalamic DA (ng/mg protein) compared to rats that did not attack.
Haney et al. 1990	Pair-housed male mice. DA measurements in the n.accumbens, corpus striatum and amygdala were obtained following 0, 1 or 10 daily attack experiences	DOPAC:DA in the n. accumbens was increased following 1 attack experience. Amine measurements in mice with repeated attack experience did not differ from behaviorally naive controls.

Killing

Barr et al. 1979	Muricide by isolated male rats. Animals were sacrificed 1 week after attack experience. DA was assayed in the hypothalamus, amygdala and olfactory bulbs.	Concentrations of DA (ng/mg protein) were similar in muricidal and non-muricidal rats in each brain region
Broderick et al. 1985	Muricide by male rats. Animals were sacrificed 1-2 hrs following the introduction of the mouse into the home cage. DA and its metabolites were assayed in the hypothalamus, thalamus, hippocampus, striatum, cortex and brain stem	Muricidal rats had increased DOPAC in the septum and DA in the anterior hippocampus compared to non-muricidal rats.

Study	Description	Results
Tani et al. 1987	Muricide by male rats with olfactory bulbectomy, n.raphe or n.accumbens lesioning. Animals were sacrificed immediately after muricide. DA and its metabolites were assayed in the frontal cortex and hypothalamic and amygdaloid nuclei	Muricidal rats had increased HVA and DOPAC in the frontal cortex and LH in comparison to intact non-muricidal rats. Type of lesion did not affect neurochemical measures.

C. Catecholaminergic Correlates of Human Aggression and Violence

CSF Correlates of Human Aggression

Inpatient studies

Study	Description	Results
Subrahmanyam 1975	Schizophrenic, manic-depressive psychotics and healthy controls were compared: 6/60 acute schizophrenics were in an "acute aggressive state". CSF MHPG and HVA and urinary NE, EPI, 5HIAA, MHPG and VMA were measured	CSF MHPG (ng/ml) in acute, aggressive schizophrenics was comparable to controls but higher than non-aggressive acute schizophrenics. Urinary amines appeared elevated in aggressive schizophrenics in comparison to both controls and non-aggressive schizophrenics. No statistics reported
Brown et al. 1979	CSF MHPG was measured in military men diagnosed with borderline personality disorder without affective illness (n=26).	CSF MHPG (ng/ml) did not differ in patients with borderline personality disorder and normal controls. If just the patients are analyzed, there is a positive correlation between scores based on the life history of aggression and CSF MHPG.

Criminal Violence

Study	Description	Results
Bioulac et al. 1980	CSF DA was measured in healthy controls (n=5) and XYY patients arrested for crimes ranging from vagrancy to assault (n=6). Probenecid was used to estimate turnover	DA levels and turnover did not differ between groups. No statistics were reported
Linnoila et al. 1983	CSF NE, MHPG, DOPAC, and HVA were assayed in men convicted of violent crime. Subjects were subdivided into categories specifying the pre-meditated or impulsive nature of the criminal act.	Catecholamine levels or turnover rates did not differ between groups.

Peripheral Correlates of Human Aggression

Outpatient Studies

Perini et al. 1986

Plasma NE and EPI in 24 borderline hypertension patients and 24 controls given psychological tests to assess suppressed aggression and anxiety.

Compared to subjects without suppressed aggression, borderline hypertensives with suppressed aggression had higher heart rates, diastolic blood pressures and plasma NE following one of the two mildly stressful conditions (mental arithmetic and the Stroop reading task).

Experimental Studies

Ekkers 1975

Urinary methyladrenaline, methylnoradrenaline, VMH and creatinine were measured in 12-17 year old males. Subjects participated in 3 laboratory measures of aggression involving the administration of aversive noise to another person.

The reliability of the aggression measures was low. Methyladrenaline weakly correlated with aggression (no units).

Criminal Studies

Wiedeking et al. 1977

Plasma NE and DBH in an XYY and an XXY male (n=2) convicted of murder and rape were compared to "normal" males (n=9) under resting conditions and during physical and emotional stress.

Levels of plasma NE and DBH (pg/ml) did not differ between violent and non-violent subjects.

Woodman et al. 1977

Urinary and plasma EPI and NE were measured in incarcerated male patients in a maximum security hospital setting. Half the men were convicted of violent personal attack and half were convicted of arson, sexual or property offenses (n=50).

Men convicted of violent personal attack, other than rape, had had lower plasma and urinary EPI and higher urinary NE than men convicted of other types of offenses.

Sandler et al. 1978

Comparison of violent and non-violent prisoners (n=10).

Plasma concentrations of free and conjugated phenylacetic acid (metabolite of phenylethylamine) were elevated in violent prisoners.

360

Reference	Method	Results
Woodman and Hinton 1978a, b; Woodman 1979	Urinary cAMP, NE, and EPI (control for differences due to urine volume) were measured in incarcerated male patients in a maximum security hospital setting. Criminal convictions were of a wide variety. Healthy men and mentally ill patients without a history of violence served as controls. Subjects were exposed to "stressful" experimental procedures.	EPI (nmol/g creatinine) was elevated in both incarcerated and mentally ill men; NE was elevated in the mentally ill group. The incarcerated offenders were subdivided into two groups based on their anticipatory response to an upcoming stressful event: Group 1 was similar to controls. Group 2 had less cortisol and EPI and more NE in their urine than the other groups. Group 2 was categorized, post hoc, as more violent. Differential response to stress in Group 2 patients was replicated 4-25 months later.
Boulton et al. 1983	Plasma phenylacetic acid, *m*-hydroxyphenylacetic acid, *p*-hydroxyphenylacetic acid and platelet MAO were measured in non-violent male offenders, healthy controls, and violent offenders (murder, rape, physical assault) that were undergoing neuroleptic drug treatment (n=23).	Violent offenders had lower unconjugated p-hydroxyphenylacetic acid (ng/ml) and conjugated phenylacetic acid than non-violent offenders ($p <$ .10); platelet MAO did not differ between groups.

D. Neuropharmacological Manipulations of Catecholamines

Animal Studies

Catecholamine synthesis manipulation

Isolation-induced Aggression

Reference	Method	Results
Kletzkin 1969; Rolinski 1973; Hodge and Butcher 1975; Herbut and Rolinski 1985; Rolinski and Herbut 1985	Similarly-treated pairs of male mice tested in a neutral arena	*L*-DOPA (200-800 mg/kg i.p.) and *DL*-DOPA (250, 500 mg/kg i.v) decreased attack frequency at doses that often increased stereotypies; whole brain DA (mg/g) was increased and n. accumbens 5-HT was decreased. *L*-DOPA (50 mg/kg) co-administered with nialamide (40 mg/kg i.p.) increased aggression compared to non-injected controls.
Miczek 1977; Miczek and O'Donnell 1978	Male rats confronting a male conspecific	*L*-DOPA (10 mg/kg i.p.) increased attack frequency; higher doses (20, 200 mg/kg i.p.) decreased attack frequency.

361

Pain-induced Aggression and Defense

Reference	Paradigm	Results
Torda 1976	Electric foots shock in pairs of similarly treated male rats	AMPT (20-50 μg) infused into the VMH decreased aggression.

Drug-Induced Aggression

Reference	Paradigm	Results
Vander Wande and Spoerlein 1962; Bryson and Bischoff 1971	DOPA-induced target biting in group-housed male mice	DOPA (400-500 mg/kg i.v.) induced catatonia, stupor and biting of inanimate objects placed near the mouth. Biting induced by *L*-DOPA (500-970 mg/kg i.p.) was associated with hyperactivity and jumping.
Troncone et al. 1988	Apomorphine-induced aggression in REM sleep-deprived male rats. **Both members of the pair were similarly treated.**	*L*-DOPA (200 mg/kg i.p.) reduced aggression.

Defensive Aggression induced by brain stimulation

Reference	Paradigm	Results
Katz and Thomas 1976	Quiet predatory aggression induced by LH stimulation	AMPT (70 mg/kg i.p.) raised the threshold of stimulation necessary to elicit attack and decreased approach and biting of attack object.

Aggression by Resident toward Intruder

Reference	Paradigm	Results
Thurmond et al. 1977, 1979, 1980	Male group-housed male mice isolated overnight confronted a naive intruder. Diets were supplemented with amino acid precursors.	Aggression and locomotion were enhanced when *l*-tyrosine (2,4%) and *l*-phenylalanine (2,4%), were administered alone or in combination for 1-2 weeks. Tyrosine increased whole brain levels of tyrosine, 5-HT and 5HIAA. Phenylalanine treatment increased phenylalanine and tyrosine levels. After 5 weeks of dietary supplement, tolerance developed to the enhancement in aggression but not to the neurochemical alterations.
Diringer et al., 1982	Male mice were fed isocaloric diets with supplemental casein or tyrosine.	High doses (50-60 mg/kg i.p.) of the DBH inhibitor fusaric acid inhibited aggression. These doses reduced brain NE and DA but increased 5-HT and 5HIAA.
Miczek and O'Donnell 1978	Male rats housed with a female confronted a male conspecific	*L*-DOPA (100-200 mg/kg i.p.) following carbidopa pretreatment decreased attack frequency.

362

Dominance-related Aggression

Benkert et al. 1973	Group-housed male rats	Pretreatment with reserpine (2 mg/kg i.p.) 16 hours before the combined administration of the dopa decarboxylase inhibitor, Ro 4-4602 (50 mg/kg i.p.) and *L*-DOPA (200 mg/kg i.p.) increased fighting.
Redmond et al. 1971a, b.	Group-housed male and female macaques.	AMPT (160-250 mg/kg p.o, (twice daily/14 days) decreased attacks, threats and social rank in 50% of animals. Facial expressions and motor activity were also suppressed. Urinary MHPG and VMA were significantly decreased.

Killing

Banerjee 1974; McLain et al. 1974	Individually housed male rats	AMPT (3x 125 mg/kg i.p.) decreased muricide and produced marked sedation; AMPT (50 mg/kg i.p. 3x daily for 3 days) increased muricide.
Schmidt 1979, 1983	Predatory aggression in ferrets	L-DOPA (30, 60 mg/kg p.o.) disrupted capture, pursuit and biting of prey.

Catecholamine Agonists

Experimenter-elicited Aggression

Maler and Ellis 1987	High frequency electric organ discharges in South American electric fish elicited by simulated electric signals.	NE (.1 μg i.c.v.) increased aggressive signalling

Isolation-induced Aggression

Hodge and Butcher 1975; Lassen 1978	Pairs of similarly treated male mice placed together in a neutral arena	Apomorphine (200-800 mg/kg i.p.) decreased attack frequency at doses that increased stereotypies. Clonidine (0.05-.15 mg/kg) and piperoxan (10,20 mg/kg) also inhibited aggression. An ineffective dose of piperoxan (5 mg/kg) reduced the anti-aggressive effect of clonidine.
Puech et al. 1974; Thor and Ghiselli 1975	Groups of male rats placed together in a neutral arena	Apomorphine (0.125-8.0 mg/kg i.v. or 20 mg/kg i.p.) decreased locomotion and increased aggression.

Baggio and Ferrari 1980	Male mice confronted an untreated male intruder within the home cage isolated 20 days	DA agonists, apomorphine (0.05-5 mg/kg i.p.) and N-n-propyl-norapomorphine (1-100 mg/kg i.p.), dose-dependently reduced aggression. These effects were reversible with haloperidol. Locomotion not reported.

Pain-induced Aggression and Defense

Geyer and Segal 1974; Torda 1976; Baggio and Ferrari 1980; Ray et al. 1983	Electric foots shock in pairs of similarly treated male rats	Aggression was increased by intraventricular administration of DA (1, 3, 6, 50 µg/µl) and decreased by NE (.5, 2, 50 µg/µl i.c.v.); pain sensitivity was not altered. The combined microinjection of NE and DA (10-50 ng/2 ml) into the VMH increased aggression. Both apomorphine (0.05-5 mg/kg i.p.) and N-n-propyl-norapomorphine (1-100 mg/kg i.p.) also increased aggression. These effects were reversible with haloperidol.

Drug-induced aggression

Maj et al. 1987	Clonidine-induced aggression in groups of male mice	Beta antagonist, adimolol (10 mg/kg) reduced aggression.
Hasselager et al. 1972	d-Amphetamine-induced aggression in groups of male mice	AMPT (350 mg/kg s.c.) decreased aggression, which is defined as abrupt locomotion, defensive posture and sound.
Troncone et al. 1988	Apomorphine-induced aggression in REM sleep-deprived male rats. Both members of the pair were similarly treated.	Bromocriptine (no dose reported) did not affect aggression
Nagy and Decsi 1974;	Intrahypothalamic carbachol-induced "sham rage" in male and female cats	NE (20-50 µg) or isoprenaline (50 µg) administered into the dorsal hippocampus reversed the effects of carbachol. Intrahippocampal DA (50 µg) or phenylephrine (50 µg) did not

364

Defensive Aggression induced by brain stimulation

Reference	Subject/Method	Results
Torda 1976	"Sham rage" elicited by electrical stimulation of the hypothalamus in male rats	Frequency of aggression was increased by NE and DA and decreased by phentolamine, propranolol and AMPT infused into the ventromedial hypothalamus. Threshold to elicit aggression was also changed.
Barrett et al. 1987, 1990	"Sham rage" and quiet biting elicited by electrical stimulation of the hypothalamus in male and female cats	Injections of NE (250-500 ng) or clonidine (.9 nmol) into the anterior hypothalamus reduced the threshold for hissing; yohimbine (775 ng) blocked the NE facilitation.
Goldstein and Siegel 1980 Crescimanno et al. 1986; Maeda and Maki 1986;	"Sham rage" and quiet biting elicited by electrical stimulation of the hypothalamus in male and female cats	Substantia nigra stimulation reduced the threshold and the latency to hiss. In contrast, ventral tegmental and n. accumbens stimulation suppressed sham rage and quiet biting. Apomorphine (1 mg/kg) reversed the inhibitory effect of amygdaloid lesions on attack induced by hypothalamic stimulation.

Aggression by Resident toward Intruder

Reference	Subject/Method	Results
Winslow and Miczek 1983; Tidey and Miczek 1991	Male pair-housed mice encountering a male conspecific	Apomorphine (.1-1.0 mg/kg i.p.) decreased attacks, aggressive threats and locomotor activity. Specific D1 (SKF 38393 3-100 mg/kg i.p.) and D2 (quinpirol .1-1.0 mg/kg i.p.) agonists also decreased aggressive behavior.

Dominance-related Aggression

Reference	Subject/Method	Results
McKenzie 1971	Pairs of male or female group-housed rats	Apomorphine (10-30 mg/kg i.p.) increased aggression between males if at least one member of the pair was dominant within its group. Apomorphine did not enhance aggression in females.

Killing

Rolinski 1975; Baggio and Ferrari 1980; Shibata et al. 1982; Berzsenyi et al. 1983; Pucilowski and Valzelli 1986; Molina et al. 1987; Isel and Mandel 1989	Muricide in male and female rats	Killing was inhibited by electrical stimulation of the locus coeruleus; inhibition was reversed by clonidine (0.15 mg/kg i.p.). NE (20, 50 µg/2 µl) infusion into the medial amygdala also inhibited muricide; phenoxybenzamine pretreatment (20 mg) reversed this effect. Both apomorphine (0.05-5 mg/kg i.p.) and *N-n*-propyl-norapomorphine (1-100 mg/kg i.p.) reduced muricide; these effects were reversible with haloperidol. DA (50 µg) infused into the medial amygdala had no effect.
Baggio and Ferrari 1980	Predatory aggression by male rats toward turtles	Apomorphine (0.05-5 mg/kg i.p.) and *N-n*-propyl-norapomorphine (1-100 mg/kg i.p.) reduced muricide; these effects were reversible with haloperidol.
Bandler 1970, 1971a,b	Ranacide and muricide in male rats	NE (3-10 µg) into the VTA decreased attack latencies in approximately 17% of rats.
Goldstein and Siegel 1980	Muricide in non-predatory female cats	Electrical stimulation of VTA or n. accumbens suppressed attack elicited by hypothalamic stimulation without altering the autonomic response.
Schmidt 1979, 1983	Predatory aggression in ferrets	Apomorphine (1 mg/kg i.m.) increased the latency to attack. Bromocriptine (8 mg/kg i.m.) decreased the latency to attack and to kill prey.

Catecholamine Antagonists

Killing

Hong et al 1987	Muricide in olfactory bulbectomized male rats	1 mg/kg of α_2 receptor antagonists (e.g. yohimbine) but not α_1 antagonist (corynanthine) reduced muricide.

366

Catecholamine Lesions

Isolation-induced Aggression

Yen et al. 1959	Male albino mice, individually housed for 3 weeks; 5 min. observation	Reserpine (3 mg/kg p.o.) decreased the percentage of mice fighting at doses that decreased activity and jumping.

Drug-Induced Aggression

Senault 1968, 1970, 1971, 1972, 1973, 1974	Apomorphine-induced aggression in pairs of male rats	Reserpine (5-10 mg/kg i.p.) sensitized non-aggressive rats to the effects of apomorphine.

Pain-induced Aggression and Defense

Chen et al. 1963; Kostowski 1966; Tedeschi et al. 1969	Electric foot shock in male mice	Peripherally administered reserpine suppressed fighting and motor activity.
Brunaud and Siou 1959; Eichelman et al. 1972; Thoa et al. 1972a; Eichelman and Thoa 1973; Sorensen and Ellison 1973; Geyer and Segal 1974; Pucilowski and Valzelli 1986	Electric foot shock in male rats	6-OHDA (i.c.v. or directly infused into the n. accumbens) increased shock induced fighting without altering jump threshold, motor behavior or spontaneous fighting. Reserpine (.5-5.0 mg/kg, route of administration not specified) had no effect on aggression.

Drug-induced Aggression

Yen et al. 1970	DL-DOPA-induced target biting in male mice	Reserpine (ED50 1.8 mg/kg i.p.) reversed biting
McKenzie 1971; Pucilowski et al. 1986, 1987	Apomorphine-induced aggression in male rats	Reserpine (10 mg/kg i.p.) pretreatment increased sensitivity to apomorphine. Locus coeruleus lesions reduced NE levels, increased striatal 5-HT, and enhanced aggression. Amygdala lesions also reduced NE levels and enhanced aggression; this was reversible with NE infusion into the amygdala.

367

Defensive Aggression induced by brain stimulation

Reference	Model	Findings
Nakamura and Thoenen 1972; Dubinsky et al. 1973; Johansson et al. 1974; Beleslin et al. 1986	"Sham rage" following electrical hypothalamic stimulation in male and female cats.	Low doses of 6-OHDA (2 mg/kg i.c.v.) increased the threshold to induce attack. Higher doses of 6-OHDA (300 µg i.c.v 2x) increased irritability and defensive biting. The degree of irritability inversely correlated with whole brain NE. Yohimbine, phenoxybenzamine, propranolol, chlorpromazine and haloperidol did not reverse the behavioral effects of 6-OHDA.

Aggression by Resident toward Intruder

Walaszek and Abood 1956	Male Siamese fighting fish confronting a male conspecific	Reserpine (10 mg/ml) decreased fighting and slightly decreased locomotor activity.

Dominance-related Aggression

Ellison 1976	Male rats living in colonies with established dominance hierarchies	Individual animals receiving 6-OHDA (25 µg/day i.c.v. for 3 days) became inactive, explored less and decreased in dominance status over a 25 day period.
Redmond et al. 1973	Free-ranging colonies of macaques	6-OHDA (2-30 mg/kg i.c.v. daily/4 days) decreased threat, attack, and other social behaviors; 33% treated individuals failed to return to their colony

Killing

Kostowski 1966	Predatory aggression in ants towards a beetle	Reserpine (.5 mg/mg p.o.) effects on aggression were time-dependent: fewer ants attacked a beetle 2-3 hrs following drug administration, while a greater percentage attacked 18-24 hours after receiving reserpine.
Karli 1959; Rolinski 1975; Pucilowski and Valzelli 1986; Molina et al. 1987	Muricide in male rats	Reserpine (10 mg/kg) or 6-OHDA (2 x 250 mg i.c.v.) increased muricide. 6-OHDA directly administered into the n. accumbens did not increase muricide.
Banerjee 1974	Muricide in individually housed male rats	6-OHDA (2x 200 µg i.c.v.) increased aggression in rats that were previously indifferent to mice.

368

| Liou et al. 1985 | Olfactory bulbectomy-induced muricide in male rats | Electroconvulsive shock (ECS) suppressed muricide. 6-OHDA or locus coeruleus lesions attenuated these effects but substantia nigra lesions did not. |
| Jimerson and Reis 1973 | Ranacide in male rats demonstrating stable attack behavior | Bilateral 6-OHDA (32 µg) administration into the LH decreased the percentage or rats killing frog. Also decreased feeding, drinking and motor activity. L-DOPA (10 mg/kg i.p.) did not restore aggression. |

E. Neuropharmacological Manipulations of Catecholamines

Human Studies

Inpatient Studies

| Goodwin et al. 1970 | L-DOPA-induced aggression in hospitalized depressed patients. Clinical state was assessed by nursing staff. Increasing doses of L-DOPA were administered blindly. | Large doses of L-DOPA (over 4 g daily p.o.) failed to alleviate depression and increased verbal and facial expressions of anger in 7 out of 11 patients; physical aggression was not noted. |

369

Table 3: Serotonin

REFERENCES	METHODS AND PROCEDURES	RESULTS AND CONCLUSIONS
A. 5-HT Correlates of Animal Aggression		
Whole brain measurements		
Isolation-induced Aggression		
Garattini et al. 1967; Valzelli and Garattini 1968	Comparison of neurochemistry in isolated and group-housed male mice.	Isolated mice had lower whole brain levels of 5HIAA (µg/g) than group-housed mice; 5-HT levels did not differ. Isolation-induced decreases in 5-HT turnover rate did not correlate with the onset of isolation-induced aggression
Welch and Welch 1968a	Pairs of male mice fought for 60 min and were sacrificed.	Fighting increased whole brain levels of 5-HT (µg/g). PCPA (360 mg/kg i.p.) pretreatment blocked this effect.
Karczmar et al. 1973; Goldberg et al. 1973	Comparison of whole brain 5-HT and 5HIAA; turnover was estimated using pargyline in various strains of male mice.	Neural steady state levels or turnover rates are not correlated with aggression levels across strains.
Lasley and Thurmond 1985	Compared NE, DA, 5HT and major metabolites in isolated and group-housed mice confronting a male conspecific in a neutral arena; not clear when amines were measured in relation to the agonisitic experience.	After 14 days of isolation, isolated and group-housed mice showed comparable levels of aggression but the isolated group had elevated 5HT turnover (5HIAA:5-HT).
Pain-induced Aggression and Defense		
Anand et al. 1985	Electric foot shock in female rats with substantia nigra, septal, or amygdala lesions. Timing of neuro-chemical measures in relation to behavior not specified	Aggression frequency was increased by substantia nigra or septum lesions and decreased by amygdala lesions. Whole brain measures of 5-HT (mg/g) were decreased in all 3 groups.
Aggression by Resident toward Intruder		
Walletschek and Raab 1982	Individually-housed male tree-shrews encountered male conspecifics	Firing rate of 5-HT-containing neurons in the dorsal raphe decreases during offensive encounters and increases during defensive fighting compared to resting animals.

Dominance-related Aggression

McIntyre et al. 1979	Comparison of whole brain 5-HT in dominant and submissive rainbow trout.	Submissive fish that are rarely attacked have 5-HT comparable to dominant fish.

Regional brain measurements

Isolation-induced Aggression

Modigh 1973	Male mice placed together in a neutral environment for 30 min. Animals were sacrificed immediately after the aggressive interaction. 5-HT and its precursors and metabolites were assayed in the striatum, cerebral hemispheres and the rest of the brain. NSD 1015 (aromatic amino acid decarboxylase inhibitor) was used to estimate turnover	In animals administered NSD 1015, fight experience resulted in increased tryptophan and 5-HTP accumulation in each brain region, compared to isolates with no fight experience.
Payne et al. 1984, 1985	Isolated and group-housed male hamsters sacrificed after the behavioral interaction. 5-HT and 5HIAA were assayed in the midbrain and hypothalamus.	More aggressive isolates do not differ from less aggressive group-housed hamsters in 5-HT and 5HIAA (ng/100 mg) in the hypothalamus and midbrain. Attack experience did not alter the percentage change in 5-HT following pargyline, but confronting an intruder did prevent a pargyline-induced decrease in hypothalamic 5HIAA when measured 20 min post-injection; attack levels did not correlate with changes in 5HIAA. Residents defeated by intruders had amine changes similar to attackers.
Hadfield and Milio 1988	Pairs of male mice tested in a neutral arena. Animals were sacrificed immediately after behavioral interactions. 5-HT and its metabolites were assayed in 10 brain regions	5HIAA:5-HT levels were not significantly altered as a function of fighting experience

Pain-induced Aggression and Defense

Lee et al. 1987	Electric foot shock in male rats. Animals were sacrificed immediately after aggressive interaction. 5-HT and 5HIAA were measured in the striatum, hippocampus and medial and dorsal raphe nucleus.	Aggressive experience was associated with decreased 5-HT (no units) in the dorsal raphe and striatum and decreased 5HIAA in the hippocampus.

Aggression by Resident toward Intruder

Reference	Methods	Results
Haney et al. 1990	Pair-housed male mice. 5-HT measurements in the nucleus accumbens, corpus striatum and amygdala were obtained following 0, 1 or 10 daily attack experiences	5HIAA:5-HT (ng/mg protein) in the amygdala was increased following 1 attack experience. Amine measurements in mice with repeated attack experience did not differ from behaviorally naive controls

Dominance-related Aggression

Reference	Methods	Results
Garris et al. 1984	Olfactory-bulbectomized group-housed male mice. Aggression determined by number of head and body lesions	Aggression and 5-HT histofluorescence in the olfactory bulb, lateral olfactory tract and pyriform cortex progressively increased over time since olfactory bulbectomy

Killing

Reference	Methods	Results
Goldberg and Salama 1969	Muricide by male rats. Forebrain and hindbrain 5-HT was measured 24 hrs following muricide.	5-HT levels did not differ between muricidal and non-muricidal rats.
Broderick et al. 1985	Muricide by male rats. Animals were sacrificed 1-2 hrs following the introduction of the mouse into the home cage. 5-HT and its metabolites were assayed in the hypothalamus, thalamus, hippocampus, striatum, cortex and brain stem	Muricidal rats had increased 5-HT (ng/g) in the amygdala and increased 5HIAA in the anterior hippocampus compared to non-muricidal rats.
Tani et al. 1987	Muricide by male rats with olfactory bulbectomy, n. raphe or n. accumbens lesioning. Animals were sacrificed immediately after muricide. 5-HT and its metabolites were assayed in the frontal cortex and hypothalamic and amygdaloid nuclei	5-HT and 5HIAA (ng/mg protein) were not different in muricidal and non-muricidal rats. (5-HT and 5HIAA were decreased in hypothalamic and amygdaloid nuclei of raphe lesioned group. 5HIAA/5-HT was increased in the LH of raphe lesioned, decreased in the central amygdaloid nucleus of n. accumbens group and decreased in LH and mammillary body of OB.
Nikulina and Popova 1988	Ranacide in minks. 5-HT and 5HIAA were measured in the amygdala and lateral and medial hypothalami	Continuous access to food was associated with increased 5HIAA in the lateral hypothalamus and amygdala and increased latency for ranacide (15 sec to 1.5 min). Food-deprived minks and those given a single meal did not differ in 5HIAA levels or ranacide latency.

372

Peripheral Correlates of Animal Aggression

Dominance related Aggression

Raleigh et al. 1983 — Group-housed male vervet monkeys — Dominant males have higher concentrations of whole blood 5-HT (ng/ml) and CSF 5HIAA than nondominant males

Elsworth et al. 1985 — Group-housed male vervet monkeys. Plasma phenylacetic acid was measured in dominant or submissive monkeys — Dominant males had higher concentrations of free and conjugated plasma phenylacetic acid (ng/ml) than lower ranking males.

CSF correlates of Animal Aggression

Dominance-related Aggression

Yodyingyuad et al. 1985 — Group-housed male talapoin monkeys. CSF 5HIAA and HVA and plasma cortisol were assayed during the formation of a social hierarchy, in established groups and in relation to the daily performance of aggressive behaviors — During the establishment of social hierarchies, males that became lowest in rank had significantly more CSF 5HIAA and HVA (ng/ml) (n=3) than before they were group-housed; 5HIAA levels decreased in certain individuals that became dominant (n=3). In established hierarchies, dominant males and females had less 5HIAA and plasma cortisol than low ranking animals. Daily variations in attacks and threats received did not correlate with CSF 5HIAA in subordinates. Dominant monkeys had higher 5HIAA on days they were overtly aggressive.

Sahakian et al. 1986 — Pairs of isolated male rats tested in a neutral arena. 5-HT turnover (nmol/ml/h) was estimated with probenecid administration. — CSF tryptophan (nmol/ml) positively correlated with attack bite frequency

373

B. Neuropharmacological Manipulations of 5-HT

Animal Studies

5-HT synthesis manipulation

Isolation-induced Aggression

Randt et al. 1975	Compared two strains of adult male mice that were undernourished in utero. Aggression directed at an olfactory-bulbectomized conspecific. 5-HT turnover estimated with pargyline.	A greater percentage (75%) of previously under-nourished DBA but not C57 mice attacked an opponent compared to controls (33%). Although 5-HT turnover was lower in the undernourished group, there was not a systematic difference between aggressive and non-aggressive individuals.
Rolinski 1975	Male mice confronting a male conspecific	PCPA (ED50: 100 mg/kg i.p.) decreases fighting.
Eichelman 1981	Male mice confronting a male conspecific	Mice fed a tryptophan-deficient diet for 4 weeks fought more than controls fed a normal diet or controls fed reduced calories with tryptophan replacement.
Weinstock and Weiss 1980	Male mice confronted a group-housed male intruder in a neutral arena	5-HTP (10 mg/kg s.c.) elicited aggression in non-aggressive mice and increased attack bite frequency in mice that already were aggressive.
Lasley and Thurmond 1985	Male mice confronted a group-housed intruder in a neutral arena	Tryptophan (.50% supplement/10 days) in diet increased aggression without affecting motor activity.
Payne et al. 1984	Male hamsters confronting a group-housed intruder	PCPA (180 mg/kg i.p.) blocked the pro-aggressive effects of prolonged isolation, while 5-HTP (40 mg/kg i.p.) enhanced it

374

Drug-induced Aggression

Rolinski and Herbut 1979	Apomorphine-induced aggression in similarly treated pairs of male rats.	Tryptophan (200 mg/kg i.p.) and 5-hydroxytryptophan (100 mg/kg i.p.) decreased fighting frequency. PCPA administered for 3 days (200, 200, 100 mg/kg i.p.) prior to apomorphine also suppressed aggression.
Carlini and Lindsey 1982	THC-induced aggression in similarly treated pairs of REM-deprived male rats	Tryptophan (200 mg/kg i.p.) potentiated aggression. PCPA (300 mg/kg i.p.) pretreatment inhibited aggression.
Fujiwara and Ueki 1974	THC-induced muricide in group-housed male rats.	PCPA (300 mg/kg i.p.) pretreatment induced muricide in 70% of non-aggressive, group-housed rats. They also exhibited hyperirritability, hypersexuality and catalepsy.

Pain-induced Aggression and Defense

Rolinski and Herbut 1981	Electric foot shock-induced fighting in male mice	l-Tryptophan (200-400 mg/kg) did not alter fighting; 5-HTP (100-200 mg/kg) reduced fighting with the lowest dose being the most effective. PCPA (200 mg/kg/day) also reduced fighting.
Eichelman 1981	Electric foot shock-induced fighting in male rats	Rats fed a tryptophan deficient diet were more aggressive than controls fed a normal diet or controls fed a low-caloric diet with tryptophan supplements. Pain sensitivity and whole brain levels of 5-HT were also reduced.
Ellison and Bresler 1974 Conner et al. 1973	Electric foot shock-induced fighting in male rats	PCPA (100 mg/kg s.c./2 days for 11 days) increased fighting and decreased locomotion, rearing and grooming. Higher doses (320-920 mg/kg s.c./6 days) had no effect on fighting behavior.

Brain lesion-induced aggression

Dominguez and Longo 1969, 1970	Septal lesion-induced aggression in male rats	PCPA (300 mg/kg) tamed hyperirritability in septal lesioned rats; When administered before septal lesions, PCPA (300-600 mg/kg) and AMPT (25-200 mg/kg) did not prevent hyperirritability, and higher doses of AMPT produced sedation.

Defensive Aggression induced by brain stimulation

Reference	Paradigm	Findings
MacDonnell et al. 1971; Dubinsky et al. 1973; Katz and Thomas 1976	"Sham rage" following electrical hypothalamic stimulation in male and female cats	PCPA (150 mg/kg i.p. daily for 3 days) or 5-HTP (5 mg/kg i.p.) had no effect on aggression. Higher doses of PCPA (250-300 mg/kg i.p. daily for 2-3 days) potentiated growling and biting inanimate objects or experimenter.

Aggression by Resident toward Intruder

Reference	Paradigm	Findings
Thurmond et al. 1979, 1980	Male group-housed mice isolated over night confronted a naive intruder. Diets were supplemented with amino acid precursors.	l-Tryptophan (0.25, 0.5% for 2 weeks) supplements increased aggression. Longer administration (2-5 weeks) of l-tryptophan (4%) decreased aggression. Whole brain levels of 5-HT and 5HIAA were increased at all concentrations of l-tryptophan, while NE and DA were decreased.

Female Aggression

Reference	Paradigm	Findings
Svare and Mann 1983	Lactating mice confronting a conspecific	DL PCPA (400 mg/kg/day for 6 days; route of administration not stated) decreased the proportion of mice attacking an intruder.
Ieni and Thurmond 1985	Maternal aggression in mice confronting a male opponent	PCPA (200-400 mg/kg, i.p.) and 5-HTP (100 mg/kg, i.p.) increased the latency to attack and reduced the number of attacks.

Dominance-related Aggression

Reference	Paradigm	Findings
Sheard 1970a	Male rats placed into a chamber with a male and female rat and a mouse.	PCPA (320 mg/kg i.p.) increased sexual and aggressive behavior, while decreasing whole brain 5-HT and 5HIAA. Chronic lithium pretreatment (5 meq/kg for 5 days) blocked PCPA effects on behavior without blocking serotonergic depletion.
Rolinski 1975	Male and female rats confronting a conspecific in a neutral environment	PCPA (400 mg/kg i.p.) increased aggressive behavior in both males and females.

Reference	Condition	Results
Raleigh et al. 1980	Group-housed monkeys	PCPA (80 mg/kg/day) administered for 14 days increased aggression in vervet monkeys; concurrent administration of 5-HTP (40 mg/kg/day) further increased aggression while tryptophan (20 mg/kg/day) had no effect. Daily PCPA administration (no dose specified) did not affect social behavior in macaques.
Chamberlain et al. 1987	Group-housed male and female vervet monkeys. Aggression occurring spontaneously and during food competition was assessed.	Males fed a tryptophan-free diet were more spontaneously aggressive than controls. During food competition, males fed a tryptophan-free diet were more aggressive, while males and females fed a diet with excess tryptophan were less aggressive than controls.

Killing

Reference	Condition	Results
McCarty et al. 1976	Cricket-killing in male and female grasshopper mice	PCPA (50 mg/kg/day for 5 days i.p.) decreased duration of predatory attack and increased attack latency
DiChiara et al. 1971; Eichelman and Thoa 1973; Conner et al. 1973 Miczek et al. 1975 Rolinski 1975; Gibbons et al. 1978; Berzsenyi et al. 1983; Pucilowski and Valzelli 1986; Isel and Mandel 1989; Molina et al. 1987	Muricide in male and female rats	PCPA decreased attack latency and increased the percentage of male and female rats that killed mice; whole brain 5-HT and 5HIAA were concomitantly decreased.
Kulkarni 1970; DiChiara et al. 1971; Bocknik and Kulkarni 1974; Gibbons et al. 1978	Isolated male rats selected for muricidal behavior	5-HTP (30, 100, 200 mg/kg i.p.) decreased the percentage of animals that killed mice. High doses (150, 200 mg/kg i.p.) of the decarboxylase inhibitor, Ro4-4602, blocked 5-HTP-induced inhibition whereas low doses (5, 10 mg/kg i.p.) enhanced it.

377

Gibbons et al. 1979	Muricide in isolated male rats.	A tryptophan-free diet (4-6 days) induced muricide in 60% of non-muricidal rats and facilitated killing in muricidal rats. Whole brain 5-HT and 5HIAA were concomitantly decreased.
Gibbons et al. 1981 Broderick and Lynch 1982	Muricide in isolated male rats. 5-HT turnover was assessed with tranylcypromine.	Acute (100-800 mg/kg i.p.) and long-term *l*-tryptophan (100 mg/kg i.p. for 7 days) decreased muricide without affecting food intake or motor function. Forebrain and hindbrain 5-HT turnover was concomitantly increased.
Copenhaver et al. 1989	Filicide in nulliparous female Sprague-Dawley rats	PCPA (400 mg/kg s.c.) induced filicide.
Nikulina and Popova 1988	Predatory aggression in the mink	5-HTP (50, 100 mg/kg i.p.) increased levels of 5-HT in the hypothalamus and midbrain and suppressed predatory aggression; 100 mg/kg also suppressed locomotor activity.

5-HT releasers

Experimenter-provoked Aggression

| Raleigh et al. 1986 | Individually-housed male vervet monkeys. Aggression was elicited by an experimentor staring at the subject. | Chronic fenfluramine (1-4 mg i.m. daily for 10 weeks) decreased plasma 5-HT and CSF 5HIAA and increased aggressive threats. |

Pain-induced Aggression and Defense

| Rolinski and Herbut 1981 | Electric foot shock-induced fighing in male mice | Fenfluramine (5-10 mg/kg) dose dependently reduced aggressive behavior |
| Sheard 1976 | Electric foot shock in rats (sex unspecified) | Acutely, parachloroamphetamine (PCA) (2.5-10 mg/kg) suppresses fighting while over time fighting frequency and intensity are increased. Neurochemically, acute PCA releases 5-HT while over time it inhibits TH and is neurotoxic. |

Dominance-related Aggression

| Raleigh et al. 1986 | Individually-housed male vervet monkeys. Aggression directed toward an inaccessible conspecific was measured. | Chronic fenfluramine (1-4 mg i.m. daily for 10 weeks) decreased plasma 5-HT and CSF 5HIAA and increased aggressive threats |

5-HT receptor agonists

Experimenter-elicited Aggression

Maler and Ellis 1987	High frequency electric organ discharges in South American electric fish elicited by simulated electric signals.	5-HT (.1 µg i.c.v.) decreased aggressive signalling

Isolation-induced Aggression

Lindgren and Kantak 1987; Olivier et al. 1989	Male mice confronting a group-housed intruder	5-HT1A agonists, 8-OH-DPAT (0.05-6.25, s.c.), 5-Me-ODMT (0.3-10 mg/kg , i.p.), ipsapirone (0.3-10 mg/kg i.p.), buspirone (0.3-10 mg/kg) and 5-methoxytryptamine (2.5-20 mg/kg, i.p.) reduced a composite measure of aggression; at higher doses, 8-OH-DPAT and 5-Me-ODMT enhanced avoidance-defensive behaviors.

Pain-induced Aggression and Defense

Rolinski and Herbut 1981	Electric foot shock-induced fighting in male mice	5-methoxytryptamine (2 mg/kg i.p.) increased the ferocity and number of attacks and decreased spontaneous motor activity; quipazine (10 mg/kg i.p.) did not alter aggression.
Rodgers 1977; Ray et al. 1983	Electric foot shock in male rats	Intraventricular 5-HT (25 µg/µl) increased fighting without affecting pain sensitivity. 5-HT (10 µg/µl) bilaterally infused into the corticomedial but not basolateral amygdala decreased attack frequency by 40%; sensitivity to footshock was concomitantly decreased.

Drug-Induced Aggression

Golebiewski and Romaniuk 1985	Intrahypothalamic carbachol-induced "sham rage" in male and female cats	5-HT (5 µg) bilaterally infused into the anterior hypothalamus decreased the frequency and duration of carbachol-induced growling.
Hahn et al. 1982	Apomorphine-induced aggression in male rats chronically treated with clonidine (5 mg/ml in drinking water, 7 days).	5-HT1B agonist mCPP (0.3-10 mg/kg, i.p.) dose dependently reduced aggression induced by apomorphine and clonidine at doses that did not alter behavior in control rats.

379

Aggression by Resident Toward Intruder

Olivier et al., 1984	Male rats	Fluprazine (5-20 mg/kg) dose dependently reduced aggressive threats and attacks.
Lindgren and Kantak, 1987	Male mice confronting a group-housed intruder.	5-HT3 agonist, quipazine (5-25 mg/kg, i.p.), reduced aggressive threats and attacks.
Haug et al., 1990	Group-housed female mice confronting a lactating female resident	8-OH-DPAT (200, 250 mg i.p.) decreased attack frequency and increased attack latency without affecting locomotor activity.

Dominance-related Aggression

| Kennett et al., 1989 | Social interaction test in pairs of similarly treated rats | 5-HT1B agonist, mCPP (0.5-1 mg/kg, i.p.) reduced boxing, biting as well as non-aggressive social behaviors whereas TFMPP (0.2-1.0 mg/kg, i.p.) did not; the antagonist cyanopindolol (6 mg/kg, s.c.) did not reverse the effects of mCPP. |

Killing

| Rolinski 1975; Applegate 1980; Berzsenyi et al. 1983; Pucilowski et al. 1985; Molina et al. 1986; Pucilowski and Valzelli 1986; Strickland and DaVanzo 1986 | Muricide in rats induced by isolation or olfactory bulbectomy | 5-HT (10 µg i.c.v) increased the latency to kill mice. 5-HT agonists, 8-OH-DPAT and 5-Me-ODM reduced the percentage of muricidal animals. Bilateral microinjection of 5-HT3 agonist, quipazine (20 µg/µl) into the cortico-medial amygdala reduced the latency to kill. |
| Applegate 1980 | Muricide in rats induced by intraventricular 5,7-DHT | 5-HT (.5, 10 µg i.c.v.) increased the latency to kill mice. |

5-HT Antagonists

Isolation-induced Aggression

| Malick and Barnett 1976 | Isolated male mice confronted another isolate within the home cage | Methiothepin (0.04 mg/kg, ED50), mianserin (0.5 mg/kg, ED50), methysergide (1.0 mg/kg, ED50), cyproheptadine (1.1 mg/kg, ED50), pizotyline (1.5 mg/kg, ED50), xylamidine (2.5 mg/kg, ED50), and cinanserine (7.3 mg/kg, ED50) prevented fighting without altering motor activity or performance in the inlined screen test. |

Weinstock and Weiss 1980	Male mice confronted a group-housed intruder within a neutral arena	Methysergide (ED50=4.1 mg/kg s.c.) decreased aggression without decreasing locomotor activity.

Pain-induced Aggression and Defense

Rodgers 1977	Electric foot shock-induced aggression in rats (sex unspecified)	Methysergide (5 µg/µl) bilaterally infused into the corticomedial but not basolateral amygdala increased attack frequency by 46%. Sensitivity to footshock was concomitantly increased.

Drug-Induced Aggression

Rolinski and Herbut 1979	Apomorphine-induced aggression in similarly treated pairs of male rats.	Cyproheptadine (1, 5 mg/kg i.p.) potentiated aggression. Rats receiving subthreshold doses of apomorphine became aggressive following either cyproheptadine or metergoline (1.5 mg/kg i.p.).
Hahn et al. 1982	Apomorphine-induced aggression in male rats chronically treated with clonidine (5 mg/ml in drinking water x 7 days)	5-HT1 antagonist, metergoline (3 mg/kg, i.p.) enhanced aggression induced by apomorphine and clonidine at doses that did not alter behavior in control rats
Golebiewski and Romaniuk 1985	Intrahypothalamic carbachol-induced "sham rage" in male and female cats	Methysergide (10 µg) bilaterally infused into the anterior hypothalamus increased carbachol-induced growling.

Aggression by Resident toward Intruder

Winslow and Miczek 1983; Lindgren and Kantak 1987; Haney and Miczek 1989	Male mice confronting group-housed intruder	Methysergide (3.0, 10.0 mg/kg i.p.), a non-specific 5-HT antagonist, decreased attacks, aggressive threats and locomotion. The 5-HT1 antagonist mianserin (0.5-5 mg/kg, i.p.) and the 5-HT2 antagonist ketanserin (1-10 mg/kg, i.p.) decreased aggression at doses which did not decrease locomotion.

Female Aggression

Ieni and Thurmond 1985	Maternal aggression in mice	5-HT1 antagonists, mianserin (2-4 mg/kg, i.p.), methysergide (4 mg/kg, i.p.) and methiopin (0.25-0.5 mg/kg) decreased attack behavior compared to controls.

Dominance-related Aggression

Kennett et al., 1989; File and Johnston 1989	Social interaction test in pairs of similarly treated rats under low light and familiar conditions	5-HT1 antagonists, mianserin (2 mg/kg, s.c.), cyproheptadine (2 mg/kg, s.c.), and metergoline (2.5 mg/kg, s.c.) prevented reduction in interaction time elicited by 5-HT1B agonists, mCPP and TFMPP (0.1-1 mg/kg, i.p.). 5-HT3 antagonists, ICS 205 930 (0.05-1 mg/kg, s.c.), GR 38032F (0.1-1 mg/kg, p.o. and zacopride (0.01-1 mg/kg, i.p.) did not alter social interactions nor did they prevent the reduction in interaction time elicited by 5-HT1B agonists, mCPP and TFMPP (0.1-1 mg/kg, i.p.).
Kennett et al., 1989	Social interaction test in pairs of similarly treated rats	5-HT1 antagonists, ritanserin (0.6 mg/kg, s.c.) and cyanopindolol (6 mg/kg, s.c.), and 5-HT2 antagonist, ketanserin, (0.2 mg/kg, s.c.) did not alter social interactions nor did they prevent the reduction in interaction time elicited by 5-HT1B agonists, mCPP and TFMPP (0.1-1 mg/kg, i.p.).

Killing

Pucilowski et al. 1985; Strickland and DaVanzo 1986	Isolation-induced muricide in male rats selected for their aggressivity	5-HT1 antagonist, mianserin (ED50=10.47) decreased muricide.

5-HT lesioning

Isolation-induced Aggression

Poschlová et al. 1975; Poschlová et al. 1977; Matte 1982	Male mice confronting a non-aggressive opponent	Administration of 5,6-DHT (1-50 µg i.c.v.) or 5,7-DHT (25 µg i.c.v) increased alert postures and reduced social behaviors, including aggressive threats and attacks.

Pain-induced Aggression and Defense

Pucilowski and Valzelli 1986	Electric foot shock in male rats	Electrolytic lesions of the dorsal but not medial raphe nucleus increased the number of fighting bouts. 5,7-DHT infused into the n. accumbens did not affect aggressive behavior.

382

Drug-induced Aggression

Palemo Neto et al. 1975	Chronic THC-induced aggression in pair-housed, ovariectomized female rats receiving hormonal replacement. Aggression toward untreated cage-mate assessed. 5-HT turnover measured with pargyline.	Chronic cannabis increased aggression while decreasing whole brain levels of 5-HT. Not clear if decreased 5-HT preceded fighting, or was a consequence of fighting.
Romaniuk et al. 1987	Carbachol-induced growling in male and female cats	Raphe lesioning increased growling, while decreasing concentrations of NE, 5-HT and 5HIAA and increasing DA.

Aggression by Resident toward Intruder

Winslow and Miczek 1983	Male, pair-housed mice encountering a male conspecific	24 post-administration, PCA (50 mg/kg i.p.) enhanced attack, aggressive threat and locomotion.

Dominance-related Aggression

Ellison 1976	Male rats living in colonies with established dominance hierarchies	Individual animals receiving 5,6-DHT (10 µg/day i.c.v. for 3 days) showed increased motor, sexual and aggressive behavior and increased in dominance status over a 25 day period

Killing

Banerjee 1974; Breese and Cooper 1975; Vergnes et al. 1973, 1988; Applegate 1980; Marks et al. 1977; Pucilowski and Valzelli 1986; Isel and Mandel 1989; Yamamoto et al. 1988	Muricide in male rats	Intraventricular 5,7-DHT or specific lesioning of the raphe nucleus or the intrahypothalamic ascending 5-HT pathways increased the percentage of rats that killed mice, especially when NE depletion was blocked by the concurrent administration of MAO-inhibitors. The combined administration of a 5-HT antagonist and an NE precursor inhibited muricide. Muricide was not induced when 5,7-DHT was limited to the n. accumbens or if rats were handled prior to the administration of 5,7-DHT.
Liou et al. 1985	Olfactory bulbectomy-induced muricide in male rats	The suppressive effects of ECS on muricide were not altered with raphe nucleus lesions

Human Studies

Criminal Violence

Bioulac et al. 1980

XYY patients arrested for crimes ranging from vagrancy to assault (n=6) were administered L-5-hydroxytryptophan (1.5-1.85 g/day) for 5 months.

Tryptophan administration increased CSF 5HIAA and resulted in a clinically observed reduction in aggression in 4 of 6 patients. No statistics were reported.

Inpatient Studies

Coccaro et al. 1989

Fenfluramine (60 mg p.o.) was administered to "normal" controls and to patients with either a personality or a major affective disorder. Plasma PRL following fenfluramine administration was used as an index of central 5-HT activity.

PRL response was significantly reduced in both sets of patients in comparison to controls. Scores on aggression scales negatively correlated with plasma PRL levels in patients with personality disorder but not major affective disorder. Past history of suicide attempts correlated with reduced PRL response in both groups of patients.

Volavka et al. 1990

Tryptophan (up to 6 g/day p.o. for 25-35 days) or placebo treatment in violent male and female psychiatric patients. Antipsychotics or sedatives were concurrently administered to control violent behavior

Tryptophan had no effect on the number of violent incidents (agitation, verbal assault, assault on others or self, and property assault) compared to the placebo group. However the tryptophan group apparently required fewer drug treatments to control violence than the placebo group.

Outpatient Studies

Fishbein et al. 1989

Fenfluramine (60 mg p.o.) was administered to male polydrug users seeking treatment for addiction. Subjects were assigned to a high and low aggressive group, based on psychometric testing. Plasma PRL and cortisol following fenfluramine administration   used as an indice of central 5-HT activity.

When adjusted for differences in baseline, PRL and cortisol in the plasma were elevated in subjects scoring high on tests of aggression and impulsivity. The neuroendocrine response to fenfluramine was better correlated with impulsivity than aggression.

C. 5-HT Correlates of Human Aggression and Violence

Peripheral correlates of aggression in humans

Inpatient studies

Subrahmanyam 1975

Schizophrenic, manic-depressive psychotics and healthy controls were compared: 6 out of 60 schizophrenics were in an "acute aggressive state". Urinary 5HIAA was measured.

Urinary 5HIAA appeared elevated in aggressive schizophrenics in comparison to both controls and non-aggressive schizophrenics. No statistics reported.

Greenberg and Coleman 1976

Plasma 5-hydroxyindole (5HI) was measured in 24 hyperactive, institutionalized mentally retarded men and women before and after drug treatment. Individual patients were compared to age- and sex-matched controls. Clinical and staff reports were used as dependent variables.

Baseline 5HI levels were depressed in 83% of patients compared to controls. In 63% of these patients, decreases in hyperactivity and aggression correlated with increases in 5HI.

Outpatient Studies

Kent et al. 1988;
Brown et al. 1989

5-HT uptake in blood platelets was measured in male outpatients seeking treatment for frequent bouts of verbal and physical aggression and age- and sex-matched healthy controls ($n=15$)

In 11/15 cases, platelet 5-HT uptake (pmol/2 x 107 platelets) was slightly lower in aggressive patients compared to controls. 5-HT uptake negatively correlated with scores on impulsivity but not anger scales.

Leckman et al. 1990

CSF 5HIAA was measured in an adult male and female obsessively preoccupied with violent thoughts. Imagery but not behavior was overtly aggressive.

CSF 5HIAA was significantly reduced in patients with violent thoughts compared to normal and psychiatrically disturbed populations.

Criminal Violence

Raisanen et al. 1984

Urinary bufotenine (5-HT metabolite) was measured in male offenders arrested for murder or attempted murder ($n=48$) and healthy controls.

Urinary bufotenine (nmol/g creatinine) was higher in violent patients compared to controls.

385

Study	Description	Findings
Virkkunen and Narvanen 1987	Plasma 5-HT and tryptophan were measured before and after a glucose tolerance test in healthy controls and violent incarcerated male offenders, diagnosed with either intermittent explosive disorder ($n=6$) or antisocial personality.	Convicts with intermittent explosive disorder had higher plasma tryptophan levels (mmol/l) than the other groups. Following glucose administration, convicts with intermittent explosive disorder had higher tryptophan and insulin (30 min post injection) than normals. 5-HT levels did not differ between groups.

CSF correlates of aggression in humans

Inpatient studies

Study	Description	Findings
Subrahmanyam 1975	Schizophrenic, manic-depressive psychotics and healthy controls were compared: 6 out of 60 schizophrenics were in an "acute aggressive state". CSF 5HIAA was measured.	CSF 5HIAA (ng/ml) in acute, aggressive schizophrenics was comparable to controls but higher than non-aggressive acute schizophrenics. No statistics reported.
Brown et al. 1979	CSF 5-HT and 5HIAA were measured in military men diagnosed with borderline personality disorder without affective illness ($n=26$).	CSF 5HIAA (ng/ml) did not differ in patients with borderline personality disorder and normal controls. If just the patients are analyzed, there is a negative correlation between scores based on the life history of aggression and CSF 5HIAA.
Brown et al. 1982	CSF 5-HT and 5HIAA were measured in military men diagnosed with borderline personality disorder without affective illness ($n=12$).	CSF 5HIAA (ng/ml) negatively correlated with psychopathic deviate scores on the MMPI (-.77) and with scores of aggression based on life history (-.53); CSF 5HIAA did not covary with the Buss-Durkee Inventory for aggression.

Criminal Violence

Study	Description	Findings
Bioulac et al. 1980	CSF 5-HT was measured in healthy controls ($n=5$) and XYY patients arrested for crimes ranging from vagrancy to assault ($n=6$). Probenicid was used to estimate turnover.	Following probenicid administration, XYY patients had less 5HIAA (ng/ml) than controls; baseline 5HIAA and HVA levels did not appear to differ between groups.
Linnoila et al. 1983	CSF 5HIAA was assayed in men convicted of violent crime. Subjects were subdivided into categories specifying the pre-mediated or impulsive nature of the criminal act.	Impulsive violent offenders had lower CSF 5HIAA (pmol/ml) than those in the premeditated group. When subjects were re-categorized based on the number of crimes committed, CSF 5HIAA was lower in the group that committed more than 1 violent act.

386

Reference		
Lidberg et al. 1985	CSF 5HIAA, HVA and MHPG were assayed in homicidal convicts, suicidal patients and healthy controls.	Suicidal patients had lower concentrations of 5HIAA (nmol/l) than controls. Overall, metabolite concentrations in homicidal convicts did not differ from controls. Post-hoc categorization of convicts indicate men who killed a sexual partner had lower levels of 5HIAA than controls.
Linnoila et al. 1989; Virkkunen et al. 1989a,b	Measured CSF 5HIAA, HVA and MHPG in male alcohol abusers, imprisoned for manslaughter or arson	CSF 5HIAA (nmol/l) was lower in imprisoned males compared to healthy controls. Prisoners having a family history of alcoholism had lower CSF 5HIAA than those that did not. Subjects with a history of suicidal attempts had lower CSF 5HIAA and MHPG than those that did not.

Behavioral Disorders in Juveniles

Kruesi et al. 1990	Measured CSF 5HIAA, HVA and MHPG in children and adolescents diagnosed with either disruptive behavioral disorder or obsessive compulsive disorder	Patients with disruptive behavior disorders had lower CSF 5HIAA (pmol/ml) and higher aggression than age-, sex-, and race-matched patients with obsessive-compulsive disorder. CSF 5HIAA negatively correlated with scores on 2/9 measures of aggression in subjects with disruptive behavior disorders.

TABLE 4: GABA

References	Methods and Procedures	Results and Conclusions
A. GABA Correlates of Animal Aggression		
Whole Brain Measurements		
Aggression by Resident toward Intruder		
Earley and Leonard 1977	Testosterone-treated (1 mg/kg) or cyproterone acetate-treated (1 mg/kg) group housed and untreated isolated male mice	Housing conditions alter brain concentrations of GABA as well as aggression. Aggressive response was inversely related to GABA content in olfactory bulb and striatum. Testosterone administration increased GABA concentrations in group-housed mice and reduced aggressive attacks received by untreated isolates
Female Aggression		
Haug et al. 1987	Confrontations between female C57, C3H, B6HEF1 and HEB6F1 strain resident mice vs lactating intruders	No relationship was found between aggressiveness and brain GABA concentration
Killing		
Mack et al. 1975	Predatory aggression in male rats	Brain GABA levels were reduced in the olfactory bulbs of mouse killing rats compared to non-killing rats
Regional brain measurements		
Isolation-induced aggression		
Simler et al. 1982	Aggression in pairs of DBA/2 and C57 strain male mice either group housed or single housed for 8 weeks	Non-aggressive C57 mice and aggressive DBA/2 mice showed decreases in GABA concentrations in septum, striatum, olfactory bulb and posterior colliculus following isolation. Compared to C57 mice, DBA/2 mice showed high levels of aggression, an increase in GABA concentrations in the amygdala, and reduced GABA concentrations in olfactory bulb and striatum

Female Aggression

Potegal et al. 1982	Confrontations between aggressive and non-aggressive ovariectomized female resident hamsters vs methotrimeprazine-treated intruder hamsters	Aggressive females had 15-25% higher GABA concentrations in midregions of brain (limbic, striatal, and diencephalic structures) than non-aggressive females

Aggression by Resident toward Intruder

Haug et al. 1984	Confrontations between three resident castrated or sham-operated males vs. lactating female intruders in three strains of mice	Brain GABA levels were increased in the hypothalamus, olfactory bulbs, and amygdala in castrates of the most aggressive strain (C57); C57 and C3H castrates displayed shorter attack latencies and increased attacks than sham-operated animals
Muñoz-Blanco et al. 1986	Comparison between aggressive and non-aggressive strains of female bulls. Synaptosomes from 7 brain regions were measured for amino acid content	Compared to the non-aggressive strain, Spanish fighting-bulls had higher GABA concentrations (120%) in the thalamus and slightly lower concentrations in the hypothalamus (69%), caudate nucleus (66%) and corpus striatum (82%). A higher ratio of excitatory (glutamate, aspartate) to inhibitory (GABA, glycine) neurotransmitter amino acids was found in the aggressive strain.

Killing

DaVanzo et al. 1986	Predatory aggression in isolated and group housed male rats.	Mouse-killing rats showed slight increases (32-34%) in muscimol binding in the amygdala compared to non-killing rats regardless of housing condition.

B. Neuropharmacological manipulations of GABA

<u>GABA Synthesis Manipulations</u>

Isolation-induced Aggression

Puglisi-Allegra and Mandel 1980	Isolation-induced aggression in DBA/2 strain male mice; aggression measured with automated bite detector.	Low doses of a GABA-T inhibitor, valproate (200 mg/kg) or GABA reuptake blocker, nipecotic acid (125 mg/kg), did not alter aggression on initial administration, but inhibited aggression on the second day. Higher doses (300 mg/kg valproate) reduced aggression all 3 days.

389

Poshivalov 1981	Isolation-induced aggression in male mice; aggression measured with automated bite detector	GABA synthesis inhibitor, thiosemicarbazide (1 mg/kg) increased probability of attacks; GABA-T inhibitor, gamma-acetylenic GABA (75-100 mg/kg) reduced attack, threat and ambivalent behaviors (e.g., tail rattle)
Simler et al. 1983	Isolation-induced aggression in DBA/2 and C57 strain male mice	A GABA-T inhibitor, valproate (300 mg/kg) blocked aggression in DBA/2 mice in a time-dependent manner peaking at 75 min. post injection; decreases in aggression paralleled increased concentrations of GABA in olfactory bulb, striatum, posterior colliculus and septum
Sulcova and Kršiak 1981	Isolation-induced aggression or timidity in male mice	GABA-T inhibitors, valproate (25-200 mg/kg) and aminooxyacetic acid (1-9 mg/kg) dose dependently reduced aggressive threats, attacks and tail-rattle in aggressive mice, and defensive escape behaviors in timid mice
DaVanzo and Sydow 1979	Isolation-induced aggression in male mice	GABA-T inhibitors aminooxyacetic acid (20-40 mg/kg) and gama-acetylenic GABA (100-150 mg/kg) dose-dependently prolonged latency to attack and increased whole brain GABA concentration.
Oehler et al. 1985	Isolation-induced aggression in male mice	Initial administration of a GABA-T inhibitor, valproate (200 mg/kg/day in drinking water) reduced viewer rated aggression, but was inactive after chronic administration (4 weeks).

Pain-induced Aggression

| Rodgers and Depaulis 1982 | Electrical foot shock-induced aggression in pairs of male rats | Valproate (100-200 mg/kg), a reversible GABA-T inhibitor, did not alter shock-induced fighting behavior; g-vinyl GABA (100-200 mg/kg), an irreversible GABA-T inhibitor dose dependently reduced fighting. |

Reference	Model	Results
Puglisi-Allegra et al. 1981	Electrical tail shock-induced aggression in C57 and DBA/2 male mice	GABA-T inhibitor, valproate (200-250 mg/kg) inhibited shock-induced biting whereas GAD (glutamic acid decarboxylase) inhibitor, DL-allylglycine (15-25 mg/kg) elicited shock-induced biting in 20-21 week old, highly aggressive C57 mice. Biting in non-aggressive DBA/2 mice was unchanged.

Drug-induced Aggression

Arnt and Scheel-Krüger 1980	GABA receptor antagonist-induced self directed aggression in male rats	Systemic g-acetylenic GABA (GABA-T inhibitor) blocked bicuculline-induced (30-60 ng/ml/rat) and picrotoxin-induced (50-100 ng/ml/rat) self-biting

Aggression by Resident toward intruder

Haug et al. 1980	Confrontations between three intact or gonadectomized male and female mice vs. lactating female intruders	GABA-T inhibitor, valproate (200-300 mg/kg) prevented attacks by intact resident males or females towards lactating intruders, and reduced aggression in castrated males; Valproate increased GABA concentrations in the hypothalamus (40-50%) in intact and gonadectomized male and female mice, and in olfactory bulbs (50%), and amygdala (70%) in intact males. Female and male gonadectomized mice showed increased (85 and 70% respectively) GABA content in posterior colliculus

Killing

Mack et al. 1975	Predatory aggression in male rats	Valproate (200 mg/kg) abolished mouse killing behavior in 90% of killer rats, but was ineffective in olfactory bulbectomized killer rats. Intraolfactory bulb microinjection of valproate (25 mg) blocked mouse killing behavior 30 minutes to 3 hours post injection. GABA (200 mg/kg) immediately abolished mouse killing behavior, but effects lasted less than 1 hour. Simultaneous injection of valproate with GABA blocked mouse killing immediately and persisted for 4 hours post injection.

391

Potegal et al. 1983	Predatory aggression and viewer-rated aggression directed at experimenter in isolated male rats	Microinjections (0.005-0.5 mg) of GABA synthesis inhibitor, thiosemicarbazide , into the septum dose-dependently decreased latency to kill mice, and increased irritability (escape, biting and vocalizations).
Depaulis and Vergnes 1984	Predatory aggression in male rats	g-Vinyl GABA (200-400 mg/kg) (irreversible inhibitor of GABA-T) and nipecotic acid (GABA reuptake blocker) reduced mouse-killing in experienced rats, but increased incidence of mouse killing in naive, food-deprived rats; reductions in mouse-killing were concurrent with sedation as measured in open field tests and with photobeam interruptions. Dipropylacetate (reversible inhibitor of GABA-T and succinic semialdehyde) was inactive
Molina et al. 1986	Predatory aggression in isolated male rats	Intraolfactory administration (0.15 mmol/rat) of GABA-T inhibitors: DPA and g-vinyl GABA reduced mouse killing animals to 5-30% of previous levels 30 minutes to 3 1/2 hours after administration. GABA reuptake blockers were less effective at reducing animals displaying mouse killing; nipecotic acid amide (1-2 hrs post administration) and guvacine (1-4 hours post administration) reduced mouse killing to 40-60% of previous levels.

392

GABA Agonists

Isolation-induced Aggression

| Puglisi-Allegra and Mandel 1980; Sulcova and Kršiak 1980; Poshivalov 1981 | Aggression in male mice; measured with an automated bite detector or by direct observation | Muscimol (0.2-1.5 mg/kg), a GABA receptor agonist, piracetam (300-1500 mg/kg), a cyclized derivative of GABA, and phenibut (50-100 mg/kg) and phenylpyrrolidon (50 mg/kg), GABA analogues, reduced attack, threat and ambivalent behaviors (e.g., tail rattle) and reduced defensive and alert postures. In timid mice, a low dose of piracetam (300 mg/kg) was ineffective, whereas the high dose (1500 mg/kg) reduced defensive postures and escapes. In another study, piracetam, increased aggression (200 mg/kg/day in drinking water), but also increased locomotor activity |

Pain-induced Aggression

| Puglisi-Allegra et al. 1981 | Electric tail shock-induced aggression in C57 and DBA/2 strain male mice | GABA agonist, muscimol (0.5-1 mg/kg), inhibited shock-induced biting in aggressive C57 mice, and did not alter biting in non-aggressive DBA/2 mice |

Brain-lesion induced Aggression

| Breese et al. 1987 | 6-OHDA lesion-induced self directed aggression in rats | Bilateral microinjections of GABA receptor agonist, muscimol (30 ng/rat), into substantia nigra produced self biting and lacerations in 2 of 11 controls and in all neonatally 6-OHDA-lesioned rats; this behavior did not occur in adult lesioned rats |

Aggression by Resident toward Intruder

| Depaulis and Vergnes 1985 | Resident-intruder confrontations in male rats | GABA agonist, THIP (1.25-2.5 mg i.c.v.) induced attack and threat behaviors compared to non-aggressive control |

393

Killing

Reference	Model	Findings
Depaulis and Vergnes 1983, 1984; Molina et al. 1986	Predatory aggression in male rats	GABAA agonist, THIP (2.5-5 mg/5ml i.c.v.) reduced latencies to attack and kill mice and induced mouse killing behavior in 60% of non-muricidal rats. THIP did not alter mouse-killing in experienced rats, but increased incidence of mouse killing in naive, food-deprived rats. Intraolfactory administration (0.15 mmol/rat) of muscimol, THIP, and isoguvacine also reduced the percentage of animals showing mouse killing to 25-50%.
Delini-Stula and Vassout 1978	Olfactory bulb ablation-induced predatory behavior in male rats	GABA agonists; baclofen (3-10 mg/kg) reduced mouse killing in normal and lesioned rats, whereas muscimol (0.75-1.5 mg/kg) and GABA-acetylester (10-50 mg/kg) reduced mouse killing only in normal rats
Arnt and Scheel-Krüger 1980	GABA receptor antagonist-induced self directed aggression in male rats	Concurrent microinjections of GABA agonists muscimol and THIP into the substantia nigra blocked bicuculline-induced (30-60 ng/ml/rat) and picrotoxin-induced (50-100 ng/ml/rat) self-biting.

GABA Antagonists

Isolation-induced Aggression

Reference	Model	Findings
Poshivalov 1981	Male mice; aggression measured with a bite detector	GABA receptor antagonist picrotoxin (1 mg/kg) increased probability of attacks and bicuculline increased number of attacks at a low dose (0.5 mg/kg) and reduced threats and attacks at higher doses(1-1.5 mg/kg)

Pain-induced Aggression

Reference	Model	Findings
Puglisi-Allegra et al. 1981	Electric tail shock-induced aggression in C57 and DBA/2 strain male mice	GABA antagonist, picrotoxin (0.25 mg/kg) elicited shock-induced biting in non-aggressive, 10 week old, C57 and 20 week old DBA/2 mice.

	Electric foot shock-induced aggression in male rats	Bicuculline (0.25-4 mg/kg), a competitive GABA receptor antagonist, reduced aggression, but effects were variable. Picrotoxin (0.125-2 mg/kg), a noncompetitive GABA receptor antagonist, had a biphasic effect on aggression, where the low dose (0.125 mg/kg) enhanced fighting and higher doses reduced it dose dependently.
Rodgers and Depaulis 1982		

Drug-induced Aggression

Arnt and Scheel-Krüger 1980	GABA receptor antagonist-induced self directed aggression in male rats	Bilateral microinjection of bicuculline (30-60 ng/ml/rat) and picrotoxin (50-100 ng/ml/rat) into the substantia nigra produced self-biting.

Aggression by Resident toward Intruder

Depaulis and Vergnes 1985, 1986	Male rats confronting a male conspecific	GABA antagonist, bicuculline (62.5-125 ng i.c.v.) reduced aggressive threats and attacks, and increased defensive postures compared to non-aggressive controls. Microinjection of GABA antagonist, picrotoxin (25-50 ng/0.25 ml/rat) into PAG produced a shift from offensive to defensive behaviors when intruder was located contralaterally from injection site.

Female Aggression

Hansen and Ferreira 1986	Maternal aggression in female rats	Bilateral microinjections of GABA receptor antagonist, bicuculline (60 ng/ml), into the VMH or amygdala reduced aggressive threats, attacks and bites directed at intruder male; defensive freezing behavior was not altered.
Potegal et al. 1983	Predatory aggression and viewer-rated aggression directed at experimenter in isolated male rats	Microinjections (0.005-0.5 mg) of GABA receptor antagonist, muscimol, into the septum dose-dependently decreased latency to kill mice, and increased irritability (escape, biting and vocalizations).

395

Killing

Depaulis and Vergnes 1983   Predatory aggression in male rats   Intracerebroventricular injections of GABA receptor antagonist, bicuculline (125 ng/5 ml/rat) prolonged attack and killing latencies, and reduced percentage of animals showing mouse killing

TABLE 5: Acetylcholine

References	Methods and Procedures	Results and Conclusions
A. Cholinergic Correlates of Animal Aggression		
Whole brain measurements		
Isolation-Induced Aggression		
Karczmar et al. 1973	Comparison between various species of male mice.	Neural levels or turnover rates of ACh are not correlated with levels of aggression across species.
Drug-Induced Aggression		
Jain and Barar 1986	Clonidine-induced aggression in pairs of male and female mice. Following the aggressive interaction, mice were treated with ether and sacrificed	ACh levels in whole brain decreased following clonidine-induced fighting
Pain-induced Aggression and Defense		
Jain and Barar 1986	Electric foot shock in male mice. Following the aggressive interaction, mice were treated with ether and sacrificed	ACh levels in whole brain decreased following shock-induced fighting
Peripheral correlates of aggression in animals		
Dominance-related Aggression		
Welch and Goldberg 1973	Isolated male mice grouped in a neutral cage daily for 1-5 days. The adrenal medulla was excised 18-20 hours following the last aggressive interaction and choline acetyltransferase activity was measured	Choline acetyltransferase activity (mmol/g protein/h) was slightly decreased following 4 days of fighting experience compared to mice transferred to an empty cage. As time since the cessation of daily fighting elapsed, enzyme activity tended to increase above control values
Stoddard et al. 1986	Stimulation of hypothalamic sites mediating sham rage in anesthetized male and female cats. Adrenal medullary output of NE and EPI was measured.	Hypothalamic stimulation increased adrenal output of NE and EPI (ng/kg/min)

397

B. Neuropharmacological Manipulations of Acetylcholine in Animals

Acetylcholinesterase Inhibitors

Isolation-induced Aggression

DaVanzo et al. 1966; Herbut and Rolinski 1985; Rolinski and Herbut 1985	Male mice tested in a neutral arena	Physostigmine (.05-1.0 mg/kg) increased aggression compared to non-injected controls.

Pain-induced Aggression and Defense

Anand et al. 1989	Electric footshock in female rats	Fenitrothion, an acetylcholinesterase inhibitor, increased aggression in rats with septum, substantia nigra or amygdala lesions. Whole brain measures of NE and 5-HT were decreased in treated rats.

Brain lesion-induced Aggression

Stark and Henderson 1972	Septal lesion-induced aggression in male rats	Physostigmine (.4, .57 mg/kg i.p.) decreased hyperreactivity (target biting, irritability toward experimenter), while the peripheral anticholinesterase inhibitor, neostigmine, had no effect. Atropine (.2, .4 mg/kg i.p.) pretreatment blocked the effects of physostigmine.

Drug-Induced Aggression

Rolinski and Herbut 1979	Apomorphine-induced aggression in similarly treated pairs of male rats.	Physostigmine (1 mg/kg i.p.) inhibited aggression and increased stereotypic gnawing.

Killing

Bandler 1969, 1970, 1971a,b	Ranacide in male rats	ACh (3-10 mg) alone had no effect but ACh with physostigmine decreased attack latencies.

Nicotinic Agonists

Pain-induced Aggression and Defense

Reference	Condition	Result
Driscoll and Baetig 1981	Electric footshock in male and female rats	Central or peripherally administered nicotine decreased fighting in pairs of male rats without altering vocal or escape behavior. Fighting in females only decreased at doses that also inhibited movement.
Emley and Hutchinson 1983	Tail-shock induced target biting in individually housed male and female squirrel monkeys	Nicotine (.32 mg/kg s.c.) elevated lever pressing while slightly decreasing target biting.

Defensive Aggression induced by brain stimulation

Kono et al. 1986	"Sham rage" in cats following electrical stimulation of the VMH	Ventral amygdalofugal but not stria terminalis lesions increased the threshold of stimulation necessary to elicit "rage." "Threshold returned to control levels 48 hrs post-lesioning. ACh (.5 pmol) microinjected into the VMH increased stimulation threshold for 24 hrs in control and lesioned animals.

Dominance-related Aggression

Silverman 1971	Pairs of male rats separated and re-introduced into their home cage; only one animal injected.	Nicotine (25 µg/kg s.c.) decreased aggression, social investigation and sexual behavior.

Killing

Bandler 1969, 1970, 1971a,b; Lonowski et al. 1973, 1975; Miczek 1976 Waldbillig 1980	Muricide in male rats	In male rats that were muricidal, nicotine (100-1000 mg/kg i.p.) decreased muricide; central (mecamylamine 30 mg/kg) but not peripheral (hexamethonium 30 mg/kg) nicotinic receptor antagonists reversed nicotine's effects.

Muscarinic Agonists

Isolation-induced Aggression

DaVanzo et al. 1966; Herbut and Rolinski 1985; Rolinski and Herbut 1985	Male mice tested in a neutral arena	Arecoline (.5-1.0 mg/kg), betanechol (.1-.4 mg/kg), carbachol (.05-.1 mg/kg), oxotremorine (.025-.05 mg/kg) and pilocarpine (.5-1.0 mg/kg) increased aggression compared to non-injected controls.

399

Drug-Induced Aggression

Rolinski and Herbut 1979	Apomorphine-induced aggression in similarly treated pairs of male rats.	Pilocarpine (2.5,10 mg/kg i.p.), oxotremorine (1 mg/kg i.p.), and choline chloride (60, 100 mg/kg i.p.) inhibited aggression and increased stereotypic gnawing.
Hernandez-Peon et al. 1963; Baxter 1968a; Romaniuk et al. 1973; 1974; Nagy and Decsi 1979; Beleslin and Samardzic 1979; Brudzynski 1981a,b	"Sham rage" in male and female cats elicited by intracerebral drug administration	Carbachol (2.5-40 mg), or ACh (no dose stated), infused into certain regions of the diencephalon, limbic forebrain, or mesencephalon elicited "sham rage", although some reports describe defensive behavior with very little aggression. Rage was blocked either by intrathypothalamic atropine (10 mg) prior to carbachol or by the concurrent administration of carbachol (.62-1.25) into the dorsal hippocampus. Muscarine (.005-.03 mg/kg i.c.v.) resulted in attack, fear, and autonomic and motoric activation that was antagonized by atropine (.2-.5 mg/kg), scopolamine (.5 mg/kg), NE (.5-1 mg/kg), DA (.5-1 mg/kg) and EPI (.5 mg/kg).

Killing

Bandler 1969, 1970, 1971a,b; Lonowski et al. 1973, 1975; Miczek 1976 Waldbillig 1980	Muricide in male rats	In muricidal rats, carbachol (2.0 mg) infused into the hypothalamus increased attack latency. Pilocarpine (12.5, 25 mg/kg/day i.p. for 3 weeks) induced muricide in 25-70% of non-muricidal rats.
Bandler 1969, 1970, 1971a,b; Smith et al. 1970; Lonowski et al. 1973, 1975	Muricide in female rats	In non-muricidal female rats (70%), killing was elicited by carbachol (20 mg) infused into the LH at night but not during the day.

Bandler 1969, 1970, 1971a,b	Ranacide in male rats	Carbachol (1-10 mg) infused into the dorsomedial thalamus or lateral hypothalamus decreased the latency to kill frogs in approximately 40% of rats. Atropine (10 mg) in the same thalamic sites blocked the facilitative effects of intrahypothalamic carbachol. Systemic atropine (10,25 mg/kg i.p.) or atropine methylnitrate (10, 25 mg/kg) also blocked carbachol's effects. Carbachol (3-10 mg) into the VTA decreased attack latencies in approximately 25% of rats.
Berntson and Leibowitz 1973; Berntson et al. 1976	Biting and killing of rats by male and female cats. Cats that were not spontaneously aggressive were selected.	Arecoline (5-12 mg/kg i.p.) and oxotremorine (90-150 i.p.) increased attack, hissing and growling; this effect was blocked by nicotine (.5 mg/kg) pretreatment.
Leaf and Wnek 1978	Male and female cats selected for spontaneous muricide	Pilocarpine (.25-.5 i.p.) decreased muricide. Methyl atropine (.125-.50 i.p.) partially blocked this effect.

Cholinergic Antagonists

Isolation-induced Aggression

| Herbut and Rolinski 1985; Rolinski and Herbut 1985 | Male mice tested in a neutral arena | Atropine (5.0-10.0) and scopolamine (.25-.5) suppressed aggression. Low doses of central (mecamylamine 2 mg/kg i.p.) and peripheral (hexamethonium bromide 2, 4 mg/kg i.p.) nicotine antagonists increased aggression, while higher doses decreased it. Nicotine could reverse the inhibitory effects of mecamylamine. |

Pain-induced Aggression and Defense

| Lapin 1967; Widy-Tyszkiewicz 1975 | Electric footshock in male and female mice | Atropine (.1 mM/kg i.p.), scopolamine (.1 mM/kg i.p.) and PAT-4 increased the shock threshold necessary to elicit fighting. |
| Powell et. al. 1973 | Electric footshock in female rats | Atropine (10 mg/kg s.c.) and scopolamine (.25-3.0 mg/kg s.c.) decreased fighting. The peripheral antagonists, atropine methyl nitrate and scopolamine methyl nitrate, did not. |

401

Brain lesion-induced Aggression

Herink et al. 1973	Septal lesion-induced aggression in male rats	Atropine (1 mg/kg i.p.) decreased aggression in rats with septal lesions but increased aggression in controls.

Drug-Induced Aggression

Rolinski 1975	PCPA-induced muricide in male and female rats	Atropine (7.5 mg/kg i.p.) and scopolamine (7.5 mg/kg i.p.) suppressed PCPA-induced muricide.
Rolinski and Herbut 1979	Apomorphine-induced aggression in similarly treated pairs of male rats.	Atropine (10,20 mg/kg i.p.) and scopolamine (1-4 mg/kg) suppressed aggression.
De Feo et al. 1983	Clonidine-induced target biting in male and female mice	Scopolamine (10-20 mg/kg p.o.) potentiated target biting and phenoxybenzamine (10-20 mg/kg p.o.) inhibited it.

Defensive Aggression induced by brain stimulation

Katz and Thomas 1975	Quiet predatory aggression induced by LH stimulation	Scopolamine (1.0 mg/kg i.p.) raised the threshold of stimulation necessary to elicit attack and decreased approach and biting of attack object.

Aggression by Resident toward intruder

Avis and Pecke 1979	Male convict cichlids confronted a male conspecific	Scopolamine (5 mg/l) decreased aggressive gill displays at doses that did not affect feeding or locomotion.
Van der Poel and Remmelts 1971	Treated male rats confronted an untreated male conspecific in a neutral environment	Scopolamine (.03-3.0 mg/kg i.p.) decreased aggressive, defensive and play behavior. Methylscopolamine (.2-30 mg/kg i.p. had no effect).

Killing

Smith et al. 1970; Lonowski et al. 1973; Malick 1976; Strickland and Da Vanzo 1986	Spontaneous and olfactory bulb-induced muricide in male and female rats.	Atropine decreased spontaneous (ED50: 2.9 mg/kg i.p.) and olfactory bulb-induced muricide (ED50: 5.9 mg/kg i.p.). Atropine (2.5 mg) unilaterally infused into the LH had no effect on muricide. Scopolamine (1 mg/kg i.p.) also decreased spontaneous muricide.

402

C. Neuropharmacological Manipulations and Human Aggression

Experimental Studies

Cherek 1981, 1984

Male and female subjects: aggressive acts entailed subtracting money or administering an aversive noise to another person; non-aggressive responses earned money.

Nicotine (.42, 2.19 mg) cigarettes decreased aggressive and increased non-aggressive responding.

Peripheral correlates of aggression in humans

Experimental Studies

Dotson et al. 1975

Groups of men observed in a social setting, where alcoholic beverages, cigarettes and cigars were available ad lib. Buss-Durkee Hostility Inventory was administered immediately prior to the social interaction.

Numbers of cigarettes smoked weakly correlated (.25) with the score on the Hostility Inventory

403

TABLE 6: Opioid Peptides

REFERENCES	METHODS AND PROCEDURES	RESULTS AND CONCLUSIONS
A. Opioid Correlates of Animal Aggression		
Regional brain measurements		
Dominance-related Aggression		
Diaz and Asai 1990	Group-housed mice were assigned an ordinal dominance rank. Dominance was re-determined by reassembling mice according to their initial dominance rank. Met-enkephalin was measured in brainstem and forebrain.	Met-enkephalin concentrations in the brainstem and forebrain inversely correlated with the first assessment of dominance. In the second dominance determination, dominant and subdominant also had less brainstem met-enkephalin than subordinate and non-aggressive mice; brainstem met-enkephalin concentrations decreased and increased after dominance acquisition and loss, respectively.
CSF correlates of aggression in animals		
Dominance-related Aggression		
Martensz et al. 1986	Group-housed male talapoin monkeys	CSF b-EP(fmol/ml) was inversely correlated with social rank: Dominant males had less b-EP than subordinate males. There was a negative correlation between the amount of aggression exhibited and the level of b-EP. Dominant males are more sensitive to the facilitatory effects of naloxone (.125-5.0 mg/kg i.m.) on plasma LH levels than intermediate and subordinate males.

404

B. Neuropharmacological Manipulations of Opioid Peptides

Opioid Agonists

Isolation-induced Aggression

Poshivalov 1982; Benton 1985	Male and female mice confronting a group-housed opponent	In males, met-enkephalin (50 μg i.c.v.) decreased the probability of attack and threats while increasing social investigation. Neo-endorphin (25 mg/kg i.p.) increased social investigation and locomotor behaviors. DAGO had no significant effect. In female mice, DAGO decreased social behavior and increased locomotion (250 μg, 1 mg/kg).

Pain-induced Aggression and Defense

Plotnikoff et al. 1976	Electric foot shock in pairs of similarly treated male mice.	Low doses of met-enkephalin (.1, 1.0, mg/kg i.p.) decreased fighting behavior, while a higher dose (10.0 mg/kg) did not
Tazi et al. 1983	Electric foot shock in pairs of similarly treated male rats.	b-EP (10 mg/kg s.c.) did not affect attack rates.

Defensive Aggression induced by brain stimulation

Shaikh et al. 1988, 1990	"Sham rage" elicited by hypothalamic and periaqueductal gray stimulation in male and female cats	DAME (.25-1.0 μg/.5 μl), morphiceptin (.4 nmol), and DPDPE (.8 nmol) microinjected into the periaqueductal gray increased the threshold of stimulation needed to elicit defensive aggression. Pretreatment with naloxone (10 μg/.5 μl) blocked the suppressive effects of DAME. b-FNA and ICI 174,864 antagonized the effects of morphiceptin and DPDPE, respectively.
Pott et al. 1987	"Sham rage" in male and female cats elicited by stimulation of the VMH	Aggression is facilitated by concurrent stimulation of ventral central gray nuclei and inhibited by concurrent stimulation of the dorsal central gray. Naloxone (1.0 μg) microinjections antagonized the suppressive effects of dorsal central gray stimulation, but did not alter the facilitatory effects of ventral central gray stimulation

405

Aggression by Resident toward Intruder

Christ 1985	Male cichlids confronting a male conspecific	Low doses of met-enkephalin (10 mg/g increased fighting. High doses (30 mg/g) non-selectively decreased fighting.

Opioid Antagonists

Isolation-induced Aggression

Poshivalov 1982; Brain et al. 1984, 1985; Benton 1985	Male and female mice confronting a conspecific	Naloxone (.25 mg/kg i.p.) increased the frequency of attack. Detailed ethological analysis (dendrograms) indicates that naloxone (.1, 10 mg/kg s.c.) increases the association between aggressive and defensive behaviors. The delta antagonist, ICI 154,129 (30, 80 mg/kg s.c.), disrupts the normal sequence of both social and non-social behaviors, in comparison to saline-treated controls. In male but not female mice, κ αππα antagonists, tifluadom (.5, 1.0 mg/kg s.c.) and U-50488 (2.5, 5.0 mg/kg s.c.), decreased attack and chase while increasing defensive and non-social behaviors.
Valenca and Falcao-Valaenca 1988	Siamese fighting fish confronting a male conspecific	Naloxone (10 mg/l) decreased the duration of gill cover erection without altering the frequency. Avoidance measures did not differ between groups

Pain-induced Aggression and Defense

Fanselow et al. 1980; Fanselow and Sigmundi 1982	Electric foot shock in pairs of similarly treated female rats.	The number of fighting encounters increased linearly with the number of shocks administered and the shock intensity. Naloxone (4 mg/kg i.p.) increased fighting in rats receiving 1-75 shocks; naloxone enhancement of aggression was pronounced at higher shock intensities.

406

Reference	Paradigm	Result
Gorelick et al. 1981; McGivern et al. 1981	Electric foot shock in isolated and group-housed male rats	Naloxone (2.5 mg/kg i.p.) pretreatment increased the duration and frequency of fighting episodes in one study, while in another it (4 mg/kg i.p.) had no effect. Naloxone injected in the middle of a session decreased fighting frequency in both isolated and group-housed rats.
Tazi et al. 1983	Electric foot shock in pairs of similarly treated male rats.	The incidence of fighting increased over time in rats treated with saline or naloxone (2 mg/kg s.c.) following 9 daily shock administrations; when baseline levels of fighting were low, naloxone increased the fighting episodes above saline.
Defensive Aggression induced by brain stimulation		
Shaikh et al. 1988, 1990	"Sham rage" elicited by hypothalamic and periaqueductal gray stimulation in male and female cats	Naloxone (.5-10.0 mg/kg) reduced the threshold of stimulation necessary to elicit aggression.
Aggression by Resident toward Intruder		
Haug et al. 1986	Castrated male mice housed in groups of 3 confronted lactating females	Naloxone (.25-2.0 mg/kg i.p.) dose-dependently decreased the frequency of attack.
Female Aggression		
Olivier and Mos 1986	Maternal aggression in female rats	Naloxone (.1-12.5 i.p.) had no effect on aggression
Dominance-related aggression		
Meller et al. 1980	Group-housed male talpoin monkeys	Naltrexone (500 mg/kg i.m.) decreased the frequency with which 1 of the top-ranking males attacked subordinates, but did not affect aggression in the other top-ranking male; the frequency of threats was not altered in either dominant male. Naloxone administered to subordinate males did not alter the amount of aggression received

Reviews of the Literature

Animal Literature

Valzelli 1967	Endocrine, anatomical and pharmacological aspects of aggression induced by brain lesion, pain, drugs, and isolation. 180 references from 1930 to 1966.	Fighting alters central catecholamine and serotonin levels and aggressive animals have a distinct drug sensitivity. Although agents from a variety of drug classes decrease fighting, none are selective for aggression.
Rothballer 1967	Older review of effects of hormones and neurotransmitters on aggression. 101 references	Some emphasis on role of catecholamines.
Kermani 1969	Neurological and biochemical data on aggression. 84 references from 1907 to 1968.	Amygdala, hypothalamus and midbrain are areas associated with hostility. NE is inversely correlated with aggressive behavior
Bryson 1971	Neurotransmitter levels in environmental conditions that alter aggressive behavior. 258 references from 1877 to 1970	NE and 5-HT levels are inversely correlated with aggression
Welch and Welch 1971	Relationship between neural NE, DA and 5-HT and aggression. 95 references from 1940 to 1969.	The initial effect of fighting is to elevate all 3 amines in whole brain; prolonged exposure to fighting may result in lowered amine concentrations in subcortical regions and elevations in the telencephalon
Allikmets 1974	Intracerebral administration of ACh, NE, DA and 5-HT and aggression in rats and cats. 75 references from 1939 to 1973	ACh or cholinomimetics in the amygdala, septum, hippocampus, hypothalamus and mesencephalon increase aggression in cats. NE, DA and 5-HT do not.
Johansson 1974	Role of 5-HT, DA and NE in predatory, shock-induced, and drug-induced aggression. 33 references from 1915 to 1974.	Inhibition of 5-HT synthesis is associated with increased predatory aggression. NE is related to irritable aggression.
Reis 1974	Relationship between neurotransmitters and affective or predatory aggression	NE and possibly DA facilitate affective aggression and inhibit predatory aggression. ACh facilitates and 5-HT inhibits both types of aggression

408

Reference	Description	Findings
Bernard 1975	Relationship between neural catecholamines and shock-induced aggression, ranacide and septal lesion-induced aggression. 38 references from 1956 to 1974.	Data suggest 5-HT inhibits and DA and NE facilitate aggression
Depue and Spoont 1986	Relationship between neural catecholamines and "irritative" aggression. 66 references from 1959 to 1986	Irritative aggression, based in the amygdala and septum, is facilitated by DA and inhibited by 5-HT and NE
Gianutsos and Lal 1976	Effects of DA, NE, ACh, and 5-HT on morphine-withdrawal aggression and chemical-induced aggression. 111 references from 1953 to 1975	DA facilitates drug-induced and morphine-withdrawal aggression. The effects of 5-HT and NE are contradictory, while the effects of ACh are unknown.
Daruna 1978	Relationship between brain catecholamines levels and different models of intraspecies aggression in rodents. 200 references from 1959 to 1977	Proposes that aggressive behavior is regulated by the balance between NE and 5-HT in a few specific brain regions: Isolation-induced aggression is often facilitated by increases in DA, NE and 5-HT activity. Shock- and chemically-induced defensive aggression are correlated with increased DA, 5-HT and decreased NE or increased DA and NE and decreased 5-HT activity in mesolimbic regions
Mandel et al. 1979 Mandel et al. 1981	Review of 39 references on the molecular basis of aggressive behavior. Review of 56 references on inhibitory amino acids and aggression	GABA concentrations were lower in the olfactory bulbs of rats that displayed mouse killing behavior. Local or systemic administration of compounds that facilitate GABAergic transmission blocks mouse killing behavior, whereas compounds that inhibit GABAergic transmission induce killing behavior in non-killing rats
McNaughton and Mason 1980	Reviewed the role of the dorsal ascending noradrenergic bundle in animal behavior, including various models of aggression. 542 references	Concentrations of neural NE are inversely related to aggression

Reference	Description	Findings
Eichelman et al. 1981; Eichelman 1986, 1987	Review of neurochemical and pharmacological aspects of aggression and violence. 1981 review: 211 references from 1939 to 1980; 1986 review: 244 references from 1937 to 1984; 1987 review: 136 references from 1939 to 1986.	Different types of aggression depend upon the interaction of different neurotransmitters: NE and DA release correlated with increased affective aggression and decreased predatory aggression. Both types of aggression are increased by ACh activation and decreased by 5-HT release. Drugs enhancing GABA activity inhibit most types of animal aggression.
Valzelli 1981	Psychopharmacology of pathological aggression, as distinct from behaviors that are within a species-normal repertoire. 84 references from 1959 to 1981.	Treatment of pathological aggression should involve techniques to increase central 5-HT activity
Pucilowski and Kostowski 1983	5-HT involvement in affective and predatory aggression. 110 references from 1957 to 1981	Mesostriatal 5-HT system inhibits both affective and predatory aggression; mesolimbic 5-HT is not directly involved in either type of aggression
Singhal and Telner 1983	Review of 75 references on psychopharmacological aspects of aggression	GABA appears inversely related to predatory and pain-induced aggression. Increased levels of GABA inhibit shock induced fighting and predatory aggression, whereas GABA-antagonist-induced reductions in GABA produce or facilitate these types of aggression.
Bell et al. 1985	Review of the relationship between cholinergic drugs and neural mechanisms of affective and predatory aggression; 83 references from 1962 to 1984	Both types of aggression are facilitated by muscarinic agents and inhibited by nicotinic agents.
Valzelli 1985	Neurochemical changes associated with various animal models of aggression. 28 references from 1962 to 1984.	Impaired 5-HT transmission, due to either genetic, dietary or chemical factors, is associated with aggressive behavior
Pucilowski 1987	Monoaminergic control of affective aggression. 140 references from 1964 to 1987	Mesolimbic DA system facilitates affective aggression, while 5-HT and NE pathways from the locus coeruleus to the amygdala and n. accumbens inhibit aggression.
Coccaro 1989	Review of 111 animal and human articles between 1957 and 1989 on central 5-HT and aggression	Reduction and augmentation of central 5-HT activity is related to increases and decreases in aggression, respectively

Reference	Description	Comments
Miczek and Donat 1989	Neurochemical and pharmacological evidence relating 5-HT to various laboratory models of aggression. Miczek and Donat: 169 references from 1957 to 1987. Miczek et al. 1989: 30 references from 1958 to 1989.	Although 5-HT systems are involved in the mediation of aggressive and defensive behaviors, there is no consistent relationship between 5-HT activity and particular types of aggressive behavior. The continuing development of 5-HT agents with specific receptor selectivity will characterize this relationship more definitively
Human Literature		
Verebey et al. 1978	Clinical psychopharmacology of opiate agonists and antagonists in psychiatry. 188 references from 1972 to 1978.	There is not a great deal of supportive evidence for the use of opiate agonists or antagonists for treatment of mental disease
Woodman 1979	Peripheral biochemistry of patients maintained in a maximum security hospital. 36 references from 1885 to 1978	Presents evidence for an abnormal adrenal response to stress in a subgroup of prisoners
Eichelman et al. 1981; Eichelman 1986	Review neurochemical and pharmacological aspects of aggression and violence. 1981 review: 211 references from 1939 to 1980; 1986 review: 244 references from 1937 to 1984; 1987 review: 136 references from 1939 to 1986.	Multiple neurotransmitter systems interact to modulate the propensity for aggressive behavior. Genetics and environmental stress can modulate the neurotransmitter systems mediating the behavior. Pharmacological treatment of aggression depends upon the etiology of the violent behavior i.e. senile dementia, sexual violence, acute schizophrenia.
Kraemer 1985	Relevance of CSF and peripheral measures of 5HIAA to aggressive behavior. 28 references from 1970 to 1985.	Human violence appears to be correlated with low levels of CSF 5HIAA paired with high levels of CSF NE; not clear if differences reflect etiology or indirect aspects of violent behavior.
Muhlbauer 1985	Role of central 5-HT in human aggression. 13 references from 1965 to 1980	Presents data indicating a negative correlation between CSF 5HIAA and history of aggressive behavior.
Brown and Goodwin 1986	Relationship between suicide and aggression and CSF 5HIAA. 67 references from 1960 to 1986	Most of the literature indicates an inverse correlation between CSF 5HIAA and aggression, alcoholism and suicide

411

Reference		
van Praag et al. 1986 van Praag et al. 1987	Relationship between 5-HT and psychopathology. 1986: 65 references from 1950 to 1985. 1987: 74 references from 1965 to 1987	Low CSF 5HIAA correlate with various measures of hostility and aggression
Roy et al. 1987	CSF 5-HT turnover and violence in a subgroup of alcoholics. 29 references from 1973 to 1986	Low CSF 5HIAA correlated with suicidal and impulsive behavior in alcoholics.
Burrowes et al. 1988	Neurochemical and other physiological correlates of violence. 70 references from 1969 to 1987	CSF 5HIAA levels inversely correlate with psychological tests of aggression or irritability
Roy and Linnoila 1988	Relationship between violent and suicidal behavior and 5-HT. 14 references from 1976 to 1987.	CSF 5HIAA is inversely correlated with aggression, alcoholism and suicide.
Coccaro 1989	Review of 111 human and animal articles between 1957 and 1989 on central 5-HT and aggression	Central 5-HT activity is reduced in mood and personality disorder patients with history of suicidal and/or impulsive aggressive behavior
Ellis 1991	Review of 173 articles on the relationship between MAO and antisocial behavior	Correlates low platelet MAO activity with high probability of criminality, psychopathy and drug abuse; no evidence of a causal relationship presented.

412

TABLE 7A: Effects of Antipsychotics on Aggression in Animals

References	Methods and Procedures	Results and Conclusions
Isolation-Induced Aggression		
Yen et al. 1959	Male albino mice, individually housed for 3 weeks; 5 min. observation	Chlorpromazine (10 mg/kg, p.o.) decreased both aggressive and motor activities.
Cook and Weidley 1960	Mice, individually-housed for at least 25 days, paired with non-isolated intruder	Chlorpromazine: ED50 for inhibiting attack behavior=11.3 mg/kg p.o.; prochlorperazine: ED50=7.4 mg/kg; trifluoperazine: ED50=5.4 mg/kg; trimeprazine: ED50=9.5 mg/kg; methiomeprazine: ED50=7.0 mg/kg.
Janssen et al. 1960	Male mice, individually-housed for 24 hours; 1 min. observation	Acetopromazine (ED50=1 umol/kg), triflupromazine (ED50=2.8 umol/kg), methopromazine (ED50=2.9 umol/kg), chlorpromazine (ED50=3.1 umol/kg), perphenazine promazine (ED50=16 umol/kg), perphenazine (ED50= 1.6 umol/kg), thiopropazate (ED50=2.1 umol/kg), prochlorperazine (ED50=7.8 umol/kg), and haloperidol (ED50=1.9 umol/kg) decreased aggression only at doses which suppressed locomotor activity.
Gray et al. 1961	Male albino mice, individually-housed; 3 min observation	Chlorpromazine (ED50=8 mg/kg) non-selectively decreased aggressive behavior.
Scriabine and Blake 1962	Male albino mice, selected for aggressive behavior and isolated for 24 hr; tested in pairs, 5 min observation	Chlorpromazine (0.5-8 mg/kg i.p.) decreased fighting time; impairment of motor activity was observed with 4-8 mg/kg.
Cole and Wolf 1966	Male and female mice, individually-housed for 3 days-3 weeks; 5 min. observation	Chlorpromazine (5 mg/kg i.p.) decreased fighting in one strain of the 2 tested (Onychomys torridus) and increased latency to attack; these animals also appeared sedated.

413

Citation	Methods	Results
DaVanzo et al. 1966	Male C57 B1/10 J or Dublin (ICR) mice, individually- or group- housed for 3 weeks; 3 min. observation	Chlorpromazine: ED50 for reducing aggressive behavior=1.54 mg/kg i.p.; perphenazine: ED50=1.84 mg/kg i.p.; butaperazine: ED50=6.7 mg/kg (C57 B1/10 J mice) and 8.5 mg/kg (Dublin mice).
Valzelli et al. 1967	Male Swiss albino mice, individually housed for 4 weeks; tested in groups of 3 for 5 min.; 5-point rating scale for aggressive behavior	Chlorpromazine (2.5, 10 mg/kg, i.p.), levomepromazine (0.3-1.2 mg/kg) and propericiazine (0.5, 1 mg/kg) decreased aggressive behavior; 10 mg/kg chlorpromazine produced overt sedation.
Boissier et al. 1968	Male CF1 mice, isolated 24-48 hours; 5 min. observation	Chlorpromazine (1-8 mg/kg p.o.) and haloperidol (.125-.5 mg/kg p.o.) dose-dependently decreased aggressiveness.
Hoffmeister and Wuttke 1969	Mice, individually-housed for several days; tested with non-isolated intruder	Chlorpromazine: ED50 for inhibition of aggression=93.23 mg/kg p.o.; ED50 for sedation=33.05 mg/kg
Le Douarec and Broussy 1969	Male and female CD and Swiss mice, individually-housed for 1 month; 5 min. observation, both animals in test drugged	Chlorpromazine (2 mg/kg) significantly reduced attacks by isolated mouse and vocalizations by intruder without affecting motor behavior; 4 mg/kg chlorpromazine decreased motor behavior with aggressiveness.
Sofia 1969b	Male Swiss Webster mice, individually-housed for 8 weeks; 3 min observation; rotarod test for neurotoxicity	Chlorpromazine (ED50=1.6 mg/kg), methotrimeprazine (ED50=3.0 mg/kg), thioridazine (ED50=4.6 mg/kg) and thiothixene (ED50=50 mg/kg) non-selectively decreased aggressive behavior.
Cole and Wolf 1970	Male and female mice (Onychomys leucogaster and O. torridus), individually housed for 3 days-3 weeks; 5 min observation	Chlordiazepoxide (O. leucogaster: 5.0 mg/kg, O. torridus: 5.1 mg/kg) significantly increased mean fighting time of mice; chlorpromazine (O. leucogaster: 27 mg/kg, O. torridus: 14.75 mg/kg) decreased fighting time in O. torridus.

414

Valzelli 1971 Valzelli and Bernasconi 1976	Male mice, individually housed for 4 weeks; tested in groups of 3; exploration measured using a hole board.	Chlorpromazine (5 mg/kg i.p.), propericiazine (1 mg/kg), haloperidol (1 mg/kg) and propericiazine (1 mg/kg) decreased fighting; propericiazine (0.1 mg/kg i.p.) decreased exploratory behavior of normal mice but not of aggressive mice.
Cairns and Scholz 1973	Male mice, individually- or group-housed for 7 weeks; 10 min. observation; non-isolated intruder was drugged	Chlorpromazine (4-16 mg/kg i.p.) administered to intruder decreased attacks by isolated mouse due to decreased reactivity of intruder.
Goldberg et al. 1973	Male and female mice, individually-housed for 5 weeks; 1 min. observation	Chlorpromazine (1.25-5 mg/kg i.p.) inhibited fighting behavior in males at doses which caused 60% inhibition in motor activity; isolated females did not become aggressive; CPZ inhibits 5-HT synthesis in isolated males.
Maengwyn-Davies et al. 1973	Male BALB/C, A/He J and C57BR/edJ mice, individually or group-housed for 2 weeks; aggressiveness toward another male measured for 10 minutes daily for 7 consecutive days	C57BR/edJ mice did not become aggressive after isolation; in these mice, chlorpromazine (4 mg/kg s.c.) did not alter defensive behavior; chlorpromazine did not affect the elevated adrenal tyrosine hydroxylase activity and was minimally effective in altering elevated phenylethanolamine-$\underline{N}$-methyltransferase in response to fighting stress.
Poshivalov 1973	Mice, individually-housed for 1-12 weeks	Droperidol (0.125-1.25 mg/kg) decreased fighting.
Barnett et al. 1974	Male CF-1 mice, individually-housed; tested in home cage; 3 min observation	Benzazepine (SCH 12679, ED50=14.6 mg/kg i.p.) selectively decreased fighting. Perphenazine (ED50=1 mg/kg i.p.) non-selectively decreased fighting.

Niemegeers et al. 1974	Male mice, individually-housed; 1 min. observation	Azaperone (.31-2.5 mg/kg, ED50=.74 mg/kg) dose-dependently inhibited fighting at doses 10 times lower than doses producing loss of righting reflex (ED50=7 mg/kg); azaperone is more selective in inhibiting aggression than haloperidol, levomepromazine, promazine or chlorpromazine.
Poshivalov 1974	Male C57BL mice, individually-housed for 1 day-12 weeks; 5 min. observation	Haloperidol (.06 mg/kg i.p.) decreased aggression without influencing motor behavior, higher doses of haloperidol (.125, 1.25 mg/kg) and chlorpromazine (1, 5 mg/kg i.p.) decreased motor behavior along with aggression.
Coscina et al. 1975	Male Swiss mice, individually-housed for 4 weeks; tested in groups of 9	Chlorpromazine: ED50 for reducing aggression=1.92 (1 hr.), ED50 for reducing spontaneous motor activity=8.1 mg/kg; piperazine derivatives RMI 61 140, RMI 61 144, RMI 61 280: ED50 for reducing aggression=.05-.11 mg/kg, ED50 for reducing motor activity=.25-.37 mg/kg.
Hodge and Butcher 1975	Male mice, individually-housed; latency, frequency, and duration of fighting measured; 15 min. observation.	Pimozide (0.1-0.8 mg/kg i.p.) decreased frequency of fighting and locomotor activity; disulfiram (45-190 mg/kg i.p.) decreased fight frequency and duration, increased latency to attack and decreased locomotor activity.
Kršiak 1975a,b	Male "timid" and aggressive mice, individually-housed for 3-6 weeks; tested with group-housed males in a neutral cage	Chlorpromazine (7.5 mg/kg) inhibited timidity without affecting motor activities; 2.5 mg/kg chlorpromazine selectively decreased tail rattling in aggressive mice.
Rolinski 1975	Male Swiss mice, individually-housed for 4 weeks; tested in pairs with another isolate; 5 min. observation	Pimozide, alpha-methyltyrosine, spiramide, trifluperazine and haloperidol all reduced aggression at doses higher than the ED50 for reducing motor activity; only phenoxybenzamine reduced aggression at doses lower than the ED50 for reducing motor activity.

Tobe and Kobayashi 1976	Male ddY mice, individually-housed for 6-7 weeks; tested in pairs, 10 min. observation	Chlorpromazine: ED50 for suppressing aggression=3.9 mg/kg p.o., 1.7 mg/kg i.p., ED50 for suppressing motor activity=6.8 mg/kg p.o., 2.0 mg/kg i.p.; TR-10 (a triazine derivative): ED50 for suppressing aggression=18 mg/kg p.o., 9.0 mg/kg i.p., ED50 for suppressing motor activity=13.5 mg/kg p.o., 12 mg/kg i.p.
Humber et al. 1979 Philipp et al. 1979	Male mice, individually-housed 4-5 weeks; 5 min. observation	Butaclamol inhibited fighting (ED50=1.9 mg/kg i.p.) and caused some catalepsy; the butaclamol analogues anhydrobutaclamol and deoxybutaclamol inhibited fighting behavior (ED50s=3.0 and 2.1 mg/kg) but only at doses which induce 100% catalepsy; isobutaclamol inhibits fighting (ED50=2 mg/kg) at a dose which induces some catalepsy (45% of max).
Olivier and van Dalen 1982	Male mice, individually-housed for 3 weeks, tested in a neutral cage with a group-housed intruder for 5 min.	Chlorpromazine (2.5, 5 mg/kg) non-specifically decreased aggressive behavior.
Poshivalov 1982	Male CC57W mice, individually-housed for 6-8 weeks; tested weekly with group-housed intruder for 4 min.	Haloperidol (1 mg/kg) decreased attack frequency, sociability and locomotion.
Benton 1984	Male TO mice, individually- or group-housed for 4-5 weeks; 23 hr. automated observation	Chlorpromazine (1-5 mg/kg s.c.) dose-dependently decreased aggression as measured by squeaking.
McMillen et al. 1989	Male mice, individually-housed for 3 weeks; 3 min. observations, measured fight frequency and locomotor activity	Inhibition of aggression by clozapine (1-5.5 mg/kg), sulpiride (3-30 mg/kg), haloperidol (.01-1 mg/kg), SCH 23390 (.01-1 mg/kg) and trifluoperazine (1, 3 mg/kg) is secondary to inhibition of locomotor activity; Chlorpromazine (3 mg/kg) inhibited fighting with less disruption of locomotor activity; BMY 20661 (ED50=.46 mg/kg) did inhibit aggression without disrupting locomotor activity.

417

Reference	Method	Results
Redolat et al. 1991	Male albino mice, isolated 5-6 weeks; tested with group-housed anosmic (zinc sulfate) opponents; 10 min observation.	Acute sulpiride (20, 50, 100 mg/kg i.p.) dose-dependently decreased attacks and threats, while increasing both immobility and non-social exploration. Chronic treatment (10, 20, 50 mg/kg sulpiride for 1 or 2 weeks) only increased immobility.

Pain-Induced Aggression and defense

In Mice

Reference	Method	Results
Tedeschi et al. 1959b	Male CF-1 mice, tested in pairs while receiving footshock; 3 min observation	Chlorpromazine (ED50=6.8 mg/kg p.o.), prochlorperazine (ED50=4.6 mg/kg) and trifluoperazine (ED50=0.85 mg/kg) decreased fighting only at sedative doses.
Chen et al. 1963	Male Swiss Webster mice, tested in pairs after receiving foot shock	Chlorpromazine (ED50=4.2 mg/kg p.o.) decreased fighting (sedative effects not tested).
Kostowski 1966	Mice, tested in pairs while receiving foot shock; 3 min observation	Chlorpromazine (5 mg/kg i.p.) decreased aggressiveness and induced ataxia.
Hoffmeister and Wuttke 1969	Male mice, tested in pairs while receiving electric foot shock	Chlorpromazine: ED50 for inhibition of aggressive behavior = 9.99 mg/kg p.o.
Sharma et al. 1969	Mice, tested in pairs while receiving mild intermittent electric foot shock, 1 min. observation	Chlorpromazine (3-10 mg/kg), phenocyloxime chlorpromazine (5, 10 mg), trifluoperazine (3-10 mg/kg), diphenacyloxime trifluoperazine (5, 10 mg/kg), perphenazine (1-10 mg/kg) and diphenacyloxime perphenazine (5, 10 mg/kg) dose-dependently suppressed fighting; these doses also significantly decreased motor activity.
Sofia 1969b	Mice, tested in pairs while receiving foot shock; 3 min observation; Rotarod test for neurotoxicity	Chlorpromazine (ED50=3.4 mg/kg), methotrimeprazine (ED50=2.6 mg/kg) and thiothixene (ED50=33.4 mg/kg) non-selectively decreased aggressive behavior; thioridazine (ED50=4.8 mg/kg) selectively decreased aggressive behavior.

418

Reference	Subjects	Results
Tedeschi et al. 1969	Mice, tested in pairs while receiving foot shock; 3 min observation	Chlorpromazine (ED50=10.8 mg/kg p.o.), promazine (ED50=16 mg/kg), prochlorperazine (ED50=5.1 mg/kg), trifluoperazine (ED50=1.8 mg/kg), perphenazine (ED50=2.5 mg/kg), thioridazine (ED50=18.7 mg/kg), and chlorprothixene (ED50=4.3 mg/kg) decreased fighting at doses which also decreased locomotor activity.
Barkov 1973	Mice, tested in pairs while receiving foot shock	Chlorpromazine: ED50 for suppression of aggression=1.1 mg/kg, carbidine: ED50=0.039 mg/kg.
Shintomi 1975	Male ddY mice, tested in pairs while receiving foot shock	Haloperidol (ED50=8.3 mg/kg p.o.), trifluperidol (ED50=8.3 mg/kg), benperidol (ED50=15.7 mg/kg), spiroperidol (ED50=12.4 mg/kg), chlorpromazine (ED50=15.7 mg/kg), trifluopromazine (ED50=8.3 mg/kg), levomepromazine (ED50=8.3 mg/kg), perphenazine (ED50=12.4 mg/kg), prochlorperazine (ED50=24.6 mg/kg), trifluoperazine (ED50=8.3 mg/kg), thioridazine (ED50=40.8 mg/kg), thiothixene (ED50=12.4 mg/kg), chlorprothixene (ED50=18.5 mg/kg), clothiapine (ED50=12.4 mg/kg) and mepazine (ED50=82.6 mg/kg) antagonized shock-induced fighting and hyperactivity.
Rolinski and Kozak 1979	Male mice, tested in pairs while receiving foot shock; also tested spontaneous motor activity	Haloperidol (0.5, 1 mg/kg i.p.) and pimozide (1 mg/kg) significantly decreased aggressive behavior at doses higher than ED50 for causing immobility.
Nakao et al. 1985	Male ddY mice, tested in pairs while receiving foot shock	Haloperidol (0.5 mg/kg i.p.) significantly decreased the number of fighting episodes.
In Rats		
Brunaud and Siou 1959	Male rats, tested in pairs while receiving foot shock	Chlorpromazine (5 mg/kg) reduced aggressive behavior without sedative effects.

419

Reference	Subjects	Results
Laverty and Taylor 1969	Male rats, tested in pairs while receiving foot shock	Chlorpromazine (1-5 mg/kg s.c.) dose-dependently reduced fighting, rotarod time and motor activity.
Barkov 1973	Rats, tested in pairs while receiving foot shock	Chlorpromazine: ED50 for suppression of aggression=3.5 mg/kg, carbidine: ED50=0.22 mg/kg.
Powell et al. 1973	Male and female Sprague-Dawley rats, pair-housed, tested in same-sex pairs while receiving foot shock	Chlorpromazine (0.5-10 mg/kg s.c.) dose-dependently decreased fighting.
Lal et al. 1975	Male rats, tested in pairs while receiving foot shock	Haloperidol (0.63 mg/kg) reduced aggression at each shock intensity level.
Rodgers et al. 1976	Male Sprague-Dawley rats, tested in pairs while receiving foot shock	Chlorpromazine and haloperidol (10 ug i.c. to corticomedial amygdala) reduced both fighting and pain sensitivity.
Bean et al. 1978	Male Long-Evans rats, tested in pairs while receiving foot shock	Benzazepine (2.5-20 mg/kg i.p.) dose-dependently reduced fighting, time on rotarod and spontaneous motor activity.
Hegstrand and Eichelman 1983	Male Sprague-Dawley rats, tested in pairs while receiving foot shock	Haloperidol (1 mg/kg/day, i.p., 14 days) significantly decreased number of attacks; DA receptor density increased 37% in caudate.

In Squirrel Monkeys

Emley and Hutchinson 1972, Emley and Hutchinson 1983	Male and female squirrel monkeys, lever pressing responses and biting responses on a latex rubber hose were measured while shock delivered to tail	Chlorpromazine (.06-2 mg/kg s.c.) dose-dependently decreased biting; low doses increased and higher doses decreased lever-pressing in response to shock.

Experimenter-Provoked Aggression

In cats

Hoffmeister and Wuttke 1969	Cats, provoked into displaying defensive-aggressive behavior using a leather glove	Chlorpromazine (2.5-10 mg/kg p.o.) inhibited aggressive behavior.

In monkeys

Das et al. 1954	Spontaneously aggressive rhesus monkeys	Chlorpromazine (0.7-2 mg/kg) eliminated aggressive behavior and produced akinesia and somnolence in 8 of the animals; 1 monkey which was previously timid became aggressive.
Barkov 1973	Spontaneously aggressive male monkeys (rhesus, green guenon)	Trifluoperazine (5-10 mg/kg) and carbidine (4-14 mg/kg) reduced aggressiveness, higher doses caused sedation; chlorpromazine (3, 10 mg/kg) and perphenazine (5-10 mg/kg) chiefly caused sedation without affecting aggressiveness.
Barnett et al. 1974	Rhesus monkeys, selected for aggressive behavior toward investigator.	Benzazepine (SCH 12679, 1.25-10 mg/kg s.c.) dose-dependently decreased aggressive behavior without producing motor impairment; perphenazine (0.25 mg/kg s.c.) non-selectively decreased aggression.

Defensive aggression induced by brain stimulation

Kido et al. 1967	Male and female cats; electrical stimulation applied to posterior hypothalamus and central gray.	Chlorpromazine (3 mg/kg i.v.) decreased rage response induced by stimulation of posterior hypothalamus and hissing induced by stimulation of central gray.
Baxter 1968b	Male and female cats; electrical stimulation applied to perifornical region of hypothalamus	Chlorpromazine (1-5 mg/kg i.p.) had inconsistent effects on hissing threshold; 5 mg/kg produced depression and ataxia.
Funderburk et al. 1970	Male and female cats; electrical stimulation applied to perifornical region of hypothalamus; hissing response measured	Chlorpromazine (2.5 mg/kg) and triperidol (2.5 mg/kg) lowered the threshold for hissing; trifluoperazine and perphenazine (1, 2 mg/kg) increased the threshold for hissing.
Dubinsky and Goldberg 1971	Male and female cats, electrical stimulation to the perifornical region of the hypothalamus, the ventral hippocampus and the dorsal medial nucleus of the thalamus	Chlorpromazine (2-8 mg/kg) did not suppress rat killing despite severe locomotor deficits.

Malick et al. 1969	Male Long-Evans rats with bilateral septal, olfactory bulb and ventral medial hypothalamus lesions; tested for responses to inanimate objects	Chlorpromazine: ED50 for inhibiting aggression=12.1 mg/kg i.p. (septal), 22.7 mg/kg (VMH), 5.9 mg/kg (OB); trifluoperazine: ED50 for inhibiting aggression=11.5 mg/kg i.p. (septal), 18.4 mg/kg (VMH), 6.7 mg/kg (OB).
Goldberg 1970	Male Long-Evans rats, received bilateral electrolytic septal lesions; tested for aggressive reactions to an inanimate object; rotarod test for neurotoxicity	Chlorpromazine (ED50=11.3 mg/kg i.p.) decreased aggression at doses causing neurotoxicity.
Ueki et al. 1972	Male rats, individually housed, received olfactory bulbectomies or electrolytic lesioning of the septum or amygdala; selected for "emotionality" and mouse-killing	Chlorpromazine (5, 10 mg/kg i.p.) decreased muricide, attack response to an inanimate object, and motor activity in olfactory bulbectomized and septal rats.
Barnett et al. 1974	Male Long-Evans rats, received electrolytic lesions to septum; tested for aggressiveness to inanimate object	Benzazepine (SCH 12679, ED50=22.2 mg/kg i.p.) selectively decreased aggressive behavior; perphenazine (ED50=4.5 mg/kg i.p.) non-selectively decreased aggression.

Aggression due to omission of reward

In Rats

| Thompson 1961, 1962 | Male rats, water deprived; trained to lever press for water reinforcement which was later withheld; aggression defined as marked increases in lever responding | Chlorpromazine (1.5 mg/kg, i.p., 4 times daily for 4 days) increased "aggressive" lever responding during the extinction period. Chlorpromazine (0.9-3.0 mg/kg, i.p.) and thioridazine (3.0-11.5 mg/kg, i.p.) increased "aggressive" responding at low doses and decreased "aggressive" responding at higher doses. |

422

Andy and Velamati 1978	Male cats; electrical stimulation to the septum, hypothalamus, amygdala, preoptic area, basal ganglia and hippocampus	Haloperidol and phenoxybenzamine reduced the production of aggressive seizures but also lowered the threshold for evoking stimulus-bound aggression.
Fukuda and Tsumagari 1983	Male and female cats; electrical stimulation to medial hypothalamus; measured hissing and attack toward experimenter	Chlorpromazine (ED50=5 mg/kg) increased the threshold for directed attack at doses which caused sedation; haloperidol (1 mg/kg) and chlorpromazine (1 mg/kg) did not change the hissing threshold.
Sweidan et al. 1990	Male and female cats; electrical stimulation to ventromedial hypothalamus; defensive behavior facilitated with apomorphine (0.1-1 mg/kg i.p.); measured hissing threshold	Pretreatment with haloperidol (0.1, 0.5 mg/kg i.p.), spiperone (0.2 mg/kg) and SCH 23390 (0.5 mg/kg) antagonized facilitatory effect of apomorphine on hissing. In cats not receiving apomorphine, haloperidol (0.1 mg/kg), spiperone (0.2 mg/kg) and SCH 23390 (0.1 mg/kg) suppressed hissing.

Brain lesion-induced irritability

In Mice

Kletzkin 1969	Male mice, individually- or group-housed, received septal lesions; tested for aggressive reactions to inanimate objects and fighting with an untreated mouse	Chlorpromazine (2 mg/kg i.p.) abolished fighting in normal, septal, and isolated mice without altering reactivity.

In Rats

Horovitz et al. 1963	Rats with bilateral septal lesions; tested for reaction to an inanimate object; rotarod test for neurotoxicity	Chlorpromazine (ED50=6.8 mg/kg i.p.) decreased irritability at sedative doses.
Beattie et al. 1969	Male Wistar rats with septal and/or hypothalamic electrolytic lesions; reactivity to handling stress was rated	Chlorpromazine (ED50=1.25 mg/kg) produced "taming" without causing neurological deficits in rats with septal lesions; in rats with hypothalamic or hypothalamic-septal lesions chlorpromazine produced taming (ED50=2.5 mg/kg) with minimal neurological deficits.

423

Drug-induced aggression

In Mice

Reference	Method	Results
Yen et al. 1970	Male ICR mice, injected with 500 mg/kg l-DOPA i.v. to induce aggression; biting response to an inanimate object measured	Chlorpromazine: ED50 to reduce aggression=2.4 mg/kg i.p., TI=49, haloperidol: ED50=0.9 mg/kg, TI=47, reserpine: ED50=0.18 mg/kg, TI=389, trifluoperazine: ED50=0.15, TI=1233.
Hasselager et al. 1972	Male NMRI mice, injected with d-amphetamine to induce aggression; measured agonistic behavior among groups of 4 mice housed under highly crowded conditions every other minute for 180 minutes after injection	Spiramide (0.05-1 mg/kg) and trifluperazine (0.15 mg/kg) inhibited aggressive behavior without causing sedation.
Rolinski 1973, 1974	Swiss mice, treated with amphetamine (15 mg/kg s.c.) and/or l-DOPA (200 or 800 mg/kg i.p.) to induce aggression; tested in groups of 4	Pimozide (0.125, 0.25 mg/kg i.p.), spiramide (0.1, 1.2 mg/kg), trifluperazine (0.15 mg/kg) and nialamide (40 mg/kg) pretreatment selectively reduced aggressive behavior induced by amphetamine or l-DOPA; FLA 63 (30 mg/kg) had no effect.
Lal et al. 1975	Mice, treated with 4 mg/kg amphetamine and 400 mg/kg dl-DOPA to induce aggression, tested in groups	Haloperidol (1 mg/kg) inhibited attacks, defensive rearing and vocalizations.

In Rats

Reference	Method	Results
Lammers and Van Rossum 1968	Male Wistar rats, treated with Ro 4-4602 (52.3 mg/kg i.p.) and amphetamine (.54-1.72 mg/kg i.p.) or l-DOPA (6.23-62.3 mg/kg i.p.) to induce aggression; tested in groups of 6	Chlorpromazine (3.55 mg/kg i.p.) and haloperidol (0.4 mg/kg i.p.) suppressed defensive rearing.
Nakamura and Thoenen 1972	Male Wistar rats; aggression induced with i.v. injection of 6-OHDA; measured responses (irritability) to non-painful stimuli	Chlorpromazine (ED50=5.09 mg/kg i.p.) and haloperidol (ED50=0.72 mg/kg i.p.) decreased irritability only at doses which caused catalepsy.
Gianutsos et al. 1974	Male Long-Evans rats, treated chronically with haloperidol (2.5-20 mg/kg/day i.p. for 16 days); tested in groups, 60 min. observation	Chronic haloperidol induced catalepsy; haloperidol withdrawal induced "wet dog shakes" and intense aggression in response to a sub-threshold dose of apomorphine.

Reference	Methods	Results
Lal et al. 1975	Male rats; aggression elicited with apomorphine (20 mg/kg) or by withdrawal from morphine	Haloperidol (0.63, 2.5 mg/kg) reduced attacks, defensive rearing, and vocalizations in apomorphine-treated and morphine withdrawn animals.
Shintomi 1975	Male ddK mice, group-housed; aggression elicited with methamphetamine (5 mg/kg s.c.)	Haloperidol (0.1, 0.5 mg/kg p.o.), trifluperidol (0.1, 0.5 mg/kg), benperidol (0.25, 0.5 mg/kg), spiroperidol (0.1, 0.25), chlorpromazine (5, 10 mg/kg), trifluopromazine (0.5, 1 mg/kg), levomepromazine (1 mg/kg), perphenazine (1 mg/kg), prochlorperazine (1, 2.5 mg/kg), trifluoperazine (0.5, 1 mg/kg), thioridazine (5, 10 mg/kg), thiothixene (1, 5 mg/kg), chlorprothixene (2.5-10 mg/kg), clothiapine (1 mg/kg) and mepazine (200 mg/kg) antagonized methamphetamine-induced fighting and hyperactivity.
Zetler and Hauer 1975	Male Wistar rats, housed in groups of 5; aggression induced by combined injection of isocarboxazid (MAO inhibitor) and imipramine; tested response to inanimate object	Chlorpromazine (1-5 mg/kg i.p.), trifluoperazine (0.15-5 mg/kg), acepromazine (1.5-5 mg/kg), and propionylpromazine (1.25-10 mg/kg) decreased biting at doses which induced catalepsy; haloperidol (5 mg/kg) decreased biting without inducing catalepsy.
Albert and Richmond 1977	Male hooded rats; tested for reactivity to presentation of pencil or gloved hand, or to prodding or grasping of the body (3-point scale) after injection of dopaminergic, alpha-adrenergic, beta-adrenergic, and cholinergic antagonists injected i.c. ventral to anterior septum	Haloperidol (5 mg/ml) did not increase reactivity over saline control levels.
Mueller et al. 1982	Male Long-Evans rats, treated with caffeine (140 mg/kg s.c.) or amphetamine (43 mg pellet implanted s.c. for 4.5 days+3 mg/kg i.p.) to induce self-injurious behavior (SB)	Haloperidol (0.2 mg/kg i.p.) pretreatment did not significantly reduce SB in caffeine-treated animals; pimozide (1.5 mg/kg i.p.) pretreatment eliminated SB in amphetamine-treated animals.

Reference	Method	Results
Pucilowski and Kostowski 1988	Male Wistar rats; aggression induced by apomorphine injection; 5 min. observation of aggression among paired rats	Rats which were originally non-responsive to apomorphine exhibited aggression after ethanol or haloperidol withdrawal (14 days, 0.5 mg/kg twice daily); co-administration with diltiazem decreased this aggression.
In chicks		
Ayitey-Smith and Addae-Mensah 1983	Male and female chicks, treated with apomorphine (0.5 mg/kg s.c.) to induce aggression; tested in pairs	Chlorpromazine (10 mg/kg i.p.) pretreatment completely antagonized aggressive behavior to self and other chicks but did not abolish pecking to inanimate objects.

Aggression by resident toward intruder

In mice

Reference	Method	Results
Kršiak and Steinberg 1969	Male TO mice, residents housed in groups of 4 and intruders individually-housed; all residents in group drug-treated; 6 min observation	Chlorpromazine (4 mg/kg, i.p.) decreased aggression and induced sedation.
Thurmond 1975	CF-1 mice, housed individually for 24 hours; tested with an intruder male; 30 min. observation	Chlorpromazine (2.5 mg/kg, i.p.) decreased the number of residents attacking, decreased number of attacks and increased latency to attack.
Dixon 1982	Male LAC mice, individually-housed for 4 weeks; tested with group-housed male intruders (intruder received drug); 6 min observation	Clozapine (0.3-1 mg/kg p.o.) selectively increased defensive behavior.

In rats

Reference	Method	Results
Silverman 1965a,b, 1972	Male rats, individually housed, tested in pairs in home cage of drugged rat, 10 min. observation	Chlorpromazine (1, 4 mg/kg, i.p.) reduced aggression by reducing responsiveness to external stimuli.
Olivier and van Dalen 1982	Male rats, tested in home cage with male intruder	Haloperidol (0.04-0.2 mg/kg) non-specifically decreased aggressive behavior.

Female aggression

Yoshimura and Ogawa 1989

Female ICR mice, housed with a male for 4 days and then housed alone; tested with male group-housed intruders on postpartum days 5 and 7; 5 min observation

Haloperidol (0.1-0.4 mg/kg i.p.) dose-dependently decreased biting and locomotion; chronic haloperidol (0.1, 0.2 mg/k.g for 20-22 days) did not significantly alter biting or locomotion.

Dominance-related aggression

In fish

Walaszek and Abood 1956

Male Siamese fighting fish, individually housed; tested for aggressive responses to conspecifics

Chlorpromazine (2 ug per milliliter tank water) caused sedation and eliminated aggression. Chlorpromazine sulfoxide (50 ug per milliliter tank water) had no effect.

Munro 1986

Female blue acaras, individually-housed; tested for aggressive responses to mirrors or models

Chlorpromazine (2.5-20 mg/10 l tank water) decreased incidence of biting and decreased displays (approach/avoidance behavior) at higher doses; motor activity was decreased.

In mice

Vessey 1967

Male and female mice (Mus musculus, C57) of three confined, freely-growing populations; 20 min. observation

Aggressive interactions increased with increases in populations size; chlorpromazine (>0.6 mg/g food) reduced aggressive behavior and increased infant survival.

Zwirner et al. 1975

Male OF-1 mice, housed in groups of 3; 10 min observation; tested aggressive behavior between 2 groups

Chlorpromazine (1-10 mg/kg p.o.) decreased aggression at doses which also decreased locomotor activity.

In rats

Heimstra 1961

Male Wistar rats, individually housed, 22 hr. food-deprived; tested in pairs, 5 min. observation

Chlorpromazine (0.5 mg/kg i.p.), when administered to either the dominant or the submissive animal caused that animal to gain control of the food source; administration of chlorpromazine to one or both animals reduced fighting behavior.

In cats		
Hoffmeister and Wuttke 1969	Male cats, tested for aggressive behavior toward an untreated male cat; 6-point aggression score	Chlorpromazine (2.5-8 mg/kg p.o.) abolished attack behavior.
In pigs		
Gonyou et al. 1988	Pigs, tested in groups of 15; 48 hour observation	Both amperozide (1 mg/kg i.m.) and azaperone (2.2 mg/kg i.m.) decreased overall aggressiveness of group; amperozide was more selective.
In monkeys		
Lister et al. 1971	Social colony of baboons	Chlorpromazine (2.5 mg/kg i.m.), administered to the dominant male, reduced aggressive behavior by about 50%, 3 mg/kg was sedative.
Miczek and Yoshimura 1982	Male squirrel monkeys, housed in social colonies, treated with d-amphetamine (1 mg/kg p.o.) or cocaine (10 mg/kg p.o.) to induce altered social behavior	Chlorpromazine (0.25-1 mg/kg p.o.) and haloperidol (0.25, 0.5 mg/kg p.o.) pretreatment antagonized locomotor effects of amphetamine and cocaine but did not reliably normalize social behavior.
Killing		
In mice		
Gay and Clark 1976	Male Onychomys leucogaster (northern grasshopper mice), induced to kill CBA mice with electric shock	Chlorpromazine (2.5, 5 mg/kg i.p.) did not alter killing but facilitated killing on postdrug trials.
Poshivalov 1980	Male CC57W mice, selected for aggressive behavior and rated for dominant status	Droperidol (2.5 mg/kg) inhibited aggression, sociability and locomotion; chronic administration (7 days) to the dominant mouse induced sedation, decreased social interaction, and caused the dominant mouse to lose this status.

428

Karli 1958, 1959a,b	Male rats, selected for muricidal behavior	Chlorpromazine (1-15 mg/kg i.p.) did not affect muricidal behavior despite the sedative effects.
Janssen et al. 1962	Male and female rats, selected for muricidal behavior	Chlorpromazine (2.5-40 mg/kg s.c.) haloperidol (1.25-5 mg/kg) and other neuroleptics non-specifically inhibited or abolished muricidal behavior.
Vergnes and Karli 1963	Male and female rats which became muricidal after olfactory bulb ablation	Hydroxyzine (30-40 mg/kg, i.p.) abolished muricidal behavior.
Horovitz et al. 1966	Female rats, selected for muricidal behavior; rotarod and conditioned avoidance tests for sedation.	Chlorpromazine (ED50=5.5 mg/kg) and fluphenazine (ED50=0.15 mg/kg) non-selectively decreased killing.
McCarthy 1966	Rats, separated into groups of spontaneous-killers, capricious killers and non-killers	Chlorpromazine was more effective in blocking pilocarpine-induced muricidal behavior in non-killers than spontaneous killing.
Loiselle and Capparell 1967	Male Hooded rats, selected for muricidal behavior; 15 min. observation; also tested motor activity	Chlorpromazine (10-50 mg/kg) slightly decreased muricidal behavior and was sedating.
Sofia 1969a	Male Long Evans rats, individually-housed, selected for muricidal behavior; rotarod test for neurotoxicity	Chlorpromazine (ED50=9 mg/kg), methotrimeprazine (ED50=120.6 mg/kg), thioridazine (ED50=76.1 mg/kg) and thiothixene (ED50=100 mg/kg) non-selectively reduced killing behavior.
Goldberg 1970	Male Long-Evans rats, individually housed for 6 weeks; selected for muricidal behavior; rotarod test for neurotoxicity	Chlorpromazine (5.6 mg/kg, i.p.) decreased killing at doses which caused neurotoxicity.
Valzelli and Bernasconi 1971, 1976	Male Wistar rats, individually housed for 6 weeks; selected for muricidal behavior	Chlorpromazine (5 mg/kg), propericiazine (1 mg/kg), haloperidol (0.5, 1 mg/kg) and triperidol (1 mg/kg) have anti-muricidal activity which may be secondary to sedative effects.

Reference	Subjects	Results
Ueki et al. 1972	Male rats, individually housed, received olfactory bulbectomies or electrolytic lesioning of the septum or amygdala; selected for "emotionality" and mouse-killing	Chlorpromazine (5, 10 mg/kg i.p.) decreased muricide, attack response to an inanimate object, and motor activity in olfactory bulbectomized and septal rats.
Kostowski and Czlonkowski 1973	Male Wistar rats, individually- or group-housed, separated into "killers" and "non-killers"	Isolated rats were less sensitive to cataleptic effects of chlorpromazine (5 mg/kg s.c.) and haloperidol (0.1, 0.2 mg/kg) than group-housed; "killers" were less sensitive to chlorpromazine, "non-killers" were less sensitive to haloperidol.
Barnett et al. 1974	Male Long-Evans rats, selected for muricidal behavior.	Benzazepine (SCH 12679, ED50=5.6 mg/kg i.p.) and perphenazine (ED50=0.9 mg/kg i.p.) selectively reduced killing.
Valzelli and Bernasconi 1976	Male Wistar rats, individually-housed, screened for muricidal behavior; subjected to a single electric shock during test	Chlorpromazine (5 mg/kg i.p.) and haloperidol (1 mg/kg) non-selectively decreased muricidal behavior; propericiazine (1 mg/kg) and triperidol (1, 7.5 mg/kg) were ineffective.

In ferrets

Reference	Subjects	Results
Schmidt and Apfelbach 1977 Schmidt 1979, 1983	Ferrets were tested for predatory behavior toward Wistar rats.	Haloperidol (0.14, 0.6 mg/kg, i.m.), metoclopramide (10 mg/kg i.m.) and tiapride (16 mg/kg i.m.) increased the efficiency of the predatory behavior. Clozapine (10 mg/kg), sulpiride (40 mg/kg i.p., 90 mg/kg i.m.) and chlorpromazine (4 mg/kg) non-specifically decreased predatory behavior by causing sedation.

In cats

Reference	Subjects	Results
Leaf et al. 1978	Male and female cats, selected for muricidal behavior	Chlorpromazine (1-8 mg/kg) and haloperidol (0.0625-0.5 mg/kg) did not inhibit muricidal behavior.

Literature Reviews

Delini-Stula and Vassout 1979	Review of the effects of drugs on aggressive behavior (51 references spanning 1956 to 1978)	Antipsychotic drugs have aggression-decreasing effects which may be secondary to their inhibitory effects on overall behavior; antipsychotics do not effectively decrease predatory aggression.
Valzelli 1979	Review of the effects of sedative and anxiolytic drugs on aggression (129 references spanning 1948 to 1979)	Thioridazine appears to be the most selectively anti-aggressive phenothiazine; among the butyrophenones, azaperone appears to be selective for aggressive behavior; other phenothiazines and butyrophenones have non-selective effects on aggressive behavior.
Miczek and Winslow 1987	Review of research on the psychopharmacology of aggression (over 350 references spanning 1928 to 1986)	The aggression-inhibiting effects of antipsychotic drugs in most pre-clinical experimental paradigms are secondary to their sedative effects.
Miczek et al. 1994	Review of current pharmacotherapeutic approaches to the management of violence and aggression (97 references spanning 1959 to 1991)	The suppressive effects of typical neuroleptic drugs, SCH 23390 and raclopride are part of an overall suppression of active behavior.

431

TABLE 7B: Effects of Antipsychotics on Aggression in Humans

References	Methods and Procedures	Results and Conclusions
Behavior Disorders in Juveniles		
Shaw et al. 1963	91 male and female emotionally disturbed children (symptoms included aggressiveness), aged 7-15; behavior evaluated by nursing staff; placebo controlled, double-crossover design	68% of subjects showed some improvement of symptoms with phenothiazine (trifluoperazine, thioridazine, triflupromazine, fluphenazine) treatment; thioridazine produced the highest percentage of improvement; side effects were drowsiness, apathy and some extrapyramidal dysfunctioning.
Alderton and Hoddinott 1964	9 male children, ages 6-12, some diagnosed with mild brain damage or behavior disorder, all diagnosed as aggressive and hyperactive; aggressive and other symptoms rated as present or absent; placebo-controlled, double-blind	Thioridazine (25 mg t.i.d.) significantly decreased aggressive and destructive symptoms as well as hyperactivity.
Vialatte 1966	51 children with behavioral disorders (including aggression in 29 cases), ages 3 months-14 years; behavior evaluated by parents; no control group	Thioridazine (1-2 mg/kg/day) decreased aggressiveness in all 29 cases.
Alexandris and Lundell 1968	21 male and female institutionalized children, ages 7-12, diagnosed as hyperactive; 5-point rating scale for hyperactive, aggressive, and other behaviors; placebo-controlled, double-blind	Thioridazine (15-150 mg/day for 6 mos.) significantly decreased hyperactivity and aggression scores.
Cunningham et al. 1968	12 children, ages 8-13, who had shown aggressive, hyperactive or destructive behavior; 3-point rating scale for behavior, rated by nursing staff and teacher; reaction time, pegboard, "twisting path", Woodworth-Wells substitution and learning test administered; placebo-controlled double-blind cross-over design	Haloperidol (0.25-1.5 mg b.i.d.), administered for four weeks with benzhexol (1 mg b.i.d., to minimize side effects) caused significant decreases in hyperactivity, destructiveness, resentfulness and aggressiveness; benzhexol alone also decreased hyperactivity; side effects of haloperidol included apathy, sleepiness and transient muscle pain, as well as slower test times.

432

Study	Sample	Results
Ucer and Kreger 1969	50 male and female retarded, emotionally disturbed children, ages 7-12; affective and behavioral symptoms including hyperactivity and aggression were rated; double-blind, no placebo group	Haloperidol (0.75-3.75 mg/day) decreased aggressiveness in 52% and increased aggressiveness in 16% of subjects; thioridazine (15-75 mg/day) decreased aggressiveness in 14.2% (not significant) and increased aggressiveness in 4.8%; minor side effects.
Faretra et al. 1970	60 male and female disturbed children, 87% of them childhood schizophrenics, ages 5-12; rated by investigator, double blind, no placebo group	Overall, haloperidol (0.25-1.25 mg t.i.d.) improved 57% and fluphenazine (0.75-3.75 mg/day) improved 67% of patients; haloperidol decreased assaultiveness scores after 4 weeks but fluphenazine did not.
Ojeda 1970	35 male and female children with symptoms including hyperactivity and aggressiveness, 21 with childhood schizophrenia and 8 with minimal brain damage syndrome, ages 6-12; 5-point rating scale; placebo controlled, double-blind	Thioridazine (100-400 mg/day for 21 weeks) improved symptoms in all 20 patients, placebo improved symptoms in 8 of 15 patients; no side effects reported.
Sprague et al. 1970	12 male emotionally disturbed children, mean age 7.8 years; tested performance on a recognition task, activity level; observed deviant behavior in the classroom, attention to school work and contact with teachers; placebo controlled	Thioridazine (0.75, 1.0 mg/kg/day) decreased accuracy and increased reaction time in the recognition task without affecting activity levels; thioridazine did not significantly affect deviant classroom behavior, contact with teachers or attention; methylphenidate increased accuracy, attention and cooperative behavior while decreasing reaction time and hyperactivity.
Le Vann 1971	61 male and female mentally retarded children exhibiting symptoms of hostility, aggressiveness and compulsiveness, ages 5-12; rated by investigator, double-blind, no placebo	Haloperidol (1-2 mg/day) improved 95% and chlorpromazine (41-140 mg/day) improved 30% of patients; haloperidol was more effective at reducing hostile, impulsive and aggressive symptoms; some side effects (drowsiness), symptoms of one patient on haloperidol worsened.

Bartunkova et al. 1972	21 male and female children with symptoms of motor restlessness, insubordination and aggressiveness, ages 10-15; 7-point rating scale; double-blind	Chlorpromazine and propericiazine reduced aggressivity, insubordination and motor restlessness.
Campbell et al. 1972	10 severely disturbed male and female children (6 schizophrenic, 1 autistic), ages 3-6, matched for motor activity and prognosis; half were administered chlorpromazine and half were administered lithium carbonate, double-blind cross-over design	Chlorpromazine (15-45 mg/day for 8-10 weeks) slightly but non-significantly decreased psychotic speech, withdrawal and stereotypy; slightly but non-significantly increased responsiveness, vocabulary, attention span and speech initiation; EEG effects inconclusive; side effects included sedation and worsening of schizophrenic symptoms; not specific to aggression.
Minde et al. 1972	Follow-up study of 91 hyperactive and aggressive children, ages 11-17, initially seen 5 years previously; were treated with chlorpromazine or d-amphetamine	Symptoms diminished with maturity but subjects' behavior still more problematic than controls; psychological adjustment did not appear to be dependent on duration of drug therapy; drug therapy alone not sufficient to improve behavior.
Aman and Werry 1975 Werry and Aman 1975	Male and female hyperactive and aggressive children, ages 4-12; placebo-controlled, double-blind	Haloperidol (.035 mg/kg, acute) had no effect on cardiovascular function beyond a slight and transient increase of heart rate; low doses (.025 mg/kg) increased cognitive functioning and high doses (.05 mg/kg) decreased functioning; improvement seen mostly in low-complexity tasks; methylphenidate appears more effective overall.
Saletu et al. 1975	62 male and female hyperactive children with symptoms which included aggressiveness, ages 6-13; Global Clinical Impression Rating Scale, Parents' Questionnaire, Teachers' Questionnaire, measurement of visual evoked potentials; double blind, placebo-controlled	Thioridazine (20-80 mg/day) improved symptomatology but was less effective than d-amphetamine; short latencies and small amplitudes of visual evoked potentials preceding treatment were predictors of good therapeutic response to thioridazine.

434

Gittelman-Klein et al. 1976	155 male and female children, ages 6-12, diagnosed as hyperactive; 4-point rating scale completed by parents, 6-point clinical evaluation scale; 8-point Global Improvement Scale by teachers, parents, clinicians; placebo-controlled, single-blind	Thioridazine (50-300 mg/day, alone or with 52 mg/day methylphenidate for 12 weeks) did not improve symptoms after 4 or 12 weeks treatment; children receiving thioridazine with methylphenidate were improved; mild side effects (decreased appetite, sleep difficulties) were noted.
Campbell 1987	Autistic and conduct disorder aggressive children	Haloperidol facilitated learning in retarded autistic children and did not induce side effects, but did slow reaction time and performance on Porteus Mazes in non-retarded aggressive children as well as induce Parkinsonian side effects.
Criminal Violence		
Jirgl et al. 1970	15 aggressive and auto-aggressive delinquents with psyphopathic personal features; tested using Clyde's mood scale, Knobloch-Hausner neuroticism questionnaire N-5, Zunge depression scale, multiple affect adjective check list and others; no control group	Propericiazine (doses not mentioned) decreased hostility, depression and aggressivity markedly; emotional lability and thinking were moderately improved; side effects were somnolence and muscular rigidity.
Scarnati 1986	47 male prisoners, ages-55, 44 had committed violent crimes; 25 diagnosed with paranoid schizophrenia, 6 diagnosed with bipolar disorder, 29 were alcohol and/or drug abusers; observations by prison psychiatrist	Most inmates (n=15) were being treated with fluphenazine deconate (24-62.5 mg/2 weeks) or thioridazine HCl (5 inmates, 100-500 mg/day); schizophrenic patients treated with antipsychotics show better cognitive functioning than untreated schizophrenics; this prison population appears to require higher than normal doses of antipsychotic medication.

Inpatient Studies

Schizophrenic patients

Overall et al. 1964	68 male schizophrenic and 77 male depressed inpatients, ages 25-76; evaluated before and 4 weeks after treatment commenced; Brief Psychiatric Rating Scale (BPRS), Manifest Depression Scale of the Inpatient Multidimensional Psychiatric Scale (IMPS), Manifest Depression Scale of the Minnesota Multiphasic Personality Inventory (MMPI)	Thioridazine (mean daily dose 507-605 mg) reduced hostility, suspiciousness and overall pathology more effectively than imipramine in both patient groups.
Kellam et al. 1967	Over 340 acute schizophrenic inpatients; assessed before and 6 weeks after drug treatment; global ratings by psychiatrist, psychologist and nurse, Inpatient Multidimensional Psychiatric Rating Scale completed by psychiatrist or psychologist, Ward Behavior Rating Scale completed by nurse; placebo controlled, double blind	Chlorpromazine, fluphenazine, and thioridazine decreased hostility and aggression scores; regardless of drug or placebo treatment, patients in low disturbed behavior wards improved more than patients in high disturbed behavior wards.
Ebert et al. 1977	36 male and female schizophrenic inpatients, ages 19-41; global ratings, Gottschalk's Five Minute Verbalization Task, Rorschach tests given before and 5 and 13 weeks after initial drug administration; behavior ratings by clinicians and nurses; douple-blind	Chlorpromazine, fluphenazine and acetophenazine decreased overall and aggressive primary process expression at 13 weeks; this was correlated with degree of illness but not with scores on the Rorschach at 5 weeks.
Yesavage 1982	58 male violent schizophrenic inpatients, ages 22-58; antipsychotic serum levels and instances of assaulted-related behavior on the ward measured for 1 week	Mean serum level after thiothixene treatment was 13.3 ng/ml; significant correlations were found between lower serum levels and higher incidence of assaultive behavior.
Appelbaum et al. 1983	Retrospective study of 93 inpatients known to have committed violent acts within 6 months of the study, diagnosed with schizophrenia, senile dementia or manic-depressive psychosis; compared medication received before and after patients committed acts of violence with drug treatment of non-violent control group	No significant differences were found in drug or doses prescribed for violent and nonviolent groups; majority of violent acts committed by patients taking low (less than 400 mg/day) doses of antipsychotic medication (chlorpromazine, haloperidol, fluphenazine deconate).

436

Herrera et al. 1988	16 male schizophrenic inpatients proven resistant to previous neuroleptic treatment, ages 25-44; global effects, motor effects and violence (Lion's Scale of Inpatinet Violence) rated by research and nursing staff; placebo-controlled, double-blind	Chlorpromazine (1800 mg/day for 6 weeks, with 6 mg benztropine), clozapine (900 mg/day), or placebo treatment resulted in significantly fewer violent episodes than haloperidol (max dose 60 mg/day for 6 weeks, with 6 mg benztropine); in a subgroup of patients, a moderately high-dose of haloperidol can increase violent behavior (correlated with akathesia).
Mentally Retarded		
Llorente 1969	Retrospective study of 65 male and female mentally retarded and aggressive inpatients, ages 7-63; behavior rated on a 7-point scale	Thioridazine (100-800 mg/day) significantly decreased aggressiveness, assaultiveness and hyperactivity.
Elie et al. 1980	253 male and female aggressive, mentally retarded inpatients; 5-point rating scale of aggressive and other symptoms and CGI assessed by chief investigator and nursing staff; placebo-controlled, double-blind	SCH 12679 (benzazepine, 100 mg 4 times daily) decreased aggressive symptoms within 2 weeks, reduced agitation and hyperactivity; thioridazine (50 mg 4 times daily) increased aggressive behavior; SCH 12679 caused side effects including sedation, anorexia, GI disturbances.
Hacke 1980	18 female mentally retarded aggressive and self-injurious inpatients, ages 14-32	Melperone (a butyrophenone, 100-300 mg/day) significantly reduced aggressive and auto-aggressive behavior; no severe side-effects were found.
De Cuyper et al. 1985	21 female mentally retarded inpatients displaying chronic aggressive behavior, ages 26-76; 2 scales for aggressive behavior; double-blind placebo-controlled	Milenperone (a benzimidazolone propylamine, 10 mg b.i.d. for 6 weeks), added as an adjunct to previously administered psychotropic medication, did not significantly reduce aggressiveness; side effects included orofacial dyskinesias.
Read and Batchelor 1986	21 male and female mentally handicapped inpatients who exhibited severe violent and/or self-injurious behavior, ages 21-54; global observations; no placebo control	Haloperidol decanoate (200-300 mg every 28 days with procyclidine, 5 mg bd or tds) controlled aggression and self-injury in those patients treated and also decreased the general level of aggression in untreated patients.

Other

Ananth et al. 1972	56 male and female psychotic inpatients, ages 18-60; drug was administered in emergency situations; 3-point aggression scale; double-blind, chlorpromazine administered as control	Both propericiazine (20 mg i.m.) and chlorpromazine (100 mg) significantly improved behavior without significant side effects; propericiazine was superior in reducing agitation and aggression, chlorpromazine was superior in reducing excitement.
Tardiff 1982, 1983, 1984	Survey of the use of drugs in treating over 5,000 male and female assaultive and nonassaultive inpatients; most medicated with neuroleptics (haloperidol, chlorpromazine, thioridazine, thiothixene) alone or in combination with anticonvulsants (phenytoin); rated using NOSIE scale of psychopathology and behavior	The use of neuroleptics, except chlorpromazine, was correlated with an increased need for emergency medication, restraint or seclusion; there is no evidence that this relationship is causal.
Brenner et al. 1984	20 male and female inpatients with significant aggressive or auto-aggressive behavior, ages 62-90; measured aggressive, auto-aggressive, spontaneous and initiative behavior as well as cognitive performance; double-blind cross-over design, pipamperone used as reference	Febarbamat (3-8 150-mg tablets/day for 8 weeks) and pipamperone (3-8 20-mg tablets/day) did not significantly differ in effects on social behavior; results of cognitive tests were inconclusive.
Gelders et al. 1984	34 male and female out- and inpatients, aged 24-56; 16 of these were epileptics and 19 were alcoholics; behavior evaluated with a 22-item rating scale; no placebo control	Milenperone (2 mg b.i.d., alone or as an adjuvant to existing medication) significantly reduced verbal and physical aggression in both in- and outpatients; side effects were observed in 2 patients.
Schulte 1985	5 cases of violent behavior associated with haloperidol treatment (in- and outpatients)	Violent reactions to haloperidol which are associated with akathisia may occur up to 24 hrs after treatment; a follow-up should be done within 24-48 hrs of treatment.
Fugham et al. 1989	48 male and female elderly demented, aggressive inpatients; symptoms rated on a 4-point scale, 1,2, and 4 weeks after treatment; Comprehensive Psychopathological Rating Scale and GBS scale for dementia	Both zuclopenthixol (max mean daily dose 5.1 mg) and haloperidol/levomepromazine (max mean daily dose 1.6/7.6 mg) reduced aggressiveness within 1 week of treatment; zuclopenthixol was more effective.

438

Outpatient Studies

Schizophrenic Patients

Cohen et al. 1968

126 male and female schizophrenic outpatients who scored highly on a measure of social aggression at home (Explicit opposition), ages 18-42; interviews of patient by psychiatrist and interviews of close relatives used to measure social aggression; double-blind placebo-controlled

Chlorpromazine (mean daily dose 100-400 mg) decreased aggressive symptoms in patients living in homes judged "low conflict and tension" but chlorpromazine and promazine were equally ineffective for patients living in homes judged "high conflict and tension".

Rada and Donlon 1972

27 male and female schizophrenic outpatients, ages 21-58; placebo-controlled, double-blind

Piperacetazine (up to 160 mg/day) and thioridazine (up to 800 mg/day) both improved anergia and thinking disorder; hostility, suspiciousness and unusual thought content were significantly reduced with thioridazine.

Mentally Retarded Patients

Deberdt 1976

188 male and female outpatients showing behavior disturbances, 50 of these mentally retarded, ages 3-100; 4-point rating scale of symptoms; no placebo control

Pipamperone (median optimum daily dose 95 mg for mentally retarded subjects, in non-mentally retarded subjects 47 mg for children and 80 mg for adults and aged subjects, 4 weeks) significantly reduced aggressive, destructive, and other behavior; few side effects reported (sleepiness).

Other

Keckich 1978

Case study of a 29 year-old outpatient man diagnosed with sociopathic personality and transvestism who had a history of drug abuse and hostility

Haloperidol treatment (2-4 mg/day) along with imipramine treatment (100 mg/day) was prescribed to control depression, violent tendencies; akathitic side effects from haloperidol provoked an urge to assault.

Clinton et al. 1987

136 male and female patients treated in an emergency department for violent behavior due to alcohol intoxication, head trauma and personality disorder, among others; response categorized as +, -, no effect

Haloperidol (mean cumulative dose=8.2 mg i.m., i.v. or p.o.) alleviated violent behavior in 113 patients, had no effect in 3 patients; complications were observed in 4 patients.

Literature Reviews

Lion 1975	Review of the use of drug therapy for aggression (47 articles spanning 1957-1975)	It is impossible to state whether either the phenothiazines, thiothixenes or butyrophenones are better for the treatment of aggression; this depends on the patient.
Winsberg et al. 1976	Discussion of drug therapy in hyperactive and aggressive children	Psychostimulant and tricyclic treatment is preferable to antipsychotic treatment because these drugs are less toxic; chlorpromazine and thioridazine can have effects within one day in children who are responders; haloperidol can be useful in children who do not respond to stimulants, tricyclics, or phenothiazines.
Eichelman 1977	Review of the use of pharmacological therapy in treating aggression (23 articles spanning 1968-1977)	Pharmacological intervention should be used to treat the underlying cause of the aggressive behavior. Antipsychotics should be used for patients whose violence is secondary to their psychoses, delusions or hallucinations.
Itil and Mukhopadhyay 1978	Review of the use of drug treatment in the management of aggressive behavior associated with various mental disorders and drug abuse (87 articles spanning 1937-1977)	Antipsychotics can be used successfully to treat aggression associated with schizophrenia, epilepsy, mental retardation, personality disorders, drug abuse, and children with minimal brain dysfunction; side effects include sedation; no drug treats aggression specifically.
Aman and Singh 1980	Review of the effectiveness of thioridazine in treating disturbed children (24 articles spanning 1958-1978)	Only 34 of 101 dimensional ratings of behavior are successfully influenced by thioridazine treatment, among those are aggressiveness, hyperactivity, and mood; drowsiness is a prominent side effect; effects on cognitive functioning are inconclusive.

440

Reference	Review description	Findings
Extein 1980	Review of the effectiveness of pharmacotherapy in emergency psychiatric treatment (14 articles spanning 1964-1979)	Symptoms which respond favorably and consistently with antipsychotic treatment include combativeness and hostility; low potency neuroleptics (chlorpromazine, thioridazine) are most likely to cause side effects of sedation and hypotension; high potency neuroleptics (haloperidol, fluphenazine, trifluoperazine, thiothixene) are most likely to produce rigidity, dystonia and akinesias.
Itil 1981	Review of the pharmacological management of human aggressive behavior (71 references spanning 1959-1980)	Chlorpromazine, triflupromazine, thioridazine, pericyazine, fluphenazine, thiothixene, and haloperidol are recommended for aggression related to psychosis, dementia, minimum brain dysfunction, mental retardation, antisocial personality and conduct disorder; antipsychotics are not recommended for aggression due to epilepsy.
Leventhal and Brodie 1981	Review of the psychopharmacology of aggression and violence (69 references spanning 1968-1979)	Antipsychotics are potent anti-aggressive agents in psychotic and nonpsychotic patients but the benefits of these substances must be weighed against their potential for provoking debilitating side effects.
Werry 1981	Review of the effects of drugs on learning in children (44 articles spanning 1937-1980)	Data on the effects of antipsychotics and learning are scarce and inconclusive.
Sheard 1983	Review of the psychopharmacology of aggressive behavior in animals and humans (60 articles spanning 1966-1981)	Careful use of neuroleptics is recommended for the management of aggression associated with violent psychotic states, alcohol withdrawal, amphetamine psychosis, mental retardation and children with minimal brain damage. Neuroleptics are less useful for treating aggression associated with epileptic seizures, personality disorders, phencyclidine psychosis and affective disorders.

Conn and Lion 1984	Review of drug therapy in treating violent patients (11 articles spanning 1972-1980)	No single drug is available to treat aggression, the pharmacotherapy used depends on the patient's symptoms; antipsychotics are useful for controlling aggressive behavior in emergency situations and in patients with thought disorders; small doses may be useful for paranoid patients but also may worsen the aggression by causing the patient to feel out of control.
Tupin 1985	Review of drug therapy in managing violent patients (35 articles spanning 1970-1983)	High potency antipsychotics, along with barbiturates and occasionally benzodiazepines, are useful for controlling violence in short-term emergency situations; for long-term management, clinical assessment should be done to determine appropriate treatment.
Cherek and Steinberg 1987	Review of the effects of drugs on human aggression (191 references spanning 1937-1986)	Further research is needed to delineate factors predicting favorable response to neuroleptics in non-psychotic patients.
Yudofsky et al. 1987	Review of the pharmacological treatment of aggressive behavior in humans	No medication specifically treats aggression, however antipsychotics are used most commonly; the effectiveness of antipsychotics is usually due to their sedative effects; because of their debilitating side effects, antipsychotics should not be used chronically to treat aggression.
Brizer 1988	Review of pharmacotherapy in managing aggressive patients (145 articles spanning 1962-1987)	Antipsychotics are reported to be useful in treating aggressive schizophrenics, behaviorally disordered children, and patients with borderline personality disorder, organic brain syndromes, alcohol- and drug-induced aggression, or psychoses; however many studies are methodologically flawed.

| Itil and Reisberg 1978 | Discussion of drug treatment in aggressive patients (30 articles spanning 1944-1978) | Chlorpromazine (25-100 mg i.m.) is favored for acute emergency use but blood pressure must be monitored; in persistent acute aggression (beyond 48 hours) ECT treatment or a combination of chlordiazepoxide and antipsychotic medication is useful. |
| Klar and Siever 1984 | Discussion of drug thearapy in treating aggression (43 articles spanning 1967-1983) | Impulsive aggressive behavior in psychotic patients is effectively treated with antipsychotic drugs. |

TABLE 8A: Effects of Antidepressant and Monoamine Oxidase Inhibitor Drugs on Aggression in Animals

References	Methods and Procedures	Results and Conclusions
A. NORADRENERGIC REUPTAKE BLOCKERS		
Isolation-induced Aggression		
Cook and Weidley 1960	Male mice; one subject drug treated; group-housed opponent	Imipramine (ED50: 12.7 mg/kg i.p.) decreased attack behavior.
DaVanzo et al. 1966 Kršiak 1975b Kršiak 1979 Kršiak et al. 1981	Male albino C57B1/10J and ICR mice; isolated subject drug treated; group-housed opponent	Neurotoxic doses of imipramine (ED50: 31.2 mg/kg i.p.) and amitriptyline (ED50: 11.2 mg/kg i.p.) inhibited fighting behavior. 25 mg/kg i.p. desipramine in conjunction with 0.5 mg/kg i.c.v. 6-OHDA increased attack behavior.
Valzelli et al. 1967 Valzelli and Bernasconi 1971	Groups of 3 male Swiss albino mice; all subjects drug treated; 5 point aggression scale	10-30 mg/kg i.p. amitriptyline inhibited aggression in 100% of the subjects. 20 mg/kg i.p. imipramine inhibited aggression in 25% of the subjects. 20 mg/kg i.p. desipramine did not inhibit aggression.
Le Douarec and Broussy 1969 Sofia 1969b Wenzl et al. 1978 Tobe et al. 1981	Male and female CD, Swiss and NMRI mice; both subjects drug treated	5-20 mg/kg i.p. imipramine dose dependently decreased attacks, exploratory and motor behavior. Thiazesim (ED50: 22.3 mg/kg i.p.) decreased fighting. Amitriptyline (ED50: 15.6 mg/kg p.o.), 31.6 mg/kg p.o. nortriptyline and MCI-2016 (ED50: 44.0 mg/kg p.o.) inhibited aggression. 75 mg/kg p.o. amitriptylinoxide inhibited biting due to sedation.
van Riezen et al. 1973	Male Swiss mice; both subjects drug treated	Imipramine (ED50: 7 mg/kg s.c.) and OI77 (5-methylaminacetyl-6-methyl-5,6-dihydro-phenanthridine-HCl; ED50: 11 mg/kg s.c.) decreased fighting without motor impairment.

Reference	Subjects/Method	Results
Delini-Stula and Vassout 1981	Male albino NMRI mice; acute and 21 day treatment; both subjects drug treated	Acute administration of 10, 25 mg/kg i.p. imipramine did not affect aggression. Chronic treatment with 10 mg/kg imipramine decreased the proportion of animals fighting by 30%. Acute administration of 10, 15 mg/kg i.p. maprotiline or 3, 7.5 mg/kg i.p. amitriptyline decreased fighting by 30-50% which showed no signs of tolerance after chronic treatment.
Sieber et al. 1982	Male C3H/HeJ mice; resident or intruder drug treated	10 mg/kg p.o. imipramine to resident or intruder decreased social investigation and increased aggressive behavior.
Yoshimura and Ogawa 1984	Male mice; resident or intruder drug treated	When administered to the resident, 5-20 mg/kg i.p. imipramine suppressed bites and aggressive postures.
Oehler et al. 1985b	Male albino AB/Jena mice; 3 point scale	20 day administration of 5 mg/kg (in water supply) desipramine had no effect on aggressivity.

Pain-induced aggression and defense

Reference	Subjects/Method	Results
Allikmets and Lapin 1967	Footshock to pairs of albino rats with and without amygdaloid lesions; both subjects drug treated	Acute administration of 20 mg/kg i.p. imipramine increased aggressiveness in lesioned animals. 10 day administration of 5-20 mg/kg lowered the threshold for aggressiveness in lesioned animals. Acute and chronic administration of 5 mg/kg i.p. desipramine slightly enhanced aggressiveness in lesioned animals.
Lapin 1967	Footshock to pairs of male and female albino, BALB, C57Br and C57Bl mice; both subjects drug treated	0.05 mM/kg i.p. desipramine and imipramine modestly increased aggressiveness.

Reference	Subjects	Results
Sofia 1969b Irwin et al. 1971 van Riezen et al. 1973 Rolinski and Herbut 1981 Tobe et al. 1981	Footshock to pairs of male and female albino mice; both subjects drug treated	Neurotoxic doses of imipramine (ED50: 22 mg/kg i.p.), desipramine (ED50: 45.6 mg/kg i.p.) and amitriptyline (50, 100 mg/kg p.o.) decreased fighting and increased escape responses. 100-200 mg/kg p.o. MCI-2016 dose dependently decreased fighting. Up to 32 mg/kg s.c. OI77 (5-methylaminoacetyl-6-methyl-5,6-dihydro-phenanthridine) and 50 mg/kg i.p. thiazesim had no effect on fighting.
Tedeschi et al. 1969	Footshock to pairs of male mice; both subjects drug treated	Amitriptyline (ED50: 39.8 mg/kg p.o.) decreased fighting and produced motor impairment. 20 mg/kg i.p. imipramine produced no effect on fighting.
Crowley 1972 Crowley and Rutledge 1974 Anand et al. 1977	Footshock to pairs of male Sprague-Dawley rats; both subjects drug treated	10, 20 mg/kg i.p. imipramine dose dependently decreased fighting duration and was without effect on motor behavior.
Burov 1975	Footshock to pairs of male albino rats; both subjects drug treated	5 mg/kg i.p. amitriptyline, imipramine increased the number of fight cycles.
Eichelman and Barchas 1975 Prasad and Sheard 1982 Prasad and Sheard 1983a,b	Footshock to pairs of male Sprague-Dawley rats; both subjects drug treated	Acute administration of 10 mg/kg i.p. desipramine did not change fighting. 2-5 day administration of 10 mg/kg i.p. b.i.d. amitriptyline, desmethylimipramine, imipramine or 10, 15 mg/kg i.p. desipramine potentiated attacks.
Sheard et al. 1977	Footshock to pairs of male albino Sprague-Dawley rats with and without 30 ug/kg i.p. LSD; both subjects drug treated	5 mg/kg i.p. desipramine pretreatment antagonized LSD-enhanced fighting while having no effect on fighting without LSD.
Bell and Brown 1979	Footshock to pairs of male Sprague-Dawley rats; both subjects drug treated	1-10 mg/kg i.p. thiazesim produced no effect on aggression.

Reference	Model	Results
Mogilnicka and Przewlocka 1981 Mogilnicka et al. 1983	Footshock to pairs of male albino Wistar rats; acute and 10 day administration; both subjects drug treated	Acute administration of 10 mg/kg i.p. amitriptyline, imipramine, mianserin, iprindole, maprotiline or (+, -, ±)oxaprotiline produced no effect on fighting. Chronic administration of 10 mg/kg i.p. b.i.d. amitriptyline, imipramine, mianserin, iprindole, maprotiline or (+, -, ±)oxaprotiline increased the frequency and duration of fighting.
Valdman and Poshivalov 1986	Footshock to pairs of male CC57W mice; 7 day pretreatment with 0.5 mg/kg i.p. reserpine; both subjects drug treated	7 day administration of 10 mg/kg i.p. trazodone reduced timid-defensive and restored aggressive behavior due to nociceptive stimulation.

Aggression due to omission of reward

Reference	Model	Results
Kamioka et al. 1977	Lever pressing in male macaque monkeys; reinforcer alternated between food delivery and tail shock to conspecific; one subject drug treated	10 mg/kg p.o. imipramine administered to either subject did not alter the number of shock deliveries.
Kampov-Polevoi 1978	Escape from water competition in pairs of male albino Wistar rats; one animal drug treated	1, 5 mg/kg i.p. amitriptyline and 0.3, 2.5 mg/kg i.p. imipramine dose dependently decreased effective attacks by the dominant rat. 1, 5 mg/kg amitriptyline and 0.15-2.5 mg/kg imipramine dose dependently increased defensive and offensive abilities of subordinate rats.

Defensive aggression induced by brain stimulation

Reference	Model	Results
Penaloza-Rojas et al. 1961	Electrical stimulation of the posterior hypothalamus in adult cats	2-5 mg/kg i.p. imipramine decreased and 8-10 mg/kg increased the response threshold for rage during hypothalamic stimulation.
Allikmets and Delgado 1968 Allikmets et al. 1968	EEG recordings and "spontaneous" behavior after electrical stimulation of the amygdala and hippocampus in restrained monkeys (*Macaca mulatta*)	100 uL imipramine and 100 uL amitriptyline increased the threshold necessary for aggressiveness and vocalizations and suppressed EEG activity in the amygdala.
Baxter 1968b	Hissing response via stimulation of the perifornical-ventromedial hypothalamus in cats	3-12 mg/kg i.p. imipramine produced no effects on the hissing response and ataxia.

447

Reference	Method	Results
Dubinsky and Goldberg 1970; Dubinsky et al. 1973	Hissing response and attack behavior via electrical stimulation of the perifornical region of the hypothalamus in adult male and female cats	Imipramine (ED50: 8.5 mg/kg i.p.) decreased attacks with inconsistent effects on the hissing response. 12 mg/kg i.p. desipramine inhibited attack in 3 of 6 cats.
Malick 1970	Hissing response via electrical stimulation of the perifornical region of the hypothalamus in cats	5 mg/kg i.p. imipramine lowered and 10 mg/kg increased the threshold to elicit the hissing response. 10, 20 mg/kg i.p. thiazesim increased "irritability" but failed to consistently decrease hiss threshold.
Funderburk et al. 1970	Hissing response via electrical stimulation of the perifornical region of the hypothalamus in adult male and female mongrel cats	5 mg/kg i.p. amitriptyline, desipramine and imipramine elevated the threshold to elicit the hissing response.

Lesion-induced aggression

Reference	Method	Results
Reis and Fuxe 1964	Sham rage in male and female adult cats produced by decerebration (lesions from the superior colliculus to the optic chiasm)	5-10 mg/kg i.v. protriptyline increased sham rage for up to 60 minutes; rage was abolished with 5 mg/kg i.v. haloperidol.
Malick et al. 1969; Sofia 1969b; Goldberg 1970	Biting of an inanimate object in male Long-Evans hooded rats with lesions of the septum, ventral-medial hypothalamus or olfactory bulbs; 3 point aggression scale	Imipramine suppressed aggression in septal (ED50: 52.7 mg/kg i.p.) and OB (ED50: 42.9 mg/kg) rats, but was inactive in VMH rats. Thiazesim decreased aggression in septal (ED50: 24.8 mg/kg i.p.), VMH (ED50: 23.5 mg/kg) and OB (ED50: 26.1 mg/kg) rats. Desipramine (ED50: 45.4 mg/kg i.p.) decreased biting in septal rats. Doses which were effective in suppressing aggression were neurotoxic.
Ueki et al. 1972	"Emotionality" (reactions to inanimate bite target, handling, tail pinching and mice) in male Wistar King A rats with lesions of the olfactory bulb, septum, or amygdala; 4 point scale	10, 20 mg/kg i.p. imipramine produced no effects on aggressive reactivity in septal rats except for a dose dependent decrease in muricide. 10, 20 mg/kg i.p. imipramine and 5, 10 mg/kg i.p. amitriptyline dose dependently decreased reactivity in OB rats. Effects were at doses which decreased ambulation and rearing.

448

Drug-induced Aggression

Lapin 1962	Groups of ten male albino mice treated with 20 mg/kg s.c. amphetamine; all subjects drug treated	0.5, 1 mg/kg s.c. imipramine potentiated amphetamine-induced motor excitation and aggressivity. 62.5 mg/kg imipramine decreased vocalizations and biting and motor activity.
Allikmets et al. 1969	Male and female cats treated with 200 ug acetylcholine to the amygdala	Pretreatment with 5 mg/kg i.m. imipramine decreased acetylcholine-induced "emotional" reactions, seizures and salivation.
Fog 1969	Groups of four male albino Wistar rats pretreated with 200 mg/kg s.c. pargyline; all subjects drug treated	50 mg/kg s.c. imipramine, desipramine, nortriptyline and amitriptyline induced vocalizations and defensive reactions without stereotypy in 100% of the rats.
Zetler and Otten 1969	Vocalizations and reaction to inanimate bite target in male Wistar rats	2.5-20 mg/kg i.p. imipramine does not induce aggression and vocalizations. 2.5-20 mg/kg imipramine dose dependently increased vocalizations and biting after 30 mg/kg i.p. isocarboxazid or 100 mg/kg i.p. iproniazid, but not 50 mg/kg i.p. phenelzine.
Schrold 1970	Pairs of 3-10 day old female White-Leghorn or male New Hampshire chicks; both subjects drug treated	6.3, 12.5 mg/kg i.p. imipramine, desipramine, 10, 20 mg/kg i.p. protriptyline and 5, 20 mg/kg nortriptyline dose dependently increased the intensity of attack pecks. 5-20 mg/kg i.p. amitriptyline produced no effect on pecking.
Yen et al. 1970	Biting of an inanimate object in male ICR mice after administration of 500 mg/kg i.v. dl-DOPA	Imipramine (ED50: 11.5 mg/kg i.p.), desipramine (ED50: 14 mg/kg i.p.) decreased DOPA-induced biting. Amitriptyline (ED50: 34 mg/kg i.p.) decreased DOPA-induced biting at sedative doses.
Zetler and Hauer 1975	Vocalizations and biting of an inanimate object in male Wistar rats pretreated with 30 mg/kg i.p. isocarboxazid	30 mg/kg i.p. imipramine increased vocalizations and attacks to the inanimate object.
Maj et al. 1979	Pairs of male Wistar rats treated with 5 mg/kg s.c. apomorphine; both subjects drug treated	14 day pretreatment with 10 mg/kg s.c. amitriptyline, imipramine, desipramine, mianserin and iprindole produced a 30-60% increased in the proportion of apomorphine-induced fighting pairs.

Maj et al. 1980 Maj et al. 1981 Maj et al. 1982 Maj et al. 1983	Groups of four male albino Swiss mice treated with 20 mg/kg clonidine i.p.; all subjects drug treated	Acute: 10 mg/kg i.p. amitriptyline, imipramine, mianserin and iprindole attenuated clonidine induced aggressiveness. 10 mg/kg i.p. (+), (-) oxaprotiline did not affect clonidine-induced aggression. Chronic: Clonidine induced aggression was enhanced at 2, but not 72 hours after 10 day treatment with 10 mg/kg i.p. b.i.d. amitriptyline, imipramine, mianserin, iprindole or (+)oxaprotiline.
Kostowski et al. 1986	Groups of 3 male albino Swiss mice administered 10 mg/kg i.p. clonidine; pairs of male Wistar rats administered 10 mg/kg i.p. apomorphine; all subjects drug treated	21 day pretreatment with 10 mg/kg i.p. desipramine increased the number of biting attacks, vocalizations in mice and increased aggression in rats.

Aggression by resident toward intruder

Avis and Peeke 1979	Male convict cichlids (*Cichlasoma nigrofasciatum*); resident drug treated	0.05, 1.0 mg/L (in aquarium water) imipramine decreased the frequency of attack displays.
Willner et al. 1981	Male Lister hooded rats; resident drug treated	7 day treatment with 5-20 mg/kg i.p. desmethylimipramine dose-dependently increased reactivity to handling. In the home cage, 20 mg/kg increased attacks to intruders. 7.5, 10 mg/kg produced aggressive behavior if period of drug treatment increased or 3-4 day withdrawal period was included.

Killing

Reference	Behavior	Results
Horovitz et al. 1965 Horovitz et al. 1966 Kulkarni 1968 Barnett et al. 1969 Sofia 1969a Goldberg 1970 Salama and Goldberg 1970 Valzelli and Bernasconi 1971 Hitchens et al. 1972 van Riezen et al. 1973 Rush and Mendels 1975 Valzelli and Bernasconi 1976 Eisenstein et al. 1982 Strickland and DaVanzo 1986	Muricidal behavior in male and female Wistar, Long-Evans, Sprague-Dawley and Holtzman rats	Amitriptyline (ED50: 5.1 mg/kg i.p.), imipramine (ED50: 8 mg/kg i.p.), desipramine (ED50: 9.8 mg/kg i.p.), mianserin (ED50: 10.5 mg/kg i.p.), trazodone (ED50: 7.2 mg/kg i.p.), bupropion (ED50: 15.3 mg/kg i.p.), aletamine (ED50: 1.0 mg/kg i.p.), thiazesim (ED50: 14.8 mg/kg i.p.), doxepine (ED50: 10 mg/kg i.p.), desmethyldoxepine ED50: 10 mg/kg i.p.), 10-30 mg/kg s.c. and p.o. OI77 (5-methylaminoacetyl-6-methyl-5,6-dihydro-phenanthridine-HCl) dose dependently decreased muricide with minimal effects on motor coordination.
Didiergeorges et al. 1968 Malick 1976 Yamamoto and Ueki 1978 Watanabe et al. 1979 Tobe et al. 1981 Hara et al. 1983 Shibata et al. 1983 Hara et al. 1984 Shibata et al. 1984 Iwasaki et al. 1986	Muricidal behavior in male Wistar and Long-Evans rats with olfactory bulbectomies	Peripheral administration: Imipramine (ED50: 25.8 mg/kg i.p.), amitriptyline (ED50: 34.0 mg/kg i.p.), desipramine (ED50: 3.3 mg/kg i.p.), MCI-2016 (ED50: 19.2 mg/kg i.p.), 10-20 mg/kg i.p. doxepine, 5-10 mg/kg i.p. maprotiline and 10-50 mg/kg i.p. lofepramine dose dependently decreased muricide in OB rats. There were no signs of tolerance after 21 day administration. Central administration: Bilateral injections of 5-20 ug imipramine, 10-30 ug amitriptyline or 10 ug doxepine into the medial amygdala, posterior lateral hypothalamus or amygdala suppressed muricide in OB rats.
Leaf et al. 1969	Muricidal behavior in male hooded Wesleyan and female albino Holtzman rats	Bilateral administration of 50 ug imipramine or thiazesim to the amygdala produced immediate inhibition of muricide for up to 2 hr. Thiazesim decreased muricide when administered to the medial n. of the amygdala in males and central or baso-lateral, -medial n. of the amygdala in females.
Rolinski 1975	Muricidal behavior in male and female Wistar rats pretreated with 400 mg/kg i.p. PCPA	20 mg/kg i.p. imipramine suppressed muricide.

451

Barr et al. 1976	Frog killing and muricide in male and female rats	12 mg/kg imipramine had no effect on attack and kill latencies of mice but inhibited frog kill latencies.
Leaf et al. 1978	Muricidal behavior in adult male and female cats	2-64 mg/kg i.p. imipramine and 4-32 mg/kg i.p. amitriptyline did not inhibit muricide.
Yamamoto and Ueki 1978	Muricidal behavior in male Wistar King A rats with midbrain raphe lesions	Imipramine (ED50: 17.8 mg/kg i.p.) and desipramine (ED50: 25.1 mg/kg i.p.) dose dependently decreased muricide in raphe lesioned rats.
Schmidt 1979 Schmidt 1980 Schmidt and Meierl 1980	Rat killing behavior in male ferrets	10, 40 mg/kg p.o. maprotiline and 17 day administration of 5, 10 mg/kg p.o. b.i.d. imipramine produced no effects on capture elicitation and attack.
Fujiwara et al. 1980	Muricidal behavior induced by chronic 6 mg/kg i.p. D^9-THC administration in Male Wistar King A rats	5-20 mg/kg i.p. imipramine, 10-30 mg/kg i.p. amitriptyline, 10-20 mg/kg i.p. doxepine, 5-10 mg/kg i.p. maprotiline, 10-50 mg/kg i.p. lofepramine and 5-20 mg/kg i.p. desipramine dose dependently decreased muricide.
Kostowski et al. 1984	Muricidal behavior in male Wistar rats with or without 3 150 mg/kg p.o. PCPA pretreatments	2.5, 5 mg/kg i.p. desipramine and nomifensine dose dependently decreased spontaneous and PCPA-induced muricide. Anti-muricide effects were greater in PCPA treated animals.
Al-Khatib et al. 1987	Muricidal behavior in adult male Wistar King-A rats with lesions of the nucleus accumbens	30 mg/kg i.p. imipramine, 15 mg/kg i.p. nomifensine and 15 mg/kg i.p. mianserin suppressed muricide.
Literature Reviews		
Valzelli 1967	Review of over 200 articles between 1934 and 1966 on drugs and aggressiveness	Antidepressants (amitriptyline, desipramine, imipramine and thiazesim) decrease aggression in isolation-, lesion-induced and muricide paradigms.

452

Karii et al. 1968	Review of 28 articles between 1956 and 1968 on brain stimulation and ablation effects on predatory aggression	Imipramine suppresses muricide through its action on the centromedial amygdala; the dose of imipramine required for suppression of muricide is doubled after deafferentiation of the olfactory bulb.
DaVanzo 1969	Review of 11 articles between 1942 and 1967 on drug effects on isolation-induced aggression	Amitriptyline and desmethylimipramine decrease isolation-induced aggression. This effect is potentiated with concurrent scopolamine administration.
Randrup and Munkvad 1969a Kršiak 1974b	Review of over 200 animal articles between 1923 and 1973 on the pharmacology of aggression; focus is on mouse aggression and stereotyped hyperactivity and "rage"	Imipramine inhibits shock-induced and isolation-induced fighting. Pretreatment with MAOI's plus tricyclics produce vocalizations and boxing postures.
Carlini et al. 1976	Review of 80 articles between 1964 and 1975 on drug and environmental factors in marihuana effects	Nomifensine induces aggression in 3 day REM sleep deprived animals.
Miczek and Barry 1976 Miczek 1987 Miczek and Donat 1989	Review of over 1500 animal articles between 1920 and 1989 on the pharmacology of sex and aggression	Imipramine decreases intraspecies aggressive behavior at high doses which also impair motor activity. Imipramine decreases killing and can increase aggressive activities in footshock and drug-induced paradigms, but the interpretation is problematic.
Pinder et al. 1977 Malick 1979	Review of over 125 animal and human articles between 1942 and 1977 on the pharmacology of aggression; focus is on maprotiline and isolation-induced aggression	Isolation-induced fighting is selectively antagonized with antidepressants (including amitriptyline, desipramine, doxepine, imipramine and maprotiline) at doses not producing neurological impairment.
Sheard 1977a Eichelman 1979 Maj 1980 Sheard 1981	Review of over 200 animal articles between 1928 and 1980 on animal models of aggressive behavior	Chronic administration of imipramine, amitriptyline, desipramine, mianserin and iprindole facilitate shock-, apomorphine- and clonidine-induced aggression by prolonging the action of NE in the synapse.
Delini-Stula and Vassout 1979	Review of 51 articles between 1956 and 1978 on the effects of psychoactive drugs on aggressive behavior in mice and rats	Predatory, but not shock-induced aggression appears to be consistently inhibited by antidepressants. Amitriptyline and maprotiline suppress isolation-induced aggression.

453

B. SEROTONERGIC REUPTAKE BLOCKERS

Isolation-induced aggression

Reference	Subject	Results
Ogren et al. 1980	Male albino mice; one subject drug treated	5 mg/kg i.p. zimelidine inhibits aggressive behavior by 60% for up to 4 hours.
Delini-Stula and Vassout 1981	Male albino NMRI mice; acute and 21 day treatment; both subjects drug treated	Acute administration of 10, 25 mg/kg i.p. clomipramine did not affect aggression. Chronic treatment with 10 mg/kg clomipramine decreased the proportion of animals fighting by 40%.
Poshivalov 1981	Male CC57W mice with or without 10 mg/kg i.p. l-DOPA or 500 mg/kg i.p. PCPA pretreatment; one subject drug treated	10 mg/kg i.p. fluoxetine decreased aggressive, social and sexual behavior while increasing defensive behavior. After PCPA or l-DOPA pretreatment, 10 or 20 mg/kg fluoxetine (respectively) decreased aggressive, social and defensive behavior.
Olivier and van Dalen 1982	Male mice; one subject drug treated	25-50 mg/kg i.p. fluvoxamine and chlorimipramine decreased aggressive and increased defensive behaviors.
Oehler et al. 1985b	Male albino AB/Jena mice; 3 point scale	20 day administration of 5 mg/kg (in water supply) clomipramine had no effect on aggressivity.

Pain-induced aggression and defense

Reference	Subject	Results
Sheard et al. 1977	Footshock to pairs of male albino Sprague-Dawley rats treated with 30 ug/kg i.p. LSD; both subjects drug treated	5 mg/kg i.p. chlorimipramine pretreatment antagonized LSD-enhanced fighting while having no effect on fighting without LSD.
Marini et al. 1979	Footshocks to pairs of male albino Sprague-Dawley rats; both subjects drug treated	2.5, 5 mg/kg i.p. chlorimipramine produced a nonsignificant decrease in fighting.
Rolinski and Herbut 1981	Footshock to pairs of male Swiss mice; both subjects drug treated	0.75, 1.25 mg/kg i.p. fluoxetine dose dependently decreased fighting episodes with a 15-26% decrease in locomotion.

454

Reference	Method	Results
Valdman and Poshivalov 1986	Footshock to pairs of male CC57W mice; 7 day pretreatment with 0.5 mg/kg i.p. b.i.d. reserpine; both subjects drug treated	7 day administration of 10 mg/kg i.p. chlorimipramine or zimelidine reduced timid-defensive and restored aggressive behavior due to nociceptive stimulation.
Vogel et al. 1988	Neonatal administration of 15 mg/kg s.c. b.i.d. clomipramine (given from postnatal days 8-21), subsequently, footshocks to pairs of adult male Sprague-Dawley rats; one subject drug treated	Clomipramine treated subjects displayed fewer offensive (offensive uprights, lateral crouch, mounting, leaps toward other subject in response to shock) and more defensive responses (defensive uprights, freezing crouch, supine) than control subjects.

Defensive aggression induced by brain stimulation

Reference	Method	Results
Dubinsky et al. 1973	Hissing response and attack behavior via electrical stimulation of the perifornical region of the hypothalamus in male and female cats	Chlorimipramine (ED50: 3.4 mg/kg i.p.) decreased attacks with inconsistent effects on the hissing response.

Drug-induced aggression

Reference	Method	Results
Schrold 1970	Pairs of 3-10 day old female White-Leghorn or male New Hampshire chicks; both subjects drug treated	5, 10 mg/kg i.p. chlorimipramine produced a moderate increase in pecking.
Maj et al. 1979	Pairs of male Wistar rats treated with 5 mg/kg s.c. apomorphine	14 day pretreatment with 10 mg/kg s.c. clomipramine produced a 60% increase in the proportion of apomorphine-induced fighting pairs.
Maj et al. 1981	Groups of four male albino Swiss mice treated with 20 mg/kg clonidine i.p.	10 mg/kg i.p. fluoxetine and zimelidine attenuated clonidine induced aggressiveness. Clonidine induced aggressiveness was enhanced 2, but not 72 hours after chronic administration of zimelidine, but not by fluoxetine (10 mg/kg i.p. b.i.d. for 10 days).

Aggression induced by REM sleep deprivation

Reference	Method	Results
Carlini and Lindsey 1982	Male albino Wistar rats undergoing 72 hour REM sleep deprivation followed by 2.5, 5 mg/kg i.p. THC; both subjects drug treated	10, 20 mg/kg i.p. fluoxetine produced aggressive behavior in non-THC treated rats. 5 mg/kg fluoxetine potentiated aggressiveness in THC-treated rats.

Female aggression

Reference	Study	Results
Haug et al. 1990	Groups of 4 female Swiss albino mice; 3 drug treated littermates and 1 untreated lactating dam	2-8 mg/kg i.p. fluoxetine produced no effect on biting attacks or attack latency.

Killing

Reference	Study	Results
Marks et al. 1978	Muricide and lesions of ascending 5HT projections (dorsal and median raphe nuclei) in adult male Wistar rats	10-25 mg/kg i.p. chlorimipramine dose dependently inhibited muricide in non-lesioned rats but produced only modest suppression of muricide in lesioned rats without motor impairment.
Yamamoto and Ueki 1978 Shibata et al. 1984	Muricide in adult male Wistar King A rats with midbrain raphe lesions or olfactory bulbectomies; acute and chronic (up to 21 day) administration	Chlorimipramine suppressed muricide in raphe lesioned (ED50: 10.0 mg/kg i.p.) and OB (ED50: 27.5 mg/kg) rats. 20 mg/kg s.c. chlorimipramine decreased muricide by 50% with slight tolerance.
Fujiwara et al. 1980	Muricide induced by chronic D^9-THC administration in Male Wistar King A rats	5-20 mg/kg i.p. chlorimipramine dose dependently decreased muricide.
Schmidt 1980 Schmidt and Meierl 1980	Rat killing behavior in male ferrets	15 mg/kg p.o. chlorimipramine and 6, 10 mg/kg p.o. fluoxetine produced no effects on capture elicitation and attack.
Berzsenyi et al. 1983 Kostowski et al. 1984 Molina et al. 1987	Muricidal behavior in male Wistar rats treated with 2-3 150 mg/kg p.o. PCPA pretreatments or raphe lesions	2.5-28 mg/kg i.p. fluoxetine, 2.5-12.5 mg/kg i.p. citalopram dose dependently decreased muricide with minimal effects on motor activity in intact, PCPA treated and raphe lesioned rats. Anti-muricide effects were greater in PCPA treated animals.
Al-Khatib et al. 1987	Muricidal behavior in adult male Wistar King-A rats with lesions of the nucleus accumbens	15 mg/kg i.p. zimelidine suppressed muricide.

Literature Reviews

Reference	Study	Results
Sheard 1977a	Review of 119 animal articles between 1928 and 1977 on animal models of aggressive behavior	Chlorimipramine has no effect on shock-elicited fighting, but antagonizes LSD-potentiation of shock-elicited fighting.

Eichelman 1979	Review of 177 articles between 1953 and 1979 on the role of the biogenic amines and aggressive behavior	Tricyclic antidepressants block muricide by blocking 5-HT reuptake or degradation. Chronic administration of antidepressants increases shock-induced fighting.
Malick 1979	Review of 49 articles between 1942 and 1977 on the pharmacology of isolation-induced aggression in mice	Isolation-induced fighting is selectively antagonized with chlorimipramine at doses not producing neurological impairment.
Maj 1980	Review of 68 articles between 1970 and 1980 on the action of antidepressants	Chronic administration of chlorimipramine enhance aggressiveness induced by apomorphine.
Miczek 1987 Miczek and Donat 1989	Review of over 1500 articles between 1920 and 1988 on the pharmacology of aggression	Tricyclic antidepressants decrease isolation-induced and predatory aggression. Inconsistent effects are observed with acute administration while chronic administration increases shock-induced aggression.

C. MONOAMINE OXIDASE INHIBITORS (MAOI)

Isolation-induced aggression

DaVanzo et al. 1966	Male C57B1/10J and ICR mice; one subject drug treated	Phenelzine sulfate (ED50: 33.4 mg/kg i.p.), isocarboxazid (ED50: 43.5 mg/kg i.p.) and etryptamine (ED50: 8 mg/kg i.p.) decreased fighting without affecting motor behavior.
Valzelli et al. 1967	Groups of 3 male Swiss albino mice; all subjects drug treated; 5 point aggression scale	20 mg/kg i.p. phenelzine did not inhibit aggression. 5, 10 mg/kg i.p. pheniprazine and 5-15 mg/kg i.p. tranylcypromine inhibited aggression in 50% of the subjects.
Welch and Welch 1968b	Male Swiss mice; both subjects drug treated	50 mg/kg i.p. pargyline increased biting contacts, whole brain NE (22%), DA (35%) and 5-HT (17%); 100 mg/kg decreased biting contacts.

Pain-induced aggression and defense

Tedeschi et al. 1969	Footshock to pairs of male mice; both subjects drug treated	200 mg/kg i.p. iproniazid increased fighting without affecting motor behavior.

457

Eichelman and Barchas 1975	Footshocks to pairs of male Sprague-Dawley rats; both subjects drug treated	3-5 day administration of 100 mg/kg nialamide i.p., 150 mg/kg i.p. iproniazid or 20 mg/kg i.p. pargyline potentiated attacks without altering pain thresholds.
Rolinski and Herbut 1981	Footshock to pairs of male Swiss mice; both subjects drug treated	30 mg/kg pargyline increased the number of fighting episodes.
Valdman and Poshivalov 1986	Footshock to pairs of male CC57W mice; 7 day pretreatment with 0.5 mg/kg i.p. b.i.d. reserpine; both subjects drug treated	7 day administration of 10 mg/kg i.p. pyrazidol reduced timid-defensive and restored aggressive behavior.

Defensive aggression induced by brain stimulation

| Malick 1970 | Hissing response via electrical stimulation of the perifornical region of the hypothalamus in cats | 4-10 mg/kg i.p. tranylcypromine increased behavioral "irritability" and decreased the threshold to elicit the hissing response. 50-125 mg/kg i.p. pargyline and 5-20 mg/kg i.p. phenelzine increased "irritability" but failed to consistently decrease hiss threshold. |

Lesion-induced aggression

| Malick et al. 1969 | Biting of an inanimate object in male Long-Evans hooded rats with lesions of the septum, ventral-medial hypothalamus or olfactory bulbs | Pargyline decreased aggression in septal (ED50: 52.2 mg/kg i.p.), VMH (ED50: 193.7 mg/kg) and OB (ED50: 139.2 mg/kg) rats. Phenelzine decreased aggression in septal (ED50: 40.2 mg/kg i.p.), VMH (ED50: 180.7 mg/kg) and OB (ED50: 38.2 mg/kg) rats. Tranylcypromine decreased aggression in septal (ED50: 5.8 mg/kg i.p.), but not in VMH or OB rats. |

Drug-induced aggression

| Scheel-Kruger and Randrup 1968 | Pairs of male Wistar rats administered pargyline with or without disodium diethyldithiocarbamate (DDC) pretreatment; both subjects drug treated | 150 mg/kg s.c. pargyline produced defensive aggressive behavior in rats receiving DDC pretreatment. DDC and pargyline reduced brain NE and normetanephrine. |
| Fog 1969 | Groups of four male albino Wistar rats; all subjects drug treated | 200 mg/kg s.c. pargyline failed to induce vocalizations and defense postures. |

458

Reference	Subjects	Results
Reis et al. 1970	Male and female mongrel cats with or without 20 mg/kg i.p. l-DOPA	20 mg/kg i.p. pheniprazine produced no effect on aggressive behavior.
Yen et al. 1970	Biting of an inanimate object in male ICR mice after administration of 500 mg/kg i.v. dl-DOPA	100 mg/kg i.p. iproniazid and 3-6 mg/kg i.p. tranylcypromine increased the intensity and duration of DOPA-induced biting.
Zetler and Hauer 1975	Vocalizations and biting of an inanimate object in male Wistar rats	30 mg/kg i.p. isocarboxazid increased vocalizations and attacks.
Rolinski and Herbut 1979	Pairs of male Wistar rats treated with 20 mg/kg apomorphine; both subjects drug treated	50 mg/kg i.p. pargyline suppressed aggressive behavior.

Aggression by resident toward intruder

Reference	Subjects	Results
Avis and Peeke 1979	Male convict cichlids (*Cichlasoma nigrofasciatum*); resident drug treated	5, 10 mg/L (in aquarium water) pargyline dose dependently decreased attack displays.
Payne et al. 1985	Male hamsters; resident drug treated	70 mg/kg i.p. pargyline decreased attacks and increased attack latency.

Killing

Reference	Subjects	Results
Horovitz et al. 1965 Horovitz et al. 1966 Hitchens et al. 1972	Muricidal behavior in female Holtzman rats	Iproniazid (ED50: 155 mg/kg i.p.) and phenelzine (ED50: 5 mg/kg i.p.) inhibited muricide with minimal effects on motor coordination.
Sofia 1969a	Muricidal behavior in male hooded Long-Evans rats	Pargyline (ED50: 127.6 mg/kg i.p.), tranylcypromine (ED50: 4.1 mg/kg i.p.), etryptamine (ED50: 9.77 mg/kg i.p.), nialamide (ED50: 158.9 mg/kg i.p.), isocarboxazid (ED50: 28.7 mg/kg i.p.) and phenelzine (ED50: 28.0 mg/kg i.p.) decreased muricide. Iproniazid (ED50: 216.9 mg/kg i.p.) decreased muricide at neurotoxic doses.
Leaf et al. 1978	Muricidal behavior in adult male and female cats	0.25-1 mg/kg i.p. tranylcypromine did not inhibit muricide.
Watanabe et al. 1979	Muricidal behavior in male Wistar King A rats with olfactory bulbectomies	Bilateral injection of 100 ug nialamide into the medial, basal or anterior amygdala, lateral septum or ventral hippocampus did not inhibit muricide.

| Isel et al. 1988
Isel and Mandel 1989 | Muricidal behavior in male Wistar rats with or without 2 day 150 mg/kg i.p. PCPA treatment or raphe lesions | 18.6, 37.2 mg/kg i.p. moclobemide moderately decreased muricide while decreasing motor activity in intact, PCPA treated and raphe lesioned rats. 5.0-11.8 mg/kg i.p. cimoxatone, 24.2-72.5 mg/kg i.p. toloxatone, 3-10 mg/kg i.p. minaprine dose dependently decreased muricide with minimal effects on locomotor activity in intact, PCPA treated and raphe lesioned rats. |

Literature Reviews

Valzelli 1967	Review of over 200 articles between 1934 and 1966 on drugs and aggressiveness	MAOI's (iproniazid, phenelzine and pheniprazine) increase aggression in isolation-, shock-induced and muricide paradigms.
Randrup and Munkvad 1969a	Review of 80 animal articles between 1923 and 1969 on the mechanisms involved in stereotyped hyperactivity and "rage"	MAOI's with and without DOPA produce aggression in pairs of rats and mice and may be related to NE effects.
Welch and Welch 1969a Welch and Welch 1973 Malick 1979	Review of over 60 animal articles between 1942 and 1977 on aggression and the biogenic amines	MAOI's (pargyline, pheniprazine, isocarboxazid, etryptamine and phenelzine) decreases isolation-induced aggression without producing neurological impairment by accelerating the release of NE, DA and 5HT. However, low doses may enhance fighting for a short period of time.
Kršiak 1974b Sheard 1977a Eichelman 1979 Sheard 1981	Review of over 200 animal articles between 1928 and 1979 on the pharmacology of aggression	MAOI's increase shock-elicited fighting and decrease muricide. MAOI's plus tricyclics or adrenergic compounds produce vocalizations and boxing postures.
Miczek 1987 Miczek and Donat 1989	Review of over 1500 articles between 1920 and 1989 on the pharmacology of aggression	MAOI's decrease isolation-induced and predatory aggression. Acute and chronic MAOI's decrease and increase shock-induced aggression, respectively.

Table 8B: Effects of Antidepressant Drugs on Aggression in Humans

REFERENCES	METHODS AND PROCEDURES	RESULTS AND CONCLUSIONS
A. NORADRENERGIC REUPTAKE BLOCKERS		
Behavioral disorders in juveniles		
Winsberg et al. 1972	32 male and female hyperkinetic and aggressive children (age: 5.3-13.6 years) in New York; 39-item behavior rating scale; double-blind, placebo control	50 mg p.o. t.i.d. imipramine decreased aggressivity, hyperactivity and inattention in 69% of the patients compared to 44% of the patients receiving 5-10 mg p.o. t.i.d. d-amphetamine. Cognitive impairment was not observed.
Rapoport et al. 1974	Conners rating scale, diary evaluation, and psychiatric evaluation of 76 male hyperactive (age: 6-12 years) outpatients; double-blind, placebo control	80 mg/day imipramine decreased hyperactivity and produced cognitive improvement in the most inhibited, anxious children. Unusually aggressive children showed no improvement. Side effects included sedation, irritability, insomnia, decreased appetite, nausea, sadness and increased blood pressure.
Waizer et al. 1974	Psychiatric and psychological evaluation of 19 male hyperactive children (age: 6-12 years) in New York; placebo control	50-75 mg t.i.d imipramine reduced hyperactivity, defiance and inattentiveness and increased sociability. Side effects included anorexia and insomnia but "represented no serious problem."
Yepes et al. 1977	BRS, teacher and parent rating of 21 male and 1 female (mean age: 9.2 years) aggressive or hyperkinetic outpatients; double-blind, placebo control	17-50 mg p.o. t.i.d. amitriptyline decreased hyperactivity, aggression and produced sedation. This was comparable to 10-30 mg p.o. b.i.d methylphenidate without producing cognitive impairment.
Pallmeyer and Petti 1979	2 male patients (age: 6, 12 years) with childhood depression in Pittsburgh	3.5-5 mg/kg p.o. daily imipramine markedly increased aggressive and hostile behavior which subsided with discontinuation.

461

Study		
Puig-Antich 1982	Unstructured observations and K-SADS-P administered to 43 prepubertal male patients (mean age: 9.6 years) with major depression including conduct disorder; open trials and double-blind, placebo control	5-13 week treatment with 5 mg/kg/day imipramine improved affect followed by an abatement of conduct disorder.
Inpatient studies		
Wood et al. 1976	MMPI, WRAT and WAIS and self-evaluation in 15 male and female patients (mean age: 28 years) with minimal brain dysfunction; comparison of methylphenidate, pemoline and imipramine; open trial	One patient not responding favorably to methylphenidate or pemoline became less anxious, irritable and angry with 10 mg/day imipramine. Adults with minimal brain dysfunction manifested the same response to medication as do children.
Rampling 1978	Tricyclic-induced aggressiveness in 3 male and 1 female patients (age: 26-52 years) in Australia	Untoward immediate aggressiveness was a rare (possibly underreported) side effect of tricyclics (25 mg b.i.d. imipramine, 25 mg t.i.d. and 150 mg h.s. amitriptyline).
George et al. 1989	3 male and female patients (age: 28-32 years) with aggressive behavior and panic disorder	100 mg daily desipramine or 200 mg daily imipramine for panic disorder reduced anger, rage and assaultive behavior.
Outpatient studies		
Panter 1977	Untreatable, verbally and physically assaultive 30 year old female in California	200 mg daily doxepin produced no clear improvement of impulsive and dangerous behavior.
Experimental studies		
Overall et al. 1964	BPRS, IMPS and MMPI administered to 68 male schizophrenic and 77 male depressive patients (age: 25-76 years); double blind	240 mg p.o. daily imipramine produced minimal or no improvement on hostility and uncooperativeness in schizophrenic and depressive patients.
Gottschalk et al. 1965	Analysis of speech for hostility in 5 patients (age: 16-52); single and double blind, placebo control	50-200 mg p.o. q.i.d. imipramine increased verbal overt outward hostility and anxiety scores.

462

Itil et al. 1972	Clinical Global Score, HZI Depression, Hamilton Anxiety, anxiety self-rating and psychosomatic rating scales administered to 25 male and female mild to moderate depressive outpatients; double-blind, placebo control	Daily administration of 60-200 mg p.o. amitriptyline or 20-72 mg p.o. mianserin decreased anxiety and irritability which was maximal after 3 weeks; mianserin produced no side effects except for slight sedation.

Literature Reviews

Itil and Seaman 1978 Miczek 1987	Review of over 1500 articles between 1920 and 1987 on the pharmacology of aggression	Antidepressants are ineffective or inconsistent in the treatment of aggression in adults. In hyperactive children, imipramine and amitriptyline can reduce hyperactivity but not always conduct disorders.
Gunn 1979 Tupin 1985	Review of over 35 articles between 1967 and 1985 on the psychopharmacology of aggression	Antidepressants (imipramine, amitriptyline) produce a 60-70% remission of violent outbursts in depressed patients; there are some reports of paradoxical rage.

B. SEROTONERGIC REUPTAKE BLOCKERS

Outpatient studies

Yaryura-Tobias and Naziroglu 1978	12 female outpatients (age: 14-39 years) with obsessive-compulsive, aggressive and self-mutilating behavior in New York	260 mg p.o. daily chlorimipramine decreased obsessive-compulsive symptoms, aggression and self-mutilation in 75% of the patients; 25% worsened and were diagnosed as schizophrenic.

Literature Reviews

Rapoport 1989	Review of clomipramine and obsessive-compulsive disorder	4% of obsessive-compulsive patients have aggressive thoughts; clomipramine appears to have selective anti-obsessive-compulsive properties.

Table 9A: Effects of Lithium on Aggression in Animals

REFERENCES	METHODS AND PROCEDURES	RESULTS AND CONCLUSIONS
Isolation-induced Aggression		
Weischer 1969	Male albino NMRI mice; one subject drug treated	30 mEq/L lithium (in drinking water) decreased aggression after 2-3 weeks in 35% of the animals with a 14% mortality rate.
Brain 1972	Male albino mice; one subject drug treated	Daily intake of 4.83 ml (0.9%) lithium chloride (in drinking water) for 15 days decreased fighting duration and number of attacks, this corresponded to an increase in adrenocortical and a decrease in gonadal function.
Eichelman et al. 1977	Male CF1 mice; both subjects drug treated	14 day administration of 1.5-6.0 mEq/kg i.p. b.i.d. lithium dose dependently decreased fight duration.
Malick 1978b	Male CF1-S mice; both subjects drug treated	Acute administration of 40-300 mg/kg i.p. lithium did not inhibit the proportion of animals fighting. 5 day repeated administration of 40-300 mg/kg lithium dose-dependently decreased fighting.
Brain and Al-Maliki 1979	Male albino TO mice; one subject drug treated	0.2, 0.4 mEq i.p. lithium decreased the proportion of animals fighting, attacks, attack duration and increased the attack latency.
Grimm and Zelikovich 1982	Male SJL/J mice; one subject drug treated	Acute administration of 5 mEq/kg i.p. lithium eliminated attacks and nonaggressive social behavior in 40% of the subjects; 4 day administration of 5 mEq/kg lithium eliminated attacks and nonaggressive social behavior in 80% of the subjects.
Oehler et al. 1985b	Male albino AB/Jena mice; 3 point scale	20 day administration of 4 nmol/kg (in water supply) lithium decreased attacks.

464

Pain-induced aggression and defense

Sheard 1970b	Footshock to pairs of Sprague-Dawley rats; both subjects drug treated	5 mEq/kg i.p. lithium for 5 days increased attack latency and shock level sustained without fighting and did not alter the pain threshold. Polydipsia noted.
Bisbee and Cahoon 1973	Footshock to male Sprague-Dawley rats; inanimate bite target	3.2-12.7 mg/kg i.p. lithium produced no effect on biting when there was no shock delivery. When shock administered, 3.2 mg/kg lithium increased and 9.5, 12.7 mg/kg decreased biting.
Marini et al. 1979	Footshock to pairs of male albino Sprague-Dawley rats; both subjects drug treated	20-60 mEq/kg (in food supply) lithium decreased fighting; lithium antagonized 20, 40 ug/kg i.p. LSD-induced increases and 2.5, 5 mg/kg i.p. chlorimipramine-induced decreases in fighting.
Mukhurjee and Pradhan 1976a	Footshock to pairs of male Walter Reed rats; both subjects drug treated; 3 point scale	1-3 mEq/kg i.p. once a week lithium dose dependently decreased attacks for up to 48 hours. 3 mEq/kg lithium antagonized d-amphetamine-induced increases and scopolamine-induced decreases in fighting.
McGlone et al. 1980	Footshock to male and female Sprague-Dawley rats with lesions to the area postrema; both subjects drug treated	5 day administration of 5 mEq/kg i.p. lithium increased attack latency (without altering attack frequency) which was abolished with area postrema lesions.
Brain et al. 1981	Footshock to pairs of male TO mice; one subject drug treated	0.2 mEq i.p. lithium produced a nonsignificant decrease in the frequency of attacks.
Prasad and Sheard 1982	Footshock to pairs of male Sprague-Dawley rats with and without 14 day treatment with 15 mg/kg i.p. desipramine; both subjects drug treated	2 week administration of 20 mEq/L lithium (in drinking water) decreased fighting; there was a greater decrease in desipramine-treated rats.

Drug-induced aggression

Delgado and DeFreudis 1969	EEG recordings and "spontaneous" behavior in restrained adult monkeys (*Macaca mulatta*)	50-100 uL i.c. lithium into the amygdala and hippocampus diminished restlessness and aggressiveness in reaction to restraint while producing high voltage, low frequency EEG waves.

465

Reference	Method	Findings
Ozawa et al. 1975	Groups of 4 male ddI mice pretreated with 40 mg/kg nialamide plus 100 mg/kg l-DOPA or 5-40 mg/kg clonidine; all subjects drug treated	100, 200 mg/kg lithium potentiated biting attacks in nialamide plus l-DOPA and clonidine pretreated rats by 200-300% without affecting whole brain NE, DA or 5-HT.
Allikmets et al. 1979	14 day treatment with 1 mg/kg i.p. haloperidol plus 2 mEq/kg lithium or 2 mEq/kg i.p. lithium to pairs of male Sprague-Dawley rats treated with 5 mg/kg i.p. apomorphine; both subjects drug treated; 4 point aggression scale	Haloperidol potentiated apomorphine-induced stereotypy and aggression which was blocked by concurrent lithium administration. Lithium alone produced no effects on aggression and decreased apomorphine-induced stereotypy.

Lesion-induced Aggression

Reference	Method	Findings
Mukhurjee and Pradhan 1976b	Hyperexcitability (including biting) in bilateral medial and lateral septal lesioned male Walter Reed and 3 NIH black rats; 4 point scale	1-3 mEq/kg i.p. once a week lithium dose dependently decreased hyperexcitability; there was an earlier onset and longer duration of action with higher doses.

Aggression by resident toward intruder

Reference	Method	Findings
Weischer 1969	(1) Male and female hamsters; resident drug treated. (2) Reactions of Siamese fighting fish (Betta splendens) to intruder or mirror image; resident subject drug treated	(1) 30 mEq/L lithium (in drinking water) decreased aggression after 8-10 days in 55% of the animals. (2) 10-30 mEq/L lithium (in aquarium water) decreased aggression in 25% of the animals for up to 8 hours.
Sheard 1970a	Male Sprague-Dawley rats in aggressive and sexual interactions with male, female rats or white mice	5 mg/kg IP lithium for 5 days abolished sexual and aggressive behavior due to subsequent PCPA administration. Forebrain levels of 5HIAA were decreased, there was no change in 5HT.
Sheard 1973	Male Sprague-Dawley rats; resident drug treated	1.5 mEq/L in drinking water for 5 days abolished territorial aggression.

Female aggression

Reference	Method	Findings
Brain and Al-Maliki 1979	Introduction of male TO mice to lactating female albino TO mice; female drug treated	0.2 mEq i.p. lithium did not decrease the proportion of animals fighting, attacks, attack duration or attack latency.

Killing

Krames et al. 1973 O'Boyle et al. 1973	Muricidal behavior in male Long-Evans and Holtzman rats	Lithium (2% body weight, 0.15 M i.p.) administered immediately, but not at 3 hr, after muricide increased (300%) latency to kill in subsequent encounters.
Gustavson et al. 1974	Rabbit and lamb killing in male and female coyotes	2.5 g i.p. lithium administered after the kill produced sickness and aversion to the eating of prey. After two trials, lithium suppressed attack behavior.
Rush and Mendels 1975 Broderick and Lynch 1982	Muricidal behavior in male Sprague-Dawley and Long-Evans rats	Lithium (0.5-12 mEq/kg i.p. or 2 mEq/kg i.p. b.i.d. for 7-10 days) dose dependently decreased muricide in conjunction with increased fore- and hind-brain 5HT turnover. Highest doses impaired motor behavior and were neurotoxic.
Mukhurjee and Pradhan 1976b	Muricidal behavior in male hooded Long-Evans, NIH black rats and 2 bilaterally septal lesioned male NIH black rats	1, 2 mEq/kg i.p. once per week lithium had no effect on muricide; 3 mEq/kg i.p. decreased muricide in 30% of the rats.
Brain and Al-Maliki 1979	Locust killing in male and female albino TO mice	0.2 mEq i.p. lithium had no effect on experienced or nonexperienced male or female "killers".
Klunder and O'Boyle 1979 Langley 1981	Cricket killing in male ICR or wild northern grasshopper mice	Lithium (0.5-2.0% body weight, 0.15 M i.p.) increased attack latency, suppressed feeding, but did not decrease the killing of live prey. Attack of dead prey was suppressed.
Yamamoto et al. 1985	Muricidal behavior in male Wistar King A rats with midbrain raphe lesions or olfactory bulbectomies	Acute administration of 100 mg/kg i.p. lithium suppressed muricide in 25% and 35% of the raphe lesioned and bulbectomized rats, respectively; this suppression increased to 66% and 80% after 5-14 administrations.

Literature Reviews

Sheard 1977a	Review of 119 animal articles between 1928 and 1977 on animal models of aggressive behavior	Lithium dose dependently decreases electrical stimulation-induced aggression. 4-5 week administration of lithium decreases shock elicited fighting.

467

Malick 1979	Review of 49 articles between 1942 and 1977 on the pharmacology of isolation-induced aggression in mice	Isolation-induced fighting is selectively antagonized with lithium at doses not producing neurological impairment.
Svare and Mann 1983	Review of 41 articles between 1971 and 1982 on hormonal influences on maternal aggression	Lithium produces no effect on postpartum aggression but suppresses intermale aggression in mice.
Müller-Oerlinghauser 1985	Review of 40 animal and human articles between 1969 and 1985 on long term lithium treatment	Lithium enhances l-tryptophan inhibition of muricide.
Miczek 1987	Review of over 1500 articles between 1920 and 1987 on the pharmacology of animal and human aggression	Lithium decreases isolation-, shock-, drug- and lesion-induced aggression although higher doses are quite toxic. At nontoxic low doses, enhanced shock- or drug-induced aggression may be seen; lithium is ineffective in decreasing muricide.

TABLE 9B: Effects of Lithium on Aggression in Humans

References	Methods and Procedures	Results and Conclusions
Behavioral disorders in juveniles		
Campbell et al. 1972	10 male and female severely hyperactive and severely disturbed children (age: 3-6 years); crossover, double-blind	0.3-1.2 mEq/L (serum) lithium moderately diminished explosivity, hyperactivity, aggressiveness and psychotic speech while producing nausea and sedation/motor excitation.
Kelly et al. 1976	4 year treatment of a physically and verbally abusive 15 year old mentally handicapped female patient diagnosed with manic-depressive illness	0.5-1.1 mEq/L (serum) lithium increased the quality of self-control and social behavior with no hypothyroid side effects.
Platt et al. 1981 Campbell et al. 1982 Platt et al. 1984a,b	DSM3 and cognitive inventories of over 70 male and female treatment-resistant aggressive 5-13 year old inpatients in New York; double-blind, placebo control	0.32-1.51 mEq/L (serum) lithium was superior to haloperidol and chlorpromazine in decreasing aggressiveness and explosiveness. Cognitive, sedating and insomnia side effects were mild.
Vetro et al. 1985	Parent, teacher interviews, hospital records and PFT administration to 17 hyperaggressive male and female children (age: 3-12 years)	0.68 mmol/L (serum) lithium successfully treated aggressivity in 76% of the patients without sedation or cognitive impairment. Weight gain and gastrointestinal symptoms were present in 41% of the patients.
Inpatient studies		
Dostal and Zvolsky 1970	14 phenothiazine resistant aggressive and severly mentally handicapped male patients (age: 11-17 years)	0.3-0.95 mEq/L (serum) lithium reduced aggressiveness and undisciplined behavior in 79% of the patients; unusually severe polydipsia and polyuria present.
Martorano 1972	2 violent male paranoid schizophrenics (ages: 21 and 25 years)	0.3-0.7 mEq/L (serum) lithium abolished aggressive and acute psychotic reactions within 3 weeks. Affective rage and hyperactivity may be associated with lithium efficacy.

Tupin 1972	Untreatable, assaultive and self-injurious behavior in 10 patients; open trials; includes case study of a 26 year old male; diagnoses varied: antisocial personality, mental deficiency, explosive-aggressive personality	Lithium suppressed belligerent angry and self-injurious behavior (with and without an episodic course).
Micev and Lynch 1974	6 male and 4 female, untreatable, aggressive and self-mutilating mentally handicapped inpatients (age: 16-58 years) in the U.K.	0.6-1.4 mEq/L (serum) lithium modestly to significantly improved aggressive outbursts in 89% of the patients. Self-injurious behavior was abolished in 75% of the patients.
Shader et al. 1974	Untreatable, aggressive and assaultive 34 year old female in Massachusetts	0.6-1.0 mEq/L (serum) lithium decreased impulsiveness, aggressiveness and feelings of anger.
Lion et al. 1975b	Untreatable, impulsive and assaultive 27 year old mentally handicapped male patient in Maryland	1.2 mEq/L (serum) lithium reduced agitation, aggressiveness and restlessness.
Altshuler 1977	Untreatable, impulsive and aggressive behavior in 9 male and female early total deafness patients (age: 16-52 years) in New York: 6 schizophrenics, 2 acting out adolescents and 1 personality disorder	0.42-0.8 mEq/L (serum) lithium decreased or completely suppressed violent, impulsive and aggressive behavior in 78% of the patients. There was no improvement in 1 schizophrenic patient and 1 other patient due to malaise.
Goetzl et al. 1977	3 aggressive male and female mentally handicapped patients (age: 16-20 years) in New York	0.6-0.9 mEq/L (serum) lithium reduced aggressive, hyperactive behavior and increased social behavior. Nausea, vomiting and diarrhea noted in one patient.
Cutler and Heiser 1978	30 year old male with 6 year history of impulsive, violent and assaultive behavior	0.95 mEq/L (serum) lithium abolished violence and impulsiveness without untoward side effects.
Dale 1980	15 untreatable and violent mentally handicapped male and female inpatients (age: 17-63 years); open trial	0.4-1.2 mmol/L (serum) lithium reduced aggressive outbursts within 2 weeks in 73% of the patients. No response and a worsening of violence in 3, 1 patients, respectively. One patient developed tardive dyskinesia.
Sovner and Hurley 1981	Untreatable, assaultive self-injurious 26 and 44 year old severely mentally handicapped females in Massachusetts	1-1.35 mEq/L (serum) lithium decreased aggressive and self-injurious behavior while producing polyuria and hypothyroidism.

470

Study	Population	Results
Buck and Havey 1986	Untreatable and violent mentally handicapped 23 year old male inpatient in Maine	0.8 mEq/L (serum) lithium with 9.5 ug/mL carbamazepine (serum) produced a near complete suppression of violent and schizophrenic behavior.
Elliott 1986	Case studies of 2 unmanageable, aggressive mentally handicapped patients (ages: 44 and 22 years)	0.8-1.0 mEq/L (serum) lithium decreased aggressiveness, temper outbursts and inappropriate sexual behavior in both patients; this was associated with decreased IQ, attentiveness and cognitive abilities.
Glenn et al. 1989	Case studies of 5 male and 5 female brain injured inpatients (age: 20-75 years) with unremitting aggressive, combative, self-injurious or severe affective instability	0.5-1.4 mEq/L (serum) lithium improved aggressive and self-injurious behavior in 70% of the patients. Neurotoxicity demonstrated in 30% of the patients, particularly when used in conjunction with neuroleptics.
Luchins and Dojka 1989	Retrospective analysis of aggressive and self-injurious behavior (SIB) of 11 male and female mentally handicapped inpatients (age: 27-55 years)	0.6-0.95 mEq/L (serum) lithium decreased aggression and SIB in 64% and 82% of the patients, respectively.

Outpatient studies

Study	Population	Results
Panter 1977	Untreatable, verbally and physically assaultive 30 year old female in California	1.3-1.5 mEq/L (serum) lithium decreased or suppressed aggressive, assaultive and self-injurious behavior.
Freinhar and Alvarez 1985	29 year old male with bipolar and intermittent explosive disorders, alcohol and cocaine abuse in California	0.5-0.8 mEq/L (serum) lithium completely suppressed violent behavior and severe mood fluctuations.

Criminal violence

Study	Population	Results
Sheard 1971	Verbal hostility, physical aggressiveness and self-rating of 12 violent, assaultive male inmates (age: 21-43 years); single-blind, placebo control	0.6-1.5 mEq/L (serum) lithium decreased rating of aggressive affect and aggressive behavior. Side effects uncommon but included mild nausea, increase in thirst and insomnia.

Study	Method	Results
Tupin et al. 1973	Psychiatric and self evaluation, prison records of 27 male convicts exhibiting violent behavior; open trial	0.82 mEq/L (serum) lithium decreased aggressiveness, prison rule infractions for violence and feelings of aggression in 78% of the inmates. Side effects included nausea, vomiting and tremor; 4 inmates discontinued lithium therapy because of ulcers, leukocytosis and psychosis.
Kerr 1976	Untreatable and violent 29 year old mentally handicapped female child batterer in Tasmania	"Therapeutic dosages" of lithium suppressed irritability and explosive rages; patient in good health after 12 months.
Marini and Sheard 1976 Marini and Sheard 1977 Sheard et al. 1976 Sheard 1977b Sheard and Marini 1978	Motor, cognitive and prison records of 66 impulsive aggressive delinquent males (age: 16-24 years); double-blind, placebo control	0.6-0.9 mEq/L (serum) lithium decreased overt aggression (prison infractions are abolished within 3 months) which was not due to: toxicity, motor impairment, cognitive deficits, hypothyroidism, reduced testosterone or manic-depression. Side effects noted were tremor, dry-mouth, polyuria and nausea.
Experimental studies		
Rifkin et al. 1972	Psychiatric evaluation, Oklahoma Behavior Rating Scale administered to 21 male and female inpatients with emotionally unstable character disorder; crossover double-blind, placebo control	0.6-1.5 mEq/L (serum) lithium controlled mood swings and "maladaptive behavior patterns," reactivity in 67% of the patients compared to 20% placebo improvement.
van Putten and Sanders 1975	Hospital record review and GAS inventory of 35 male and female inpatients with untreatable, intractable mental illness; double-blind, placebo control	0.7-1.7 mEq/L (serum) lithium decreased agitated depression, mania, and unstable-aggressive behavior in 31% of the patients.
Worrall et al. 1975	8 non-manic depressive assaultive mentally handicapped female inpatients; 7-point aggression scale; double-blind, placebo control	0.74-1.38 mEq/L (serum) lithium decreased aggressiveness within 2 weeks of administration: 3 improved, 2 showed no change and 2 signs of neurotoxicity (at 1.16-1.38 mEq/L), 1 worsened. Note: Other psychotropic medication was given concurrently.

472

Reference	Study Description	Findings
Tyrer et al. 1984	17 male and 9 female mentally handicapped inpatients (age: 14-50 years) with assaultive histories in the U.K.; 20 point behavioral inventory; double-blind, placebo control	0.5-0.8 mmol/l (serum) lithium decreased destructiveness, self-assault and aggression in 68% of the patients without side effects. Factors associated with a positive response include female gender, epilepsy, stereotypic behavior, low initial aggressivity and overactivity.
Craft et al. 1987	20 male and 22 female mentally handicapped inpatients (mean age: 33 years) with assaultive histories in the U.K.; 5 point behavioral inventory; double-blind, placebo control	0.7-1.0 mmol/L (serum) lithium decreased aggression scores and aggressive outbursts in 73% of the patients. No change and increased aggression were noted in 17 and 9% of the patients. "Transitory" polydipsia, polyuria, tremor, drowsiness and vomiting were reported in 36% of the patients.

Literature Reviews

Reference	Study Description	Findings
Tupin 1972 Shader et al. 1974 Sheard 1975 Marini and Sheard 1977 Sheard 1977b Sheard 1978 Gunn 1979 Lena 1979 Schou 1979 Lion 1981 Sovner and Hurley 1981 Campbell et al. 1982 Jefferson 1982 Sheard 1983 Sheard 1984 Tupin 1985 Mattes 1986 Cherek and Steinberg 1987 Kazdin 1987 Miczek 1987 Wickham and Reed 1987 Yudofsky et al. 1987 Sheard 1988	Review of over 1500 human and animal articles between 1920 and 1987 on the treatment and pharmacology of aggression	Lithium decreases assaultiveness in normals and in patients suffering from a lack of impulse control (anger, rage and irritability are easily triggered), schizophrenia, personality disorders, self-injurious behavior. It is effective in mentally handicapped children and adults as well as prison populations. Lithium very rarely exacerbates aggressive behavior. Renal, thyroid and cognitive functioning should be considered and monitored.

473

Table 10A Effects of Acute Anxiolytics on Animal Aggression

REFERENCES	METHODS AND PROCEDURES	RESULTS AND CONCLUSIONS
A. Benzodiazepine Receptor Agonists		
Experimenter Provoked Aggression		
Randall et al. 1960 Randall et al. 1965	Aggression directed at experimenter in rhesus monkeys, dogs, cats, and septal or cortical-lesioned rats; no assessment of behavioral specificity.	Chlordiazepoxide produced taming effects in rats (11-21 mg/kg), dogs (10 mg/kg), cats (10 mg/kg) and monkeys (1-20 mg/kg); ataxia observed in dogs and at high dose in monkeys.
Heuschele 1961	Aggression directed at observer in zoo animals	Chlordiazepoxide prevented aggression in various species of vicious zoo animals [e.g., a lynx (6 mg/kg), dingoes (3 mg/kg), a baboon (13 mg/kg), and a macaque (5 mg/kg)].
Heise and Boff 1961 Scheckel and Boff 1968	Aggression directed at experimenter in cynomolgus monkeys	1,4-benzodiazepines including chlordiazepoxide (1-5 mg/kg), diazepam (1 mg/kg), or nitrazepam (0.125 mg/kg) prevented aggressive behavior; oxazepam reduced behavior only slightly at the highest dose (40 mg/kg).
Scheckel and Boff 1966	Aggression directed at observer in squirrel monkeys	Chlordiazepoxide (1-2 mg/kg) or diazepam (1-2 mg/kg) prevented aggressive behavior.
Hoffmeister and Wuttke 1969	Aggression directed at experimenter in cats	Chlordiazepoxide (10 mg/kg p.o.) prevents defensive aggression directed at experimenter approaching cat with leather glove without producing ataxia.
Bauen and Possanza 1970	Aggression directed at the experimenter in minks	Diazepam (15 mg/kg), chlordiazepoxide (50 mg/kg) and oxazepam (25 mg/kg) produced taming, and prevented lunges and attacks without producing muscle relaxation.
Langfeldt and Ursin 1971	Aggression directed at the experimenter in feral cats	Diazepam (1 mg/kg i.p.) reduced hissing and attack behaviors in response to experimenter's approach, prodding with a pole or handling.

474

Tsumagari et al. 1978	Aggression directed at the experimenter in rhesus monkeys	Diazepam (2.5 mg/kg) and Y-7131 (10 mg/kg) reduced aggressive displays and attacks in monkeys.
Tompkins et al. 1980	Aggression directed at experimenter in rhesus monkeys	Diazepam (5-20 mg/kg) dose dependently reduced observer-rated aggressive response to pole prodding; ataxia was observed at higher doses.
Kalin and Shelton 1989	Defensive aggression in infant rhesus monkeys when approached by experimenter	Diazepam (1 mg/kg) reduced frequency of aggressive vocalizations (e.g., barking and defensive freezing behavior, but did not alter distress calls (e.g., cooing).
Blanchard et al. 1989	Aggression directed at the experimenter in rats	Chlordiazepoxide (10-20 mg/kg), midazolam (5-10 mg/kg), and diazepam (5 mg/kg) reduced viewer rated defensive aggression. Chlordiazepoxide (10 mg/kg) reduced defensive threats, whereas low doses of midazolam (1-5 mg/kg) reduced biting and a high dose (10 mg/kg) reduced jump attacks and defensive threats.

Isolation-induced aggression

Scriabine and Blake 1962	Aggression in male mice single housed from 24 hours up to 7 weeks	Chlordiazepoxide (5-10 mg/kg, i.p.) dose dependently reduced time spent fighting. Impairment of motor activity occurred at the highest dose only (20 mg/kg).
DaVanzo et al. 1966	Aggression in male C57 B1/10J and Dublin (ICR) mice single housed 1-21 days	Chlordiazepoxide (16-35 mg/kg i.p. ED_{50}) reduced fighting in 2 strains of mice 1, 2 and 4 hours post injection; C57 were more sensitive to anti-aggressive effects. Sedative effects were observed at same or lower doses than ED_{50} that reduced aggression.
Cole and Wolf 1966	Aggression in male mice single housed for 6 weeks	Chlordiazepoxide (52 mg/kg p.o. ED_{50}) reduced percentage of mice fighting at doses that were substantially lower than those that produced neuromuscular impairment.

475

Reference	Model	Results
Valzelli 1967	Aggression in male mice single housed for 4 weeks then placed together in novel cage in groups of 3; aggressive behavior evaluated at 0.5, 1, 2, 4, 6, 8, and 24 hours after administration of saline or drug	Chlordiazepoxide (7.5-30 mg/kg i.p.), diazepam (7.5-20 mg/kg i.p.), oxazepam (10-15 mg/kg i.p.), nitrazepam (7.5-30 mg/kg i.p.) reduced viewer-rated aggression scores for 2-6 hours; some evidence of neuromuscular impairment.
Hoffmeister and Wuttke 1969	Aggression in male mice single housed for 3-5 weeks; confronts group housed intruder in home cage	Chlordiazepoxide (93.23 mg/kg ED_{50} p.o.) reduces attack behavior in isolated mice, but sedation is observed at lower doses (33 mg/kg ED_{50} p.o., in mice climbing an inclined screen) than those that reduced aggression.
Le Douarec and Broussy 1969	Isolation-induced aggression in male mice	Chlordiazepoxide (20-40 mg/kg) reduced frequency of attacks by the resident and vocalizations by the intruder; chlordiazepoxide prolonged the intervals between attacks.
Sofia 1969b	Aggression in pairs of male mice single housed for 8 weeks; confront each other in home cage of one of the isolates	Chlordiazepoxide (23.5 mg/kg i.p. ED_{50}) and diazepam (11.1 mg/kg i.p. ED_{50}) inhibited fighting episodes, but only at doses that were higher than those that altered motor performance (rotorod test).
Robichaud et al. 1970	Aggression in pairs of male mice single housed for 3-5 months	Diazepam (6 mg/kg ED_{50} p.o.), chlordiazepoxide (15 mg/kg ED_{50} p.o.) and prazepam (27 mg/kg ED_{50} p.o.) prevented fighting; motor performance on an inclined screen was impaired only at significantly higher doses in female mice tested separately.
Weischer and Opitz 1972	Aggression in male mice single housed for 4 weeks	Chlordiazepoxide (10-54 mg/kg. i.p. ED_{50}) produced taming, and muscle relaxation, but did not alter species-specific aggressive behavior.
Barzaghi et al. 1973	Aggression in pairs of male mice single housed for 4 weeks	Clobazam or chlordiazepoxide (5-20 mg/kg i.p.) prevented aggressive behavior, but reduced spontaneous motor activity dose dependently. However, ataxia (in rotorod test) was only observed at doses 3-5 times higher than those that prevented aggression.

Reference	Study	Results
Valzelli 1973	Review of references on isolation-induced aggression in male mice	Chlordiazepoxide (10 mg/kg i.p.), oxazepam (10 mg/kg i.p.) and midazolam (10 mg/kg i.p.) inhibited fighting in isolates for 3-5 hours.
Heilman et al. 1974	Aggression in male mice single housed for 3 weeks; confronted group housed intruder in neutral cage	Aggression was reduced at high doses of triflubazam (29.5 mg/kg ED_{50} i.p.), chlordiazepoxide (28.5 mg/kg ED_{50} i.p.) and diazepam (14.8 mg/kg ED_{50} i.p.); motor performance (rotorod test) was impaired at lower doses of diazepam (5.2 mg/kg ED_{50} p.o.) and triflubazam (15 mg/kg ED_{50} p.o.) and similar doses of chlordiazepoxide (30 mg/kg ED_{50} p.o.).
Ferrini et al. 1974	Aggression in pairs of male mice isolated for 4 weeks	10-20 mg/kg i.p. medazepam and SB 5833 reduced observer rated aggression when tested at 1, 2, 4 and 24 hours post injection; temazepam reduced aggression at a lower dose (2.5-5 mg/kg), but muscle relaxation was apparent.
Fernandez-Tome et al. 1975	Aggression in pairs of male mice single housed for 4 weeks	Chlordiazepoxide or QM-6008 (10-90 mg/kg p.o.) reduced observer-rated aggression in mice. In separate tests, male mice showed evidence of motor impairment at 60 mg/kg (ED_{50} i.p.) QM 6008 and at 14 mg/kg (ED_{50} i.p.) chlordiazepoxide.
Kršiak 1975a	Aggression in male mice single housed 3-6 weeks	Chlordiazepoxide (5 mg/kg p.o.) increased aggressive threats and attacks; diazepam (1 mg/kg p.o.) did not alter aggressive threats or attacks but did reduce ambivalent behavior (e.g., tail rattle). Other behaviors were unaffected.
Kršiak 1974a, 1975b	Timidity in male mice single housed 3-6 weeks	Chlordiazepoxide (20-50 mg/kg p.o.) and diazepam (5 mg/kg p.o.) reduced defensive-flight behaviors in timid mice; diazepam increased social behaviors. Walking, rearing and other motor activities were not altered.
Sulcova et al. 1976	Timidity in pairs of male mice single housed for 2 weeks	Diazepam (5 mg/kg) inhibited defensive flight behaviors and increased social behaviors without altering motor activities.

Poshivalov 1978	Timidity and aggression in pairs of male single housed mice	Low doses of medazepam (0.1 mg/kg) increased aggressive behavior in aggressive mice; higher doses of medazepam (5 mg/kg) or diazepam (5 mg/kg) suppressed aggression.
Malick 1978a	Aggression in pairs of male mice single housed for 4 weeks	Diazepam (10 mg/kg i.p.) reduced percentage of mice that displayed fighting, and produced ataxia.
Kršiak 1979	Aggression in male mice single housed for 3-5 weeks	Chlordiazepoxide (50 mg/kg p.o.) and diazepam (10 mg/kg p.o.) reduced attacks and aggressive threats; lower doses of chlordiazepoxide (5 mg/kg) and diazepam (1 mg/kg) reduced ambivalent behavior (e.g., tail rattling). Walking, rearing and other motor activities were unaffected.
Fielding and Hoffman 1979	Aggression in male mice single housed for 4 weeks	Clobazam (10 mg/kg ED_{50} p.o.) and chlordiazepoxide (14 mg/kg ED_{50} p.o.) reduced percentage of isolated mice exhibiting fighting. Motor performance (rotorod test) was only impaired at substantially higher doses (50 mg/kg p.o. ED_{50} clobazam) and (30 mg/kg p.o. ED_{50} chlordiazepoxide)
Sulcova et al. 1979 Donat and Kršiak 1985 Sulcova and Kršiak 1984, 1986	Aggression and timidity in pairs of male mice isolated for 3 weeks	Diazepam (3-4 mg/kg p.o.) reduced aggressive threats in aggressive mice and defensive flight behaviors in timid mice without altering motor behaviors; diazepam increased social behaviors in both groups.
Kršiak et al. 1981	Aggression or timidity in pairs of male mice single housed for 3-6 weeks	Diazepam (10 mg/kg p.o.) and chlordiazepoxide (50 mg/kg p.o.) reduced attacks in aggressive mice. Diazepam (5 mg/kg) and chlordiazepoxide (20, 50 mg/kg) reduced defense and escape behaviors in timid mice without altering walking or rearing.
Olivier and van Dalen 1982	Aggression in male mice single housed for 3 weeks; confronted group housed intruder in neutral cage	Chlordiazepoxide (5-7.5 mg/kg p.o.) increased social, aggressive and defensive behaviors without altering nonsocial behaviors or time spent inactive.

Reference	Condition	Findings
Sulcova 1985 Sulcova and Kršiak 1989	Aggression and timidity in pairs of male mice single housed for 3 weeks	At low doses, oral administration of alprazolam (0.05-0.25 mg/kg), oxazepam (2.5 mg/kg), and diazepam (3 mg/kg) reduced defensive behaviors in timid mice; higher doses of alprazolam (1.25 mg/kg), diazepam (10 mg/kg), oxazepam (22.5 mg/kg) reduced attacks without reducing motor activity or social behaviors. Other 1,4 benzodiazepines reduced these behaviors but also produced sedation.
Grimm and Zelikovich 1982	Aggression in pairs of male mice single housed for 40 days; prescreened for attack, only more aggressive animals were treated; confronted each other in neutral cage	Diazepam (2.5 mg/kg i.p.) reduced aggression, but also reduced social behaviors. Diazepam-treated animals were attacked by previously submissive untreated opponents.
Oehler et al. 1985a,b	Aggression in male mice single housed for 3 weeks	Diazepam (5 mg/kg i.p.) produced no changes in aggression or motor behaviors.
Skolnick et al. 1985	Aggression in male mice single housed 28 days; confronted group housed intruder in neutral test environment	Diazepam (4 mg/kg s.c.) reduced viewer-rated aggression scores without producing ataxia (rotorod test).
Beck and Cooper 1986	Aggression in pairs of male rats single housed for 2 weeks. Pretested for aggression, and more aggressive animal treated; confronted each other in neutral familiar environment	Chlordiazepoxide (5 mg/kg i.p.) increased the duration of aggressive behaviors without altering social or exploratory behavior.
Poshivalov et al. 1987	Aggression in pairs of male mice single housed 3-6 weeks	Diazepam (2.5-4 mg/kg) reduced aggressive threats and attacks.
Plummer and Holt 1987	Aggression in male rats single housed for 3 weeks; tested in novel environment	Alprazolam (0.5-2 mg/kg), triazolam (0.25-1 mg/kg) reduced aggressive threats and attacks, but bites were increased at the higher doses. Defensive postures were reduced; crouching and freezing increased.

| Kršiak and Sulcova 1990 | Aggression and timidity in male mice single housed for 3 weeks; confronted group housed opponent in novel cage | Alprazolam (0.05-2.5 mg/kg p.o.) and oxazepam (2.5-22.5 mg/kg p.o.) reduced attacks in aggressive mice and defensive postures and escapes in timid mice. However, increases in walking were observed in timid animals. Triazolam (0.04-0.75 mg/kg p.o.), nitrazepam (0.25-5 mg/kg), and lorazepam (0.2-1.8 mg/kg) also reduced aggression, but only at doses that altered motor behaviors. |

Pain-induced Aggression and Defense

Kostowski 1966	Electric foot shock-induced aggression in pairs of male mice	Chlordiazepoxide (5 mg/kg i.p.) prevented fighting in mice, but calming and ataxia were observed.
Hoffmeister and Wuttke 1969	Electric foot shock-induced aggression in pairs of male mice	Chlordiazepoxide (10 mg/kg ED_{50} p.o.) reduced aggressive behavior; sedation was observed in mice climbing an inclined screen only at higher doses (33 mg/kg ED_{50} p.o.).
Sofia 1969b	Electric foot shock-induced aggression in pairs of male mice	Chlordiazepoxide (4.2 mg/kg i.p. ED_{50}) and diazepam (0.9 mg/kg i.p. ED_{50}) inhibited fighting episodes, at doses that were significantly lower than those that altered motor performance (rotorod test).
Robichaud et al. 1970	Electric foot shock-induced aggression in pairs of male mice	Prazepam (13 mg/kg ED_{50} p.o.) prevented fighting. Motor performance on an inclined screen was impaired at high doses (74 mg/kg ED_{50}) in female mice tested separately.
Christmas and Maxwell 1970	Electric foot shock-induced aggression in mice and rats	1,4-benzodiazepines including chlordiazepoxide (4.7 mg/kg), diazepam (2.1), oxazepam (4.5 mg/kg), and nitrazepam (1.2 mg/kg) reduced percent of animals displaying fighting behaviors in mice and rats at doses well below those that reduced motor activity.
Irwin et al. 1971	Electric foot shock-induced aggression in pairs of male mice	Chlordiazepoxide (10-20 mg/kg) reduced leaping and fighting behaviors, but only at doses that reduced muscle tone and produced ataxia.

480

Emley and Hutchinson 1971, 1983	Electric tail shock-induced target biting in male and female squirrel monkeys on a fixed interval schedule	Chlordiazepoxide (1-32 mg/kg s.c.) selectively reduced target biting. At low doses (0.5-8 mg/kg) lever press responding was elevated. Diazepam (0.06-2 mg/kg s.c.) reduced biting nonspecifically as similar doses also reduced lever press responding.
Manning and Elsmore 1972	Electric foot shock-induced aggression in pairs of male rats	Chlordiazepoxide (10-40 mg/kg i.p.) dose dependently reduced percentage of fights. Sedation observed at high dose (40 mg/kg).
Barzaghi et al. 1973	Electric foot-shock-induced aggression in pairs of male mice	Clobazam or chlordiazepoxide (5-20 mg/kg i.p.) reduced the number of fights at doses well below those that reduced motor performance (rotorod test).
Goldberg et al. 1973	Electrical foot shock-induced aggression in mice	A low chlordiazepoxide dose (6.4 mg/kg ED_{50}) produced taming and reduced fighting in mice.
Heilman et al. 1974	Electric foot shock-induced aggression in pairs of male mice	Triflubazam (9 mg/kg ED_{50} i.p.), chlordiazepoxide (4.2 mg/kg ED_{50} i.p.) and diazepam (1.3 mg/kg ED_{50} i.p.) reduced shock-induced aggression without producing motor impairment (rotorod test, inclined screen, climbing apparatus).
Quenzer et al 1974	Electric foot shock-induced aggression in pairs of male rats	Chlordiazepoxide (5-30 mg/kg i.p.) dose dependently reduced shock-induced fighting. Animals were more easily handled, but no noticeable motor deficits were observed.
Robichaud and Goldberg 1974	Electric foot shock-induced aggression in pairs of male rats and pairs of male mice	Chlordiazepoxide reduced fighting in mice (6.4 ED_{50} i.p.) and rats (10 mg/kg MED i.p.); doses that produced muscle relaxation in mice were substantially higher (32.3 mg/kg). Chlordiazepoxide metabolites were not as effective as the parent compound; compounds were effective in the following order: chlordiazepoxide > demethylated chlordiazepoxide > deaminated chlordiazepoxide (demoxepam).

481

Reference	Paradigm	Results
Kamioka et al. 1977	Lever response in macaques being threatened by another monkey exposed to electric shock; drugs administered to the aggressor monkey	Oxazepam (2 mg/kg s.c.), chlordiazepoxide (0.5 mg/kg s.c.), cloxazolam (0.5 mg/kg s.c.), diazepam (0.5 mg/kg s.c.) and CS-386 (0.5 mg/kg s.c.) increased suppressed response in monkeys being threatened by another monkey exposed to electric shock; drugs administered to the aggressor monkey did not increase suppressed response in the monkey being threatened. Some sedation and slight ataxia were observed at highest doses of diazepam and cloxazolam.
Vassout and Delini-Stula 1977	Electric foot shock-induced aggression in male rats	Chlordiazepoxide (1-3 mg/kg) did not alter aggression or motor behaviors.
Tsumagari et al. 1978	Electric foot shock-induced fighting in mice	Diazepam (1.8 mg/kg ED_{50} i.p.) and Y-7131 (0.4 mg/kg ED_{50} i.p.) reduced fighting episodes in mice. Motor activities (rotorod test) were diminished at higher doses of Y-7131 (2.1 mg/kg) and diazepam (5.3 mg/kg).
Fielding and Hoffman 1979	Electric foot shock-induced aggression in male mice	Clobazam (2.6 mg/kg ED_{50} p.o.), chlordiazepoxide (14 mg/kg ED_{50} p.o) and diazepam (0.16 mg/kg ED_{50} p.o.) reduced percentage of animals fighting. No motor disturbances for any drug dose <25 mg/kg; ataxia at higher doses.
Delini-Stula and Vassout 1979	Electric foot shock-induced aggression in male rats	Diazepam (1-5 mg/kg i.p.) and oxazepam (10-50 mg/kg i.p.) reduced number of fighting bouts; similar doses of both chlordiazepoxide (2.5-10 mg/kg i.p.) and oxazepam (10-25 mg/kg i.p.) reduced locomotor activity.
Renzi 1982	Electric tail shock-induced aggression in pairs of restrained male mice tested 6 hours post injection	Chlordiazepoxide (2.5-5 mg/kg;i.p.) did not alter bites.
Jarvis et al. 1985	Electric tail shock-induced aggression in pairs of male mice	Chlordiazepoxide (2-16 mg/kg i.p.) dose dependently reduced target biting in confined mice.

482

Nakao et al. 1985	Electric foot shock-induced aggression in pairs of male mice	Diazepam (0.5 mg/kg), carbamazepine (10-20 mg/kg i.p.) reduced shock-induced frequency of fighting bouts; GABA receptor antagonists, picrotoxin (0.3 mg/kg s.c.) and bicuculline (0.5 mg/kg s.c.) blocked this effect.
Traversa et al. 1985	Electric foot shock-induced aggression in pairs of male mice	Chlor-desmethyldiazepam (0.04-0.08 mg/kg i.p.) increased shock-induced fighting in mice that showed no fighting in pre-test screening trials; a higher dose (1.25 mg/kg) reduced fighting.

Aggression Induced by Omission of Reward

Miczek 1974	Aggression in pairs of male rats confronting each other after omission of food reward	Chlordiazepoxide(5-20 mg/kg) produced a biphasic effect on aggression and submissive behaviors. Low doses increased attacks and threats in dominant rats (5 mg/kg), and increased submissive behaviors in intruder males (5-10 mg/kg); the high dose reduced these behaviors.
Moore et al. 1976	Extinction-induced aggression directed at a mirror in pigeons	Aggressive mirror response was suppressed by chlordiazepoxide (5 mg/kg), without altering key pecking response.
Arnone and Dantzer 1980	Extinction-induced aggression in pairs of pigs: Operant behaviour	Diazepam (1-2 mg/kg i.m.) increased the frequency of biting and fighting as well as total duration of aggressive bouts; diazepam reduced plasma cortisol levels that were elevated following extinction.

Defensive Aggression Induced by Electrical Brain Stimulation

Baxter 1964	Defensive aggression in cats following electrical stimulation in hypothalamus	Chlordiazepoxide (5-20 mg/kg i.p.) produced taming, but did not alter electrical threshold required to elicit hissing.
Malick 1970	Defensive aggression in cats following electrical stimulation in hypothalamus	Chlordiazepoxide (10-15 mg/kg i.p.), diazepam (4-7.5 mg/kg i.p.) or oxazepam (12 mg/kg i.p.) elevated the electrical threshold required to elicit hissing.

Funderburk et al. 1970	Defensive aggression in cats following electrical stimulation in hypothalamus	Chlordiazepoxide (10 mg/kg i.p.) elevated the electrical threshold required to elicit hissing without producing ataxia.
Otsuka et al. 1973	Defensive aggression and startle response in cats following electrical brain stimulation in hippocampus, amygdala, ventromedial hypothalamus or midbrain reticular formation	Nitrazepam (1 mg/kg i.p.) reduced hissing and attacks in response to air puffs and pole prodding, but also produced marked ataxia; diazepam (1 mg/kg i.p.) reduced hissing, but did not alter attacks.
Delgado 1973	Defensive aggression in male and female rhesus monkeys following electrical brain stimulation in thalamus and central grey	Chlordiazepoxide (8 mg/kg i.m.) prevented defensive aggression (staring, ear flattening, piloerection, barking and attacks) for 6 hours in restrained and free moving animals; chlordiazepoxide (5 mg/kg i.m.) prevented aggression directed at observers or object placed in cage. No noticeable motor deficits were observed.
Murasaki et al. 1976	Defensive aggression in cats following electrical stimulation in hypothalamus	Diazepam (1 mg/kg i.p.) elevated electrical threshold required to elicit hissing and further elevated threshold to attack.
Tsumagari et al. 1978	Defensive aggression in cats following electrical stimulation in hypothalamus	Diazepam (0.36 mg/kg ED$_{50}$ p.o.) and Y-7131 (0.06 mg/kg ED$_{50}$ p.o.) reduced rage response in cats. Sedation and ataxia were observed at 0.39 mg/kg p.o. diazepam and at 0.25 mg/kg p.o. Y-7131.
Kruk et al. 1987	Aggression in rats following electrical stimulation in hypothalamus	Chlordiazepoxide (5-20 mg/kg p.o.) did not alter threshold for attack or locomotion, but animals shifted to a less violent form of attacks (strong bite to mild bite); oxazepam (5-20 mg/kg p.o.) treated animals displayed a similar shift and a slight increase in attack threshold.
Polc et al. 1981	Defensive aggression in cats following electrical stimulation in hypothalamus	Diazepam (1 mg/kg i.p.) elevated the threshold for eliciting attack response; elevations were prevented by Ro 15-1788 (5 mg/kg i.p.).

484

| Fukuda and Tsumagari 1983 | Defensive aggression in cats following electrical stimulation in hypothalamus | Diazepam (0.25-1 mg/kg i.p.), nitrazepam (0.1-0.2 mg/kg i.p.), and lorazepam (0.05-0.1 mg/kg i.p.) elevated electrical threshold for direct attack dose dependently; hissing threshold was reduced at higher doses of diazepam (1-5 mg/kg). Muscle relaxation observed at 0.48 mg/kg diazepam, 0.61 mg/kg nitrazepam, and 0.14 mg/kg lorazepam (ED_{50}). |

Drug-induced Aggression

Yen et al. 1970	dl-DOPA-induced aggression in mice	Chlordiazepoxide (13 mg/kg ED_{50}), diazepam (2.5 mg/kg ED_{50}), oxazepam (8.8 mg/kg ED_{50}) reduced target biting without producing ataxia.
Nakamura and Thoenen 1972	6-OHDA-induced aggression in rats	Diazepam (1 mg/kg every 2 hours for 6 hours) produced taming and reduced aggression directed at the experimenter.
Mueller and Nyhan 1982	Pemoline-induced self-directed biting in rats	Diazepam (5 mg/kg i.p.) tended to reduce licking and self-biting behaviors in a time dependent manner. Slight reductions in locomotor behaviors were observed.

Brain Lesion-induced Aggression

Blyther and Marriott 1969	Hypothalamic lesion-induced aggression in rats	Chlordiazepoxide (9 mg/kg ED_{50} i.p.) reduced hyper-reactivity dose dependently.
Sofia 1969b	Septal lesion-induced aggression in male mice; "aggressiveness" was measured for attacks on inanimate objects in tail-restrained animal	Chlordiazepoxide (23.8 mg/kg i.p. ED_{50}) and diazepam (17.2 mg/kg i.p. ED_{50}) inhibited viewer-rated "aggression", but only at doses that were 3 and 4 times higher than those that altered motor performance (rotorod test).
Horovitz et al. 1963 Loizzo and Massotti 1973	Septal lesion-induced aggression in rats	Chlordiazepoxide (10-20 mg/kg) reduced hyperirritability (measured as startle response, vocalization, object attacks and bites), but only at doses 2 times higher than those that produced motor deficits in septal rats and in rats tested in a separate experiment (rotorod test).

Reference	Model	Findings
Goldberg et al. 1973	Septal lesion-induced aggression in rats	Chlordiazepoxide (16 mg/kg ED_{50}) reduced attacks at inanimate objects (pencil and glove).
Fernandez-Tome et al. 1975	Aggression in septal lesioned rats	Chlordiazepoxide or QM-6008 (10-90 mg/kg i.p.) reduced observer-rated aggression. However, in rats tested under different conditions, QM-6008 (16.3 mg/kg i.p. ED_{50}) and chlordiazepoxide (8.9 mg/kg i.p. ED_{50}) reduced motor performance (rotorod test).

Aggression by Resident Toward Intruder

Reference	Model	Findings
Le Douarec and Broussy 1969	Resident-intruder confrontations in male mice	Chlordiazepoxide (10-40 mg/kg) reduced attacks and ambivalent behavior (e.g., tail rattles) directed at untreated intruder.
Hoffmeister and Wuttke 1969	Aggression in pairs of cats confronting each other in cage where partition is removed	Chlordiazepoxide (10-20 mg/kg p.o.) does not alter attack behavior in pairs of male cats confronting one another; ataxia observed at high doses (20 mg/kg).
Olivier and van Dalen 1982	Resident male rats confront single housed intruder	Chlordiazepoxide (2.5 mg/kg i.p.) increased aggressive threats and attacks without altering time spent inactive; social behavior was slightly enhanced.
Olivier et al. 1984	Resident-intruder confrontations in male rats	Chlordiazepoxide (2.5-10 mg/kg) dose dependently reduced aggression and increased social interactions without altering exploratory behaviors or motor activities.
Miczek 1985 Weerts et al. 1988 Mos et al. 1990	Resident-intruder confrontations in male rats	Diazepam increased frequency of attacks and threats and duration of aggressive postures at low doses (0.1-1 mg/kg, i.p.) and decreased aggressive behaviors at moderate to high doses (3-17 mg/kg).
Yoshimura and Ogawa 1984 Yoshimura 1987	Single housed resident male mice confronts intruder in home cage	Chlordiazepoxide (5-20 mg/kg i.p.) reduced aggressive threats and attack bites in resident males without altering motor behaviors. Chlordiazepoxide-treated intruders received more attacks from untreated residents and displayed less defensive behaviors in a dose dependent manner.

486

Gardner and Guy 1984	Aggression in the social interaction test in male rats	Chlordiazepoxide (2-12 mg/kg), oxazepam (10-20 mg/kg), lorazepam (2 mg/kg) and nitrazepam (0.2-4 mg/kg) reduced composite measures of aggression and increased social behaviors. Locomotor behaviors were unaffected.
Mos and Olivier 1988	Dominant and subordinate resident-intruder confrontations in male rats	Chlordiazepoxide (5, 10 mg/kg) increases the duration of aggressive threats and attacks in both residents and the increases were even higher in the subordinate animal. Motor behaviors were not altered.
File 1982	Aggression in untreated male rats in an established colony (n=12) directed at a drug-treated male intruder	Chlordiazepoxide (5 mg/kg i.p.) or lorazepam (0.25 mg/kg i.p.) administration in intruder rats did not alter defensive interactions or aggression received from an untreated resident male.
Dixon 1975	Aggression in group-housed male mice	Chlordiazepoxide (no dose specified) increased aggression and social interactions, changing the social structure of mice living in territories.
Zwirner et al. 1975	Aggression in group-housed male mice	Chlordiazepoxide (3 mg/kg p.o.) increased attacks between group members. Lower doses (0.3-1 mg/kg p.o.) did not alter aggression. No effects on motor activity were observed at any dose tested.
Female Aggression		
Olivier et al. 1985 Mos et al. 1987	Maternal aggression in lactating rats	A low dose of chlordiazepoxide (5-10 mg/kg) increased attack bites dependent on baseline levels of aggression without altering time spent inactive.
Yoshimura 1987	Maternal aggression in female mice that confronted male intruder in home cage on 3, 5, 7, 9 and 11 days post parturition	Chlordiazepoxide (10 mg/kg i.p.) increased frequency of attack bites without altering motor behaviors; 5 mg/kg dose did not alter aggression.
Yoshimura and Ogawa 1989	Maternal aggression in lactating rats	Chlordiazepoxide (5-15 mg/kg i.p.) produced a biphasic effect on aggression; 10 mg/kg increased bites and 15 mg/kg decreased bites. Locomotor behaviors were not affected.

487

Mos and Olivier 1989	Maternal aggression in lactating rats	At low doses, chlordiazepoxide (5-20 mg/kg p.o.), diazepam (1.25-5 mg/kg p.o.), oxazepam (1.25-20 mg/kg i.p.) and alprazolam (1.25 mg/kg i.p.) enhanced aggression. The magnitude of pro-aggressive effects varied with drug. Oxazepam increased more behavioral elements (threats, bites to head and bites to body, total time spent on aggression) over a wider dose range than chlordiazepoxide or diazepam. Only alprazolam (2.5-5 mg/kg i.p.) reduced aggression; possibly due to stronger muscle relaxant properties.
Olivier et al. 1990	Maternal aggression in lactating rats	Low doses of chlordiazepoxide (5-10 mg/kg p.o.) and oxazepam (1.25-10 mg/kg p.o.) enhanced aggression and higher doses (20 mg/kg) reduced it. Diazepam (1.25-5 mg/kg i.p.) and alprazolam (1.25-5 mg/kg p.o.) also were similar but increased aggression was observed at a narrower dose range (1.25 mg/kg only). Drugs did not significantly alter exploratory behaviors or time spent inactive, with the exception of reduced exploration with oxazepam (2.5-20 mg/kg).

Dominance-related Aggression

Delgado et al. 1976	Dominance-related aggression in socially housed rhesus monkeys	Diazepam (0.1-10 mg/kg p.o.) reduced aggressive behaviors initiated by dominants; chlordiazepoxide-treated subordinates received less aggression from untreated dominants. Mobility was reduced at higher doses (3-10 mg/kg) in dominants and at all doses in subordinates.
Poole 1973	Dominance-related aggression in pairs of male hamsters	Chlordiazepoxide (50 mg/kg p.o.) reduced incidence of pursuit and attack in dominant animals; chlordiazepoxide-treated subordinate animals showed reductions in defensive behaviors and an increase in social investigation. No adverse effects on mobility or coordination were observed.

488

Fielding and Hoffman 1979	Aggression in pairs of hamsters	Clobazam (30-60 mg/kg ED_{50} p.o.) administered 1-6 hours prior to testing reduced bites at doses up to 50 mg/kg without producing motor disturbances. Chlordiazepoxide (10 mg/kg) reduced biting 50%, but also produced ataxia.

Killing

Horovitz et al. 1965	Predatory aggression in rats	Chlordiazepoxide (30 mg/kg ED_{50}) reduced mouse killing, but at doses that were 6 times the dosage that reduced rotorod performance.
Kostowski 1966	Predatory aggression in ants	Chlordiazepoxide (0.5 µg/mg) did not alter the number of ants that attacked beetles.
Sofia 1969a	Predatory aggression in isolated rats selected for mouse-killing behavior	Chlordiazepoxide (20.2 mg/kg i.p. ED_{50}) and diazepam (219.0 mg/kg i.p. ED_{50}) inhibited mouse-killing, but only at doses that were substantially higher than those that altered motor performance (rotorod test).
Cole and Wolf 1970	Predatory aggression in grasshopper mice	Chlordiazepoxide (15, 27 mg/kg i.p. ED_{50}) at doses that were half the TD_{50}, increased the latency to attack and the duration of fighting in *Onychomys torridus* mice, but had no effect on aggression in the more aggressive *O. leucogaster* strain.
Panksepp 1971	Predatory aggression in rats following electrical stimulation in hypothalamus	Chlordiazepoxide (5 mg/kg i.p.) reduced affective attack, but not quiet biting attacks (did not attack other rats, bit dead mice) or escape behaviors; sedation was not apparent.
Langfeldt 1974	Predatory aggression in cats	Diazepam (1-4 mg/kg) dose dependently prolonged latency to kill by inducing play behavior.
Quenzer and Feldman 1975	Predatory aggression in rats	Chlordiazepoxide (25, 50, 75 mg/kg i.p.) suppressed mouse killing.

Reference	Behavior	Findings
Leaf et al. 1975 Leaf et al. 1984	Predatory aggression in rats	Oxazepam (2.5-80 mg/kg), diazepam (1.25-10 mg/kg) and chlordiazepoxide (2.5-80 mg/kg) increased the percentage of animals displaying mouse killing behavior.
Valzelli and Bernasconi 1976	Isolation-induced predatory aggression in rats	When administered with a single electric shock, chlordiazepoxide (10 mg/kg) and medazepam (10 mg/kg) blocked mouse killing behavior in only 33-37% of killer rats. Lorazepam (0.03 mg/kg), diazepam (10 mg/kg), and oxazepam (15 mg/kg) were not effective.
Vassout and Delini-Stula 1977	Predatory aggression in normal and bulbectomized rats	Chlordiazepoxide (1-3 mg/kg) did not alter mouse killing behavior. No motor deficits were observed at these doses.
Apfelbach 1978	Predatory behavior in ferrets	Chlordiazepoxide (1 mg/kg) increased killing efficiency; shorter latency to attack larger prey (but received more bites) and less bites required to complete kill
Leaf et al. 1978	Predatory aggression in cats	Diazepam (4 mg/kg) and chlordiazepoxide (16 mg/kg) did not alter mouse killing.
Delini-Stula and Vassout 1979	Predatory aggression in male rats	Diazepam (3-5 mg/kg i.p.) reduced percentage of animals showing mouse killing behavior in killer rats; diazepam (2.5-10 mg/kg i.p.) dose dependently reduced locomotion counts in mice.
Hirose et al. 1981	Predatory aggression in rats	Diazepam (11.8 mg/kg ED$_{50}$) and 45-0088-S (3 mg/mg ED$_{50}$) reduced mouse killing.
Kostowski et al. 1983	Suppression of predatory behavior in rats during continual electrical stimulation in locus coeruleus	Chlordiazepoxide (5 mg/kg i.p.) prevented suppression of mouse killing.
Kozak et al. 1984	Suppression of predatory aggression in rats during continual electrical brain stimulation in locus coeruleus	Chlordiazepoxide (5 mg/kg), aprazolam (2.5 mg/kg), 1-pyrimidinepiperazine (10 mg/kg) and MJ-13805 (10 mg/kg) reversed suppression of mouse killing.

Valzelli and Galateo 1984	Review of isolation-induced and electrical brain stimulation-induced predatory aggression in rats	Chlordiazepoxide (5 mg/kg), 1-pyrimidine-piperazine (10 mg/kg), medazolam (10 mg/kg) and diazepam (5 mg/kg) increased muricidal behavior in isolates, and prevented electrical brain stimulation induced suppression of mouse killing.
Pellis et al. 1988	Predatory aggression in cats	Oxazepam (1-3 mg/kg) and diazepam (4 mg/kg) reduced the latency to kill and escalated levels of predation from play to killing.

B. 5-HT$_{1A}$ Receptor Agonists

Experimenter Provoked Aggression

Tompkins et al. 1980	Ratings of aggression directed at experimenter in rhesus monkeys	Buspirone (20-160 mg/kg) dose dependently reduced observer-rated aggressive response to pole prodding without producing ataxia.

Isolation-induced Aggression

Valzelli and Galateo 1984	Review of isolation-induced and electrical brain stimulation-induced predatory aggression in rats	Buspirone (10 mg/kg) increased muricidal behavior in isolates, and prevented electrical brain stimulation induced suppression of mouse killing
McMillen et al. 1987	Aggression in male mice single housed in suspended cages for 3 weeks that were trained to attack group housed intruder in 5 daily sessions. 1st test saline trial, followed by drug test (2nd) 30 minutes after	Gepirone (1.25-10 mg/kg i.p.) dose dependently reduced the number of mice displaying fighting behavior without altering motor performance (rotorod test).
Cutler and Dixon 1988	Isolation-induced aggression in pairs of male mice	Ipsapirone (0.3-3 mg/kg, i.p.) reduced aggressive behaviors as well as defensive and ambivalent behaviors. No motor deficits observed.
McMillen et al. 1988	Aggression in male mice single housed for 3 weeks that were elected for reliable attack in confrontations with group housed mouse in home cage	Buspirone (1.25-10 mg/kg i.p.), and ipsapirone (2.5-10 mg/kg i.p.) reduced the number of mice that attacked. Doses that reduced aggression did not alter ability of treated animals to orient towards the intruder or reduce motor functions in treated group housed mice (rotorod tests).

Reference	Model	Results
Olivier et al. 1989	Isolation-induced aggression in male mice	Eltoprazine (0.5-20 mg/kg, PO), ipsapirone (0.3-10 mg/kg, IP) and buspirone (0.3-10 mg/kg, IP) reduced a composite measure of aggression; eltoprazine (1-20 mg/kg) increased social interactions.

Aggression by Resident toward Intruder

Reference	Model	Results
Olivier et al. 1984	Resident/intruder confrontations in male rats	Buspirone (2-8 mg/kg) dose dependently reduced aggressive threats and attacks. Avoidance inactivity and exploration behaviors were decreased.

Killing

Reference	Model	Results
Kozak et al. 1984	Suppression of predatory aggression in rats during continual electrical brain stimulation in locus coeruleus	Buspirone (10 mg/kg) reversed suppression of mouse killing.

C. β-Noradrenergic Blockers

Isolation-induced Aggression

Reference	Model	Results
Delini-Stula and Vassout 1979	Aggression in male mice single housed for 4-8 weeks	(-)-propranolol (10-20 mg/kg i.p.) reduced percentage of mice fighting without reducing locomotor activity.
Weinstock and Weiss 1980	Aggression in male mice single housed for 4 weeks that confronted group housed male in novel test chamber for blind observations	(±)-propranolol (1 mg/kg ED_{50}) and pindolol (1 mg/kg ED_{50}) reduced the number of attacks, and (±)-propranolol increased attack latency. (+)-propranolol (20/mg/kg) did not alter aggressive behavior. Doses up to 2 times higher than ED_{50} for reducing aggression did not alter locomotor activity. Higher doses of (±)-propranolol (5 mg/kg) actually increased activity. Drugs that reduced aggression also blocked 5-HT-induced head twitching suggesting serotonergic control.

Aggression by Resident toward Intruder

Miczek 1981 Miczek and DeBold 1983	Resident male mice confronted a group-housed intruder in home cage; either the resident or the intruder was treated.	l-propranolol (10-20 mg/kg) reduced attack and threat behaviors in treated resident mice facing untreated intruders and did not alter defensive behaviors in treated intruders confronting an untreated resident.
Yoshimura and Ogawa 1985	Male mice were tested in a resident intruder paradigm for aggression. Either resident or intruder was treated	dl propranolol (5-20 mg/kg), oxprenolol (30-75 mg/kg) and carteolol (30-75 mg/kg) reduced offensive sideways, attack bite and tail rattle in treated residents, but did not alter aggression in treated intruder mice.
Yoshimura 1987 Yoshimura et al. 1987	Male mice were tested in a resident-intruder paradigm for aggression. Subjects received 0, 5, 10, 20 mg/kg d, l, or dl propranolol or practolol PO 60 min before testing.	l or dl propranolol suppressed resident aggressive behaviors. The inactive d isomer had no effect. Only the highest dose l-propranolol (20 mg/kg p.o.) suppressed locomotor activity.
Kennett et al. 1989	Social interaction test in pairs of treated rats under low light and familiar conditions	(-)-propranolol (16 mg/kg, s.c.) did not alter social interaction when administered alone, nor did it reverse suppression of social and aggressive behaviors produced by 5-HT$_{1B}$ agonist mCPP (0.5-1 mg/kg, IP).

Pain-induced Aggression

Vassout and Delini-Stula 1977	Electric foot shock-induced aggression in pairs of male rats	(-)-propranolol (10 mg/kg i.p.) reduced the number of animals displaying fighting behaviors at doses that did not alter motor capacities. (+)-propranolol was without effects.
Delini-Stula and Vassout 1979	Electric foot shock-induced aggression in male rats	(-)-propranolol (10-20 mg/kg i.p.) reduced number of fighting bouts without altering locomotor activity.
Prasad and Sheard 1983b	Electric foot shock-induced aggression in male rats; all subject received desipramine (10 mg/kg i.p.) for 2 days plus dl propranolol (20 mg/kg i.p.) or saline; 2 tests occurred on the 3rd day (preinjection and dl propranolol); rats were retested on the 4th day (no injection)	Preinjection testing on the 3rd day showed an increase in percent fighting in both rats treated with desipramine and rats treated with desipramine plus dl propranolol. In the second test, injection of dl propranolol alone reduced fighting in both groups.

Lesion-induced Aggression

Bainbridge and Greenwood 1971	Experimenter provoked "aggression" in septal lesioned rats as measured by viewer-rated aggression scores	dl- and d-propranolol (10-100 mg/kg s.c.) dose dependently reduced aggression scores. Ataxia observed at highest dose (100 mg/kg) in some animals.

Killing

Vassout and Delini-Stula 1977	Predatory aggression in intact and bulbectomized male rats	(-)-propranolol (10-20 mg/kg i.p.) inhibited mouse killing behavior without reducing motor capacities.
Delini-Stula and Vassout 1979	Predatory aggression in male rats	(-)-propranolol (10-20 mg/kg i.p.) reduced percentage of animals showing mouse killing behavior without altering locomotor activities.
Shibata et al. 1983	Predatory aggression in olfactory bulbectomized male rats	Systemic administration (10–20 µg/kg s.c.) of β-blockers, sotalol and propranolol did not alter mouse killing behaviors, or block DMI-suppressed mouse killing. Microinjection of β-blockers into amygdala (2 µg/2 µl) did not alter mouse killing behavior nor did it alter mouse killing suppressed by DMI (20 µg/2 µl).

494

TABLE 10B: Effects of Chronic Anxiolytics on Animal Aggression

References	Methods and Procedures	Results and Conclusions
A. Benzodiazepine Receptor Agonists		
Isolation-induced Aggression		
Sulcova et al. 1976	Timidity in pairs of male mice single housed for 3-5 weeks	Diazepam (5 mg/kg p.o.) inhibited defensive flight behaviors and increased social behaviors. Motor behaviors were not altered. Chronic treatment (5 mg/kg for 8 days) lessened behavioral effects; 2 days after withdrawal, defensive behaviors were not changed.
Dixon 1982	Aggression in untreated male mice single housed for three weeks that confronted drug-treated group housed opponent in home cage	Diazepam-treated (0.125 mM in drinking solution x 14 days) intruder males showed reductions in flight behaviors and increases in attacks received from nondrugged isolates
Malick 1978a	Aggression in pairs of male mice single housed for 4 weeks confronted each other in home cage of 1 animal; both animals were treated	Diazepam (10 mg/kg i.p.) reduced percentage of mice that displayed fighting, and produced ataxia in subsequent inclined screen test; mice became tolerant to ataxic effects following chronic administration (5 days), but reductions in fighting persisted.
Grimm and Zelikovich 1982	Aggression in pairs of male mice single housed for 40 days that were prescreened for attack. Only the more aggressive animals were treated; confrontations were in a neutral cage	Diazepam (2.5 mg/kg i.p.) reduced aggression initially and remained low with chronic treatment (7 days)
Valzelli 1972	Review of 7 references on isolation-induced aggression in male mice	Chlordiazepoxide (10 mg/kg), oxazepam (10 mg/kg) and midazolam (10 mg/kg) inhibit fighting in isolates for 3-5 hours
Pain-induced Aggression and Defense		
Quenzer et al., 1974	Electrical foot shock-induced aggression in pairs of male rats	Chlordiazepoxide (5-30 mg/kg x 10 days) reduced shock-induced fighting

495

Reference	Model	Findings
Renzi 1982	Electrical tail shock-induced aggression in restrained mice	Chronic administration of chlordiazepoxide (2.5-5 mg/kg/day for 10 days) induced a dose dependent increase in biting, whereas acute administration of the same doses produced no changes

Aggression Induced by Omission of Reward

Reference	Model	Findings
Moore et al. 1976	Extinction-induced aggression directed at a mirror in pigeons	Chlordiazepoxide (5 mg/kg) suppressed aggressive mirror response and remained low during chronic administration (60 days). Key peck responses were not reduced.

Drug-induced Aggression

Reference	Model	Findings
Kostowski et al. 1986	Clonidine- or apomorphine-induced aggression in mice and rats	In mice, chronic administration (5 mg/kg/day for 21 days) of diazepam, alprazolam prevented clonidine-induced (10 mg/kg) attack bites and vocalizations, whereas similar administration of adinazolam potentiated these behaviors. In rats, apomorphine-induced (10 mg/kg) aggression was increased only by chronic diazepam treatment

Aggression by Resident towards an Intruder

Reference	Model	Findings
File and Tucker 1983	Resident-intruder interactions in male rats	Chronic administration (from day 7-21 postnatally) of lorazepam (0.25-1.25 mg/kg/day) increased submissive behaviors in adult intruder males towards a non-treated resident
File 1982	Aggression in untreated male rats in an established colony (n=12) directed at a drug-treated male intruder	5 days chronic chlordiazepoxide (5 mg/kg/day i.p.) or lorazepam-treated (0.25 mg/kg/day i.p.) intruders received less aggression from residents, and initiated aggressive behaviors towards established residents; intruders displayed less submissive behaviors.
File 1986b	Resident-intruder confrontations in male rats	When tested 9 days after cessation of drug treatment, chronic (21 days) neonatal administration of clonazepam (1-5 mg/kg/day) increased aggressive behavior in neonatally treated resident males and defensive behaviors in neonatally treated intruder males

Reference	Description	Result
File 1986a	Resident-intruder confrontations in male rats	14 days after cessation of chronic (21 days) neonatal administration of diazepam (1 mg/kg/day), increased aggressive interactions were seen in resident animals, whereas cessation of chronic lorazepam (2.5 mg/kg/day) increased submissive behaviors directed at the non-treated intruder; intruders neonatally treated with diazepam (10 mg/kg/day) displayed an increase in wrestling with a non-treated resident, whereas neonatal lorazepam resulted in reductions in kicking
File and Mabbutt 1990	Aggression in rats single housed for 1 week that confronted group-housed intruder in home cage	Chronic administration of chlordiazepoxide (10 mg/kg/day, i.p. x 4 weeks) did not alter aggression in drug treated residents or intruders
Guaitani et al. 1971	Aggression in group-housed and isolated male mice	Chronic administration (2-6 months) of diazepam (1-50 mg/kg/day), N-desmethyldiazepam (50 mg/kg/day), N-methyloxazepam (50 mg/kg/day) and oxazepam (50 mg/kg/day) increased fighting leading to skin lesions and higher rates of mortality
Fox and Snyder 1969 Fox et al. 1972	Aggression in group-housed male mice	Mice consuming a diet containing diazepam (0.1 mg/g food), nitrazepam (0.05 mg/g food) or flurazepam (0.2 mg/g food) for 2-16 weeks showed increased inter-group aggression and reduced defensive behaviors; mortality and wounding were increased

Female Aggression

Reference	Description	Result
Yoshimura et al., 1987	Maternal aggression in lactating female mice that confronted male intruder in home cage day 5 post parturition; chronic treatment with chlordiazepoxide (5-10 mg/kg i.p./day) beginning day 3 of cohabitation and terminated postpartum day 3	Chronic treatment tended to enhance aggression, but not significantly; motor performance was not altered.
Yoshimura and Ogawa 1989	Maternal aggression in lactating rats	Chlordiazepoxide (5-10 mg/kg/day x 20-22 days) did not alter aggression

Killing

Quenzer and Feldman 1975

Predatory aggression in rats

Chlordiazepoxide (25, 50, 75 mg/kg) suppressed mouse killing following acute administration, but this effect diminished with repeated administration (11 days)

Literature Reviews

DiMascio 1973

Review of 40 references on experiments of aggression in animals and man

Depending on species, tests, dosage and time course, benzodiazepines can reduce or increase aggression. Generally, there are quantitative differences among chlordiazepoxide, diazepam and nitraxepam, but oxazepam appears to be ineffective or produce opposite effects than the other BNZ.

TABLE 11A: Effects of Acute Anxiolytics on Human Aggression

References	Methods and Procedures	Results and Conclusions
A. Benzodiazepine Receptor Agonists		
Treatment of Inpatients		
Lion et al. 1975a	Case report on paradoxical rage reaction in a 25 year old male following diazepam (5 mg x 3/day) for 3 days; patient had history of temper outbursts	Patient became increasingly anxious, argued with wife, and attacked her, fracturing her jaw.
Salzman 1988	Case study on administration of lorazepam in violent psychotic patients maintained on low doses of haloperidol.	Lorazepam (1-2 mg, i.m.) administered with haloperidol (5 mg) when a patient became disruptive effectively reestablished behavioral control.
Bond et al. 1989	Case reports on aggression and violence in 3 mentally retarded patients	Midazolam (5-10 mg) rapidly stopped violent and aggressive symptoms (e.g., agitation, temper tantrums, assault, and self-injurious behaviors).
Experimental Studies on Aggression		
McDonald 1967	84 paid female college students were assessed with inventory scales for personality and anxiety and grouped according to action and non-action orientation of high and low anxiety; subjects received placebo or diazepam (5 or 10 mg)	Action-oriented subjects showed an increase on the hostility scale (anger) after diazepam; subjects that were more hostile with diazepam were more anxious and depressed pre-drug and increased anger and anxiety post-drug.
Wilkinson 1985	60 male psychology undergraduates grouped as high, medium or low anxiety according to self reported anxiety inventories (A-Trait, A-State, MAACL), received placebo or diazepam (10 mg) and were told they had received a tranquilizer. Subjects were studied in a competitive reaction time task which consists of setting a shock level against an increasingly provocative opponent	Diazepam increased aggression (shock intensity) in the low anxiety group under low provocation and increased measures of depressive affect. In the high anxiety group aggression increased only under high provocation, concurrent with reductions in anxiety.

499

| Cherek et al. 1987
Cherek et al. 1990 | 9 normal male subjects were evaluated with structured psychiatric interview (SADS-L); subjects were administered placebo or diazepam (2.5-10 mg/70 kg), and completed a competitive response task, POMS and Buss-Durkee Hostility questionnaires. Each subject received each dose 3 times | Diazepam reduced aggressive responding in 7 out of 9 subjects and slightly increased "escape" responding. In 2 subjects, diazepam increased aggressive responding; aggression increases were correlated with high assaultive and hostility scores in Buss-Durkee, but not with POMS. |
| Cherek et al., 1991 | 5 normal male subjects were evaluated with a structured psychiatric interview (SADS-L). Subjects received placebo and triazolam (0.125, 0.25 and 0.5 mg/70 kg of body weight) under double blind conditions, and completed a competitive response task, POMS and Buss-Durkee Hostility questionnaires. Triazolam doses were administered 1st in ascending order and then randomly, so that each subject received each dose twice. | Triazolam effects on aggressive responding differed according to scores of hostility and anger; 2 subjects with low Buss-Durkee scores showed decreases, and 2 subjects with high Buss-Durkee scores showed increases. Triazolam reduced reinforced responding and escape responding in all subjects. |

Literature Reviews

| Itil 1981 | Review of 71 references on drug therapy management of aggression | Acute aggressive states associated with organic syndrome, isolated explosive disorders, and alcohol withdrawal, can be controlled with acute intramuscular administration of chlordiazepoxide (up to 100 mg); aggressive syndrome associated with epilepsy is also treatable with benzodiazepines. |

500

TABLE 11B: Effects of Chronic Anxiolytics on Human Aggression

References	Methods and Procedures	Results and Conclusions
A. Benzodiazepine Receptor Agonists		
Treatment of Inpatients		
Tobin and Lewis 1960	135 women and 77 men with behavioral disturbances including anxiety, phobia, obsessive/compulsive behavior, depression, and hysterical acting-out behavior were administered chlordiazepoxide (25-175 mg/day for 1 week to 13 mo.); subjects were evaluated by physician and interviews during psychotherapy	18 of 23 patients with hysterical acting-out behavior showed reductions during chlordiazepoxide treatment; acute rage reactions occurred in 3 patients.
Ingram and Timbury 1960	9 outpatients with phobic anxiety and 6 with obsessional neuroses were administered chlordiazepoxide (10-25 mg 3x/day)	Side-effects included hyperactivity in 1 patient, increased irritability in 2 patients and assaultive behavior in 1 patient.
Boyle and Tobin 1961	Case reports of 25 behaviorally disruptive patients with chronic brain syndrome, schizophrenia, psychosis or mental retardation	15 patients showed improvement with chlordiazepoxide treatment (10-500 mg for 4 days-10 mo.); 6 slightly and 8 marked reduction in aggressiveness, agitation, hostility and assaultiveness. One patient showed increased aggressiveness following chlordiazepoxide.
Murray 1962	Case history of a psychiatric patient with chronic depressive disorder and anxiety administered chlordiazepoxide (75 mg/day)	Patient's depressive state improved but he also became verbally aggressive and agitated, with numerous behavioral outburst that continued despite lowering of dosage to 25 mg/day
Feldman 1962	55 anergic and hyperanergic patients, 17 patients with depressive reactions, 70 anergic schizophrenics received chlordiazepoxide or diazepam (ED_{50} 20-30 mg/day)	Combativeness, hostility, and hallucinations were unaffected by diazepam, but were reduced by chlordiazepoxide; diazepam relieved hypo-activity, but facilitated assaultive behavior in one patient and feelings of "hate".

501

Monroe and Dale 1967	10 chronically hospitalized schizophrenic patients; 8 with "activated EEG's" received chlordiazepoxide (20-50 mg 4x/day) was added to drug treatments for month 1, then chlordiazepoxide was administered alone for month 2, and placebo alone for month 3. If chlordiazepoxide reduced symptoms, it was resumed for month 4 to determine if response could be repeated	8 of 10 patients that exhibited impulsive acting out behavior (physical attacks, suicide attempts, or inappropriate sexual behavior) showed dramatic improvement or complete disappearance of impulsiveness. 5 out of 10 were more "sociable" and communicative.
Goddard and Lokare 1970	16 epileptic patients were administered diazepam (5-10 mg 3x/day) and assessed with WPRS (Wittenborn psychiatric rating scale)	Scores for psychotic belligerence were markedly reduced.
Guldenpfennig 1973	45 epileptic patients aged 1-58 years were treated for control of seizures with clonazepam (0.5-1 mg/day) initially, and increased to a maximum of 12 mg/day in adults, 4-6 mg/day in children	3 patients discontinued treatment due to induction of aggression and temper tantrums.
Kocur et al. 1984	Case report of aggressive behavior in 35 boys with behavioral disturbances of neurotic or encephalopathic origin	Bromazepam only slightly reduced aggressive behavior in some cases; no change in 6 patients, and increased aggression in 4 cases.
Rosenbaum et al. 1984	Case reports of patients treated with alprazolam (0.5-8 mg) for panic disorders, depression, and compulsive disorders	8 out of 80 patients treated with alprazolam became hostile (physical assault, reckless driving and verbal aggression); only 1 patient had a history of hostile outbursts.
Strahan et al. 1985	Case reports of 3 patients with bipolar depression, generalized anxiety, and panic disorders were administered alprazolam (0.25-5 mg t.i.d.); all showed poor impulse control	Alprazolam increased anxiety and panic attacks and produced irritability, agitation, and interpersonal conflicts; symptoms disappeared with discontinuation of alprazolam.
Pecknold and Fleury 1986	Case reports of 2 patients with panic disorder and agoraphobia that were treated with alprazolam (4-6 mg/day)	1-4 weeks of alprazolam treatment induced manic episodes and increased irritability.
Ward et al. 1986	Case report of a paranoid schizophrenic	Lorazepam (2-4 mg) reduced agitation and aggressive behavior.
Keats and Mukherjee 1988	Case study of a seizure-prone paranoid schizophrenic	Clonazepam (2 mg/kg 4x/day) reduced violent hallucinations and postictal aggressive outbursts.

502

Crime

Kalina 1964

62 patient-inmates were administered diazepam (5-10 mg 2-3x/day for approx. 6 mo.); 40 of the inmates had behavioral problem including violent, destructive, assaultive and aggressive behavior, agitation and active paranoia

Violence (homicidal), destructiveness, assaultiveness, belligerence, and abusiveness were completely controlled in 33 out of 40 inmates with behavioral problems.

Gleser et al. 1965

46 adolescent males in a juvenile detention center were administered placebo or chlordiazepoxide (10 mg/kg 3x day 1 and 20 mg/kg day 2); aggression was assessed during recorded conversations and test period with alarm bell, flashing light, or threat of faradic shock

Anxiety and ambivalent hostility were reduced over time in the chlordiazepoxide group; chlordiazepoxide group showed a trend for reduction in overt hostility.

Simonds and Kashani, 1979

Interviews of 109 delinquent boys aged 12-18 years that were committed to a training school to determine relationship between drug abuse and criminal offense. Boys were divided according to crime against property (burglary, auto theft, vandalism or stealing) vs crime against persons (murder, rape, assault or robbery). Drug abuse was determined according to DSM-III criteria

Diazepam was used or abused (1 or more times alone or in combination with other drugs) by 31 person offenders and by 21 property-only offenders, but no significant effects in relation to criminal offense were found for this drug.

Experimental Studies on Aggression

Gardos et al. 1968

45 paid college students were grouped according to scores on TMAS (Taylor manifest anxiety scale) as low, medium or high anxiety and administered placebo, oxazepam (45 mg/kg/day in 3 doses) or chlordiazepoxide (30 mg/kg/day in 3 doses) for 1 week; hostility was determined with Buss-Durkee and Gottschalk-Gleser hostility inventories

Oxazepam did not alter hostility in any group, but chlordiazepoxide increased hostile-aggressive tendencies in the medium and highly anxious groups.

DiMascio et al. 1969

In a double blind study, 55 normal male volunteers received placebo, chlordiazepoxide (15 or 30 mg/kg/day) or oxazepam (45 mg/kg/day) for one week; anxiety and hostility were assessed with Scheier and Cattel anxiety battery and Buss-Durkee hostility scale

Chlordiazepoxide increased hostility scores, whereas oxazepam did not.

Salzman et al. 1969	40 male volunteers that had low anxiety and hostility ratings were administered placebo or chlordiazepoxide (10 mg 3x/day for 1-2 weeks). 1/2 of the subjects received instructions that the drug would produce a pleasant, friendly, more relaxed feeling. Subjects were evaluated with Buss-Durke Hostility inventory	Both the placebo group that received instructions about the drug and the unaware chlordiazepoxide group showed increased hostility ratings.
Podobnikar 1971	36 patients with anxiety related symptoms and neurotic hyperaggressiveness were administered placebo or chlordiazepoxide (10 mg 2x/day)	11 of 22 patients showed no signs of aggressiveness compared to placebo after 2-4 weeks treatment.
Salzman et al. 1974	48 male volunteers assigned to 3 person groups were evaluated with Buss-Durkee hostility inventory and self-rated and group member-rated questionnaires. Subjects were administered placebo or chlordiazepoxide (30 mg/day) for 7 days	Chlordiazepoxide produced an increase in self-rated hostile affect, but not in behavioral hostility. When a frustration stimulus was presented to the group, chlordiazepoxide increased interpersonal behavioral hostility; rage reaction was observed in 1 subject.
Rickels and Downing 1974	225 neurotic outpatients (majority were women) from 3 clinical settings administered placebo or chlordiazepoxide (40 mg/day for 4 weeks); patients were evaluated by physicians and by patient symptom checklist and grouped according to low, medium or high anxiety	All symptoms of hostility, irritability and anxiety were reduced by chlordiazepoxide treatment in all groups. No evidence for increased aggressiveness or "paradoxical rage" reactions.
Kochansky et al. 1975	33 paid volunteers, mean of 24.5 years old, responded to newspaper add that scored greater than 12 but less than 26 on TMAS (Taylor Manifest Anxiety Scale) in discussion groups of 3 were and administered the BDHI (Buss-Durkee Hostility Inventory) before and after group interaction; self administered 15 mg/kg oxazepam, 10 mg/kg chlordiazepoxide, or placebo 3x/day for 1 week tested again 8th day	Following "frustration" stimulus (telling subjects they had performed inadequately and would have to repeat task) chlordiazepoxide increased verbal hostility whereas oxazepam reduced verbal hostility compared to placebo.

504

Kochansky et al. 1977	32 paid volunteers, 21-29 years old, responding to newspaper ads were classified as medium or highly anxious with TMS (Taylor Manifest Anxiety Scale); discussed TAT (Thematic Apperception Test) cards in groups of 3 in predrug condition; subjects self administered 15 mg/kg oxazepam, 10 mg/kg chlordiazepoxide, or placebo 3x/day for 1 week	"Frustration" was induced by telling subjects they had performed inadequately and would have to repeat task; chlordiazepoxide reduced total verbal units initially, but following frustration stimulus, increased verbal hostility as measured by viewer-rated aggression scale. Oxazepam reduced verbal hostility even after frustration stimulus.
Zisook et al. 1978	51 outpatients with neurotic anxiety were administered placebo or halazepam (40 mg 2-4x/day) in a double blind study; patients were evaluated with Hamilton Anxiety Scale, MMPI and patient symptom check list	Of the 20 patients that completed the study, halazepam did not alter hostility or anger scores over a 6 week period.
Lion 1979	45 outpatients with histories of temper tantrums, assaultive behavior, and impulsiveness associated with irritability and hostility were administered placebo (4x/day), chlordiazepoxide (25-50 mg 4x/day) or oxazepam (30-60 mg 4x/day); patients were evaluated by physicians and with scored questionnaires using Buss-Durkey Hostility scale and Scheir-Cattell anxiety scale	Oxazepam significantly reduced irritability and hostility measures when compared to placebo or chlordiazepoxide.
Griffiths et al. 1983	12 men with histories of abusing barbiturates and benzodiazepines; 3 subjects also on methadone treatment. Subjects received placebo and two high doses of diazepam (50 and 100 mg/day for 5 days) in a double blind random block design. Subjects filled out questionnaires for drug effect, drug liking, ARCI (Addiction Research Center Inventory) and POMS (Profile of Mood States)	Diazepam decreased social interactions and increased ratings of hostility by staff (but not by subject); carry over effects observed in 2 week washout period.
Gardner and Cowdry 1985	16 female outpatients with borderline personality disorder and histories of dyscontrol (suicide attempts, self-abuse, assaults) were administered alprazolam (1-6 mg) or placebo for 6 weeks in a double blind random crossover design	Alprazolam produced episodes of dyscontrol in 7 out of 12 patients (58%) compared to 1 out of 13 patients taking placebo; episodes were more severe, frequent and unpredictable than previous episodes prior to drug treatment.

505

Lipman et al. 1986	387 outpatients with depressive and anxiety disorders between 18 and 70 years old that answered newspaper add completed a double blind study of 5 medications including chlordiazepoxide (20-60 mg/day for 8 weeks). Subjects completed self-rating check lists, POMS (Profile of Mood State). Physicians evaluated patients progress with HAM-A, HAM-D (Hamilton Anxiety Scale and Depression Scale), anxiety depression check list, Covi Anxiety, and Raskin Depression Screen, and Global Improvement Scale	Patients that received chlordiazepoxide scored consistently higher on measures of anger and hostility than the placebo controls throughout treatment period.

Literature Reviews

Maletzky 1973	Review of case histories of 22 patients with episodic dyscontrol syndrome evaluated from interviews of patient, family and friends	Relatives and patients noted an increase in violent episodes by chlordiazepoxide and diazepam in 5 patients.
Salzman et al. 1975	Review of 28 references of clinical reports and research on oxazepam and aggression	No clinical observation of increased hostility from oxazepam administration. In laboratory, oxazepam reduced aggression or hostile mood even in presence of frustration. Oxazepam differs from chlordiazepoxide or diazepam in reference to increased aggression.
Greenblatt et al. 1975	Review of 88 references on clinical pharmacology of flurazepam	In animals, flurazepam produces taming effects in some cases and in other cases produces an increase in aggressive hostile behavior, possibly by releasing anxiety-bound aggression.
Azcarate 1975	Review of 43 references on treatment of aggression	Clinical trials of the efficacy of minor tranquilizers have revealed results similar to preclinical animal studies; some studies report increases in hostility and a paradoxical rage reactions; Variations in results may be attributed to dose, specific compound administered, acute vs. chronic administration, individual baseline levels of anxiety and/or hostility, and personality type.

506

Zisook and Devaul 1977	Review of case studies	Chlordiazepoxide and diazepam can produce rage attacks, whereas oxazepam does not; chlordiazepoxide can increase interpersonal hostility and frustration.
Bond and Lader 1979	Review of 49 references on benzodiazepines and hostility in normal and violent individuals	Evidence for rage attacks in patients administered benzodiazepines is based on uncontrolled clinical studies and few case histories. Generally, outbursts occur in patients who received doses in excess of 50 mg/day, perhaps due to drug toxicity. Increased hostility by benzodiazepines has been observed in normal subjects in controlled laboratory experiments; chlordiazepoxide increases hostility after 1 week administration, but oxazepam does not produce these results.
Valzelli 1979	Review of 129 references on the effects of sedatives and anxiolytics on aggression	Benzodiazepines reported to increase and decrease aggression in man and animals. Suggests drugs that are capable of lowering aggression are equally capable of enhancing it.
Gunn 1979	Review of 20 references on the use of drugs in the violence clinic	Increases in hostile aggressive tendencies, and in some cases, aggression and violence, have been observed in some individuals after acute chlordiazepoxide; oxazepam not implicated in paradoxical "rage" response.
Lion 1981	Review of 22 references on medical treatment of violent patients	Benzodiazepines have little anti-aggressive activity except in paranoid patients where benzodiazepines reduce hypervigilance. BZD often produce a paradoxical "rage" response in alcoholic patients, possibly by disinhibitory action.
Atkinson 1982	Review of 18 references on managing violent hospital patients	Concern with increased aggression in some patients administered diazepam and chlordiazepoxide, suggests shorter acting benzodiazepines like oxazepam because of lack of active metabolites.

507

Sheard 1983	Review of 60 references on psychopharmacology of aggression	Violent states associated with personality disorders are controlled with diazepam (0.15 mg/kg/hour, i.v.); chlordiazepoxide or oxazepam (10 mg 3x/day) is especially useful in epileptic patients. Evidence for benzodiazepine-induced paradoxical "rage" reactions may be explained by toxic reactions, or benzodiazepine withdrawal. No reports of paradoxical "rage" with oxazepam.
Sharon 1984	Review of 42 references on the use of benzodiazepines in correctional facilities	The concern of paradoxical rage and increased aggression induced by benzodiazepines in the prison populations is unsubstantiated. Very few studies exist to warrant removal of a potentially helpful agent from an anxiogenic setting. Studies that report increases in aggression fail to consider individuals that are already very aggressive prior to benzodiazepine treatment, as well as predisposing conditions (borderline personality disorder). Suggests care in prescribing benzodiazepines in these individuals, but encourages use in individuals with disabling anxiety.
Sheard 1984	Review of 59 references on the clinical pharmacology of aggression	Increases in rage and aggressive outbursts are not strongly supported by clinical data. Reductions in hostility and anxiety in double blind studies in delinquent boys, veteran outpatients and anxious out patients. Increases in hostility and paradoxical rage reactions have been associated with chlordiazepoxide but not oxazepam. In addition to antipsychotic medications, benzodiazepines are useful in treating aggressive and combative behavior related to psychosis.

508

Tupin 1985	Review of 36 references on psychopharmacology and aggression in clinical settings	Anxiolytic substances are used to treat anxious, agitated patients, but have been shown to aggravate violence reactions in some patients, and in a few cases produce paradoxical rage reactions. Suggests importance of treating the basis for aggression, namely the underlying psychiatric and/or medical problem, instead of a symptom. Benzodiazepines are useful in treating outbursts associated with borderline personality disorders, but are not as effective in treating serious panic and combativeness associated with psychosis.
Yudofsky et al. 1987	Review of 30 references on pharmacologic treatment of aggression	Benzodiazepines' effect on aggressive behavior is inconsistent. Benzodiazepines can produce paradoxical "rage" in some patients; reductions in aggression occur at higher doses that can produce sedation. Benzodiazepines are helpful for acute management of violence, but chronic use not recommended.
Eichelman 1987	Review of 136 references on neurochemical and pharmacological aspects of aggressive behavior	Benzodiazepines (chlordiazepoxide, diazepam, oxazepam) are claimed to reduce aggressive behavior in psychotic patients, prisoners with schizophrenic and personality disorders, as well as patients with episodic dyscontrol and hostile outbursts. However, rage reaction and enhanced aggressive behavior have been reported in some patient populations in open clinical trials. Oxazepam less associated with increases in aggression than chlordiazepoxide, but notes need for blind clinical trials.

Brizer 1988	Review of 145 references on psychopharmacology and the management of violent patients	Evidence for the efficacy of benzodiazepines for controlling aggression is inconclusive. Although in open clinical trials, chlordiazepoxide, diazepam and oxazepam have been shown to reduce measures of hostility and/or aggression in schizophrenics, epileptics, and patients with organic brain dysfunction and episodic behavioral outbursts, some anecdotal reports indicate an increase of paradoxical rage reaction may be associated with benzodiazepines. Notes the lack of adequate controls for concurrent medications, medication blood levels, and psychiatric and neurologic diagnosis.
Dietch and Jennings 1988	Review of case reports and experimental studies	Clinically used BZDs increase irritability, verbal aggression, assaultiveness and self mutilation; incidence of aggression is estimated at 1% of patients treated with BZDs, with differential effects with different BZD compounds. Clonazepam most likely to induce aggression, and oxazepam least likely.

B. 5-HT$_{1A}$ Receptor Agonists

Treatment of Inpatients

Ratey et al. 1989	Case reports in mentally retarded patients	Buspirone (5-15 mg 3x/day) reduced aggression, self-injurious behaviors and maladaptive behavior in 9 out of 14 patients.
Ratey and O'Driscoll 1989	Case reports in mentally retarded patients	Buspirone (5-15 mg/kg 3x/day) reduced agitation in 10 patients; however some showed an increase in hyperactivity and agitation at higher doses.

C. β-Blockers

Treatment of Inpatients

Polakoff et al. 1986	Case study of an extremely violent 36 year old retarded man on mesoridazine (120 mg/d) + propranolol (120-200 mg/d) or nadolol (80 mg/d).	β-blocker in combination with neuroleptic treatment stopped assaultive behavior and allowed outpatient status after 26 years of institutionalization.

510

Study	Description	Results
Luchins and Dojka 1989	Evaluation of mentally retarded with aggressive and self-injurious behaviors; patients received propranolol (90-410 mg/day) or lithium (600-1800 mg/day)	Aggression and self-abuse were controlled by either propranolol (83% reduction in both behaviors) or lithium (64% reduction for aggression, 82% reduction for self-abuse).
Elliott 1977	Case reports of 7 belligerent patients (2 exhibited explosive rage responses) with acute brain damage who received propranolol (60-320 mg/day).	Propranolol reduced irritation and anger, prevented aggressive outbursts and violent rage reaction. When propranolol was discontinued, symptoms reappeared in most cases.
Schreier 1979	Case report of 12 year old boy with postencephalitic psychosis administered 20 mg propranolol b.i.d. increased to 100 mg over 2 days with other medications; maintained on propranolol for 2 weeks	Propranolol reduced agitation and verbal aggression over the 2 week treatment period. The day after the last dose, he became increasingly aggressive and destructive (breaking pictures, wrecking room, tearing clothes); propranolol was reintroduced and symptoms disappeared.
Yudofsky et al. 1981	Case reports of 4 inpatients with Chronic Brain Syndrome and episodes of aggressive and violent outbursts; propranolol (320-520 mg/day) administered with other medications	Propranolol eliminated rage and violent outbursts and improved social ability with no adverse effects when carefully monitoring vital signs.
Williams et al. 1982	Case reports of 26 male and 4 female patients (9 were inpatients) ranging in age from 7 to 35 years with organic brain dysfunction; all had ongoing psychiatric and/or neurological disturbance since childhood or adolescence and prior pharmacological intervention. Patients received 10-20 mg propranolol 3-4x/day initially, and were titrated upwards to achieve a maximal dosage of 50-1600 mg/day	12 patients showed marked improvement and 12 patients showed moderate improvement in control of rage outbursts following propranolol treatment; side effects included hypertension, somnolence and lethargy. One patient showed bradycardia when taking dose twice (=320 mg).
Ratey et al. 1983	Case reports of 3 brain damaged or mentally retarded patients with episodes of provoked and unprovoked rage. Propranolol (90-300 mg/day) was administered with other medications.	All three patients that had been unresponsive to other medications, showed improvement in control of temper tantrums and rage outbursts following propranolol treatment. When propranolol dosage was reduced rage episodes returned. Symptoms subsided with reinstitution of propranolol. One patient showed bradycardia at 300 mg/day.

Study	Description	Results
Greendyke et al. 1984	Case reports of 6 assaultive patient with organic brain disease administered propranolol (200-520 mg/day for 40-80 days); assault behavior recorded by shift nurses and observations were conducted 15 minutes every 2 hours 7 times a day	Results indicate a minimum of 1 month administration for effective treatment; propranolol decreased assaultive behaviors, pacing, agitation, and resistiveness.
Yudofsky et al. 1984	Case report of 40 year old male alcoholic with Korsakoff's psychosis; extremely violent, physical assaults included injury to nursing staff and self; required physical restraints (ankle and hand or camisole). 20 mg x 4/day increased to 150 mg x 4/day in addition to other medications (haloperidol, phenytoin, pentobarbitol)	Rage attacks were markedly reduced allowing removal of physical restraints; when propranolol dosage was reduced rage attacks returned, but disappeared with reinstitution of propranolol. No adverse effects were observed.
Ratey et al. 1986	19 institutionalized mentally retarded patients given propranolol (40-200 mg/d) along with current medications	16 of 19 showed less assaultive and self-injurious behaviors when on β-blocker. Attribute effects to a lowered level of arousal.
Sorgi et al. 1986	Retrospective chart review of 7 assaultive chronic schizophrenics given β-blocker, nadolol (40-160 mg/d) or propranolol (80 mg bid), in addition to their normal antipsychotic medication.	Six of the seven patients improved. Four had > 70% reduction in assaultive behavior. Average peak effect was seen after 12 weeks of β-blocker.
Whitman et al. 1987	Three chronically aggressive psychotic patients treated with doses of propranolol up to 600 mg/d	Treatment with β-blocker plus neuroleptics resulted in remission and prevented assaultive behavior in one of three patients. Site of action is uncertain.
Ratey et al. 1987	8 autistic adults given propranolol (120-420 mg/d) and/or nadolol (120 mg/d) and behavior evaluated over 2-19 months.	β-blocker treatment resulted in reduction or cessation of self-abuse and assaultive behavior in all 8 patients. Emphasizes possible soothing effect of β-blockers.

Experimental Studies on Aggression

Study	Description	Results
Lindem et al. 1990	22 mentally retarded patients received pindolol or placebo in a double blind study for 16 wks. Destructive behaviors assessed with the Modified-Overt Aggression Scale.	Frequency of destructive acts decreased by 30% with β-blocker, the patients' communication (47%) and socialization (149%) also improved markedly.

512

Ratey et al. 1990	Chronic psychiatric patients with histories of aggression received nadolol (n=20) or placebo (n=26) in a double-blind study.	Nadolol (beta-blocker) significantly reduced aggressive outbursts and decreased severity of illness. Effects required 1-2 weeks.

Literature Reviews

Sheard 1984	Review of 59 references on clinical pharmacology of aggressive behavior	Propranolol (60-320 mg/day) has been used successfully in treating irritability, temper outbursts, and explosive rage responses, particularly in patients with organic brain dysfunction. Notes return of symptoms when propranolol is withdrawn. Improvement does not include primary symptoms of disease (disorientation, memory impairment, or psychotic thinking). Side effects include low blood pressure, headaches, dizziness, fatigue, insomnia, and depression.
Mattes 1986	Review of 100 references on the pharmacological treatment of temper outbursts	Propranolol treatment has been successful in controlling temper outbursts in patients with severe organic brain disease, brain-damage, belligerence, Korsakoff's psychosis, schizophrenia, and in violent elderly individuals, yet no predictors of benefit are found. Mechanism of action in controlling outbursts is uncertain; may be related to membrane stabilizing effect, alteration of brain cetacholamines and/or indoles, elevation of seizure thresholds, or action on serotonergic systems.
Eichelman 1987	Review of 136 references on neurochemical and pharmacological aspects of aggressive behavior	Propranolol has been reported to effectively reduce aggressive behavior in patients with organic brain injury, Korsakoff's psychosis, schizophrenics, and children with organic impairment in open clinical trials.

Hom 1987	Review of 58 references on control of disruptive, aggressive behavior in brain-injured patients	Although FDA (Food and Drug Administration) has only approved the use of β-blockers for cardiovascular disorders, they have been used with success in patients with anxiety disorders and for control of violent and disruptive behaviors; highlights lack of a specific symptom complex, EEG finding, injury location or temporal relationship to guide clinicians in treating patients as well as difficulty in determining length of treatment and control of side effects. Suggests treatment range starting at 60-129 mg/day divided in 2-3 dosages and gradual increases to a maximum of 800 mg/day.
Brizer 1988	Review of 145 references on psychopharmacology and the management of violent patients	Propranolol and other β-blockers have been successful in controlling aggressive patients with organic brain syndromes, Korsakoff's psychosis, viral encephalitis, schizophrenia, autism, episodic dyscontrol and explosive disorders. Most patients were previously refractory in multiple medication trials, but treatment is particularly effective in patients with organic brain disease. Exact mechanism of action unclear as patients often receive β-blockers with other medications (e.g. neuroleptics). Side effects include hypertension, bradycardia, and depression, but are not frequent with careful monitoring at suggested doses (up to 800 mg/day).

514

Nutrition and Violent Behavior

Robin B. Kanarek

INTRODUCTION

The concept that nutrition can affect behavior is not new. For thousands of years, people have believed that the food they eat can have powerful effects on their behavior. Some foods have been blamed for physical and mental ills, whereas others have been valued for their curative or magic powers. Within this framework, a variety of ideas about the association between food and antisocial behavior have arisen. For example, many primitive societies believe that an individual takes on the characteristics of the food that he/she consumes. Thus, eating aggressive animals (e.g. lion) is associated with belligerent behavior, whereas eating timid creatures (e.g. rabbit) is identified with less hostile acts.

The belief that certain foods can lead to antisocial or aggressive behavior is not limited to primitive societies. In this country, the idea that food affected behavior was an integral part of the nineteenth century health reform movement. The concept that "you are what you eat" was fundamental to the movement. Diet was believed to determine not only health and disease, but also spirituality, mental health, intelligence, and temperament. The health reform movement produced persuasive leaders who charmed

Robin Kanarek is at the Department of Psychology, Tufts University.

their followers with their oratory and their own brand of proselytism. Two of the most prominent leaders of this movement, Sylvester Graham (remembered best for the development of the graham cracker) and John Harvey Kellogg (recognized for the introduction of pre-cooked breakfast cereals), lectured widely throughout this country promoting the use of natural foods and decrying the ingestion of meat, which they believed would lead to the deterioration of mental functioning and the arousal of animal passions (Whorton, 1982). Kellogg further concluded that the breakdown products of meat acted as dangerous toxins that, when absorbed from the colon, produced a variety of symptoms including depression, fatigue, headache, and aggression. Kellogg wrote that "the secret of nine-tenths of all chronic ills from which civilized human beings suffer" including "national inefficiency" and "moral and social maladies" could be traced to the meat eater's sluggish bowels (Kellogg, 1919:87). In keeping with his puritanical background, Kellogg also warned his followers that spicy or rich foods would lead to moral deterioration and acts of violence (Kellogg, 1882:244-245).

The foregoing historical information should not be seen as simply humorous background material. Ideas about food and behavior continue to be prevalent. The last decade has witnessed an explosion of interest in the field of nutrition and behavior. The current obsession with health and fitness, as well as the desire to use diet as a panacea, has led to a myriad of dietary "self-help" books. Unfortunately, the consumer's desire for simple answers to complex questions has often led to misinterpretation or even misrepresentation of scientific data. Correlational data have been interpreted as signifying cause and effect relations. With the public spotlight focused so strongly on the area of nutrition and behavior, it is crucial that research in this area be based on proper methodology and careful interpretation of data. This is particularly true for studies examining the relationship between diet and antisocial behavior because policy decisions may be made on the basis of this research.

RESEARCH ON DIET AND BEHAVIOR

METHODOLOGICAL ISSUES

One of the more difficult problems in research on diet and behavior is how to separate nutritional from nonnutritional factors. Because food is so intimately involved with other aspects of our daily lives, it contains much more than its obvious nutritional value. Food is an intrinsic part of social functions, religious observations, and cultural rituals. Because food is a "loaded" variable, both experimenters and subjects may harbor biases about expected research outcomes. To minimize the confounding effects of these biases, double-blind procedures in which neither the experimenter nor the subjects know what treatment is given must be used.

In general, research on nutrition and behavior would benefit from the methodological controls used in psychopharmacology (Dews, 1986). Two variables that are important in drug studies are dose and length of treatment. Because consumption of a small amount (dose) of a dietary component may produce behavioral effects that differ from consumption of a larger amount, several doses of a dietary component should be tested whenever feasible. By using different doses, researchers can determine if there is a systematic relationship between the dietary variable and behavior. The lack of a systematic effect should be taken as a danger sign either that the apparent effect is spurious or that the variability is greater than expected.

The duration of dietary treatment is also critical. Although short-term (acute) studies permit evaluation of the immediate effects of a dietary treatment, they cannot provide information about long-term (chronic) exposure. Because the behavioral effects of dietary components (e.g., food additives) may only appear with extended exposure, both acute and chronic studies should be used to assess the nutrition-behavior interaction.

Diurnal variations in subjects' responses to nutrients should also be considered in diet-behavior studies. For example, it was recently observed that a snack (candy bar or yogurt) significantly improved subjects' ability to pay attention to relevant stimuli when it was eaten in the late afternoon, but not when it was consumed in the late morning (Kanarek and Swinney, 1990).

Prior nutritional status also has the potential of influencing the results of acute experiments. The types and amounts of foods previously consumed can affect the metabolism of a test nutrient. Standardizing dietary intake prior to evaluating the behavioral consequences of a test nutrient can eliminate this source of variation.

Another challenge in planning nutrition and behavior experiments is choosing appropriate subjects. Differences in nutritional history, socioeconomic background, and other environmental factors create subject heterogeneity that poses a threat to the internal validity of the research. Internal validity concerns the ability to conclude that a causal relationship exists between an independent and a dependent variable. Because of subject heterogeneity, alternative explanations may exist for the observed experimental effects, which lowers internal validity. For example, research on the dietary treatment of hyperactive children has shown that the home environment can affect the results of a study. Children from an unsupportive home environment show much less improvement with dietary treatment than children from a more supportive home environment (Rumsey and Rapoport, 1983).

Finally, the external validity of experimental findings must be considered. For example, the behavioral effects of a nutritional variable, observed in male college students, and tested in a laboratory during a single test session, may have little to do with the behavior of the general public in its everyday lives. Researchers face a dilemma in trying to choose between a controlled but artificial laboratory setting and a "real" or naturalistic setting that may be full of confounding factors. "Quasi-natural" studies, which could capture the advantages of both the laboratory and the real world, should be considered (Kanarek and Orthen-Gambill, 1986).

EXPERIMENTAL STRATEGIES

Three primary strategies have been used in research on diet and behavior. *Correlational studies* have been employed for generating hypotheses about diet-behavior relationships. The major objective of these studies is to define a link between dietary intake and behavior, with the specific expectation that statistical associations will be derived between the two variables. This type of research can provide important insights for experimental evaluation of the diet-behavior relation.

There are, however, several conditions that must be met before the validity of a correlational study can be accepted. First, reliable and valid measures of dietary intake must be made. One of the most widely used approaches for assessing dietary intake is the 24-hour recall in which subjects are asked to record everything that they have consumed during the preceding day. However, because there are wide day-to-day variations in an individual's food intake, a 24-hour record may not provide an accurate determination of average food intake. As a result, it has been suggested that a minimum of seven 24-hour recalls be used (Anderson and Hrboticky, 1986).

Second, proper subject sampling techniques must be used. In general, the larger the number of subjects, the better. If the number of subjects is too small, the probability of observing a significant relation between a dietary variable and behavior is reduced, and a false negative association may be assumed. On the other hand, correlation studies using large numbers of subjects risk the possibility of false positive associations. For example, when correlations are made between several dietary variables and behavior, the chance of achieving statistically significant results increases with the number of subjects and with the number of correlations made. In addition, when large numbers of subjects are used, small correlations can become statistically significant, making it necessary for the researcher to decide on the clinical importance of such results (Anderson and Hrboticky, 1986).

A common method of subject selection used in studies of diet and behavior involves the placement of media advertisements. Although this method is convenient, such sampling increases the probability of including self-selected members of the general population. For example, if a researcher wants to test the hypothesis that sugar influences hyperactivity in children and advertises the study as such, the subjects may be derived predominantly from families in which parents believe such an association exists.

Finally, cause and effect relations cannot be established from correlational data. For example, positive correlations have been reported between sugar intake and hyperactive behavior in children. These results have been interpreted by some (especially in the popular media) as demonstrating that sugar causes hyperactivity. However, it is just as possible that high levels of activity increase sugar intake or that a third unidentified variable influences both sugar intake and hyperactivity.

In contrast to correlational studies, experimental studies have the potential of identifying causal links between diet and behavior. The manipulation of a specific dietary component (the independent variable) may alter the occurrence of a behavioral measure or cognitive function (the dependent variable). Two major paradigms have been used in these studies. In *dietary replacement studies*, the behavioral effects of two diets, one containing the food component of interest (e.g. food additives) and the other not containing that food component, are compared over some period of time (e.g. two to three weeks). One obvious advantage of this method is that chronic dietary effects can be assessed. However, because making two diets equivalent in all factors except the food component being studied is difficult, it may be impossible to use appropriate double-blind techniques. Another limitation is that it is not feasible to test more than one dose of the dietary variable. There is also evidence that the order of diet presentation can influence the results of replacement studies. Finally, replacement studies are expensive with respect to both time and money.

Dietary challenge studies are used to evaluate the acute effects of dietary components. In these studies, behavior is usually rated for several hours after an individual has consumed the food component of interest or a placebo. Double-blind procedures are relatively easy to institute. The food component and the placebo can be packaged so that neither the subjects nor the experimenters can detect which is being presented. A crossover procedure in which the subjects are given the food component on one day and the placebo on another, with the order of presentation varied among the subjects, can also be employed. In addition, although not often done, more than one dose of the dietary variable can easily be tested in challenge studies.

The obvious disadvantage of challenge studies is that they do not provide information on the possible cumulative effects of a food component.

SUGAR AND BEHAVIOR

Of the many components in our diets, none has been condemned as frequently and as vehemently as sugar. Studies reviewed by the federal government indicate that sugar is the food people most consistently want to avoid and the one they look for most often on a food label's list of ingredients (Lecos, 1980). The use of sugar in our food has become a controversial issue involving scientists, dietitians, physicians, government officials, and private citizens.

The public strongly believes that sugar has negative effects on behavior. This belief has been fostered by popular reports blaming sugar for a variety of adverse behavioral outcomes including hyperactivity, depression, mental confusion, irritability, drug and alcohol addiction, and antisocial behavior (e.g. Dufty, 1975; Ketcham and Mueller, 1983; Schoenthalar, 1985). One of the most celebrated examples of our negative views of sugar is the case of San Francisco City Supervisor Dan White who shot and killed the city's mayor and another city supervisor. White's lawyers argued that their client acted irrationally and suffered from "diminished mental capacity" as a result of his overconsumption of sugar-containing "junk" foods. On the basis of this argument, which has become known as the "Twinkie defense," White was convicted of manslaughter rather than first-degree murder.

Is there a scientific basis for our attitudes about sugar? Before this question can be answered, the term sugar must be defined. Although many different types of sugar are found in our foods, most people use the word sugar to describe the simple carbohydrate sucrose. Sucrose, the sugar on our tables and typically used in cooking, is a disaccharide composed of the monosaccharides fructose and glucose. Sucrose is broken down into its monosaccharide components in the digestive tract and absorbed across the small intestine. After absorption, glucose and fructose are carried by the blood to the liver and other tissues. Because fructose is rapidly metabolized to glucose in the intestinal mucosa and the liver, any discussion of carbohydrate metabolism is essentially a discussion of glucose. Glucose is the metabolic fuel for most cells in the body and the primary energy source for cells in the central nervous system. The critical role of glucose in the normal functioning of the central nervous system has helped to foster the belief that sugar can affect behavior.

SUGAR, HYPOGLYCEMIA, AND BEHAVIOR

Sugar intake has been condemned as the cause of a large number of psychological problems, including alterations in mood, irritability, aggression, and violent behavior. One "physiological" explanation for sugar's adverse effects is hypoglycemia or "low blood sugar." Unfortunately, the term hypoglycemia has frequently been misused. Many doctors, as well as patients, are confused about the condition (Yager and Young, 1974; Nelson, 1985).

Clinically, hypoglycemia is defined by (1) low circulating blood glucose levels—50 milligrams per deciliter (mg/dl) or less; (2) symptoms including sweating, tremors, anxiety, headaches, weakness and hunger; and (3) amelioration of symptoms when blood glucose is restored to normal levels by food intake (Nelson, 1985; McFarland et al., 1987). Hypoglycemia can occur in diabetics after the administration of insulin. Additionally, other drugs such as antibiotics, antiinflammatory agents, and antidepressants; insulin-secreting tumors; and renal disease can lead to hypoglycemia.

It has been suggested that sugar consumption is a causal factor in hypoglycemia. The rationale for this idea begins with the assumption that simple sugars are more rapidly digested and absorbed than complex carbohydrates and thus cause a greater increase in blood glucose levels. This rapid rise in blood glucose levels stimulates insulin secretion, which has the effect of decreasing blood glucose levels. This regulatory effect has been called reactive or functional hypoglycemia.

There are several problems, however, with the idea that sugar intake can cause reactive hypoglycemia. First, recent studies have shown that a simple distinction cannot be made between sugars and more complex carbohydrates with respect to blood glucose and insulin responses (Crapo, 1985). Foods high in sugar can actually lead to smaller increases in blood glucose levels than foods containing complex carbohydrates. Thus, the assumption that sugar-containing foods uniformly lead to wide fluctuations in blood glucose values must be viewed with caution. Another related problem is that low blood glucose levels are not consistently associated with clinical signs of hypoglycemia. Additionally, symptoms of hypoglycemia are frequently reported in the absence of low blood glucose levels (McFarland et al., 1987).

In many cases, a diagnosis of hypoglycemia is made on the basis of symptoms without appropriate laboratory evidence (Nelson, 1985; McFarland et al., 1987). To make a diagnosis of hypoglycemia, a relationship between low blood glucose levels and the symptoms of the disease must exist. The most common ways of doing this are to conduct an oral glucose tolerance test (OGTT) or to measure blood glucose levels after a normal meal. In either case, for a diagnosis of hypoglycemia, clinical symptoms must be associated with blood glucose levels of less than 50 mg/dl. This association is rarely observed. Patients who have glucose levels lower than 50 mg/dl are infrequent (Yager and Young, 1974; Nelson, 1985; McFarland et al., 1987).

Given the relative rarity of functional hypoglycemia, why has the disease become so popular? For individuals with psychological complaints, a diagnosis of hypoglycemia may have certain benefits. First, the disease is socially acceptable. Rather than endure a "psychological" or otherwise stigmatizing condition, the patient can suffer from a respectable metabolic illness. Second, hypoglycemia gives individuals a way of easily and actively dealing with their complaints. By following certain dietary prescriptions, the patient believes that his symptoms can effectively be eliminated. In many cases, the act of attributing psychological problems to hypoglycemia and altering one's diet in response to this condition may provide some relief. Finally, hypoglycemia may be preferable to facing the possibility of a more serious condition.

Sugar and Violent Behavior

During the past decade, theories relating sugar intake to violent behavior have received increasing attention. Once relegated to articles and books directed at food faddists, such theories are now discussed at meetings of criminologists, and are found in books and articles aimed at personnel in the correction and criminal justice systems. Moreover, on the basis of these theories, correctional facilities in several states have revised their dietary policies in an effort to reduce sugar intake and control violent behavior (Gray, 1986).

Interest in the relationship between sugar and violent behavior was sparked by studies by Virkkunen and colleagues suggesting that hypoglycemia was common in criminals and delinquents displaying habitually violent behavior (Virkkunen and Huttunen, 1982; Virkkunen, 1982, 1983a, 1986a,b; Roy et al., 1986; Linnoila et al., 1990). These studies compared glucose and insulin levels during an OGTT between violent male offenders and male controls matched for age and relative body weight. In comparison to controls, men diagnosed as having either antisocial personality disorder or intermittent explosive disorder (American Psychiatric Association, 1987) initially displayed greater increases in blood glucose concentrations during the OGTT, followed by rapid declines in glucose values to levels indicative of reactive hypoglycemia. Comparisons between the two groups of offenders revealed that individuals with intermittent explosive disorder displayed a more rapid decline of glucose levels following the initial hyperglycemia, as well as a more rapid return from hypoglycemic levels to the original baseline values, than individuals with antisocial personality disorder. Men with antisocial personality disorder also demonstrated enhanced insulin secretion compared to controls. This increase in insulin secretion could act to augment glucose uptake by the cells and thus contribute to hypoglycemia. In contrast to men with antisocial personality disorder, men with intermittent explosive disorder did not have significantly elevated insulin values compared to controls. Subsequent correlational analyses suggested that a positive relationship existed between the duration of hypoglycemia during the OGTT and behavioral or sleeping problems, truancy, stealing, number of crimes against property, and multiple prison sentences (Virkkunen, 1982; Virkkunen and Huttunen, 1982; Virkkunen, 1986a,b; Roy et al., 1986).

One explanation hypothesized for the relationship between hypoglycemia and aggression is that a functional deficit in serotonergic neurons in the central nervous system may lead to abnormalities in glucose metabolism that can be conducive to violent behavior (Roy et al., 1988; Linnoila et al., 1990). Unfortunately, the details of this hypothesis remain to be elucidated.

Although individuals with a history of violent behavior had a greater tendency toward hypoglycemia than controls, this finding cannot be viewed as unequivocal evidence of an association between hypoglycemia and aggression. First, no determination of nutritional status was made in any of the studies examining this association. However, in several papers, the authors noted that habitually violent men generally had poor appetites and may not have consumed food for many hours prior to an act of violence. It is possible that the nutritional status of these men was not adequate. This seems particularly likely because all of the violent offenders in the studies by Virkkunen and colleagues had a history of alcohol abuse. Chronic alcoholics frequently substitute alcohol for much of their normal food intake, and therefore often consume insufficient amounts of protein and essential vitamins and minerals (Shaw and Lieber, 1988). Inadequate nutrition can lead to abnormal glucose responses. Thus, it may not be that hypoglycemia results in violent behavior, but rather that a lifestyle that encompasses alcohol abuse and other behaviors that contribute to inadequate nutrition results in hypoglycemia.

A second problem is that in some studies, violent offenders were given their normal diet for three days preceding the OGTT (e.g. Virkkunen, 1983a), and in others, a hospital diet containing a minimum of 48-55 percent calories as carbohydrate (e.g. Virkkunen, 1986a). The diet of control subjects was not manipulated in any of the studies. It is thus conceivable that differences in dietary intake immediately preceding the OGTT contributed to the differences in blood glucose and insulin responses observed between violent offenders and normal controls. Future research exploring hypoglycemia and aggressive behavior should include assessments of nutrient intake for all subjects.

Another difficulty with this research is that recent work has indicated that the OGTT may not be a good indicator of the changes in blood glucose levels that occur after a normal meal (Crapo, 1985). Thus, the finding that individuals with a history of violent behavior have lower glucose levels during an OGTT does not imply that this occurs under normal feeding conditions. Measurements of glucose and insulin levels in subjects following standard meals would be useful for determining the relationship between hypoglycemia and violent behavior.

An additional problem is that although positive associations were reported between the duration of hypoglycemic responses during the OGTT and a number of measures indicative of behavioral problems, there is no evidence that violent behavior actually occurred when insulin secretion was enhanced or low blood sugar levels were experienced. Examination of mood changes and other experimental behavioral indices of aggressive impulses during the OGTT could help to resolve this problem.

Finally, as previously mentioned, all of the violent offenders studied by Virkkunen and colleagues had a history of alcohol abuse. Although, as pointed out by these investigators, alcohol may enhance insulin secretion and thus lead to a reduction in blood glucose levels, it has a variety of other effects on the central nervous system. These other actions certainly play a role in alcohol's effects on aggressive behavior (see Miczek, Haney, et al., in this volume).

In a series of studies employing a dietary replacement strategy, Schoenthaler (1982, 1983a-c, 1985) investigated the effects of reducing sugar consumption on the behavior of inmates in juvenile detention facilities. A similar experimental approach was used in all studies. At a specific point in time, the institution modified its food policy in an effort to reduce sugar intake. Typical changes in the diet included substituting honey for table sugar; molasses for white sugar in cooking; fruit juice for Kool-aid; unsweetened cereal for presweetened cereal; and fresh fruit, peanuts, coconut, popcorn, or cheese for high-sugar desserts. The dependent variable in all of these studies was the number of disciplinary actions recorded by staff members before and during the change in food policy. On the basis of these studies, Schoenthaler (1982, 1983a-c, 1985) claimed that antisocial behavior in juvenile offenders could be decreased by 21 to 54 percent if sugar intake was reduced. Because this claim has important policy implications, it warrants careful scrutiny.

The first problem posed by Schoenthaler's work is identification of the independent variable. Sugar intake is reported to be the independent variable. However, one does not have to be a nutritionist to appreciate that the dietary manipulations used were of dubious value in limiting sugar intake. Many of the dietary changes merely replaced one sugar for another (e.g. honey for sucrose). Moreover, no measurements of actual sugar intake were made in any of these studies. Thus, it is impossible to determine if the dietary alterations actually led to a reduction in sugar consumption. Intake data are essential to establish if dietary manipulations have any effects and if the independent variable is operative (Hirsch, 1987). Moreover, even if sugar intake was reduced, the diets consumed during the two periods varied sufficiently in nutrient content to make it difficult to attribute any behavioral change to reduced sugar intake.

A second problem with these studies is that appropriate behavioral techniques were not used. None of these studies used standard double-blind procedures. Both subjects and institutional officials were aware of the dietary changes. The subjects' awareness of the changes, and also simply knowing that they were in a study, could have led them to alter their behavior. Furthermore, because the observers were aware of the nature of the studies, their expectations may have influenced their observations.

The nature of the dependent variable also poses a problem. Official records of disciplinary actions were used to assess changes in violent behavior. In many institutions, the staff have the discretion to, or not to, record an incident, and variation over time in the proportion of incidents reported may lead to erroneous results (Gray, 1986). Also, in some of these studies the dietary changes were made during the last portions of the subjects' institutional stay (Schoenthaler, 1983a). One might expect that the number of disciplinary actions would decrease as the juveniles either learned the rules or learned not to get caught.

In some of these studies, concern also must be expressed for the changing nature of the subject population. Some of the juveniles were included in both the control and the treatment condition, whereas others were in one condition but not the other. Finally, questions have been raised with respect to the statistical methods used in these studies (Gray, 1986; Pease and Love, 1986).

Taken together, the studies by Schoenthaler provide little convincing evidence for the claim that sugar intake contributes to antisocial behavior. These studies are flawed by faulty experimental design and inappropriate statistical analyses, and leave open the question of whether nonspecific factors were responsible for the changes attributed to diet.

RESEARCH NEEDS FOR ASSESSING THE EFFECTS OF SUGAR ON VIOLENT BEHAVIOR

Dietary replacement studies conducted in correctional institutions could contribute valuable information about the effects of dietary variables on antisocial behavior. However, these studies must be rigorously conducted and carefully controlled. The dependent and independent variables must be adequately specified, and valid and reliable methods for measuring these variables must be determined. The starting point in this research should be accurate measurements of nutritional intake. Appropriate dietary changes can then be instituted to reduce sugar intake. It is important that double-blind procedures be used. To accomplish this goal, artificially sweetened foods and beverages could be substituted for sugar-containing items. To eliminate the possibility of order effects, a crossover procedure in which half of the subjects receive sugar-containing foods first and then artificially sweetened items—whereas the remaining subjects are tested in the reverse order—should be employed. All subjects should be examined in both treatment conditions. Food intake must be measured to determine if dietary manipulations actually reduce sugar intake.

The second major question that must be addressed in this type of research is what is the dependent variable. An objective criterion of antisocial or violent behavior must be established. Additionally, by whom and when the behaviors will be measured must be adequately specified.

It must also be realized that a number of potent variables (e.g. overcrowding, drug use) may override the behavioral effects of dietary modifications. Diet is only one of the many variables that can influence human behavior. Therefore, it is important that studies on diet and behavior be conducted in institutions offering the least number of extraneous variables. Only then can definitive conclusions about the role of diet in antisocial behavior be reached

SUGAR AND ATTENTION DEFICIT DISORDER WITH HYPERACTIVITY

Attention deficit hyperactivity disorder (ADHD) is characterized by developmentally inappropriate inattention, impulsive behavior, and significantly elevated levels of motor activity. At home, attentional difficulties may be manifested by failure to follow through on tasks and the inability to stick to activities, including play, for appropriate periods of time. In school, the child with ADHD is inattentive and impulsive, and has difficulty organizing and completing work. In approximately half of the cases, the onset of ADHD is before the age of 4. As many as 3 percent of preadolescent children may suffer from ADHD, with the disorder six to nine times more common in boys than in girls (American Psychiatric Association, 1987).

With respect to violent behaviors, children with ADHD frequently have difficulties interacting with their peers and may be seen as overly aggressive. Additionally, many children with ADHD develop oppositional defiant disorder or conduct disorder later in childhood. Among those with conduct disorder, a substantial number have antisocial personality disorder as adults. Follow-up studies of children with ADHD have revealed that approximately one-third display some symptoms of the disorder as adults (American Psychiatric Association, 1987).

One pervasive idea about sugar is that it can lead to hyperactivity in children. This idea has been accepted by both educational professionals and parents (McLoughlin and Nall, 1988). Evidence for this association came first from a correlation study by Prinz and colleagues (1980) who compared sugar intake and behavior in hyperactive and normal 4- to 7-year-old children. Seven-day dietary records were obtained for all children. Trained observers, blind to the nature of the experiment, then rated the children during play for a variety of behaviors including destructive-aggressive acts (attempts to damage, strike, kick, or throw objects in the room); restlessness (repetitive arm, leg, hand, or head movements); and overall movement. Hyperactive and normal children consumed equivalent amounts of sugar-containing foods. However, for children with ADHD the amount of sugar products consumed, the ratio of sugar to nutritional foods (foods containing neither sugar nor refined carbohydrates), and the carbohydrate/protein ratio were all positively correlated with destructive-aggressive and restless behaviors. In contrast, in the normal group, sugar intake was not associated with destructive-aggressive acts but was correlated with total body movements. Although Prinz and coworkers (1980) carefully interpreted their work as only suggestive evidence of a role for sugar in ADHD, the data were rapidly interpreted in the popular press as demonstrating a causal relation between sugar and hyperactivity. Although the majority of studies on sugar and hyperactive behavior in the last decade have not confirmed the work of Prinz et al. (1980), the original interpretation of this work continues to be part of the folklore about diet and behavior.

During the last 10 years, both correlational and dietary challenge studies have assessed the effects of sugar on hyperactive behavior. For example, in an attempt to replicate the findings of Prinz et al. (1980), Wolraich and coworkers (1985) examined the association between sugar intake and performance on 37 behavioral/cognitive variables in hyperactive and normal boys. In contrast to the Prinz et al. (1980) study, sugar intake was not reliably related to destructive-aggressive behavior in children with ADHD. One obvious problem with correlational studies is that it is impossible to establish cause and effect relations. To determine causality, experimental studies that systematically manipulate sugar intake and observe its effects on behavior are required. The most common procedure used to accomplish this goal has been the dietary challenge. In the majority of these studies, the behavior of hyperactive and normal children has been rated for several hours after they have consumed a sugar-containing food or beverage, or a placebo containing an artificial sweetener. For example, Behar and colleagues (1984) investigated the effects of sucrose on behavior in 6- to 14-year-old boys whose parents believed that they had adverse responses to sugar. After an overnight fast, the boys were given a beverage containing either glucose, sucrose, or saccharin. Motor activity, spontaneous behavior, and performance on psychological tests were then measured for 5 hours. The negative behavioral effects reported by parents to occur after sugar intake were not observed. In fact, the only significant finding was that the boys were less active 3 hours after consuming the sugar-containing beverages than after consuming the saccharin-containing beverage. Similar negative findings have been reported in a number of studies (e.g., Wolraich et al., 1985; Ferguson et al., 1986; Mahan et al., 1988).

One problem with the preceding studies is that they were conducted in a laboratory environment. This rather unnatural setting may mask sugar's effects on behavior. To circumvent this problem, Milich and Pelham (1986) incorporated a dietary challenge into a treatment program for hyperactive boys. The boys were fasted overnight and at 8 a.m. given Kool-aid containing either glucose or aspartame. A double-blind crossover design was used. Beginning one-half hour after drinking the Kool-aid, the boys were evaluated for positive (e.g. following rules, sharing) and negative (e.g. noncompliance, teasing, name calling) behaviors and for classroom performance. Sugar did not adversely affect behavior or classroom performance. Similar results were obtained by Kaplan et al. (1986) who examined the behavioral effects of breakfasts in which dietary protein, fat, and carbohydrate were controlled while sucrose and total calories were varied.

Because the original study by Prinz and colleagues (1980) suggested that younger children might be more susceptible to the adverse effects of sucrose, Kruesi and coworkers (1987) performed a dietary challenge study in preschool-aged boys with alleged sugar reactivity. Again, no significant differences were observed as a function of sugar intake.

Taken together, the results of dietary challenge studies do not support the idea that sugar plays a major role in ADHD. In studies using hyperactive and normal children of varying ages and employing a range of experimental situations, sugar intake had no effects on behavior. However, although experimental evidence is weak, parents or teachers continue to supply anecdotal reports of the deleterious consequences of sugar. How do we reconcile these differences?

Several factors could contribute to the differing views of scientists and parents or teachers. One limitation of dietary challenge studies is that in most, only a single dose of sucrose was used. The amount of sucrose used in these studies may have been too small relative to the children's normal daily intake. Larger amounts might have produced negative reactions. Similarly, challenge studies can be criticized because they do not allow for the assessment of chronic sugar intake. Cumulative sugar intake may produce behavioral effects not detectable in single challenge tests. To help solve these problems, precise dietary histories and dose-response determinations of sucrose's effects on behavior are required.

Another difficulty with challenge studies is the choice of an appropriate placebo. In most studies, either aspartame or saccharin has been used as the placebo. Although this procedure successfully blinds the subjects to the item they are consuming, it does not control for the fact that the challenge food not only contains sugar, but also provides substantial calories. This presents the possibility that any changes in behavior could be attributed to calories rather than to sugar. Additionally, it has recently been proposed that aspartame may have negative effects on behavior.

Time parameters may also be important in determining sugar's effect on behavior. Most experiments have limited behavioral observations to one time after sugar intake and may have missed the critical period for its effects on behavior.

To overcome the objections raised about dietary challenge studies, and to further investigate the hypothesis that sugar intake has negative behavioral consequences, Wolraich and his colleagues (1994) evaluated hyperactive behavior and performance on cognitive tests of school-age and preschool children placed on diets high in sucrose, aspartame, or saccharin. To ensure that the subjects ate only the specified diets, all food was removed from the children's home prior to the study, and experimental diets were provided to the subjects and their families on a weekly basis by the research team. The children were fed each of the diets for three weeks with the order of diet presentation varied among the children. Results of the study were resoundingly negative. For the school-age children, all of whom were reported by their parents to be sensitive to sugar, none of the 39 behavioral or cognitive variables measured differed as a function of dietary conditions. For the normal preschool children, 4 of 31 variables did differ significantly among dietary conditions. However, there was no consistent pattern as a function of diet in the differences that were observed.

Parents and teachers may be misperceiving a relation between sugar and behavior. Hyperactive children have difficulties in altering their behavior to changing environmental demands. Thus, in school these children have trouble changing their behavior from the relatively unstructured nature of a snack or party period to the highly structured demands of class work. Because many of the foods children consume at snack time contain sugar, it may be that the association teachers report between sugar and behavior represents these children's difficulties in getting back on task following an unstructured activity. Similarly, parents often note behavioral deterioration after their child has consumed sugar in a party situation. Hyperactive children are known to have more difficulty in groups, and the effects parents observe may be more a function of the situation than of the consumption of sugar-containing foods. Finally, if parents believe that sugar intake has negative consequences, they may be more sensitive to their child's behavior after the child has consumed a sugar-containing food.

FOOD ADDITIVES AND ATTENTION DEFICIT DISORDER WITH HYPERACTIVITY

In the early 1970s, Dr. Benjamin Feingold, a pediatrician and allergist, called attention to the fact that more than 2,000 additives are part of our food supply and hypothesized that these additives played a causal role in childhood hyperactivity (Feingold, 1973, 1975). To test this hypothesis, Feingold began treating children with ADHD with a diet free of food additives. Additionally, as a result of a presumed cross-reactivity of yellow food dye with acetylsalicylic acid (aspirin), Feingold also advocated the removal of foods containing natural salicylates (e.g. almonds, apples, all berries, oranges, raisins, tomatoes, and green peppers) from the diet of hyperactive children. On the basis of his clinical work, Feingold claimed that a diet free of food additives and natural salicylates led to dramatic improvements, with 50 to 70 percent of the hyperactive children placed on this diet displaying complete remission. To obtain success, Feingold insisted that adherence to the diet was mandatory: any infringement could lead to a return of symptoms. He also proposed that successful treatment required the entire family to be on the diet and that an individual sensitive to food additives must avoid them for life. Feingold's ideas were widely publicized and rapidly gained popularity among the public (Lipton et al., 1979; Conners, 1984).

Open clinical trials, in which parents or physicians placed children on an additive-free diet supported Feingold's claims. However, carefully controlled double-blind studies have generally yielded more negative results (e.g. Conners et al., 1976; Harley et al., 1978; Goyette et al., 1978; Weiss et al., 1980). Using a dietary replacement paradigm, Harley and colleagues (1978) compared the behavioral effects of Feingold's diet with an ordinary diet containing additives. Food for families in the study was provided by the experimenters, and neither the researchers nor the family knew which diet was being consumed at any particular time; diet phases were alternated so that all families ate both diets. No significant improvements in behavior were noted by teachers or objective raters in the 36 school-aged hyperactive boys in the study. However, some parents reported improvement on the Feingold diet, but this occurred only when the diet was given after the control diet. When 10 preschool children were tested in the same situation, all of their mothers and most of their fathers rated the children as more improved on the additive-free than on the control diet. Harley and colleagues (1978:827) concluded, "While we feel confident that the cause-effect relationship asserted by Feingold is seriously overstated with respect to school-age children, we are not in a position to refute his claims regarding the possible causative effect played by artificial food colors on preschool children."

The effects of food additives on hyperactive behavior have also been examined by using dietary challenges. In these studies, children reported by their parents to respond positively to the Feingold diet were blindly "challenged" by the addition of food additives to the diet. Although some challenge studies have demonstrated a decrement in behavior when children are given food additives (Conners, 1980; Swanson and Kinsbourne, 1980), others revealed no detrimental effects of food additives (Conners, 1980; Weiss et al., 1980; Mattes and Gittelman, 1981). Several factors may account for these discrepancies. First, the types of food additives used in these studies have varied substantially. Second, a wide ranges of doses have been used. It has been argued that the dose of food additives used in some experiments was too low to produce adverse behavioral effects. However, it should be noted that the results of studies using larger doses have been both positive and negative. Finally, age may alter sensitivity to food additives. In general, younger children have been found to be more sensitive to food additives than older children (Harley et al., 1978; Conners, 1980; Weiss et al., 1980).

The preceding data allow several inferences to be drawn about the effects of food additives on hyperactive behavior. First, Feingold's claims and those from other open trials have been overstated. At best, only a small percentage of hyperactive children may be adversely affected by food additives. Second, younger children may be more sensitive to food additives than older children. Third, there may be a dose-response curve for food additives, just as there is for any toxic substance, but this has yet to be demonstrated.

In conclusion, the data on food additives and behavior are such to preclude any major legislative or administrative action to remove food additives or severely limit their use. Further studies of those few children who appear to respond negatively to food additives seem warranted. Additionally, research with experimental animals examining whether food additives have any biological activity in the central nervous system is recommended.

RELATIONSHIP BETWEEN BLOOD CHOLESTEROL AND VIOLENT BEHAVIOR

Over the past 10 years, the results of several types of experiments have suggested that an inverse relationship may exist between blood cholesterol concentration and violent behavior (Virkkunen, 1983b; Virkkunen and Penttinen, 1984; Kaplan and Manuck, 1990; Muldoon et al., 1990). For example, Virkkunen and colleagues (Virkkunen, 1983b; Virkkunen and Penttinen, 1984) measured fasting serum cholesterol levels in male homicidal offenders and found that those with antisocial personality disorder or intermittent explosive disorder with habitually violent tendencies had lower cholesterol levels than other offenders. This difference was particularly pronounced in men under the age of 30. It was hypothesized that the lower cholesterol levels in the violent offenders may be a consequence of enhanced insulin secretion.

Although the proposed relation between cholesterol levels and violent behavior is provocative, it suffers from the same problems as the proposed association between hypoglycemia and violence. No information on dietary history is provided in any of the studies by Virkkunen and colleagues. Additionally, the role of alcohol in influencing both cholesterol level and violent behavior is not addressed. Moreover, other differences between the violent offenders and other offenders (e.g. activity levels) are never explored.

To further investigate the relation between cholesterol level and aggressive behavior, Muldoon and colleagues (1990) compared the causes of mortality for subjects in intervention groups and control groups in six large primary prevention trials for reducing cholesterol levels. Cholesterol reduction in the intervention groups was accomplished in two studies by nutritional manipulations aimed at reducing dietary cholesterol and saturated fat intake and in the remaining four studies by pharmacological treatment. Causes of mortality were divided into three categories: coronary heart disease, cancer, and causes not related to illness, which included deaths due to accident, suicide, or homicide. In all studies, the treatments led to significant reductions in cholesterol levels. Compared to controls, the average cholesterol concentration of participants in the intervention groups was reduced by approximately 10 percent. This reduction in cholesterol was not associated with a significant decline in total mortality. However, cholesterol reduction was associated with a lower mortality rate from coronary heart disease and, in some studies, with a slightly higher mortality rate from cancer. With regard to mortality not related to illness, the chance of dying from suicide or violence was approximately twice as great in the intervention groups than in the control groups. The association between lower cholesterol levels and increased mortality from causes other than illness was found regardless of whether lipid lowering was based on dietary or pharmacological treatment.

Although the results of a number of studies have suggested an inverse relation between cholesterol levels and deaths due to accidents or violence, this association has not been universally observed. For example, Pekkanen et al. (1989) found little evidence of an inverse association between serum cholesterol values and increased mortality due to accidents and deaths in their 25-year follow-up of 1,580 Finnish men.

Additional epidemiological investigations and experimental research are clearly required to assess the relationship between blood cholesterol level and violent behavior.

LEAD TOXICITY AND ANTISOCIAL BEHAVIOR

It has recently been found that lead poisoning during childhood can have long-term detrimental effects on behavior (Needleman, 1989, 1990). Exposure to lead, which most frequently occurs when young children consume lead-based paints, has been associated with ADHD. As previously mentioned, ADHD is a well-established risk factor for later antisocial behavior. The rate of later delinquency in children who display ADHD and conduct disorder has been estimated to be 0.58. The attributable risk for hyperactivity in children with elevated levels of lead is 0.55. Multiplying the lower 95 percent confidence limits for these two proportions produces a joint probability of .2 for delinquency, given excess exposure to lead. The relation between lead exposure and delinquency has not yet been systematically studied, but clues suggest that this relationship should be given serious consideration (Needleman, 1989).

CONCLUSIONS

The study of the relationship between diet and behavior is still in its infancy. Within this growing field, a number of hypotheses have been developed about the role of dietary variables in determining violent behavior. Although experimental studies have been initiated to test these hypotheses, it is too early to draw definitive conclusions from this research. Better-controlled experiments employing appropriate research methodology are required. Additionally, it is important to remember that diet is but one of the many factors that could contribute to violent behavior. Research conducted thus far suggests that it may make a relatively minor contribution.

REFERENCES

American Psychiatric Association
  1987   *Diagnostic and Statistical Manual of Mental Disorders, Third Edition—Revised.* Washington, D.C.: American Psychiatric Association.
Anderson, G.H., and N. Hrboticky
  1986   Approaches to assessing the dietary component of the diet behavior connection. *Nutrition Reviews* 44(suppl.):42-51.
Behar, D., J.L. Rapoport, A.A. Adams, C.J. Berg, and M. Cornblath
  1984   Sugar challenge testing with children considered behaviorally "sugar reactive." *Nutrition and Behavior* 1:277-288.

Conners, C.K.
   1980   *Food Additives and Hyperactive Children.* New York: Plenum Press.
   1984   Nutritional therapy in children. Pp. 159-192 in J. Galler, ed., *Nutrition and Behavior.* New York: Plenum Press.
Conners, C.K., C.H. Goyette, D.A. Southwick, J.M. Lees, and P.A. Andrulonis
   1976   Food additives and hyperkinesis: A controlled double-blind experiment. *Pediatrics* 58:154-166.
Crapo, P.A.
   1985   Simple versus complex carbohydrates in the diabetic diet. *Annual Review of Nutrition* 5:95-114.
Dews, P.B.
   1986   Dietary pharmacology. *Nutrition Reviews* 44(Suppl.):246-251.
Dufty, W.
   1975   *Sugar Blues.* New York: Warner Books.
Feingold, B.F.
   1973   *Introduction to Clinical Allergy.* Springfield, Ill.: Charles C. Thomas.
   1975   Hyperkinesis and learning disabilities linked to artificial food flavors and colors. *American Journal of Nursing* 75:797-803.
Ferguson, H.B., C. Stoddart, and J.G. Simeon
   1986   Double-blind challenge studies of behavioral and cognitive effects of sucrose-aspartame ingestion in normal children. *Nutrition Reviews* 44(Suppl.):144-150.
Goyette, C.H., C.K. Conners, and T.A. Petti
   1978   Effects of artificial colors on hyperkinetic children: A double-blind challenge study. *Psychopharmacology Bulletin* 14:39-40.
Gray, G.E.
   1986   Diet, crime and delinquency: A critique. *Nutrition Reviews* 44(suppl.): 89-94.
Harley, J.P., R.S. Ray, L. Tomasi, P.L. Eichman, C.G. Matthews, R. Chun, C.S. Cleeland, and E. Traisman
   1978   Hyperkinesis and food additives: Testing the Feingold hypothesis. *Pediatrics* 61:818-828.
Hirsch, E.
   1987   Sweetness and performance. Pp. 205-223 in J. Dobbing, ed., *Sweetness.* New York: Springer-Verlag.
Kanarek, R.B., and N. Orthen-Gambill
   1986   Complex interactions affecting nutrition-behavior research. *Nutrition Reviews* 44(suppl.):172-175.
Kanarek, R.B., and D. Swinney
   1990   Effects of food snacks on cognitive performance in male college students. *Appetite* 14:15-27.
Kaplan, H.K., F.S. Wamboldt, and M. Barnhart
   1986   Behavioral effects of dietary sucrose in disturbed children. *American Journal of Psychiatry* 143:944-945.

Kaplan, J.R., and S.B. Manuck
1990   The effects of fat and cholesterol on aggressive behavior in monkeys. *Psychosomatic Medicine* 52:226-227.

Kellogg, J.H.
1882   *Plain Facts for Old and Young.* Burlington, Iowa.
1919   *The Itinerary of Breakfast.* New York: Funk & Wagnalls.

Ketcham, K., and L.A. Mueller.
1983   *Eating Right to Live Sober.* New York: Signet.

Kruesi, M.J.P., J.L. Rapoport, M. Cummings, C.J. Berg, D.R. Ismond, M. Flament, M. Yarrow, and C. Zahr-Waxler
1987   Effects of sugar and aspartame on aggression and activity in children. *American Journal of Psychiatry* 144:1487-1490.

Lecos, C.W.
1980   Food labels and the sugar recognition factor. *FDA Consumer* April:3-5.

Linnoila, M., M. Virkkunen, A. Roy, and W.Z. Potter
1990   Monoamines, glucose metabolism and impulse control. Pp. 218-241 in H.M. Van Praag, R. Plutchik, and A. Apter, eds., *Violence and Suicidality: Perspectives in Clinical and Psychobiological Research.* New York: Brunner/Mazel.

Lipton, M.A., C.B. Nemeroff, and R.B. Mailman
1979   Hyperkinesis and food additives. Pp. 1-27 in R.J. Wurtman and J.J. Wurtman, eds., *Nutrition and the Brain*, Vol. 4. New York: Raven Press.

Mahan, L.K., M. Chase, C.T. Furukawa, S. Sulzbacher, G.G. Shapiro, W.E. Pierson, and C.W. Bierman
1988   Sugar "allergy" and children's behavior. *Annals of Allergy* 61:453-458.

Mattes, J.A., and R. Gittelman
1981   Effects of artificial food colorings in children with hyperactive symptoms. *Archives of General Psychiatry* 38:714.

McFarland, K.F., C. Baker and S.D. Ferguson
1987   Demystifying hypoglycemia. *Postgraduate Medicine* 82:54-65.

McLoughlin, J.A., and M. Nall
1988   Teacher opinion of the role of food allergy on school behavior and achievement. *Annals of Allergy* 61:89-91.

Milich, R., and W.E. Pelham
1986   Effects of sugar ingestion on the classroom and playgroup behavior of attention deficit disordered boys. *Journal of Consulting and Clinical Psychology* 54:714-718.

Muldoon, M.F., S.B. Manuck, and K.A. Matthews
1990   Lowering cholesterol concentrations and mortality: A quantitative review of primary prevention trials. *British Medical Journal* 301:309-314.

Needleman, H.L.
1989   The persistent threat of lead: A singular opportunity. *American Journal of Public Health* 79:643-645.

1990   The long-term effects of exposure to low doses of lead in child-hood. *New England Journal of Medicine* 322:83-88.

Nelson, R.L.
1985   Hypoglycemia: Fact or fiction. *Mayo Clinic Proceedings* 60:844-850.

Pease, S.E., and C.T. Love
1986   Optimal methods and issues in nutrition research in the correctional setting. *Nutrition Reviews* 44(suppl.):122-132.

Pekkanen, J., A. Nissinen, S. Punsar, and M.J. Karvonen
1989   Serum cholesterol and risk of accidental or violent death in a 25-year follow-up. *Archives of Internal Medicine* 149:1589-1591.

Prinz, R.J., W.A. Roberts, and E. Hantman
1980   Dietary correlates of hyperactive behavior in children. *Journal of Consulting and Clinical Psychology* 48:760-769.

Roy, A., M. Virkkunen, S. Guthrie, R. Poland, and M. Linnoila
1986   Monoamines, glucose metabolism, suicidal and aggressive behavior. *Psychopharmacology Bulletin* 22:661-665.

Roy, A., M. Virkkunen, and M. Linnoila
1988   Monoamines, glucose metabolism, aggression towards self and others. *International Journal of Neuroscience* 41:261-264.

Rumsey, J.M., and J.L. Rapoport
1983   Assessing behavioral and cognitive effects of diet in pediatric populations. Pp. 101-162 in R.J. Wurtman and J.J. Wurtman, eds., *Nutrition and the Brain*, Vol. 6. New York: Raven Press.

Schoenthaler, S.J.
1982   The effect of sugar on the treatment and control of anti-social behavior: A double-blind study of an incarcerated juvenile population. *International Journal of Biosocial Research* 3:1-9.

1983a   Diet and crime: An empirical examination of the value of nutrition in the control and treatment of incarcerated juvenile offenders. *International Journal of Biosocial Research* 4:25-39.

1983b   Diet and delinquency: A multi-state replication. *International Journal of Biosocial Research* 5:70-78.

1983c   The Los Angeles probation department diet-behavior program: An empirical analysis of six institutional settings. *International Journal of Biosocial Research* 5:88-98.

1985   Nutritional policies and institutional anti-social behavior. *Nutrition Today* 20:16-25.

Shaw, S., and C.S. Lieber
1988   Nutrition and diet in alcoholism. Pp. 1423-1449 in M.E. Shils and V.R. Young, eds., *Nutrition in Health and Disease*. Philadelphia: Lea & Febiger.

Swanson, J.M., and M. Kinsbourne
1980   Food dyes impair performance of hyperactive children on a laboratory learning test. *Science* 207:1485-1486.

Virkkunen, M.
1982 Reactive hypoglycemia tendency among habitually violent offenders. *Neuropsychobiology* 8:35-40.
1983a Insulin secretion during the glucose tolerance test in antisocial personality. *British Journal of Psychiatry* 142:598-604.
1983b Serum cholesterol levels in homicidal offenders. *Neuropsychobiology* 10:65-69.
1986a Insulin secretion during the glucose tolerance test among habitually violent and impulsive offenders. *Aggressive Behavior* 12:303-310.
1986b Reactive hypoglycemic tendency among habitually violent offenders. *Nutrition Reviews* 44(Suppl.):94-103.
Virkkunen, M., and M.O. Huttunen
1982 Evidence for abnormal glucose tolerance test among violent offenders. *Neuropsychobiology* 8:30-34.
Virkkunen, M., and H. Penttinen
1984 Serum cholesterol in aggressive conduct disorder: A preliminary study. *Biological Psychiatry* 19:435-439.
Weiss, B., J.H. Williams, S. Margen, B. Abrams, B. Caan, L.J. Citron, C. Cox, J. McKibben, D. Ogar, and S. Schultz
1980 Behavioral response to artificial food colors. *Science* 207:1487-1488.
Whorton, J.C.
1982 *Crusaders for Fitness: The History of the American Health Reforms.* Princeton, N.J.: Princeton University Press.
Wolraich, M., R. Milich, P. Stumbo, and F. Schultz
1985 Effects of sucrose ingestion on the behavior of hyperactive boys. *Journal of Pediatrics* 106:675-682.
Wolraich, M.L., S.D. Lindgren, P.J. Stumbo, L.D. Stegink, M.I. Appelbaum, and M.C. Kiritsy
1994 Effects of diets high in sucrose or aspartame on the behavior and cognitive performance of children. *New England Journal of Medicine* 330:301-307.
Yager, J., and R.T. Young
1974 Non-hypoglycemia is an epidemic condition. *New England Journal of Medicine* 291:907-908.

Index

A

Academic achievement, *see* School performance and failure

Acetylcholine (ACh), 246, 248, 249, 266-267

Acetylsalicylic acid, 531

Activational hormonal effects, 5, 6-7

Adaptive behavior
and brain damage, 14
models, 12, 21, 78-79

Additive genetic variance, 23, 42

Adolescents, *see* Juvenile offenders; Puberty

Adoption studies, 3, 27, 29, 30, 31, 34-39, 42-43, 45, 49, 94
alcohol abuse, 4, 37, 38, 40-41, 44
standardization of, 46
of unrelated siblings, 47

Adrenal hormones, 7, 8, 184, 189, 222-225

Adrenal medulla, 184, 193, 222-225

Adrenocorticotropic hormone (ACTH), 7, 182, 184, 211, 218

Affective defense behavior, 12, 61, 63, 69-78

Age, 79, 90
and schizophrenia, 81

Aggressiveness
animal models, 12, 62-79, 174-179, 209-214, 247-248

definitions of, 173-179, 247-248
drug effects on, 9
and epilepsy, 82-85
and fetal steroid exposure, 6
genetic influences, 2, 21-22
heterogeneity in, 179-180
hormonal influences, 6-7, 8, 180-229
and hypoglycemia, 16
neuroanatomical factors, 12
neurochemical influences, 9, 10, 11

Agonistic behavior, 21-22, 175, 192, 211

Alcohol use and abuse, 4, 82, 275, 520, 534
adoption studies, 4, 37, 38, 40-41, 44
and benzodiazepine receptors, 11
genetic influences, 3, 4, 24, 46
and hormonal mechanisms, 224, 226, 228
neurochemistry, 10, 256, 261
and neuroimaging results, 16
neuropsychologic factors, 13
in pregnancy, 14, 94-95
and sugar metabolism, 17, 524, 525
among violent offenders, 86
withdrawal management, 10, 270

Alprazolam, 277, 278

Aminergic hormones, 182

541